Emerging Design Solutions in Structural Health Monitoring Systems

Diego Alexander Tibaduiza Burgos
Universidad Santo Tomás, Colombia

Luis Eduardo Mujica
Universitat Politecnica de Catalunya, Spain

Jose Rodellar
Universitat Politecnica de Catalunya, Spain

A volume in the Advances in Civil and Industrial Engineering (ACIE) Book Series

ENGINEERING SCIENCE REFERENCE
An Imprint of IGI Global

Published in the United States of America by
Engineering Science Reference (an imprint of IGI Global)
701 E. Chocolate Avenue
Hershey PA, USA 17033
Tel: 717-533-8845
Fax: 717-533-8661
E-mail: cust@igi-global.com
Web site: http://www.igi-global.com

Library of Congress Cataloging-in-Publication Data

Emerging design solutions in structural health monitoring systems / Diego Alexander Tibaduiza Burgos, Luis Eduardo Mujica, and Jos? Rodellar, editors.

pages cm

Includes bibliographical references and index.

ISBN 978-1-4666-8490-4 (hardcover) -- ISBN 978-1-4666-8491-1 (ebook) 1. Structural health monitoring. 2. Artificial intelligence--Engineering applications. 3. Structural analysis (Engineering) 4. Structural design. I. Tibaduiza Burgos, Diego Alexander, 1980- editor. II. Mujica, Luis Eduardo, 1976- editor. III. Rodellar, Jos?, editor.

TA656.6.E44 2015

624.1'71--dc23

2015010307

This book is published in the IGI Global book series Advances in Civil and Industrial Engineering (ACIE) (ISSN: 2326-6139; eISSN: 2326-6155)

British Cataloguing in Publication Data

A Cataloguing in Publication record for this book is available from the British Library.

Advances in Civil and Industrial Engineering (ACIE) Book Series

ISSN: 2326-6139
EISSN: 2326-6155

Mission

Private and public sector infrastructures begin to age, or require change in the face of developing technologies, the fields of civil and industrial engineering have become increasingly important as a method to mitigate and manage these changes. As governments and the public at large begin to grapple with climate change and growing populations, civil engineering has become more interdisciplinary and the need for publications that discuss the rapid changes and advancements in the field have become more in-demand. Additionally, private corporations and companies are facing similar changes and challenges, with the pressure for new and innovative methods being placed on those involved in industrial engineering.

The **Advances in Civil and Industrial Engineering (ACIE) Book Series** aims to present research and methodology that will provide solutions and discussions to meet such needs. The latest methodologies, applications, tools, and analysis will be published through the books included in **ACIE** in order to keep the available research in civil and industrial engineering as current and timely as possible.

Coverage

- Materials Management
- Quality Engineering
- Ergonomics
- Production Planning and Control
- Productivity
- Optimization Techniques
- Earthquake engineering
- Construction Engineering
- Operations Research
- Hydraulic Engineering

IGI Global is currently accepting manuscripts for publication within this series. To submit a proposal for a volume in this series, please contact our Acquisition Editors at Acquisitions@igi-global.com or visit: http://www.igi-global.com/publish/.

The Advances in Civil and Industrial Engineering (ACIE) Book Series (ISSN 2326-6139) is published by IGI Global, 701 E. Chocolate Avenue, Hershey, PA 17033-1240, USA, www.igi-global.com. This series is composed of titles available for purchase individually; each title is edited to be contextually exclusive from any other title within the series. For pricing and ordering information please visit http://www.igi-global.com/book-series/advances-civil-industrial-engineering/73673. Postmaster: Send all address changes to above address.

Titles in this Series

For a list of additional titles in this series, please visit: www.igi-global.com

Robotics, Automation, and Control in Industrial and Service Settings
Zongwei Luo (South University of Science and Technology of China, China)
Engineering Science Reference • copyright 2015 • 338pp • H/C (ISBN: 9781466686939) • US $215.00 (our price)

Using Decision Support Systems for Transportation Planning Efficiency
Ebru V. Ocalir-Akunal (Gazi University, Turkey)
Engineering Science Reference • copyright 2016 • 377pp • H/C (ISBN: 9781466686489) • US $215.00 (our price)

Contemporary Ethical Issues in Engineering
Satya Sundar Sethy (Indian Institute of Technology Madras, India)
Engineering Science Reference • copyright 2015 • 343pp • H/C (ISBN: 9781466681309) • US $215.00 (our price)

Emerging Issues, Challenges, and Opportunities in Urban E-Planning
Carlos Nunes Silva (University of Lisbon, Portugal)
Engineering Science Reference • copyright 2015 • 380pp • H/C (ISBN: 9781466681507) • US $205.00 (our price)

Technology and Practice in Geotechnical Engineering
Joseph B. Adeyeri (Federal University Oye-Ekiti, Nigeria)
Engineering Science Reference • copyright 2015 • 836pp • H/C (ISBN: 9781466665057) • US $225.00 (our price)

Fracture and Damage Mechanics for Structural Engineering of Frames State-of-the-Art Industrial Applications
Julio Flórez-López (University of Los Andes, Venezuela) María Eugenia Marante (Lisandro Alvarado University, Venezuela) and Ricardo Picón (Lisandro Alvarado University, Venezuela)
Engineering Science Reference • copyright 2015 • 602pp • H/C (ISBN: 9781466663794) • US $225.00 (our price)

Computer-Mediated Briefing for Architects
Alexander Koutamanis (Delft University of Technology, The Netherlands)
Engineering Science Reference • copyright 2014 • 321pp • H/C (ISBN: 9781466646476) • US $180.00 (our price)

Technologies for Urban and Spatial Planning Virtual Cities and Territories
Nuno Norte Pinto (The University of Manchester, UK) José António Tenedório (Universidade NOVA de Lisboa, Portugal) António Pais Antunes (University of Coimbra, Portugal) and Josep Roca Cladera (Technical University of Catalonia, BarcelonaTech, Spain)
Information Science Reference • copyright 2014 • 349pp • H/C (ISBN: 9781466643499) • US $200.00 (our price)

IGI GLOBAL
DISSEMINATOR OF KNOWLEDGE
www.igi-global.com

701 E. Chocolate Ave., Hershey, PA 17033
Order online at www.igi-global.com or call 717-533-8845 x100
To place a standing order for titles released in this series, contact: cust@igi-global.com
Mon-Fri 8:00 am - 5:00 pm (est) or fax 24 hours a day 717-533-8661

Editorial Advisory Board

Table of Contents

Detailed Table of Contents

Direct stress and sub-stress caused by fire, temperature variation and external loading in a structure are most important for the development of cracks. The chemical reactions of natural healing in the matrix was not been established conclusively. The most significant factor that influences the self-healing is the precipitation of calcium carbonate crystals on the crack surface. The mechanism which contribute autogenic healing are: (a) Continued hydration of cement at cracked surface as well as continued hydration of already formed gel and also inter-crystallization of fractured crystals; (b) blocking of flow path by water impurities and concrete particles broken from the crack surface due to cracking; (c) expansion of concrete in the crack flank (swelling) and closing of cracks by spalling of loose concrete particle are also reported as the sealing mechanism by researchers. The recovery of mechanical as well as physical property was discussed by different researchers. An experimental investigation was carried out to study the autogenic healing of fire damaged fly ash and conventional cementitious mortar samples subjected to steam followed by water curing at normal atmospheric pressure. The micro cracks are generated artificially by heating the 28 days aged mortar samples at 800 Deg. C. The effect of fly-ash replacing ordinary Portland cement by 0 and 20% was studied. Recovery of compressive strength and physical properties i.e. apparent porosity, water absorption, ultrasonic pulse velocity and rapid chloride ion penetration test confirm the self-healing of micro cracks. Such healing is more prominent for fly ash mortar mix. Optical as well as scanning electron microscopy With EDAX analysis and X-ray diffraction study of the white crystalline material formed in the crack, confirms formation of calcium carbonate.

This chapter is devoted to present novel techniques in Structural Health Monitoring (SHM). These techniques are based on different statistical and signal processing methods that are used in other fields but their performance and capability in SHM is presented and tested for the first time in this work. This work is dedicated to the first level of SHM, which might be considered the main and most important level. Piezoceramic (PZT) devices are chosen in this work to capture the signals due to their special characteristics such as high performance, low energy consumption and reasonable price. Suggested techniques are tested on different laboratory and real scale test benchmarks. Moreover, this work considers the effect of environmental changes on performance of the presented techniques. This work shows that although those techniques have a significant result in normal conditions, their performance can be affected by any environmental discrepancy such as temperature change. As such, there is a vital need to consider their effect. In this work, temperature change is chosen, as it is one of the main environmental fluctuation factors.

An important task in structural health monitoring (SHM) is that of damage detection under varying environmental and operational conditions. Structures, under varying environmental conditions, change their mass, elasticity and damping properties whereas changing operational conditions cause changes to excitations. A damage detection methodology implemented in these circumstances faces serious challenges since changes to structural behaviour imparted by environmental or operational conditions could be wrongly attributed to damage. The part of a damage detection decision algorithm that removes environmental and operational effects is called normalization. In this chapter a normalization methodology that is based on the similarity between continuous wavelet transform maxima decay lines is presented. This methodology is implemented on both simulated and experimental data. Simulated data were obtained from a three degree of freedom system. Varying environmental conditions were simulated by temperature dependent stiffness parameters and operating conditions by changing the colour of random excitation. Experimental data were obtained from damaged cantilever beams that were subjected to random excitations of different colour and varying temperatures.

Considering the possibility that from the area of electronics can be provided feasible solutions in the field of non-destructive testing, this chapter present a prototype and methodology that allows energize an ultrasound transducer. This system is used to evaluate for detecting weld failures at the junctions of metallic parts. Subsequently, in order to validate the design quality of that source, a computer system that allows control of a card developed ultrasound. Finally, it is implemented the ultrasonic imaging by time of flight diffraction technique, in order to obtain an objective comparison methodology to both systems.

Structural damage assessment methodologies allow providing knowledge about the current state of the structure. This information is important because allows to avoid possible accidents and perform maintenance tasks in the structure. This chapter proposes the use of an artificial immune system to detect and classify damages in structures by using data from a multi-actuator piezoelectric system that is working in several actuation phases. In a first step of the methodology, principal component analysis (PCA) is used to build a baseline model by using the collected data. In a second step, the same experiments under similar conditions are performed with the structure in different states (damaged or not). These data are projected into the different baseline models for each actuator, in order to obtain the damages indices and build the antigens. The antigens are compared with the antibodies by using an affinity function and the result of this process allows detecting and classifying damages.

Extreme learning machine (ELM) is a learning algorithm for single-hidden layer feedforward neural networks. In theory, this algorithm is able to provide good generalization capability at extremely fast learning speed. Comparative studies of benchmark function approximation problems revealed that ELM can learn thousands of times faster than conventional neural network (NN) and can produce good generalization performance in most cases. Unfortunately, the research on damage localization using ELM is limited in the literature. In this chapter, the ELM is extended to the domain of damage localization of plate structures. Its effectiveness in comparison with typical neural networks such as back-propagation neural network (BPNN) and least squares support vector machine (LSSVM) is illustrated through experimental studies. Comparative investigations in terms of learning time and localization accuracy are carried out in detail. It is shown that ELM paves a new way in the domain of plate structure health monitoring. Both advantages and disadvantages of using ELM are discussed.

This chapter presents an expert monitoring algorithm approach to detect, locate and quantify stiffness variations in structures. The algorithm is based on pattern recognition and artificial intelligence techniques that emulate knowledge based on human reasoning. The expert system (ES) uses time-frequency information about dynamics of structure, which is processed by using discrete wavelet transform (DWT), self-organizing maps (SOM), case-based reasoning (CBR) and principal component analysis (PCA). In addition, two applications are considered in order to evaluate the effectiveness of vibration analysis methodology and CBR in damage detection. The first application (Camacho 2010) uses the environmental excitation to detect and quantify damage in a Mechanical UBC ASCE Benchmark. The second one (Sandoval 2010) uses a predesigned signal to detect geometric damages on a gas pipeline. In both cases, a finite element model (FEM) is used to simulate different damages scenarios, which correspond to stiffness variations in different location.

This chapter presents a statistical methodology for structural damage diagnosis (detection, localization and estimation), in the context of continuous online monitoring. There are several sources of uncertainty such as physical variability, measurement uncertainty and model errors that affect structural damage diagnosis, and therefore, it may not be possible to precisely detect, localize, and estimate damage. Hence, a statistical approach can help to identify these sources of uncertainty, quantify their combined effect on diagnosis, and thereby, provide an estimate of the confidence in the results of diagnosis. Damage detection is based on residuals between nominal and damaged system-level responses, using statistical hypothesis testing whose uncertainty can be captured easily. Localization is based on the comparison of damage signatures derived from the system model. Both classical statistics-based methods and Bayesian statistics-based methods are investigated to quantify the uncertainty in all the three steps of diagnosis, i.e. detection, localization, and quantification. While classical statistics-based methods use the concept of least squares-based optimization, Bayesian methods make use of likelihood function and Bayes theorem. The uncertainties in damage detection, isolation and quantification are combined to quantify the overall uncertainty in diagnosis. The proposed methods are illustrated using the example of a structural frame.

Structural Health Monitoring (SHM) is an emerging discipline that aims at improving the management of the life cycle of industrial components. The scope of this chapter is to present the integration of nonlinear ultrasonics with the Bayesian inverse problem as an appropriate tool to estimate the updated health state of a component taking into account the associated uncertainties. This updated information can be further used by prognostics algorithms to estimate the future damage stages. Nonlinear ultrasonics allows an early detection of damage moving forward the achievement of reliable predictions, while the inverse problem emerges as a rigorous method to extract the slight signature of early damage inside the experimental signals using theoretical models. The Bayesian version of the inverse problem allows measuring the underlying uncertainties, improving the prediction process. This chapter presents the fundamentals of nonlinear ultrasonics, their practical application for SHM, and the Bayesian inverse problem as a method to unveil damage and manage uncertainty.

Chapter 10

Shankar Sankararaman, SGT Inc., USA & NASA Ames Research Center, USA
You Ling, General Electric Global Research Center, USA
Sankaran Mahadevan, Vanderbilt University, USA

This chapter describes a computational methodology for fatigue crack growth analysis and damage prognosis in structures. This methodology is applicable to a variety structural components and systems with complicated geometry and subjected to variable amplitude multi-axial loading. Finite element analysis is used to address complicated geometry and calculate the stress intensity factors. Multi-modal stress intensity factors due to multi-axial loading conditions are combined to calculate an equivalent stress intensity factor using a characteristic plane approach. Crack growth under variable amplitude loading is modeled using a modified Paris law that includes retardation effects. During cycle-by-cycle integration of the crack growth law, a Gaussian process surrogate model is used to replace the expensive finite element analysis, thereby significantly improving computational effort. The effect of different types of uncertainty – physical variability, data uncertainty and modeling errors – on crack growth prediction is investigated. The various sources of uncertainty include, but not limited to, variability in loading conditions, material parameters, experimental data, model uncertainty, etc. Three different types of modeling errors – crack growth model error, discretization error and surrogate model error – are included in analysis. The different types of uncertainty are incorporated into the framework for calibration and crack growth prediction, and their combined effect on crack growth prediction is computed. Finally, damage prognosis is achieved by predicting the probability distribution of crack size as a function of number of load cycles, and this methodology is illustrated using a numerical example of surface cracking in a cylindrical component.

Chapter 11

J. Chiachío, University of Granada, Spain
M. Chiachío, University of Granada, Spain
S. Sankararaman, SGT Inc., USA & NASA Ames Research Center, USA
A. Saxena, SGT Inc., USA & NASA Ames Research Center, USA
K. Goebel, NASA Ames Research Center, USA

The chapter describes the application of prognostic techniques to the domain of structural health and demonstrates the efficacy of the methods using fatigue data from a graphite-epoxy composite coupon. Prognostics denotes the in-situ assessment of the health of a component and the repeated estimation of remaining life, conditional on anticipated future usage. The methods shown here use a physics-based modeling approach whereby the behavior of the damaged components is encapsulated via mathematical equations that describe the characteristics of the components as it experiences increasing degrees of

Preface

Structural Health Monitoring (SHM) consists of the acquisition and the analysis of data from a set of permanently installed sensors to determine the health of a structure in a non-destructive way. Since it is possible to assess the structure on demand or continuously, it is feasible to verify the integrity of the structure and determine whether a structure can work properly or whether it needs to be repaired or replaced, with the subsequent maintenance cost saving. In general, SHM includes different levels such as damage detection, damage localization, damage classification and prognosis of remaining lifetime. This book provides relevant theoretical frameworks and some of the latest research and developments in the SHM field. It is no focused specifically on data acquisition technologies neither strategies for signal processing or damage detection. It is focused as a whole system capable to monitor damages in structures. In addition, the book provides examples of applications and results using data-driven approaches in different areas such as civil, aerospace, mechanical engineering, among others.

The book is addressed to students, engineers and researchers involved in the design and development of solutions for structural health monitoring systems. Besides, we expect that it will be a reference for understanding the importance of the field and also for encouraging new solutions in real applications.

Chapter 1 of the book is prepared as an introduction to Structural Health Monitoring (SHM), so that fundamental concepts and main levels of detection are described. Besides, a brief review of the main methodologies based on patter recognition is also presented. Chapter 2, named "Self-healing properties of conventional and fly ash cementitious mortar exposed to high temperature", demonstrates that heating at high temperature deteriorates the physical and mechanical properties of concrete with and without fly ash addition. Existence of partial micro cracks due to the thermal exposure is detected since the ultrasonic pulse velocity decreases. In addition, these micro cracks are confirmed by the decrease in the compressive strength and increase in apparent porosity or water absorption.

The next two, Chapters 3 and 4, named "New features for damage detection and their temperature stability" and "Wavelet transform modulus maxima decay lines: damage detection in varying operating conditions", respectively, are devoted to the first level of SHM, which might be considered by some researchers as the main and most important but also the easiest level. However, if the structure is subjected to variations in the environmental and/or operational normal conditions, the detection of any damage is a challenge since changes in the structural behavior due to these variations can mask patterns that indicate damages, or in opposite, they can wrongly suggest damages. The methodologies presented in these chapters use the response time series of vibrating structures and are based on Wavelet Transform (WT), complemented with Fuzzy Similarity measurements in the first case and decay lines in the second case. In the first of the two chapters, the vibration response is measured by means of piezo-ceramic transductors (PZT's). In the latter one, vibration signals are gathered by means of a high-speed camera. In both cases, structures have been subjected to variations in temperature in different ways.

The following chapters (5 to 9) are focused on the development of systems and strategies for damage detection, classification, localization and estimation of its severity. Chapter 5, entitled "Development of a system for detecting weld failures", presents a prototype of an electronic device capable to excite, efficiently, ultrasound transducers used for detecting weld failures at the junctions of metallic parts. This system can compete on the world market due to its easy implementation and small cost. To validate its efficacy, several kinds of defects on welds are tested (porosity, lack of penetration and lack of fusion) and the time of fly is analyzed comparing pristine specimens with the damaged ones. Chapter 6, "Structural damage assessment using an artificial immune system", proposes to use an artificial immune system to detect and classify damages in structures. The strategy employs data from a multiactuator piezoelectric system, which works in several actuation phases, and it is permanently attached to the surface of the structure under test. The advantage of using a multiactuator system lies in the analysis of the specimen across the surface covered by the network of transducers from different points of view. A known elastic wave is generated in one place of the structure and sensed in different locations, later on, the same wave is generated in other position and sensed in the rest of sensors. All the time series signals are organized and preprocessed by means of Principal Component Analysis (PCA) and finally used to define the antigens that are compared with antibodies by using some affinity function. Chapter 7, "Structure impact localization using emerging artificial intelligence algorithms", give an introduction of a novelty learning algorithm named Extreme Learning Machine (ELM) to locate impact damages in plate structures. The technique is compared with typical neural networks such as back-propagation neural network (BPNN) and least squares support vector machine (LSSVM) and illustrated through experimental studies. Results show that the introduction of the new technique decreases the training and testing execution time with comparable accuracy.

Chapter 8, "Case based reasoning for stiffness changes detection in structures: Numerical validation by using finite element models", presents an expert monitoring system to detect, locate and quantify stiffness variations. The methodology is based on pattern recognition and artificial intelligence techniques that emulate knowledge based on human reasoning. As previous chapters, this strategy uses time-series dynamic responses of the structure. They are preprocessed by means of Wavelet Transform and Principal Component Analysis (PCA). The knowledge of the expert is organized and retrieved following a reasoning based on cases (CBR) supported by self-organizing maps (SOM). To evaluate the accuracy, dynamical responses of the structures to different damage scenarios were simulated and collected by solving its finite element model. Chapter 9, "Systematic statistical approach to structural damage diagnosis", also presents a methodology for detecting, locating and estimating damages in structures. In this case, uncertainties such as physical variability, measurement uncertainty and model errors that affect structural damage diagnosis are considered. A statistical approach is used to identify these sources of uncertainty, quantify their combined effect on diagnosis, and thereby, provide an estimate of the confidence in the results of diagnosis. Damage detection was based on statistical hypothesis testing. While the classical hypothesis testing procedure calculated the uncertainty in detection based on the significance level of testing, the Bayesian hypothesis testing procedure calculated the likelihood of "no damage" scenario to "damage" scenario, thereby providing a robust Bayesian metric to assess the confidence in detection. The uncertainty in localization was quantified using heuristic measures in the classical statistics-based approach while the Bayesian method directly quantified the probability that a given damage parameter is faulty. The uncertainty in damage quantification was expressed in statistical confidence intervals in the classical statistics-based approach while the Bayesian method gives the entire probability distribution of the damage parameter.

The next three chapters (10 to 12) concern to the last level in SHM (Damage Prognosis), which includes the quantification of the damage to determine the useful lifetime remaining for the structure and the conditions to continue operating, The chapter "Nonlinear ultrasonics for early damage detection" suggests that nonlinear mechanical properties may be a key factor to quantify changes in an incipient state. It gives an introduction of the fundamentals of nonlinear ultrasonics, focusing primarily on classical and non-classical hysteretic nonlinearities. Next, nonlinear ultrasonic techniques of the main interest for industrial applications are described. An explanation of nonlinear evidences in metals and composites are also included, because of their relevance in industries with high security standards. Right after, the need for computational modeling to solve the inverse problem for evaluation is addressed; acquired data and theoretical models can be integrated strengthening the evaluation process to allow the solution of cases with weak damage signatures. Finally, the Bayesian inverse problem is explained as the suitable link to prognosis tools. Chapter 11, entitled "Fatigue crack growth analysis and damage prognosis in structure", presents a methodology for modeling crack growth in structures with complicated geometry and multi-axial variable amplitude loading conditions. During cycle-by-cycle integration of the crack growth law, a Gaussian process surrogate model is used to replace the expensive finite element analysis, thereby significantly improving computational effort. The effect of different types of uncertainty on crack growth prediction is investigated. Three different types of modeling errors – crack growth model error, discretization error and surrogate model error – are included in the analysis. The different types of uncertainty are incorporated into the framework for calibration and crack growth prediction, and their combined effect on crack growth prediction is computed. Finally, damage prognosis is achieved by predicting the probability distribution of crack size as a function of number of load cycles. This methodology is useful for design of structural components and systems where the focus is to maximize life expectancy and minimize cost through efficient and accurate reliability analysis. Straightaway, Chapter 12, "Prognostics design for structural health management", details a physics-based modeling approach whereby the behavior of the damaged components is encapsulated via mathematical equations that describe the characteristics of the components as it experiences increasing degrees of degradation. Mathematical rigorous techniques are used to extrapolate the remaining life to a failure threshold. Additionally, Bayesian state estimation is used to compute the probability distribution of the system states by accounting the different sources of uncertainty with making predictions.

Finally, the chapter entitled "An implementation of a complete methodology for wind energy structures health monitoring" is addressed to show a complete methodology for SHM damage detection solution in a particular application, wind energy plants. Several methodologies are proposed for the typical entire process of SHM. Starting with sensor placement (the best possible sensor locations are found), selecting the more representative data, classifying the different environmental and operational conditions and, applying a damage detection methodology, including sensor fault detection. The damage detection methodology is based on Stochastic Subspace Identification (SSI). This identification applies Null-Space analysis of the Hankel matrix, which is able to detect changes in the dynamic response of the structure.

In general, the solutions presented in this book for damage detection, localization and classification are based on data models instead on analytical models. In some cases, due to the lacking of well-equipped laboratory facilities, or to the novelty of the methodology, data are generated and collected by means of detailed physical modeling, typically of the Finite Element type. In other cases, real although not realistic structures are used to generate the data. Original signals are usually preprocessed with the help of Wavelet Transform (WT), Principal Component Analysis (PCA), among others. On the contrary, solutions focused on damage prognosis need the help of an analytical/physical/mathematical model to predict the behavior of the damage and to schedule maintenance. Regarding to examples presented through the book, the reader can find satisfactory results using simulated data as: three degree of freedom mass, spring and damper system, metallic frames, pipeline and, two-radius hollow cylinder; experimental data from simple structures as: cantilever beams, metallic plates, metallic welded specimens; and even more complexes structures as: composite plates, skin panel of the torsion box of the wing provided by stringers and ribs, a wind energy tower model and, graphite-epoxy composite.

Diego Alexander Tibaduiza Burgos
Universidad Santo Tomas, Columbia

Luis Eduardo Mujica
Universitat Politecnica de Catalunya, Spain

Jose Rodellar
Universitat Politecnica de Catalunya, Spain

Introduction

Structural Health Monitoring (SHM) defines a well-known engineering area whose main issue is the verification of the state or the health of structures in order to ensure proper performance using non-destructive tests, involving sensors permanently attached to the structure and computational algorithms. SHM brings different benefits such as: knowledge about the structural behavior under different loads and different environmental changes, knowledge on the current state in order to verify the integrity of the structure and determine whether a structure can work properly or whether it needs maintenance or replacing, with the corresponding maintenance cost saving. Basically, the use of SHM can be extended to any system with structures, this means, a wide range of applications in engineering areas such as civil, aerospace, aeronautics, mechanical and others.

As an example, it is possible to mention aeronautics, where the currently daily passengers flow, merchandise and operations in airports in a global scale suggest increasing the safety in the daily operations in all the elements involved. According to the International Civil Aviation Organization (ICAO), in 2012 the total world volume of scheduled commercial flights began to edge over 31 million per year (ICAO, 2013) and still increasing. However, despite this trend, the global accident rate (accidents per million departures) has changed compared with the year 2011, when it was of 4.2, decreasing in 2012 up to 3.2. This result implies an improving safety and the implementation of best solutions for monitoring. To ensure safety, the ICAO promotes the systematic implementation of Standards and Recommended Practices (SARPs) for the aviation safety through the following activities:

- Policy and standardization initiatives;
- Safety monitoring;
- Safety analysis;
- Regional safety;
- Implementing programs to address safety issues.

At airports level, Safety Management Systems (SMS) are defined in order to contribute to the airports to detect and correct safety problems before they result in aircraft accidents or incidents (FAA, Federal aviation administration, 2012; International Civil Aviation Organization, 2013). These management systems are very important because the risk and the probability of an accident are present in the daily tasks. In addition, there is an increment in this factor due to the higher number of operations, which are significant and still rising. As example, in Spain airports during 2011, the total number of passengers was 204.373.288 and the number of merchandise transportation was 671.722.190 according to AENA ("Aeropuertos Españoles y Navegación Aérea") (AENA, 2012). These quantities associated with the value of what is transported daily, provide important reasons for increasing the safety in airport operations and in the involved elements (airplanes, helicopters, etc.). In a low level, each flight company requires to ensure the reliability of its aircrafts during the different phases of the flight (pre-flight, departure and climb, route, cruise, descent and landing). To do that, the company needs to guarantee the proper performance of their aircraft structures, navigation systems, communication systems, among other elements. According to (FAA, Aircraft inspections, 2012): "when an aircraft is being designed and produced, the aviation authority, the manufacturer, and selected industry participants form groups called Maintenance Steering Groups (MSG) and industry steering committees (ISC). These groups, through numerous meetings determine the frequency and scope of aircraft inspections to be performed. This information is provided to another group called the Maintenance Review Board (MRB) which will issue their final recommendations to the manufacturer on how an aircraft should be maintained". In general, the inspection of any civil aircraft is determined by operation type. The aircraft must also be maintained in an airworthy condition (referred to as continued airworthiness) between those required inspections (Bureau, 2015). Additionally, a preflight inspection is conducted before each flight in ramp, which consists of checking the aircraft by visual examinations and operational tests to detect defects and maladjustments (Navyaviation, 2012). Many times these inspections have revealed faults and damages in the structures. Recently, for instance some small cracks were discovered in the world's biggest passenger aircraft (Airbus A380) during a routine inspection. To correct this possible problem 20 aircrafts were inspected and according to the vice-president of AIRBUS (Tom Williams): "This is not a fatigue problem, but a problem during the manufacturing process" (Staff, 2012). Unfortunately, failures of this type have traditionally been detected during routine inspection periods and normally with the use of various non-destructive techniques, but in the case of visual inspections, it is sometimes impossible to detect small structural damages (for instance between each flight). In this sense, Structural Health Monitoring (SHM) has appeared as a solution to provide tools for early structural damage detection using non-destructive techniques and algorithms.

In a general way, it is possible to compare a SHM system with the human nervous system as in Figure 1. In both cases a sensor network is connected to a central system, which allows apply excitation signals and measure the responses from the sensors distributed along the structure. Additionally there is a system for processing the data, which defines the state or the health of the structure.

In SHM different classes exist for damage diagnosis (Rytter, 1993) in a general way, which can be grouped in four levels (Figure 2). The first level corresponds to the damage detection. In this level it is important to know whether there is any change in the structure and if this change is due to damage. In the second level, after detecting damage, using proper techniques, the position of the damage can be determined. The third level considers the definition of the type of damage and its size. Finally, in the level 4, the remaining lifetime is determined. Recently, an extra level is considered in order to include the capability of auto-healing in smart structures (Inman & Grisso, 2007).

Most common applications in SHM are concentrated in the first three levels and there are many applications using different techniques. The majority of these implementations include the use of Non-Destructive inspections by means of sensors attached to the structure. These experimental setups normally require knowing the structure in order to define which sensors can be used and their distributions in the structure. The variety of sensors and configurations for data acquisition is quite broad as will be shown in the literature review.

Figure 1. Analogy of a SHM system and the human nervous system

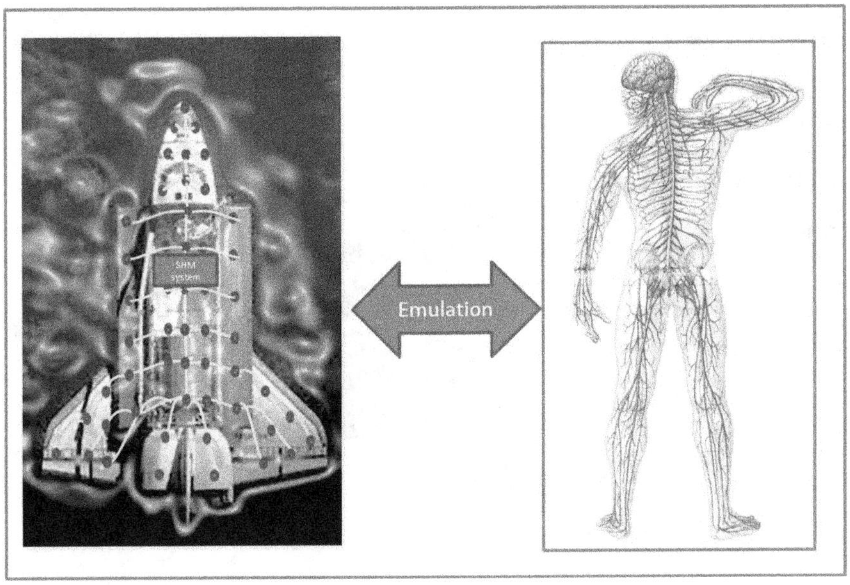

In aeronautic and astronautic areas, it is very common the use of aluminium and composite materials for building the structures (Ye, Lu, Su, & Meng, 2005). Since some years ago (probably since the first introduction into commercial use in 1944 as fuselage skin for Vultee BT-15 trainer plane (Hoskin & Baker, 1986; Ye, Lu, Su, & Meng, 2005), the trend in the design of the structures has been directed towards the use of composite materials because the advantages compared with traditional materials as the aluminum allowing the weight-saving among others benefits. There are many examples useful to show the diversity of the materials currently used in military and commercial applications among, others.

Figure 2. Levels in SHM

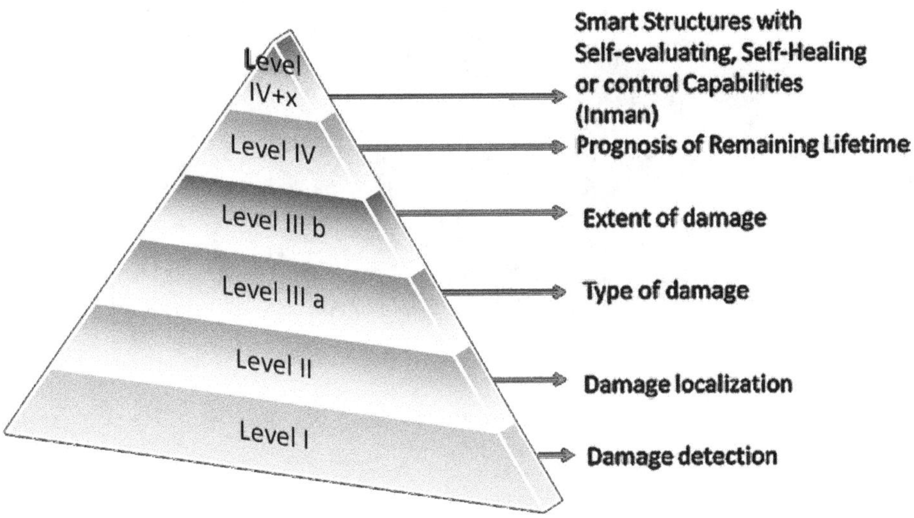

Figure 3a. Fighter aircraft F-18 E/F: F/A-18F Super Hornet *(Daily, 2012).*

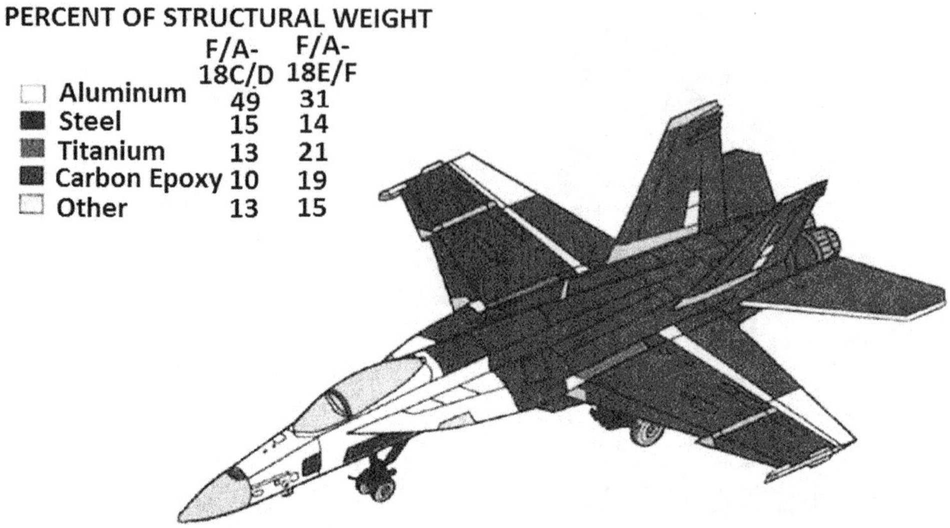

Figure 3b. Fighter aircraft F-18 E/F: schematic diagram with the percentage of structural weight *(Baker, 2004).*

Figure 4. Utility of composite structures on A380: monolithic CFRP and thermoplastics (Ye, Lu, Su, & Meng, 2005).

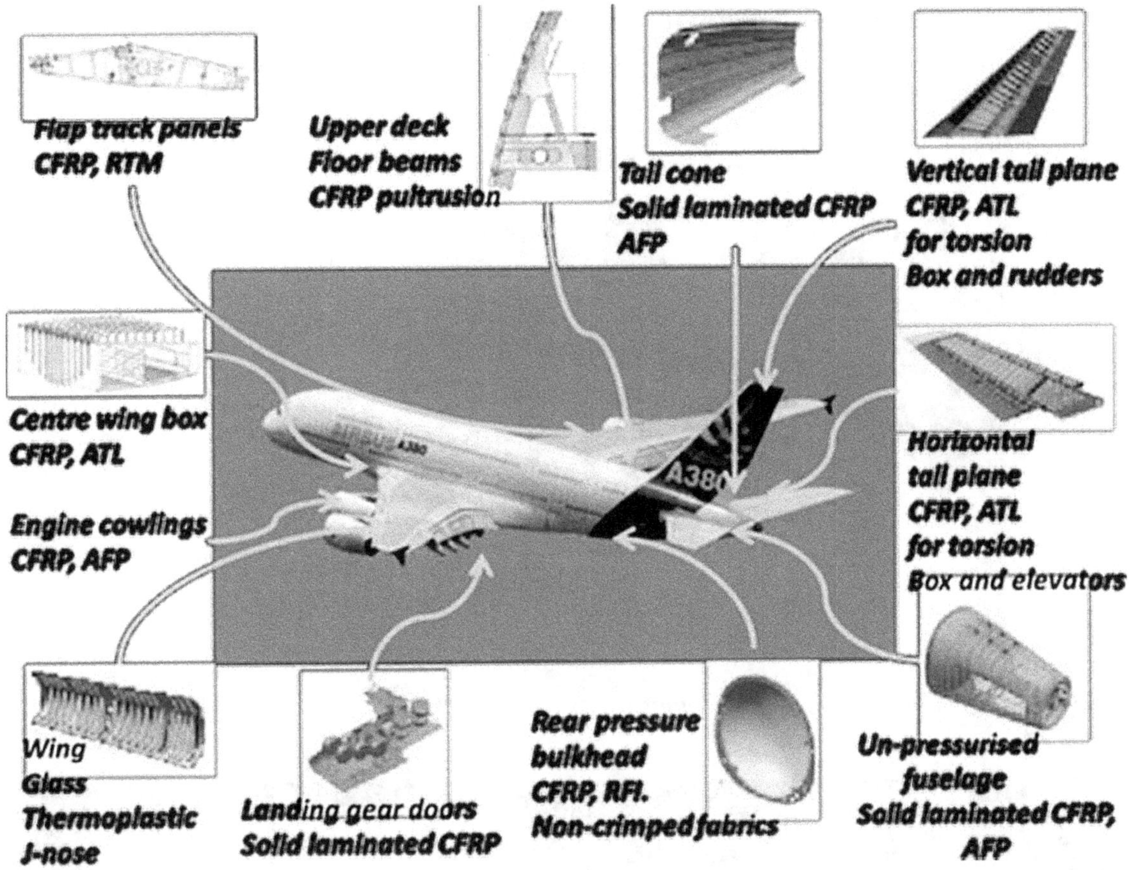

For instance, in the U.S. Navy, the F-18 E/F fighter has 31% of aluminium of the total weight, while the carbon epoxy has the 19% as can be seen in the Figure 3. Other example can be found in the AIRBUS A380, which is one of the biggest commercial aircraft. Although the use of composite materials has increased, the aluminum is still widely used (Figures 4 and 5).

The SHM systems are currently in a development stage and the majority of the applications are available in a research level, especially in the aeronautic and astronautic areas. To perform the inspection of the structures in the majority of applications, the structure is isolated and the monitoring is performed under special conditions. In areas as aeronautical and aerospace it is really important to evaluate the health of the structures in normal operational conditions when the element is integrated into the system (aircraft, helicopters, satellites, space shuttle, among others). This reason has motivated the development of new methodologies. An additional motivation is the reduction in the cost of maintenance to avoid that the aircraft goes out of operation for periodical maintenance and to increase the safety in the normal operation of the structures.

Other examples about the need of SHM systems can be found in civil engineering. This area presents big interest since large infrastructures imply big quantity of humans that use daily these civil developments. One example in this area can be found in the building called SPACE in Medellin-Colombia, where a bad design and the lack of effective state inspection, resulted in loss of life and the destruction of one of its buildings (Figure 6).

Figure 5. Utility of composite structures on A380: materials distributions (weight breakdown) *(Ye, Lu, Su, & Meng, 2005).*

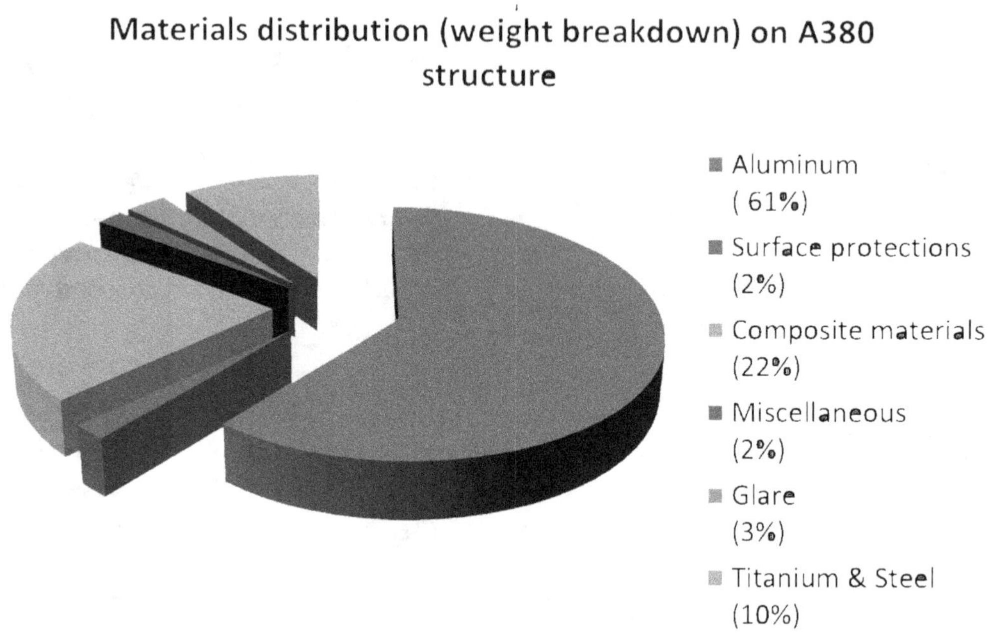

Figure 6. Building Space, Medellín-Colombia *(Tiempo, 2014).*

A BRIEF REVIEW OF STRUCTURAL HEALTH MONITORING AS A PATTERN RECOGNITION APPROACH

The problem of structural monitoring can be tackled from different points of view. Some authors have built mathematical and numerical models based on physical laws to describe the characteristics of the structure. In this kind of approaches, a high-fidelity model of the structure is required to perform a reliable damage identification. On the other hand, other authors use techniques based on data gathered by experiments or by numerical simulations. These approaches usually require a statistical model representation of the system to perform the structural state analysis (Worden & Manson, 2007). In this case, the system does not use physically based models and the problem of damage identification can be approached as a pattern recognition application, where some features of the collected signals are used as reference pattern. In general, it exploits the fact that the vibrational response of a structure is deferent when it is healthy or damaged. In this way, if some defect exists in the structure, its vibrational response will change and these changes can be analyzed.

Several reviews have been carried out in Structural Health Monitoring (SHM). Among them, Chang, Flatau, and Liu (2003) presented a review of SHM for civil infrastructures. Fritzen (2005) presented an overview of the developments of vibration based methods in the year 2005, also in 2006, Lynch and Loh (2006) presented a summary review of wireless sensors and sensor networks. Farrar and Worden (2007) performed a brief historical review of SHM technology development. The same year, Brownjohn (2007) presented a review of SHM applications to various forms of civil infrastructure, including a discussion about the damage diagnosis procedure in terms of instrumentation, data acquisition, communication systems and data mining. The next year, Chiang, Lee, and Shin (2008) presented a review focused in damage detection methods for a wind turbine system. In 2010, Mujica, Rodellar, and Vehi (2010) performed a review of impact damage detection in structures using strain data, which included sensors, specimens and impact sources used for developing and testing strategies. Recently, in 2011, Fan and Qiao (2011) presented a review and comparative study of vibration-based damage identification methods. The review included methods based on modal parameters, natural frequencies, mode shapes as well as curvature/strain mode and shape-based methods. These reviews have proved that the interest in the development of algorithms and methodologies in SHM has been growing. As result of this interest, SHM has been applied in different areas, which include applications in civil, aeronautics and astronautics structures. Many works has been reported for more than three decades (Chang, Flatau, & Liu, 2003) with promising results, for instance, tests in bridges (Azevedo, et al., 1996; Lee, Kim, Yun, Yi, & Shim, 2002; Riveros, 2007; Kawano, Mikami, & Katsuki, 2010; Panetsos, Ntotsios, Papadimitriou, Papadioti, & Dakoulas, 2010), buildings (Garziera, Amabili, & Collini, 2007; Serino & Spizzuoco, 2010), wind turbine blades (Park, Taylor, Farinholt, & Farrar, 2010), and other structures (Law & Sohn, 2000; Mujica et al., 2005; Sohn, Farrar, Hunter, & Worden, 2001; Jaques, Adams, Doyle, & Reynolds, 2010).

The next subsections give a brief overview of the SHM levels in order to show from a general point of view the wide range of methods and applications developed in SHM.

Brief Review of the Structural Health Monitoring Levels

As it was mentioned, SHM includes different levels starting by the detection of the damage and following with the localization, classification, identification and the prognosis of damages (prediction) (Worden & Manson, 2007).

Damage Detection

The damage detection corresponds to the first level in SHM, whose goal is to detect defects or damages in structures when they are still in their nascent state using non-destructive techniques and algorithms. As was previously introduced, damages can be defined as changes in the structure when it is compared with a baseline obtained from the structure in a healthy state. These changes include variations in the material or in the geometric properties.

The need of damage detection to prevent an accident is an essential factor to ensure the proper work of a structure in service. There are many applications in different areas, but it is important to note that in some areas as aeronautical and aerospace engineering the continuous monitoring of the structural health is very important to guarantee its proper performance. Probably, the use of SHM in these two areas is more important than in other areas since, normally, in areas like civil engineering, some damages are really dangerous when they start to have a considerable size. However, in aeronautical and aerospace engineering, some damages that are imperceptible when are subjected to extreme changes in their working conditions can cause catastrophes. In the aeronautical industry, for instance, the majority of inspections are performed visually (Brand & Boller, 1999; Mujica et al., 2010). However, many research groups around the world are focusing efforts to develop techniques that allow the inspection of structures in service. These investigations are focused on developing techniques for making the best detection of the different possible damages. Some developments at the damage detection level are reviewed below.

In 1992, Kudva, Munir, and Tan (1992), in 1993, Gunter, Wang, Fogg, Starr, Murphy, and Claus (1993), and Hann, Wilkerson, and Stuart (1994), Sirkis, Berkoff, Kersey, Fribele, and Jones (1994) in 1994, and Schindler, May, Claus, and Shaw (1995) in 1995 worked on impact damage detection by using advanced signal processing with Artificial Neural Networks (ANN). Friswell, Penny, and Garvey (1998) in 1998 used a genetic algorithm for damage detection based on vibration data in order to identify the position of one or more damage sites in a structure and to estimate the extent of these damages. Yap and Zimmerman (1998) used genetic algorithms for damage detection. In difference with the classical coding of the genetic algorithm, this work proposed the use of two coding enhancement strategies. In 2001, Ganguli (2001) proposed a fuzzy logic system for health monitoring of a helicopter rotor blade when this is on ground. The rules of the fuzzy system were defined in order to consider four different levels of damages in the output. The measurements used were the first four flap (transverse bending) frequencies of the rotor blade. Hao and Xia (2002) in 2002 used a genetic algorithm with real number encoding to identify the structural damage by minimizing the objective function, which directly compares the changes in the measurements before and after damage. They used three different criteria, namely, the frequency changes, the mode shape changes and a combination of both. This methodology was tested in a cantilever beam and a frame. The same year, Staszewski, Biemans, Boller, and Tomlinson (2002) used passive acousto-ultrasonic sensors to impact damage detection in composite structures. In this work, a study of different signal processing methods for passive damage monitoring was performed. In 2004, Sun and Chang (2004) presented a methodology to classify some statistical patterns using the Wavelet Packet Transform (WPT). The vibration signal collected from the structure was decomposed into wavelet packets. Later on, the signal energy of the packets were calculated and ordered according to the magnitudes. The most important magnitudes were considered and the rest were rejected, using these magnitudes and defining some thresholds and confidence limits to detect abnormal behaviors. This methodology was tested in a beam and four damage cases were studied involving line cuts of different severities in the flanges at one cross section. In 2006, Menendez, Fernandez, and Guemes (2006) used an active piezoelectric system and Fibre Bragg Grating (FBG) to detect de-bonding of subcomponents in monolithic composite parts. In the methodology, the power spectrum was used to find the damage. In the piezoelectric active system the maximum amplitude of every piezoelectric sensor is used in order to measure the energy loss of the input pulse. The same year, Fernandez, Guemes, Fritzen, and Mengelkamp (2006) performed a comparison between the use of FBG and piezoelectric sensors for damage detection. To do that, the Hankel matrix was used to obtain the damage indicator values. They proposed, as an alternative, the use of combined hybrid piezoelectric/FBG sensors. Wildy, Kotousov, and Codrington (2007) in 2007 proposed a passive method of damage detection based on the concept of strain field. Huang, Ghezzo, Rye, and Nemat-Nasser (2008) used acoustic emission in thin glass/epoxy composites plates in order to detect damages. The same year, Mujica, Vehi, Staszewski, and Worden (2008) used a hybrid methodology which combined Case Base Reasoning (CBR), wavelet transform and Self Organizing Maps (SOM) in order to detect impact damages in a wing flap. Piezoceramic sensors attached over the surface of the flap were used to collect time varying strain response data. In 2009, Chandrashekhar and Ganguli (2009) applied a fuzzy logic system with a sliding window defuzzifier using modal curvatures for damage detection. The methodology fuzzified the changes due to a damage in modal curvature using Gaussian fuzzy sets and mapped to damage location and size. This methodology was applied to a cantilever beam. Recently, Dervilis, Barthorpe, Staszewski, and Worden (2012) presented a scheme for damage detection in carbon fiber materials using novelty detection methods. These methods were applied to FRF measurements from a stiffened composite plate, which was subjected to incremental levels of impact damage.

Damage Localization

The localization of damages is the next task after damage detection. Its complexity depends on the structure, the type of sensors and their distribution. Some of the most common strategies for the location of damages include triangulation processes (Mujica et al., 2010). According to Salehian (2003), three sensors are enough for determining impact location for isotropic materials, but in anisotropic materials as composite materials these approaches must be different (Mujica et al., 2010). Other researches have combined different strategies combined with the processing of sensor data for locating damages. Some of them are discussed below.

Staszewski et al. (2000) and Worden and Staszewski (2000), used ANNs as regressor to predict the impact location and energy in composite materials. The approach used a multilayer perceptron which was trained with experimental data using back propagation.

Additionally, these works presented the combined use of the genetic algorithms and ANN to find near-optimal sensor distributions for damage detection.

In 2001, Chou and Ghaboussi (2001) presented a methodology based on genetic algorithms, which uses static measures of displacement for damage localization. In 2003, Coverley and Staszewski (2003) proposed a methodology for damage location using triangulation methods and genetic algorithms. Gorinevsky, Gordon, Beard, Kumar, and Chang (2005) presented a design as a subsystem to the Integrated Vehicle Health Management (IVHM) system of an aircraft. This system used SMART Layer sensors from Acellent Technologies. Using the signals obtained from the sensors, the mean signal amplitude was calculated and compared with the scatter signal to obtain a damage index for each actuator-sensor path. Using these values, they obtained a representation of the location structural changes and one measure of the damage size. In 2010, Mujica, Tibaduiza, and Rodellar (2010) used Principal Component Analysis and some statistic indices to localize different damages using contribution plots in an aircraft turbine blade. Recently, Hackmann, Sun, Castaneda, Lu, and Dyke (2012) in 2012 presented a holistic approach for damage localization, which integrated a decentralized computing architecture using wireless sensor networks. In the approach the damage localization algorithm used post-processed natural frequency data.

Damage Classification

The damage classification corresponds to the level which, according to some characteristics obtained from the structure in different structural states and some algorithms, a classification of each state can be performed according to their features. In general, a typical classifier is used to define which damage is present in the structure. Some examples are shown below.

Using a combination of time series analysis, neural networks and statistical inference techniques, Sohn, Worden, and Farrar (2002) in 2002 classified damages under environmental changes. First, an AR-ARX model was developed to extract damage-sensitive characteristics, then a neural network was used to normalize the data and to separate the effect caused by the environmental changes. Finally, a sequential probability ratio test was performed to define the state of the system. The methodology was tested using a numerical example of a computer hard disk and an experimental study of an eight degree-of-freedom spring-mass system. In 2007 and 2008, Mujica et al. (2007) and Mujica, Vehi, Staszewski, and Worden (2008) used Wavelet Transform in a hybrid methodology to detect, quantify and localize damages. This methodology used the wavelet transform in order to extract different characteristics from the measured signal and subsequently applied a neural network to classify the damages. Zhou, Chakraborty, Kowali, Papandreou, and Cochran (2007) proposed an algorithm for the classification of structural damage based on the use of continuous Hidden Markov Modeling (HMM) technique. It was used to model time-frequency damage features extracted from structural data using the matching pursuit decomposition algorithm.

Many works have shown the usefulness of neural networks for classification (Bakhary, Hao, & Deeks, 2007). For instance, clustering algorithms based on Self-Organizing Maps -SOM- (attempting to organize feature vectors into clusters) have been used for the classification of acoustic emissions (Yan, Holford, Carter, & Brandon, 1999; Godin, Huguet, Gaertner, & Salmon, 2004) and for active sensing damage classification (Tibaduiza, Mujica, Gemes, & Rodellar, 2010). Dua, Watkins, Chandrashekhara, and Akhavan (2001) in 2001 used an ANN with backpropagation algorithm and Finite Element Analysis (FEA) to classify impacts on composite plates. A 503,10,3 ANN was used for training and simulating the data: 503 elements in the input layer which were excited by strain profiles obtained from FEA, 10 neurons in the hidden layer and 3 neurons in the output layer. A total of seven classification groups were performed inspecting the composite plates and the kinetic energy of the falling mass. This classification was coded using Gray Code.

In 2006, Kolakowski, Mujica, and Vehi (2006) presented two approaches for damage identification. One of them was based on Virtual Distortion Method (VDM). The other methodology involved the use of Case Based-Reasoning (CBR) applying wavelet transform in order to extract features and reduce the variables to introduce into a Self-Organizing Map (SOM) for damage identification. These techniques were tested in an aluminum beam. In 2007, Bakhary, Hao, and Deeks (2007) applied a two-stage ANN system for damage location and damage severities. In the first stage, an ANN was used to identify the substructures with damage and the secondary ANN identified the damaged elements and its severity. Inputs in the first ANN were modal frequencies and mode shapes of the full structure and the outputs were modal frequencies of substructures, these were the inputs to the second ANN where the final analysis to locate the damage was performed. For testing the approach, a numerical example was used which consisted of a two-span concrete slab. Dobrzanski, Sroka, and Dobrzanski (2007) used a multilayer perceptron 9-6-5 for the classification of internal damages in steel during creep services using metallographic images. Also in 2007 and 2008, Mujica et al. (2007; Mujica, Vehi, Staszewski, & Worden, 2008) presented a methodology to detect, quantify and localize damages and impacts in several structures, among them a wing aircraft section and an aluminum beam. This methodology used wavelet transforms to extract different characteristics from the collected signal and a SOM to classify them. In 2008, Kabir, Rivard, and Ballivy (2008) presented an algorithm for damage classification using a multilayer perceptron. The methodology included the use of analysis of texture of surface deterioration using optical imagery in concrete structures. The data in the perceptron included three different datasets: spatial, spectral and a combination spatial-spectral dataset. Iskandarani (2010) in 2010 applied neural networks to classify composite structure conditions. Resin injection molded (RIM) samples response to impact damage was carried out using low frequency tapping, visual imaging, low temperature thermo imaging and tensile strength. Recently, Zhou, Tang, Zang, and Zhou (2012) presented an Artificial Inmune Pattern Recognition (AIPR) approach for the damage classification in structures. The approach included the development of an immune learning algorithm and a ARX algorithm to compress the data.

Type and Extent of Damage

The knowledge of the type of damage is a task that provides information about the severity of damage. This is performed after the detection and localization and provides information on whether the structure can still do its job or must be replaced. Some works related to the definition of the damage type and extent are outlined next.

Mares and Surace (1996) in 1996 presented a strategy based on genetic algorithms and residual force method (modal analysis theory) to detect, quantify and obtain the extent of damages in elastic structures. Friswell, Penny, and Garvey (1998) in 1998 used a genetic algorithm for damage detection in vibration based data to identify the position of one or more damages in a structure and to estimate their length. Hao and Xia (2002) in 2002 used a genetic algorithm with real number encoding to identify the structural damage by minimizing the objective function, which directly compares the changes in the measurements before and after damage. They used three different criteria, namely, the frequency changes, the mode shape changes, and a combination of them. This methodology was tested in a cantilever beam and a frame. In 2003, Mujica et al. (2003) showed the use of Case Based Reasoning as tool for damage diagnosis. A wavelet transform was also used to obtain some characteristics and a Self-Organizing Map (SOM) was used as method for handling the case base. This methodology was tested in a cantilever truss. The same year, Chang and Sun (2003) proposed a novel structural condition index for locating and quantifying structure damage based on Wavelet Packet Signature (WPS). This year, Shan and King (2003) presented a methodology to locate impacts and estimate impact magnitude on smart composites using fuzzy clustering for feature selecting and adaptive neuro-fuzzy inference system for impact locating and magnitude estimation. In 2005, Mujica et al. (2005) extended the methodology presented in 2003 to define the severity and the dimension of the damage. Recently in 2011, Gul and Catbas (2011) presented a time series methodology to detect, locate and estimate the extent of the structural changes. The approach used ARX models, which were obtained for different sensor clusters by using the free response of the structure. The approach considered also obtaining the ARX model fit ratios or the ARX coefficients as damage features.

Damage Prognosis (DP)

Damage prognosis is the last level in SHM, which includes the quantification of the damage to determine the useful lifetime remaining for the structure and the conditions to continue operating. The publications in this kind of applications are lesser in number than those existing for the previous SHM levels. Some of them are discussed below.

Staszewski, Worden, Wardle, and Tomlinson (2000) presented a discussion focused on extraction and data pre-processing in pattern recognition procedures for the diagnosis of location and severity of damage. In 2003, Farrar, Park, Robertson, Sohn, and Williams (2003) presented an approach for damage prognosis by integrating advanced sensing technology, data interrogation procedures for state awareness, novel model validation and uncertainty quantification techniques, and reliability-based decision-making algorithms. Later on, in 2007, Farrar and Lieven (2007) presented some general concepts of damage prognosis, such as the problem definition, the motivation and the process of DP. Also a review of emerging technologies with potential impact on the damage prognosis process was included. In 2010, Zhang, Zhou, and Li (2010) presented an approach that included the use of flexible piezo paint sensor and a probabilistic fracture mechanics based framework for on-line assessment and updating of the remaining fatigue life of steel bridges. In 2012, Ling and Mahadevan (2012) presented a Bayesian probabilistic methodology to integrate model-based fatigue prognosis with online and offline SHM data, considering various sources of uncertainty and errors. The methodology was tested in a numerical example, considering the surface cracking in a rotorcraft mast under service loading.

ORGANIZATION OF THE BOOK

This book aims to cover the SHM levels above discussed, considering different applications. The book is divided into 14 chapters, starting with this introduction. Chapter 2 presents an experimental investigation on self-healing properties of conventional and fly ash cementitious mortar exposed to high temperature. The study shows the self-recovery of significant physical properties to micro cracks generated in the material. Chapter 3 considers new features for damage detection in the presence of temperature uncertainties, using fuzzy similarity and wave clustering. The stability of these features is tested on different experimental case studies with increasing complexity. Chapter 4 approaches the problem of damage detection under varying operating conditions using wavelet transform, in particular the modulus maxima decay lines. The proposed methodology is evaluated on both simulated and experimental case studies under varying temperatures. Chapter 5 lies within the context of ultrasonic based non-destructive welding inspection. A prototype and a methodology for detection of weld failures are presented and tested. Chapter 6 presents a methodology for structural damage classification using an artificial immune system. The approach is implemented to an aircraft skin panel equipped with active piezoelectric devices. Chapter 7 provides a practical comparison of several learning machine algorithms for impact localization in structures. Both accuracy and execution time are considered in the experimental evaluation, where the so-called extreme learning machine exhibits a good potential for impact detection due to the fast learning speed. Chapter 8 contributes with a case based reasoning scheme to detect stiffness changes in structures. The approach is validated through numerical experiments using finite element models. Chapter 9 proposes a statistical methodology to structural damage diagnosis (detection, localization and estimation) under uncertainties such as physical variability, measurement noise and modeling errors.

The proposed methods are illustrated using two types of example problems, a structural frame and a hydraulic actuation system. Chapter 10 discusses on nonlinear ultrasonic approaches for early damage detection, considering fundamentals, experimental techniques, damage evaluation methods and future trends. Chapter 11 is devoted to analysis of fatigue crack growth and damage prognosis in structures. The effect of different types of uncertainty – physical variability, data uncertainty and modeling errors – on crack growth prediction is investigated.

Chapter 12 deals with prognostics design for structural health management. Using physics-based mathematical models, the authors propose methods to estimate the remaining life of structures to failure thresholds using fatigue data. Chapter 13 describes a complete methodology for health monitoring of wind energy structures. It includes sensor location, data selection, classification of environmental and operational conditions, damage detection and sensor fault detection. Only output data are used to implement the methodology, which is experimentally tested on a laboratory tower model. Finally, Chapter 14 closures the book with some concluding remarks.

Diego Alexander Tibaduiza Burgos
Universidad Santo Tomas, Columbia

Luis Eduardo Mujica
Universitat Politecnica de Catalunya, Spain

Jose Rodellar
Universitat Politecnica de Catalunya, Spain

REFERENCES

AENA. (2012). *Aena aeropuertos*. Retrieved Jan 10, 2015, from estadísticas tráfico: http://www.aena.es/csee/Satellite?pagename=Estadisticas/Home

Azevedo, S., Mast, J., Nelson, S., Rosenbury, E., Jones, H., McEwan, T., & Wieting, M. et al. (1996). Hermes: A high-speed radar imaging system for inspection of bridge decks. nondestructive evaluation techniques for aging infrastructure and manufacturing. *SPI*, *294*(6), 1954204.

Baker, A. (2004). Chapter introduction and overview. In A. A. Baker, S. Dutton, & D. Kelly (Eds.), *Composite materials for aircraft structures* (pp. 1–21). Reston, VA: AIAA Education Series.

Bakhary, N., Hao, H., & Deeks, A. (2007). *Neural network based damage detection using substructure technique*. Paper presented at the 5th Australian Congress on Applied Mechanics (ACAM 2007), Brisbane, Australia.

Brand, C., & Boller, C. (1999). Identification of life cycle cost reductions in structures with self-diagnostic devices. In *Proceedings of NATO RTO Symposium on Design for Low Cost Operation and Support* (pp. 1-8). NATO.

Brownjohn, J. (2007). Structural health monitoring of civil infrastructure. *Philosophical Transactions A: Royal Society A*, *365*, 589-622.

Bureau, A. S. (2015). *Required aircraft inspections*. Retrieved January 10, 2015, from Aviation Safety Bureau: http://www.aviation-safety-bureau.com/aircraft-inspections.html

Chandrashekhar, M., & Ganguli, R. (2009). Structural damage detection using modal curvature and fuzzy logic. *Structural Health Monitoring*, *8*, 267–282.

Chang, C., & Sun, Z. (2003). Locating and quantifying structure damage using spatial wavelet packet signature. In Proceedings of Smart Structures and Materials 2003: Smart Systems and Nondestructive Evaluation for Civil Infrastructures (vol. 5057, pp. 97-105). SPIE.

Chang, P. C., Flatau, A., & Liu, S. (2003). Review paper: Health monitoring of civil infrastructure. *Structural Health Monitoring*, *2*, 257–267.

Chiang, C. C., Lee, J., & Shin, H. (2008). Structural health monitoring for a wind turbine system: A review of damage detection methods. *Measurement Science & Technology*, *19*(12).

Chou, J.-H., & Ghaboussi, J. (2001). Genetic algorithm in structural damage detection. *Computers & Structures*, *79*(14), 1335–1353.

Coverley, P., & Staszewski, W. (2003). Impact damage location in composite structures using optimized sensor triangulation procedure. *Smart Materials and Structures*, *12*, 795–803.

Daily, D. I. (2012). *The USA's 2005-2009 multi-year hornet procurement contract*. Retrieved July 2012, from Defense Industry Daily: http://www.defenseindustrydaily.com/the-usas-2005-2009-multi-year-hornet-procurement-contract-03882/

Dervilis, N., Barthorpe, R., Staszewski, W., & Worden, K. (2012). Structural health monitoring of composite material typical of wind turbine blades by novelty detection on vibration response. *Key Engineering Materials*, *518*, 319–327.

Dobrzanski, L., Sroka, M., & Dobrzanski, J. (2007). Application of neural networks to classification of internal damages steel working in creep service. *Journal of Achievements in Materials and Manufacturing Engineering*, *20*(1–2), 303–306.

Dua, R., Watkins, S., Chandrashekhara, D., & Akhavan, F. (2001). *Detection and classification of impact-induced damage in composite plates using neural networks*. Paper presented at the International Joint Conference on Neural Networks 2001, Washington, DC.

FAA. (2012a). *Aircraft inspections*. Retrieved from Federal aviation administration (FAA): http://www.faa-aircraft-certification.com/aircraft-inspections.html

FAA. (2012b). *Federal aviation administration (FAA)*. Retrieved July 12, 2012, from Airport Safety: http://www.faa.gov/airports/airport_safety/

Fan, W., & Qiao, P. (2011). Vibration-based damage identification methods: A review and comparative study. *Structural Health Monitoring*, *10*(1), 83–111.

Farrar, C., & Lieven, N. (2007). Damage prognosis: the future of structural health monitoring. *Philosophical Transactions A: Royal Society A*, *365*, 623–632.

Farrar, C. H., Park, G., Robertson, A., Sohn, H., & Williams, T. (2003). A coupled approach to developing damage prognosis solutions. *Key Engineering Materials*, *245-246*, 289–306.

Farrar, C. R., & Worden, K. (2007). An introduction to structural health monitoring. *Philosophical Transactions A: Royal Society A*, *365*, 303–315.

Fernandez, A., Guemes, A., Fritzen, C., & Mengelkamp, G. (2006). *Comparison of health monitoring systems with fiber bragg grating and piezoelectric sensors*. Paper presented at the Third European Workshop on Structural Health Monitoring, Granada, Spain.

Friswell, M., Penny, J., & Garvey, S. (1998). A combined genetic and eigensensitivity algorithm for the location of damage in structures. *Computers & Structures*, *69*(5), 547–556.

Fritzen, C. (2005). Vibration-based structural health monitoring – Concepts and applications. *Key Engineering Materials*, *293-294*, 3–20.

Ganguli, R. (2001). A fuzzy logic system for ground based structural health monitoring of a helicopter rotor using modal data. *Journal of Intelligent Material Systems and Structures*, *12*, 397–407.

Garziera, R., Amabili, M., & Collini, L. (2007). *Structural health monitoring techniques for historical buildings*. Paper presented at the IV Conferencia panamericana de END, Buenos Aires, Argentina.

Godin, N., Huguet, S., Gaertner, R., & Salmon, L. (2004). Clustering of accoustic emission signals collected during tensile tests on unidirectional glass/polyester composite using supervised and unsupervised classifiers. *NDT & E International*, *37*, 253–264.

Gorinevsky, D., Gordon, G., Beard, S., Kumar, A., & Chang, F. (2005). *Design of integrated SHM system for commercial aircraft applications*. Paper presented at the Fifth International Workshop on Structural Health Monitoring, Stanford, CA.

Gul, M., & Catbas, N. (2011). Structural health monitoring and damage assessment using a novel time series analysis methodology with sensor clustering. *Journal of Sound and Vibration*, *330*, 1196–1210.

Gunter, M., Wang, A., Fogg, R., Starr, S., Murphy, K., & Claus, R. (1993). Fiber optic impact detection and location system embedded in a composite material. In *Proceedings of Meeting on Fiber Optic Smart Structures and Skins V* (pp. 262-269). Academic Press.

Hackmann, G., Sun, F., Castaneda, N., Lu, C., & Dyke, S. (2012). A holistic approach to decentralized structural damage localization using wireless sensor networks. *Computer Communications*, *36*(1), 29–41.

Hann, H., Wilkerson, B., & Stuart, J. (1994). An artificial neuronal network for low-energy impact monitoring. *Journal of Thermoplastic Composite Materials*, *7*, 344–351.

Hao, H., & Xia, H. (2002). Vibration-based damage detection of structures by genetic algorithm. *Journal of Computing in Civil Engineering*, *16*(3), 222–229.

Hoskin, B. C., & Baker, A. (1986). *Composite materials for aircraft structures*. New York: American Institute of Aeronautics and Astronautics.

Huang, Y., Ghezzo, F., Rye, P., & Nemat-Nasser, S. (2008). Passive damage detection in composite laminates with integrated sensing networks. In *Proceedings of SPIE - The International Society for Optical Engineering*. SPIE.

ICAO. (2013). *International civil aviation organization – ICAO*. Retrieved January 05, 2015, from State of global aviation safety 2013: http://www.icao.int/safety/State%20of%20Global%20Aviation%20Safety/ICAO_SGAS_book_EN_SEPT2013_final_web.pdf

Inman, D., & Grisso, B. (2007). Adaptive structures for structural health monitoring. In D. Wagg, I. Bond, P. Weaver, & M. Friswell (Eds.), *Adaptive structures: Engineering applications* (pp. 1–30). Chichester, UK: John Wiley & Sons, Ltd.

Iskandarani, M. (2010). Application of neural network to damage classification in composite structures. *Latest Trends on Computers*, *1*, 109–113.

Jaques, J., Adams, D., Doyle, D., & Reynolds, W. (2010). *Using impact modulation to identify loose bolts in a satellite structure*. Paper presented at the Fifth European Workshop on Structural Health Monitoring, Sorrento. Italy.

Kabir, S., Rivard, P., & Ballivy, G. (2008). Neural-network-based damage classification of bridge infrastructure using texture analysis. *Canadian Journal of Civil Engineering*, *35*(3).

Kawano, Y., Mikami, T., & Katsuki, F. (2010). *Health monitoring of a railway bridge by fiber optic sensor (sofo)*. Paper presented at the Fifth European Workshop on Structural Health Monitoring, Sorrento, Italy.

Kolakowski, P., Mujica, L., & Vehi, J. (2006). Two approaches to structural damage detection: VDM vs CBR. *Journal of Intelligent Material Systems and Structures*, *16*, 63–79.

Kudva, J., Munir, N., & Tan, P. (1992). Damage detection in smart structures uisng neuronal networks and finite-element analysis. *Smart Materials and Structures*, *1*, 108–112.

Law, K., & Sohn, H. (2000). Application of load-dependent ritz vectors to bayesian probabilistic damage detection. *Probabilistic Engineering Mechanics*, *15*, 139–153.

Lee, J. W., Kim, J. D., Yun, C. B., Yi, J. H., & Shim, J. M. (2002). Health monitoring method for bridges under ordinary traffic loadings. *Journal of Sound and Vibration*, *257*, 247–264.

Ling, Y., & Mahadevan, S. (2012). Integration of structural health monitoring and fatigue damage prognosis. *Mechanical Systems and Signal Processing*, *28*, 89–104.

Lynch, J., & Loh, K. (2006). A summary review of wireless sensors and sensor networks for structural health monitoring. *The Shock and Vibration Digest*, *38*(2), 91–128.

Mares, C., & Surace, C. (1996). An application of genetic algorithms to identify damage in elastic structures. *Journal of Sound and Vibration*, *195*(2), 195–215.

Menendez, J., Fernandez, A., & Guemes, A. (2006). *SHM with embedded fibre bragg gratings and piezoelectric devices*. Paper presented at the Third European Workshop on Structural Health Monitoring, Granada, Spain.

Mujica, L., Rodellar, J., & Vehi, J. (2007). *A case based reasoning approach for damage assessment in smart structures*. Paper presented at the III ECCCOMAS Thematic Conference on Smart Structures and Materials, Gdansk, Poland.

Mujica, L., Rodellar, J., & Vehi, J. (2010). A review of impact damage detection in structures using strain data. *International Journal of COMADEM*, *13*(1), 3–18.

Mujica, L., Tibaduiza, D., & Rodellar, J. (2010). *Data-driven multiactuator piezoelectric system for structural damage localization*. Paper presented at the Fifth World Conference on Structural Control and Monitoring, Tokyo, Japan.

Mujica, L., Vehi, J., Rodellar, J., & Kolakowski, P. (2005). A hybrid approach of knowledgebased reasoning for structural assessment. *Smart Materials and Structures*, *14*, 1554–1562.

Mujica, L., Vehi, J., Rodellar, J., & Kolakowsky, P. (2003). *Damage identification by using softcomputing techniques*. Paper presented at the Smart Materials and Structures-AMAS Workshop-SMART'03, Jadwisin, Poland.

Mujica, L., Vehi, J., Staszewski, W., & Worden, K. (2008). Impact damage detection in aircraft composites using knowledge-based reasoning. *Structural Health Monitoring, an International Journal*, *7*(30), 215-230.

Navyaviation. (2012). *Aircraft inspections*. Retrieved from http://navyaviation.tpub.com/14022/css/14022_41.htm

Panetsos, P., Ntotsios, E., Papadimitriou, C., Papadioti, D., & Dakoulas, P. (2010). *Health monitoring of metsovo bridge using ambient vibrations*. Paper presented at the Fifth European Workshop on Structural Health Monitoring, Sorrento, Italy.

Park, G., Taylor, S., Farinholt, K., & Farrar, C. (2010). *SHM of wind turbine blades using piezoelectric active sensors*. Paper presented at the Fifth European Workshop on Structural Health Monitoring, Sorrento, Italy.

Riveros, C. (2007). Structural health monitoring methodology for simply supported bridges: Numerical implementation. *Revista Facultad de Ingeniera Universidad de Antioquia*, *39*, 42–55.

Rytter, A. (1993). *Vibration based inspection of civil engineering structures.* (PhD thesis). Department of Building Technology and Structural Engineering, Aalborg University, Aalborg, Denmark.

Salehian, A. (2003). *Identifying the location of a sudden damage in composite laminates using wavelet approach.* (Master's thesis). Worcester Polytechnic Institute, Worcester, MA.

Schindler, P., May, R., Claus, R., & Shaw, J. (1995). Location of impacts in composite panels by embedded fibre optic sensors and neuronal network processing. In *Proceedings of SPIE Conference: Smart Structures and Materials Symposium; Smart Sensing, Processing and Instrumentation* (vol. 2444, pp. 481-490). SPIE.

Serino, G., & Spizzuoco, M. (2010). *The health monitoring system of an isolated religious buiding in Italy*. Paper presented at the Fifth European Workshop on Structural Health Monitoring, Sorrento, Italy.

Shan, Q., & King, G. (2003). Fuzzy techniques for impact locating and magnitude estimating. *Insight-Non-Destructive Testing and Condition Monitoring*, *45*(3), 190–195.

Sirkis, J. S., Berkoff, T., Kersey, A., Fribele, E., & Jones, R. (1994). Development of an impact detection technique using optical fiber sensor ans neuronal networks. In *Proceedings of SPIE on SMART Structures and Materials Symposium; Smart Sensing, Processing and Instrumentation* (pp. 158-165). Orlando, FL: SPIE.

Sohn, H., Farrar, C. R., Hunter, N., & Worden, K. (2001). *Applying the lanl statistical pattern recognition paradigm for structural health monitoring to data from a surface-effect fast patrol boat.* Technical Report LA-13761-MS. Los Alamos National Laboratory.

Sohn, H., Worden, K., & Farrar, C. (2002). Statistical damage classification under changing environmental and operational conditions. *Journal of Intelligent Material Systems and Structures*, *13*(9), 561–574.

Staff, C. W. (2012). *CNN International.* Retrieved Feb 08, 2012, from 20 airbus a380s inspected for cracks: http://edition.cnn.com/2012/01/25/travel/airbus-a380-cracks/

Staszewski, W. (2000). Advanced data preprocessing for damage identification based on pattern recognition. *International Journal of Systems Science*, *31*(11), 1381–1396.

Staszewski, W., Biemans, C., Boller, C., & Tomlinson, G. (2002). Impact damage detection in composite structures using passive acousto-ultrasonic sensors. *Key Engineering Materials*, *221-222*, 389–400.

Staszewski, W., Worden, K., Wardle, R., & Tomlinson, G. (2000). Fail-safe sensor distributions for impact detection in composite materials. *Smart Materials and Structures*, *9*, 298–303.

Sun, Z., & Chang, C. (2004). Statistical wavelet-based method for structural health monitoring. *Journal of Structural Engineering*, *130*, 1055–1062.

Tibaduiza, D., Mujica, L., Gemes, A., & Rodellar, J. (2010). *Active piezoelectric system using PCA.* Paper presented at the Fifth European Workshop on Structural Health Monitoring, Sorrento, Italy.

Tiempo, E. (2014). *Diario el tiempo.* Retrieved Dec 30, 2014, from http://www.eltiempo.com/colombia/medellin/cdo-se-niega-a-recibirle-el-lote-de-space-a-la-alcaldia-de-medellin/15031215

Wildy, S., Kotousov, A., & Codrington, J. (2007). *New passive defect detection technique.* Paper presented at the 5th Australasian Congress on Applied Mechanics, ACAM, Brisbane, Australia.

Worden, K., & Manson, G. (2007). The application of machine learning to structural health monitoring. *Philosophical Transactions A: Royal Society A, 365*, 515–537.

Worden, K., & Staszewski, W. (2000). Impact location and quantification on a composite panel using neural networks and a genetic algorithm. *Strain*, (36): 61–70.

Yan, T., Holford, K., Carter, D., & Brandon, J. (1999). Classification of accoustic emission signatures using a self-organization neuronal network. *Journal of Acoustic Emission*, *17*(1-2), 49–59.

Yap, K., & Zimmerman, D. (1998). The effect of coding on genetic algorithm based structural damage detection. In *Proceedings of 16th International Modal Analysis Conference* (pp. 165–171). Santa Barbara, CA: Society for Experimental Mechanics, Inc.

Ye, L., Lu, Y., Su, Z., & Meng, G. (2005). Functionalized composite structures for new generation airframes: A review. *Composites Science and Technology*, 1436–1446.

Zhang, Y., Zhou, C., & Li, Z. (2010). *Piezo paint acoustic emission sensor and its application to online structural health prognosis*. Paper presented at the Fifth European Workshop on Structural Health Monitoring, Sorrento, Italy.

Zhou, W., Chakraborty, D., Kowali, N., Papandreou, A., & Cochran, D. A. C. (2007). Damage classification for structural health monitoring using time-frequency feature extraction and continuous hidden markov models. In Proceedings of Signals, System and Computers. ACSSC.

Zhou, Y., Tang, S., Zang, C., & Zhou, R. (2012). An artificial inmune pattern recognition approach for damage classification in structures. In *Advances in information technology and industry applications* (pp. 11–17). Springer Berlin Heidelberg.

Chapter 1
Self-Healing Properties of Conventional and Fly Ash Cementitious Mortar, Exposed to High Temperature

Shaswata Mukherjee
Jadavpur University, India

Saroj Mondal
Jadavpur University, India

ABSTRACT

Direct stress and sub-stress caused by fire, temperature variation and external loading in a structure are most important for the development of cracks. The chemical reactions of natural healing in the matrix was not been established conclusively. The most significant factor that influences the self-healing is the precipitation of calcium carbonate crystals on the crack surface. The mechanism which contribute autogenic healing are: (a) Continued hydration of cement at cracked surface as well as continued hydration of already formed gel and also inter-crystallization of fractured crystals; (b) blocking of flow path by water impurities and concrete particles broken from the crack surface due to cracking; (c) expansion of concrete in the crack flank (swelling) and closing of cracks by spalling of loose concrete particle are also reported as the sealing mechanism by researchers. The recovery of mechanical as well as physical property was discussed by different researchers. An experimental investigation was carried out to study the autogenic healing of fire damaged fly ash and conventional cementitious mortar samples subjected to steam followed by water curing at normal atmospheric pressure. The micro cracks are generated artificially by heating the 28 days aged mortar samples at 800 Deg. C. The effect of fly-ash replacing ordinary Portland cement by 0 and 20% was studied. Recovery of compressive strength and physical properties i.e. apparent porosity, water absorption, ultrasonic pulse velocity and rapid chloride ion penetration test confirm the self-healing of micro cracks. Such healing is more prominent for fly ash mortar mix. Optical as well as scanning electron microscopy With EDAX analysis and X-ray diffraction study of the white crystalline material formed in the crack, confirms formation of calcium carbonate.

DOI: 10.4018/978-1-4666-8490-4.ch001

1. INTRODUCTION

Direct stress and sub-stress caused by fire, temperature variations and external loading in a structure are the most important reasons for the development of cracks. There are many literatures which demonstrate the degradation of concrete at different temperatures. At 100 to 200° C evaporation of the water starts, high steam pressure can leads to spalling of the cementitious composites. When the temperature reaches 250° C dehydration starts, degradation of concrete can be observed at this stage. On reaching the temperature at 400- 550° C oxidation of volatile material starts and dehydration of $CaCO_3$ occurs (Georgali & Tsakiridis, 2005). Silica undergoes volume changes at different temperatures, specifically at 300° C there is a huge volume change of silica polymorph (Chu, 1978; Powers-Couche, 1994), this difference in thermal expansion/ contraction of aggregate with rich silica, is one of the major causes for degradation of concrete, which eventually spread and may cause the failure of structure. It is reported that at 550° C reduction in compressive strength would typically range from 55% to 70% of its original value (Guise, 1999; Powers-Couche, 1992; Xu, Wong, Poon & Anson, 2003). At 800° C the shrinkage of hardened cement paste was found to be 2.2% by Y. Xu. et al. (2003).

Fly ash is produced simultaneously in thermal power plant and is used significantly with cement clinker for structural applications. In India, as per IS 1489 (part-1): 1991 the code allows 10-25% of fly ash addition with cement clinker for making Portland pozzolana cement (Indian Standard, 1991). Addition of certain amount of fly ash with OPC was found to be resource-efficient, environment friendly, durable and economical (Crouch, Hewitt & Byard, 2007). The durability, workability and long term mechanical property has been found to be more for fly ash concrete than ordinary concrete (Sivasundaram, Carette & Malhotra, 1990; Mehta, 2004). Few literature deals with the self-healing property of fly ash- cement system. Pipat Termkhajornkit et al. reported that fly ash–cement system has the self-healing ability for cracks that occur from shrinkage (Termkhajornkit, Nawa, Yamashiro, Saito, 2009). In an experimental programme Mustafa Şahmaran et al. studied the self-healing ability of self-consolidating concretes incorporating high volumes of fly ash (HVFA–SCC) subjected to mechanical load for degradation (Sahmaran, Keskin, Ozerkan & Yaman (2008). Both the articles concluded that the cracks were filled by newly formed C–S–H gels (secondary) because of the pozzolanic reactions.

Takeshi Watanabe et al. used ultrasonic pulse velocity to measure the self-healing of cement- fly ash system and concluded that fly ash has the ability towards self-healing (Watanabe, Fujiwara, Hashimoto & Ishimaru, 2011). In an experimental study Zong hui zhou et al. quantify the influence of slag and fly ash on the self-healing ability of concrete by strength measurement (Zhou, Li, Xu, Yu, 2011). Both the analytical processes were found to be important to quantify the self-healing of cementitious system.

The chemical reactions of self-healing process to heal the cracks was not been established conclusively. The most significant factor that influences the self-healing is the precipitation of calcium carbonate crystals on the crack surface (Mukherjee, Mandal & Adhikari, 2012; Heide, 2005; Edvardsen, 1999). The mechanism which contribute autogenic healing are:

1. Continued hydration of cement at cracked surface as well as continued hydration of already formed gel and also inter-crystallization of fractured crystals (Biscoping, 1998);
2. Blocking of flow path by water impurities and concrete particles broken from the crack surface due to cracking (Ramm & Biscoping, 1998; Saharam, Keskin, Ozerkan & Yaman (2008);
3. Expansion of concrete in the crack flank (swelling) and closing of cracks by spalling of loose concrete particle are also reported as the sealing mechanism by researchers (Guise, 1999).

The recovery of mechanical as well as physical property was discussed by different researchers (Edvardsen, 1999; Mukherjee, Mandal & Adhikari, 2012).

This paper reports the results of the recovery of physical and mechanical properties of mortar with and without fly ash subjected to a temperature of 800° C. The properties studied include ultrasonic pulse velocity (UPV), rapid chloride ion penetration test (RCPT), apparent porosity, water absorption and compressive strength. Material characterisation of the product formed during the autogenic healing was done using SEM and XRD.

2. EFFECT ON HIGH TEMPERATURE EXPOSURE OF CEMENTITIOUS COMPOSITE

It is well known that structural materials are exposed to severe environment in use. High temperature heating is one of such extreme condition (caused by accidental fire), where the material (concrete) is lead to severe deterioration of its properties. More specifically the deteriorations are more pronounced by loss in compressive strength, cracking and spalling of concrete, destruction of the bond between the cement paste and the aggregates and the gradual deterioration of the hardened cement paste (Georgali & Tsakiridis, 2005).

Visual observation of fire exposed concrete different colour at different temperature shows. On heating above 300°C the color of concrete can change from normal to pink (300–600°C) to whitish gray (600–900°C) and buff (900–1000°C) [20].

It is evident that concrete mass on heating at high temperature undergoes a number of transformation and chemical reactions which causes progressive breakdown of cement gel structure (Handoo, Agarwal & Agarwal, 2002). On the other hand concrete is a nonhomogeneous materials. The co- efficient of thermal expansion/contraction of aggregate and hardened cement paste (hcp) varies and this affect the interfacial transition zone of aggregate and hcp. Which decreases the mechanical properties of concrete as well as durability property.

3. EXPERIMENTAL METHODS

3.1 Materials

Ordinary Portland cement (OPC) was used in the experiment. Fly ash was collected from the National Thermal Power Corporation, Farakkha, West Bengal, India and is of Class F. The chemical composition and physical properties of fly ash and cement are given in Table 1. Fine aggregate used was river sand having specific gravity of 2.5 and fineness modulus of 2.65. Potable water was used for casting and curing.

3.2 Mix Proportions

Table 2 shows the mix proportions. The standard mix proportions for 0% fly ash represents F0 mix. The mix having cement replaced with fly ash (by weight) by 20% is designated as F20. The water binder ratio for the mixes was fixed at 0.5, while the aggregate binder ratio was kept constant at 3:1.

Table 1. Chemical and physical properties of fly ash and OPC

Components		Fly Ash	OPC
1	% SiO_2	64.58	18.62
2	% Al_2O_3	25.89	4.75
3	% Fe_2O_3	5.27	3.02
4	% CaO	0.59	61.42
5	% MgO	0.26	3.21
6	% Na_2O	0.027	1.51
7	% K_2O	0.041	1.42
8	% SO_4	0.31	2.29
Loss on Ignition		2.40	3.55
Specific Gravity		2.42	3.11

OPC was replaced with fly ash by only 0 and 20% for the entire experiment, as it is already proved from previous experimentation that upto 20% replacement the changes in mechanical and physical property is negligible (Mukherjee, Mandal & Adhikari, 2012; Mukherjee, Mandal & Adhikari, 2013). So, to see the effect of temperature and healing effect with minimum fly ash addition, the replacement level was restricted upto 20%.

3.3 Specimen Preparation and Curing

Specimen size was taken as 70.6mm x 70.6mm x 70.6 mm throughout the experiment, with an exception of 51 mm (height) X 100 mm (diameter) for rapid chloride ion penetration test, which was cut from 200 mm X 100 mm Cylinders. Mortar cubes/cylinders were compacted in vibration table. All the specimens were cured in potable water after removing them from moulds for a period of 28 days. After 28 days water curing, the compressive strength, UPV and apparent porosity of six samples (as a set) of each of the mixes was measured. Another set of specimens from each of the mixes was subjected to high temperature heating (800° C) for 2 hours to generate partial micro-cracks.

Compressive strength, UPV, apparent porosity and water absorption were measured just after the crack generation by heating, as well as at different time intervals (3, 7, 14 and 28 days) during the healing process. The heated samples were subjected to steam curing at 100° C for 3 days at normal atmospheric pressure followed by normal tap water curing until testing. RCPT was conducted after different interval as given in Table 6. Heating cycle is given in Figure 1.

Table 2. Mix proportions

Mix ID	Fly Ash (%)	OPC (%)	Aggregate: Binder	W/B
F0	0	100	3:1	0.5
F20	20	80	3:1	0.5

Figure 1. Heating cycle

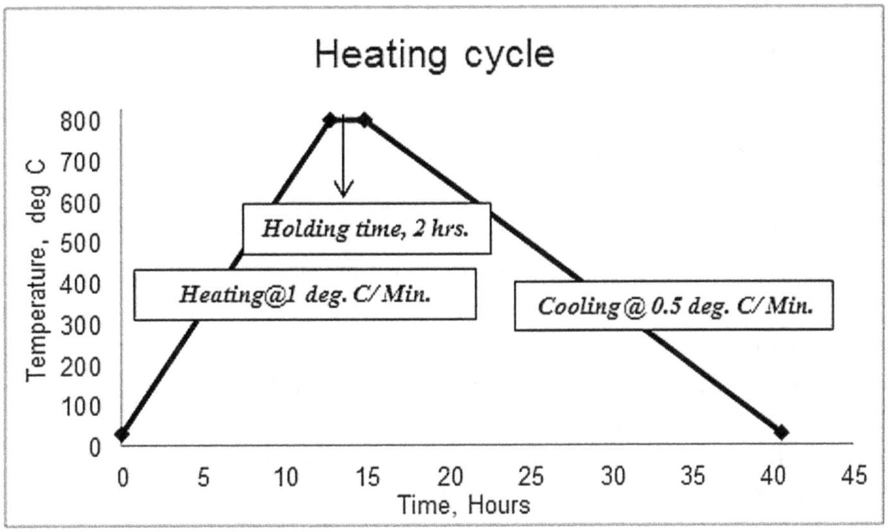

3.4 Test Methods

A digital compression testing machine was employed to determine the compressive strength of the specimen at regular intervals. Ultrasonic pulse velocity was measured according to IS: 1331 (Part 1): 1992 (Indian Standard, 1992). To determine the apparent porosity and water absorption, cubes from each mixes were dried in hot air oven at 110° C for 24 hours and their weights were taken as the dry weight (W0). The specimens were then boiled in water for 2 hours and kept for another 24 hours in the same warm water to penetrate water in the pores. Specimens were then suspended in water with copper wire to take the suspended weight (W1). The soaked weight (W2) was also recorded by removing the surface water and the copper wire. The following equations were used to find out the apparent porosity and water absorption of the specimens.

$$Apparent\ Porosity(\%) = \left\{ \frac{(W2 - W0)}{(W2 - W1)} \right\} \times 100 \quad (1)$$

$$Water\ Absorption(\%) = \left\{ \frac{(W2 - W0)}{W0} \right\} \times 100 \quad (2)$$

Two very similar standards (AASHTO T 277 and ASTM C 1202) (AASHTO,2005; ASTM, 1997) describe the rapid chloride permeability test. The tests were performed using 51mm long and 100mm diameter cylindrical specimens cut from 100mm diameter cast cylinders, and the test values were normalised using the ratio of the standard to the actual cross-sectional areas. The current across the specimen is measured every 30 minutes during the 6-hour test.

Microstructure of the white crystalline materials developed within the crack flank was examined with the help of JSM-6360 Scanning Electron Microscope. Phillips X-ray Powder Diffracto-meter 1840 was used to study the crystalline phases of the material.

4. RESULTS AND DISCUSSION

The loss and recovery in compressive strength for 0 and 20% fly ash samples are shown in Figure 2. The variations in the apparent porosity and the water absorption at different conditions are shown in Figure 3. Existence of partial micro cracks due to the thermal exposure of 800°C is confirmed by the decrease in the compressive strength and increase in apparent porosity or water absorpotion (Figures 2 and 3). The loss in compressive strength is more prominent for F0 specimen than F20 samples. It indicates that the utilisation of fly ash at 20% in mortar mix increased its resistance against thermal effect. The recovery of compressive strength after steam curing for 3 day is almost same for both F0 and F20 specimens which maintained almost thoroughly up to 28 day (healing period). The sample without fly ash did not achieved its previous 28 day compressive strength, whereas the 20% fly ash content mortar specimens crossed that line after 7 day water curing followed by 3 day steam curing. This may be due to formation of secondary C-S-H gel at later stage and lesser porosity.

The results of UPV test are presented in Table 3. The loss for UPV and gain in apparent porosity and water absorption is obvious for heated samples for both F0 and F20 samples. The loss was recovered mostly just after steam curing for 3 day. This recovery of physical property supports the self (autogenic) healing of cement based composite. Table 4 represents the rapid chloride ion penetration test result of F0 and F20 concrete specimens at different interval. In terms of permeability, 20% fly ash content specimens is less permeable than specimens containing no fly ash. Increment in RCPT value is obvious after heating at 800°C, which was found to be recovered after 3 day steam curing and 28 day water curing respectively.

Figure 2. Variations in compressive strength

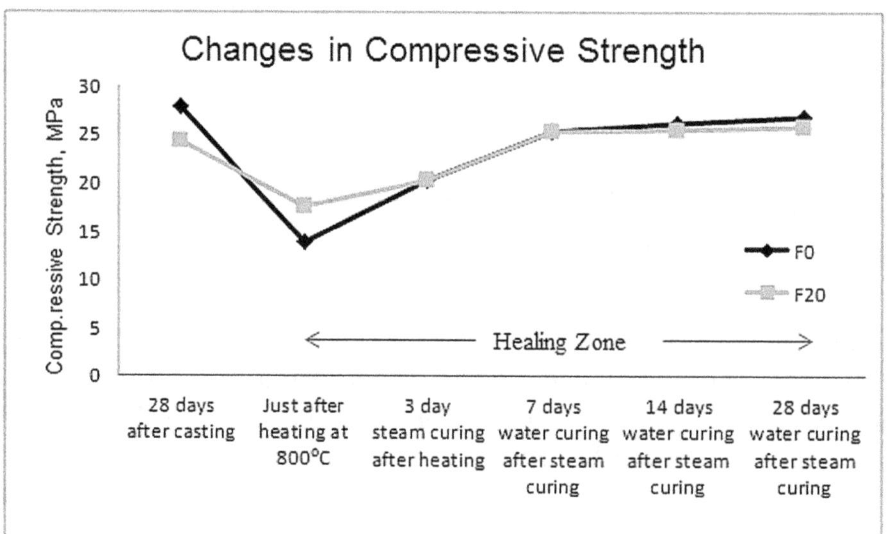

Figure 3. Variations in apparent porosity and water absorption

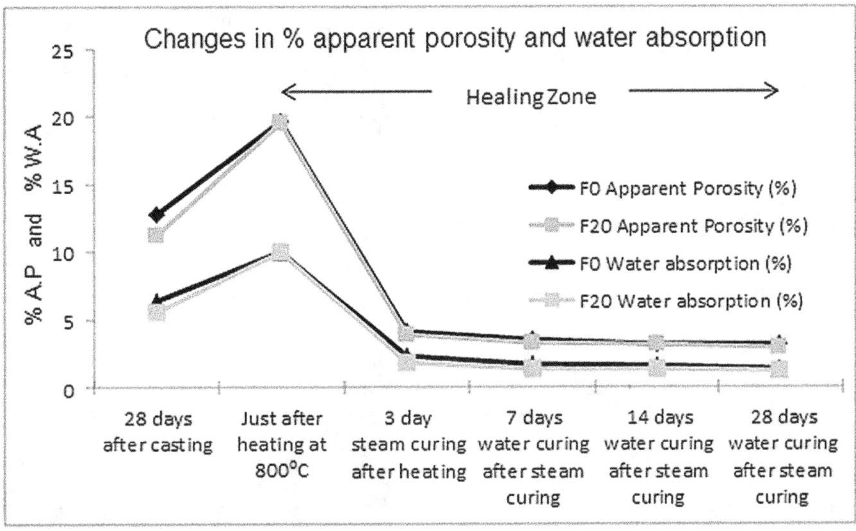

A crack detection microscope was used to find out the width of the cracks at the surface after heating. It was found that the widths were in the range of 0.05- 0.3 mm for both types of specimens. A white powdered material was also found to be deposited at the crack opening during the healing period as can be seen in Figure 4. The powder was collected from a number of specimens (from both F0 and F20) during the normal tap water curing and dried in hot air oven. The microstructure of this white powder (healing product) was investigated using SEM and XRD test. SEM image (Figure 5) indicates the agglomeration of fine powder and formation of large crystal. From XRD analysis (Figure 6) of the white

Table 3. Ultrasonic pulse velocity at different condition (Km/Sec)

Specimens	28 Days after Casting	Just after Heating at 800°C	3 Day Steam Curing after Heating	Water Curing after Steam Curing		
				7 Days	14 Days	28 Days
F0	4.95	1.75	3.74	3.74	3.75	3.77
F20	5.07	1.76	3.75	3.81	3.86	3.88

Table 4. Variations in rapid chloride ion permeability

	28 Day after Casting		Just after Heating at 800°C		3 Day Steam Curing after Heating		28 Days Water Curing after Steam Curing	
	Charge Passed (Coulombs)	Chloride Permeability	Charge Passed (Coulombs)	Chloride Permeability	Charge Passed (Coulombs)	Chloride Permeability	Charge Passed (Coulombs)	Chloride Permeability
F0	1319	Low	3123	Moderate	2730	Moderate	2340	Moderate
F20	315	Very Low	2499	Moderate	1865	Low	1436	Low

Figure 4. Deposition of white crystalline material in cracks

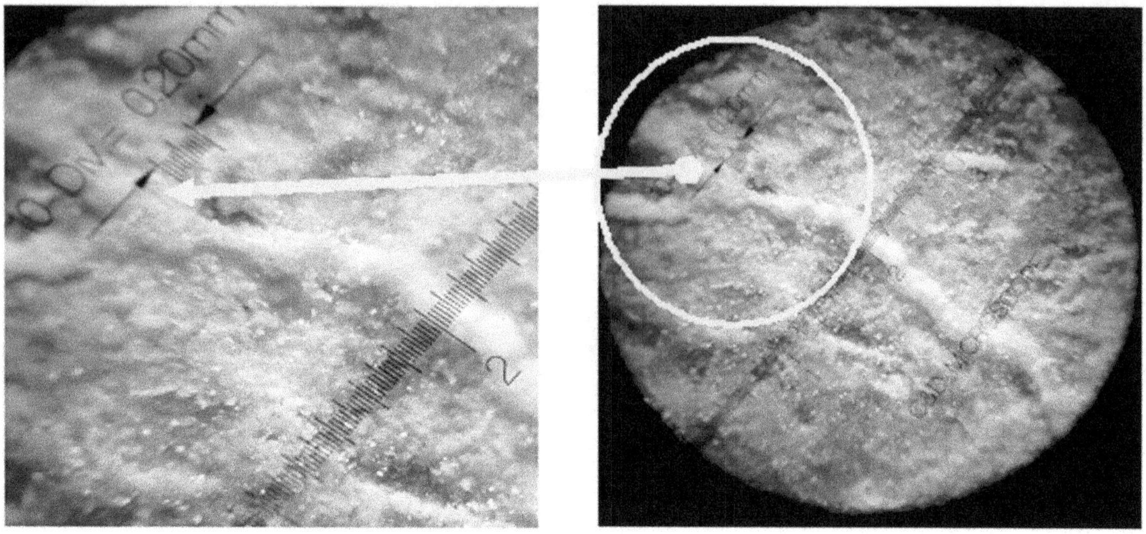

Figure 5. SEM image of white crystalline product

deposition it was found that the major phase present was calcium carbonates ($CaCO_3$). Deposition of calcium carbonate in the crack opening was also observed by other researchers (Mehta, 2004; Watanabe, Fujiwara, Hashimoto & Ishimaru, 2011).

Figure 6. XRD patterns of white crystalline product

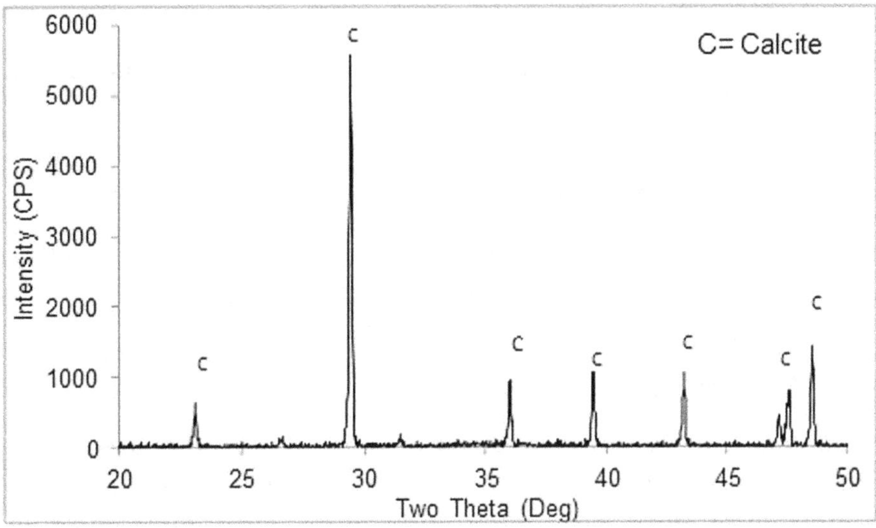

5. CONCLUSION

Heating at high temperature deteriorates the physical and mechanical properties of concrete with and without fly ash addition. Based on the experimental investigation the following conclusions can be drawn as far as natural healing property is concerned:

- After heating at 800°C, the physical, mechanical and permeability properties were found to be deteriorating for both the mixes of 0 and 20% fly ash content.
- Mechanical properties were found to be recovered partially after 3 day steam curing and fully recovered after 7 day water curing for 20% fly ash content specimens.
- UPV test, RCPT, apparent porosity and water absorption tests confirmed the recovery of physical properties after 3 day steam curing. This recovery is more pronounced for mix containing fly ash.
- XRD confirmed the self-healing product in the cracks to be calcite ($CaCO_3$).

REFERENCES

American Association of State and Highway Transportation Officials (AASHTO). (2005). *T 277: Standard Method of Test for Electrical Indication of Concrete's Ability to Resist Chloride Ion Penetration*. Washington, DC: American Association of State and Highway Transportation Officials.

American Society for Testing and Materials (ASTM). (1997). *C 1202: Test method for electrical indication concrete's ability to resist chloride ion penetration*. American Society for Testing and Materials.

Chu, T. Y. (1978). *Radiant heat evaluation of concrete—A study of the erosion of concrete due to surface heating*. Research paper SAND 77-0922. Sandia Laboratories.

Crouch, L. K., Hewitt, R., & Byard, B. (2007). *High Volume Fly Ash Concrete, Proceedings of World of Coal Ash* (pp. 1–14). WOCA.

Edvardsen, C. (1999). Water permeability and Autogenous Healing of Cracks in Concrete. *ACI Materials Journal*, *96*, 448–455.

Georgali, B., & Tsakiridis, P. E. (2005). Microstructure of fire-damaged concrete. A case study. *Cement and Concrete Composites*, *27*(2), 255–259. doi:10.1016/j.cemconcomp.2004.02.022

Guise, S. E. (1999). Petrographic and color analysis for assessment of fire damaged concrete. In *Proceedings of the 19th International Conference on Cement Microscopy.*

Gustafero, A. H. (1983). *Experiences from evaluating fire-damaged concrete structures – Fire safety of concrete structures.* American Concrete Institute.

Handoo, S. K., Agarwal, S., & Agarwal, S. K. (2002). Physicochemical, mineralogical, and morphological characteristics of concrete exposed to elevated temperatures. *Cement and Concrete Research*, *32*(7), 1009–1018. doi:10.1016/S0008-8846(01)00736-0

Indian Standard (IS). (1991). *1489 (Part 1): Portland pozzolona cement- specification (Part 1- Fly ash based).* New Delhi, India: Bureau of Indian Standards.

Indian Standard (IS). (1992). *1331 (Part 1). Nondestructive testing of concrete: Part 1- Ultrasonic pulse velocity.* New Delhi: Bureau of Indian Standards.

Kumar Mehta, P. (2004) *High-performance, high-volume fly ash concrete for sustainable development.* Paper presented at the International Workshop on Sustainable Development and Concrete Technology, Beijing, China.

Mukherjee, Mandal, & Adhikari. (2012). Study on the Autogenic Healing Property of High Volume Fly Ash Concrete. *International Journal of 3R's, 3*, 491-494.

Mukherjee, S., Mandal, S., & Adhikari, U. B. (2012). Study on the physical and mechanical property of ordinary portland cement and fly ash paste. *International Journal of Civil & Structural Engineering*, *2*(3), 731–736.

Mukherjee, S., Mandal, S., & Adhikari, U. B. (2013). Comparative study on physical and mechanical properties of high Slump and zero slump high volume fly ash concrete (HVFAC). *Global NEST Journal*, *20*(10), 1–7.

Powers-Couche, L. (1992). Fire damaged concrete-up close. *Concrete Repair Digest*, *1*, 241–248.

Powers-Couche, L. (1994). Observations of concrete exposed to very high temperature. In *Proceedings of the 16th International Conference on Cement Microscopy.*

Ramm, W., & Biscoping, M. (1998). Autogenous healing and reinforcement corrosion of water-penetrated separation cracks in reinforced concrete. *Nuclear Engineering and Design*, *179*(2), 191–200. doi:10.1016/S0029-5493(97)00266-5

Şahmaran, M., Keskin, S. B., Ozerkan, G., & Yaman, I. O. (2008). Self-healing of mechanically-loaded self consolidating concretes with high volumes of fly ash. *Cement and Concrete Composites*, *30*(10), 872–879. doi:10.1016/j.cemconcomp.2008.07.001

Sivasundaram, V., Carette, G. G., & Malhotra, V. M. (1990). Long-term strength development of high-volume fly ash concrete. *Cement and Concrete Composites*, *12*(4), 263–270. doi:10.1016/0958-9465(90)90005-I

Ter Heide, N. (2005). *Crack healing in hydrating concrete.* (Master thesis). TU Delft, Delft, The Netherlands.

Termkhajornkit, P., Nawa, T., Yamashiro, Y., & Saito, T. (2009). Self-healing ability of fly ash–cement systems. *Cement and Concrete Composites*, *31*(3), 195–203. doi:10.1016/j.cemconcomp.2008.12.009

Watanabe, Fujiwara, Hashimoto, & Ishimaru. (2011). Evaluation of self healing effect in fly-ash concrete by ultrasonic test method. *International Journal of Modern Physics B, 25.*

Xu, Y., Wong, Y. L., Poon, C. S., & Anson, M. (2003). Influence of PFA on cracking of concrete and cement paste after exposure to high temperatures. *Cement and Concrete Research*, *33*(12), 2009–2016. doi:10.1016/S0008-8846(03)00216-3

Zhou, Z. H., Li, Z. Q., Xu, D. Y., & Yu, J. H. (2011). Influence of Slag and Fly Ash on the Self-Healing Ability of Concrete, Advanced. *Materials Research*, *306-307*, 1020–1023. doi:10.4028/www.scientific.net/AMR.306-307.1020

Chapter 2
New Features for Damage Detection and Their Temperature Stability

Fahit Gharibnezhad
Universitat Politècnica de Catalunya, Spain

Luis Eduardo Mujica Delgado
Universitat Politècnica de Catalunya, Spain

Jose Rodellar
Universitat Politècnica de Catalunya, Spain

ABSTRACT

This chapter is devoted to present novel techniques in Structural Health Monitoring (SHM). These techniques are based on different statistical and signal processing methods that are used in other fields but their performance and capability in SHM is presented and tested for the first time in this work. This work is dedicated to the first level of SHM, which might be considered the main and most important level. Piezoceramic (PZT) devices are chosen in this work to capture the signals due to their special characteristics such as high performance, low energy consumption and reasonable price. Suggested techniques are tested on different laboratory and real scale test benchmarks. Moreover, this work considers the effect of environmental changes on performance of the presented techniques. This work shows that although those techniques have a significant result in normal conditions, their performance can be affected by any environmental discrepancy such as temperature change. As such, there is a vital need to consider their effect. In this work, temperature change is chosen, as it is one of the main environmental fluctuation factors.

WAVE CLUSTERING

The aim of data-clustering methods is to group the objects in databases into meaningful subclasses. Clustering tries to detect groups and assign labels to the objects based on the cluster that they belong to. A good clustering algorithm should be time efficient, order-insensitive and be able to identify clus-

DOI: 10.4018/978-1-4666-8490-4.ch002

ters irrespective of their shapes and relative position. Clustering algorithms are categorized into four main groups: partitioning algorithms (Kaufman & Rousseeuw 2005), hierarchical algorithms (Zhang, Ramakrishnan, & Livny 1996), density based algorithms (Ester et al. 1996) and grid based algorithms (Sheikholeslami, Chatterjee, & Zhang 1998; Xu & Ester 1998).

Partitioning methods relocate instances by moving them from one cluster to another, starting from an initial partitioning. To do this, some iterative control strategy is used to optimize an objective function such as Error Minimization Algorithm (k – means) (Kaufman & Rousseeuw 2005), Graph-Theoretic Clustering (Urquhart 1982), and others.

Hierarchical algorithms construct the clusters by recursively partitioning the instances. This algorithm iteratively splits the database into smaller subsets until some termination condition is satisfied. Comparing with partitioning algorithm, hierarchical algorithms do not need k as input parameter. This is an advantage but in contrary the termination condition need to be declared (Rokach & Maimon 2010).

Density based algorithm allocate the observation with the similar probability distribution to the same cluster. The overall distribution of the data is assumed to be a mixture of several distributions. This algorithm aims to identify the cluster and its distribution parameter (Ester et al. 1996).

Grid based methods partition the space into a finite number of cells that form a grid structure on which all of the operations for clustering are performed. The main advantage of the approach is its fast processing time because it is independent of the number of objects. In this work we apply a spatial data-mining method termed Wave Cluster developed by (Sheikholeslami, Chatterjee, & Zhang 1998, 2000) that belongs to this category.

Wave Cluster is an efficient method with low computational complexity. The results are less affected by noise and the method is not sensitive to the order of input objects. According to (Sheikholeslami, Chatterjee, & Zhang 2000) the main idea of Wave Cluster is to transform the original feature space by applying wavelet transform and then find the dense regions in the new space. According to them, in this procedure the process of finding the connected components in the transformed space is easier than in the original feature space, because the dense regions in the feature space will be more salient. Applying wavelet transform on a signal decomposes it into different frequency sub-bands. The high frequency parts of the signal are related to the regions where there is a rapid change such as boundaries of cluster, while the low frequency part corresponds to the areas that the objects are more concentrated, the cluster themselves. In another words, observations in a *2D* feature space are considered as an image where each pixel of image corresponds to one cell in the feature space. To achieve this goal, the discrete wavelet transform (DWT) is used. Using DWT, details and approximation of signal are decomposed for different steps. Approximation or average sub-band carries information about content of cluster and details sub-bands have information about the boundaries of clusters.

The multidimensional spatial data objects can be represented in a d – dimensional *feature space*. For an object with d numerical attributes, the feature vector will be one point in d – dimensional feature space. Wave Cluster proposes to look at the feature space from signal processing perspective. It means that the high frequency parts of the signal correspond to the regions of the feature space where there is a rapid change in the distribution of objects, that is, the boundaries of clusters. The low-frequency parts of the d – dimensional signal that have high amplitude correspond to the areas of the feature space where the objects are concentrated, in other words, the clusters themselves. For instance, Figure 1 presents a *2D* feature space where each point of the image represents the feature values of one object in spatial dataset. Thanks to the properties of wavelet transform for detecting rapid changes in the signal, we can

Figure 1. A sample 2D feature space

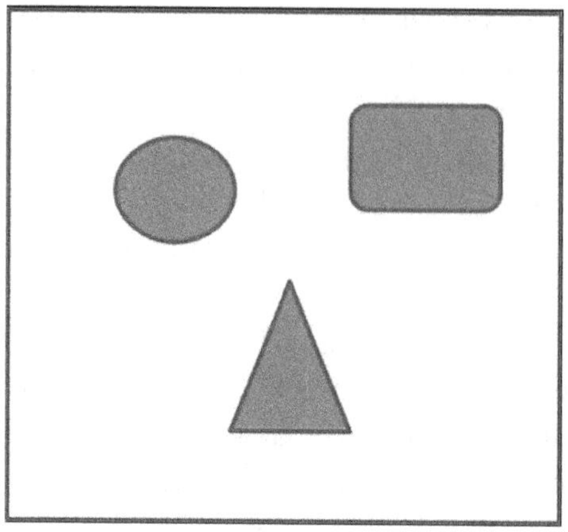

apply wavelet transform to find high-frequency and low-frequency parts of the $d-$ dimensional signal that bring about clusters detection.

As all data should be at least read, the required time for this algorithm to detect the clusters is linear in terms of number of input. After reading the data, the processing time will be only a function of number of cells in the feature space. Therefore, applying PCA as a primary step to reduce the dimension of the data is a beneficial step based on the fact that Wave Cluster will be very efficient for low number of feature space dimension.

Wavelet transform can be generalized to $d-$ dimensional signal in which a $1D$ transform can be applied multiple times. The key point of Wave Cluster is to apply wavelet on feature space instead of the objects themselves to find the clusters. Wavelet decomposes the original feature space into an *approximation* which has information about the content of clusters, a *details* which have information about the boundaries of the clusters.

According to Sheikholeslami et al (Sheikholeslami, Chatterjee, & Zhang 1998, 2000), WaveCluster algorithm is implemented through the following steps:

Input: Multidimensional data objects' feature vectors.
Output: Clustered objects.

1. Quantize feature space, and then assign objects to the cells.
2. Apply wavelet transform on the quantized feature space.
3. Find the connected components (clusters) in the sub bands of transformed feature space, at different levels.
4. Assign labels to the cells.
5. Make the look-up table.
6. Map the objects to the clusters.

To describe the algorithm in details, let $A_1 \times A_2 \times \ldots \times A_d$ be a $d-$ dimensional feature space. The input dataset is a set of $d-$ dimensional points $= \{O_1, O_2, \ldots, O_N\}$, where $O_i = (O_{i1}, O_{i2}, \ldots, O_{id})$, $1 \leq i \leq N$. The first step is to quantize A_i feature space to m_i number of intervals. For sake of simplicity, m_i could be equal for all dimensions. Each cell is defined as an intersection of one interval from each dimension. It has the form $< c_{i1}, c_{i2}, \ldots, c_{id} >$ where $c_{ij} = [l_{ij}, h_{ij})$. We say that a point $o_k =< o_{k1}, \ldots, o_{kd} >$ is contained in a cell c_i if $l_{ij} \leq o_{ki} \leq h_{ij}$ for $1 \leq j \leq d$. We count only the number of observations belong to the cells as statistic to be used.

In the next step, discrete wavelet transform will be applied on the quantized (each cell now is represented by the number of observation belong to it) feature space. Applying wavelet transform on the cells results in a new feature space and hence new units. There are many well-known algorithms for finding connected components in image and Wave Cluster uses the algorithm in (Horn 1986) to find the connected components in the 2 – dimensional feature space. The founded clusters are in the transformed feature space and are based on wavelet coefficients. Thus, they cannot be directly used to define the clusters in the original feature space. Wave Cluster makes a hookup table to map the units in the transformed feature space to the units in the original feature space. Finally, Wave Cluster assigns the label of each unit in the feature space to all the objects whose feature vector is in that unit, and thus the clusters are determined (Sheikholeslami, Chatterjee, & Zhang 1998, 2000).

As all data should be at least read, the required time for this algorithm to detect the clusters is linear in terms of number of input. After reading the data, the processing time will be only a function of number of cells in the feature space. Therefore, aFpplying PCA as a primary step to reduce the dimension of the data is a beneficial step based on the fact that Wave Cluster will be very efficient for low number of feature space dimension. It is recommended to readers to refer to (Sheikholeslami, Chatterjee, & Zhang 1998, 2000) for detailed information about Wave Clustering method.

Damage Detection Using Wave Clustering

Composites are widely used in aerospace structures because of special properties they offer. Due to complexity of wave propagation in composites, any kind of damage detection and localization seems to be more sophisticated. In this case study a composite plate is used as the specimen (see Figure 2. and Figure 3.). The plate is made of 4 equal layers and stacking of [0/90/0/90] with dimensions: $200mm \times 250mm$ and a thickness of approximately $1.7mm$. Nominal material parameters of the unidirectional (UD) layers are $E1 = 122GPa, E2 = 10GPa, v_{12} = 0.33, v_{13} = 0.3, v_{23} = 0.34, G_{12} = G_{13} = 7.4GPa$ and $G_{23} = 5.4GPa$. The density is about $1700kg/m^3$.

Nine PZT transducers PIC-151 from PI Ceramics were attached to the surface of the structure with equidistant spacing. The piezo transducers have a diameter of $10mm$ and a thickness of $0.5mm$. The excitation signal to the actuators was generated using the arbitrary signal generation capability of a HandyScope HS3.

Damages are simulated by adding appropriate mass powered by magnet to hold them and apply enough pressure on the plate (see Figure 3.).

A 10 V Hann-windowed tone-burst signal is used as an actuating signal with different carrier frequencies (30, 50, 60 KHz) with 3 cycles.

For sake of simplicity and according to the symmetry of the structure, transducers number 4, 5, 7 and 8 are used for this experiment (the rest of transducers was allocated because of other experiments in the local laboratory). The four transducers are used as actuator and 13 observations are recorded per sensor. Actuating the structure is performed for any status of structure including pristine and structure with different simulated damages. Signals are de-noised and decimated using Matlab® software. Obser-

Figure 2. Composite plate for damage detection, schematic

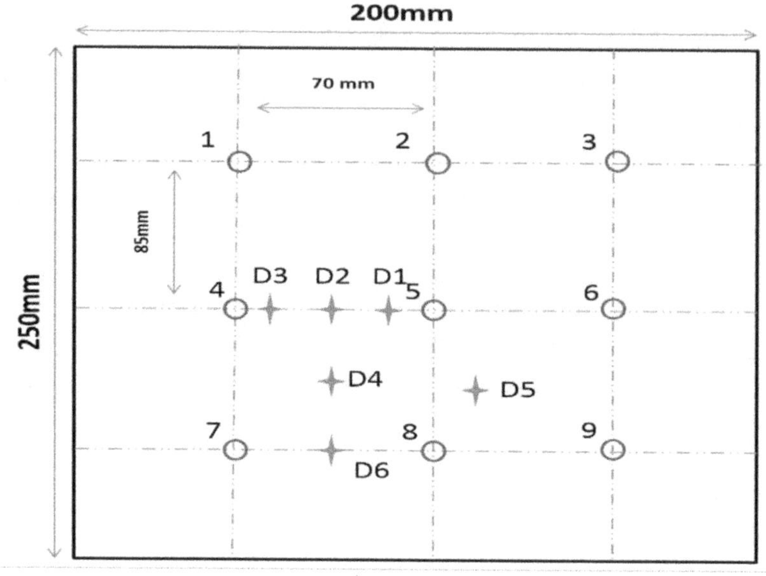

vations are used in two different ways. Primary, damage detection is performed per each route. A route is defined as the path between one transducer as an actuator and another as a sensor; hence there are 3 different routes for each actuator. Secondly, data from each actuator to all other sensors are folded in a unique matrix and the same procedure is repeated for damages in the area surrounded by 4 transducers.

Figure 3. Composite plate for damage detection, snapshot

Figure 4. Damage Classification in route $5 \rightarrow 4$

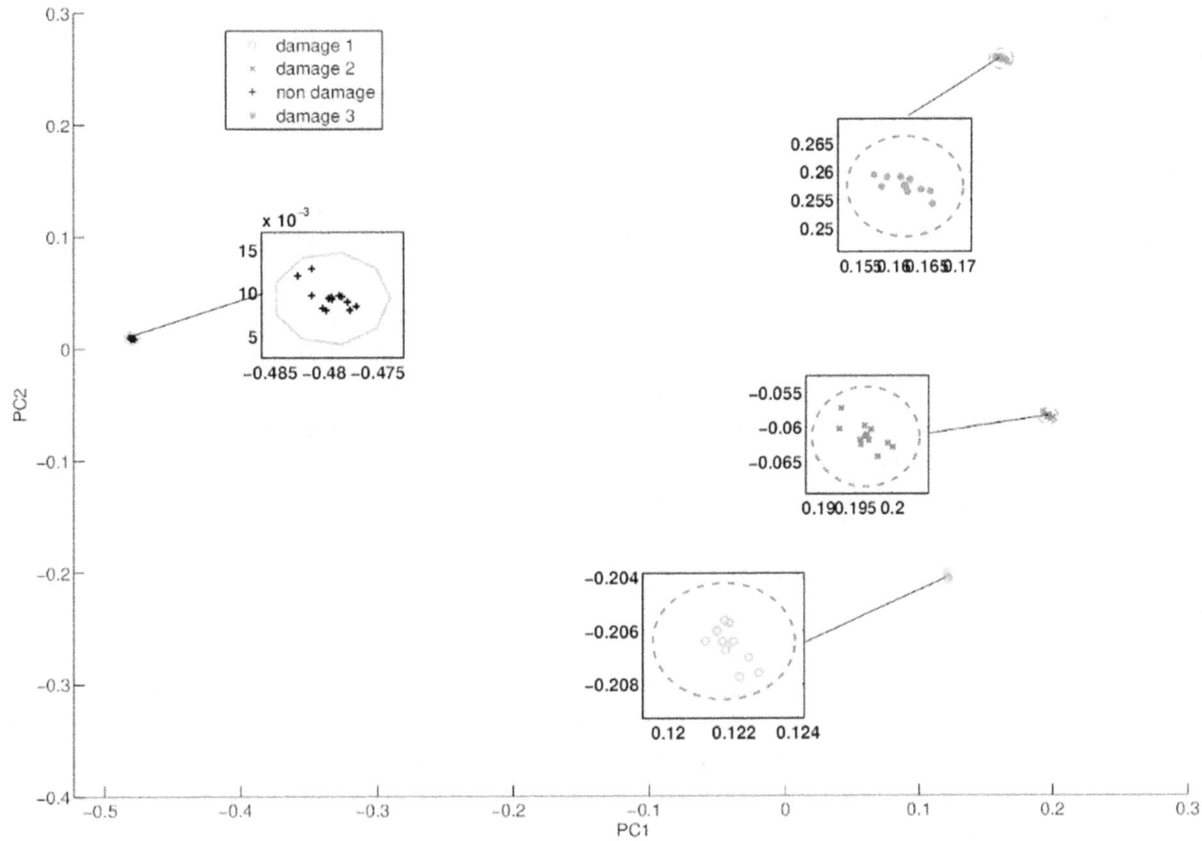

Damage Detection Results

Although Principal Component Analysis (PCA) (Jolliffe 2002; Smith 2002) and its robust counterpart show a successful role in damage detection (Gharibnezhad, Mujica, & Rodellar 2014), they still need a complimentary step for automatic damage classification. It means that a human effort still is required to classify different existing clusters. In previous works, self-organizing map (SOM) (Tibaduiza 2013) was proposed as a classifier method. In this work another classifier method based on wavelet is proposed that possess several merits. This section is dedicated to use this classifier concentrating on automatic classification of damages with different severity. To do this, PCA is used as a tool for dimensionality reduction and then a wavelet classifier is applied on the result to classify different patterns in the structure each of them associated to significant states of the structure.

To start detection and classification procedure, route between transducer 5 and 4 is chosen as a primary route to be analyzed. To do this, signals from actuator 5 to sensor 4 are collected in 4 different states of structure including pristine, damage 1, 2 and 3 (see Figure 2.) Above all, PCA is applied on collected data associated to this route contain of 4 packages (1 by status) each one consist of 13 observations (totally 52 observations). As a result the dimension of data is reduced to the first two principal components $X_{52 \times 2}$.

In the second step, Wave Cluster is applied on the data with reduced dimension to classify the clusters. As a result, Wave Cluster algorithm indicates 4 different clusters in data set. In Figure 4, all observations are depicted based on the Wave Cluster labeling. As it can be seen in this figure, observations belong to different clusters including pristine (connected circle) and three different damages (dashed circles) are clearly distinguished and classified clearly. It should be highlighted that Wave Cluster releases the need for visually exploring the groups depicted on the new dimension space (PC1-PC2) by automatically labeling the observations belonging to the same group. Here PC1 and 2 stands for first and second principal component.

For more comprehensive test, the collected data from all sensors associated with a significant actuator are folded in a single observation. Mujica et al (L. Mujica et al. 2010) suggest a method to organize recorded data somehow that response over a whole time period excited by an actuator and captured from all sensors are folded in a 2×D matrix where rows show the observation and columns are consists of package of data from each sensor laid beside each other. This method highlights the correlations in time for each signal and the correlations between sensors. Therefore, for the first actuator, 3 packages of data from each sensor containing 3000 recorded samples per observation are used. Figure 5 shows the result of clustering applied on actuator-based data. As it can be seen from the figure, all clusters related to different condition of structure are distinctly separated and automatically categorized.

FUZZY SIMILARITY

The concept of similarity is of fundamental importance in different fields and it can be considered as a many valued generalization of the classical notation of equivalence. For instance, similarity plays an essential role in taxonomy, recognition, case-based reasoning and many other fields. Fuzzy similarity is an equivalence relation that can be used to classify multivalued objects. In this work, Fuzzy Similarity approach is used and applied in SHM as a new damage detection index. A brief review on this approach is presented here. Readers are suggested to refer to (Pasi Luukka & Leppälampi 2006) for more information.

Definition: A *lattice* L is a set in which for any $x, y \in L$, $x \wedge y$ and $x \vee y$ exist in L. $x \wedge y$ is called conjunction (meet, infimum) $x \vee y$ is called disjunction (join, supremum) of x and y.

Definition: A lattice is called *residuated* if it contains the greatest element 1, and binary operations $\odot$ (called multiplication) and $\rightarrow$ (called residuum) such that following conditions hold:

$\odot$ is associative, commutative and isotone

$a \odot 1 = a$ for all elements $a \in L$ and

for all elements $a, b, c \in L$, $a \odot b \leq b$ if and only if $a \leq b \rightarrow c$.

Definition: Let L be a residuated lattice and X is a non-empty set. L– valued binary relation S defined in X is a *fuzzy similarity* if it fulfills the following conditions [84]:

Reflexivity: $\forall x \in X: S\langle x, x\rangle = 1$

Symmetricity: $\forall x, y \in X: S\langle x, y\rangle = S\langle y, x\rangle$

Transitivity: $\forall x, y, z \in X: S\langle x, y\rangle \odot S\langle y, z\rangle \leq S\langle x, z\rangle$.

Definition: Łukasiewicz structure:

$$a \odot b = max\left\{a + b - 1, 0\right\}, a \rightarrow b = min\left\{1, 1 - a + b\right\}, a \leftrightarrow b = 1 - \left|a - b\right|$$

Figure 5. Clustering using Wave Cluster applied on folded data belonging to one actuator

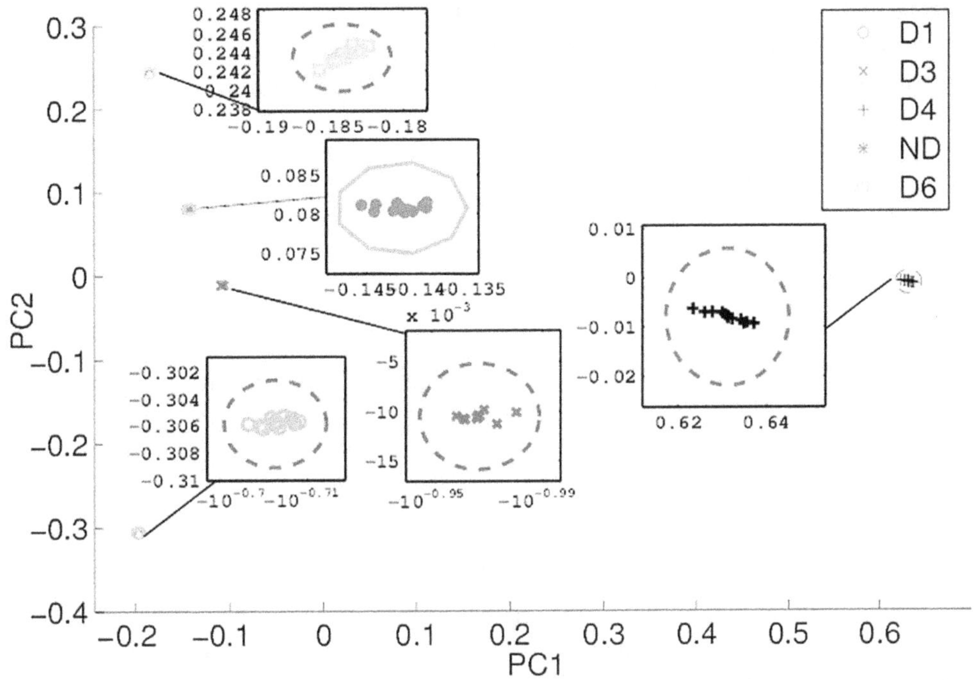

In generalized Łukasiewicz structure, the equivalence relation can be defined as

$$a \leftrightarrow b = 1 - \left|a^p - b^p\right|^{\frac{1}{p}}$$

Here p is an arbitrary number.

If the value for the magnitude of each different feature of one object can be expressed between [0, 1] we can use fuzzy similarity for finding similar pairs. Let $X_{n \times m}$ be a set of n objects each of which having m features (variable). If we know the similarity value of the features $f_1, \ldots, f_m$ between the objects, it is possible to choose the objects that have the highest similarity value. The problem lies in finding, for object x_i, a similar object x_j, where $1 \le i, j \le n$ and $i \ne j$. By choosing the Łukasiewicz structure for the features of the objects we obtain m fuzzy similarities, one for each feature, on comparing two objects (x_1, x_2) (P. Luukka, Saastamoinen, & Kononen 2001):

$$S_{fi}\left\langle x_1, x_2 \right\rangle = x_1\left(fi\right) \leftrightarrow x_2\left(fi\right). \tag{1}$$

where $x_1, x_2 \in X$ and $i \in 1, \ldots, m$. As the Łukasiewicz structure is chosen for the feature of objects, we can define the fuzzy similarity as follows (Pasi Luukka 2009):

$$S\left\langle x_1, x_2 \right\rangle = \frac{1}{m}\sum_{i=1}^{m} x_1\left(fi\right) \leftrightarrow x_2\left(fi\right) \tag{2}$$

It is very important to realize that this is true only in the Łukasiewicz structure. Readers are recommended to read (Lukasiewicz 1970) for more information about Łukasiewicz structure. Moreover, in the Łukasiewicz structure we can assign different non-zero weights ($W_1, \ldots, W_m$) to the different features and obtain the following equation, which again meets the definition of the fuzzy similarity:

$$S\left\langle x_1, x_2 \right\rangle = \frac{\sum_{i=1}^{m} W_i \left(x_1\left(fi\right) \leftrightarrow x_2\left(fi\right) \right)}{\sum_{i=1}^{m} W_i} \tag{3}$$

A weighted similarity measure in the generalized Łukasiewicz structure can be defined as follows:

$$S\left\langle x_1, x_2 \right\rangle = \frac{1}{m} \sum_{i=1}^{m} W_i \sqrt[p]{1 - \left(x_1\left(fi\right)\right)^p - \left(x_2\left(fi\right)\right)^p} \tag{4}$$

This formula uses an arithmetic mean but it is clear that other means can also be used. This can be implemented into the similarity measure based on the Łukasiewicz structure. The formula of the similarity based on the generalized mean and the generalized Łukasiewicz structure attains the following form (Pasi Luukka & Leppälampi 2006):

$$S\left\langle x_1, x_2 \right\rangle = \left[\frac{1}{m} \sum_{i=1}^{m} W_i \left(\sqrt[p]{1 - \left(x_1\left(fi\right)\right)^p - \left(x_2\left(fi\right)\right)^p} \right)^m \right]^{1/m} \tag{5}$$

This formulation above is used as a new damage detector index to compute the similarity between features from any observation captured from current status of structure (probably damaged) and the pristine structure (Baseline). The features are selected using PCA and are fed to the formulation above.

A Methodology for Damage Detection using Fuzzy Similarity

The proposed damage detection methodology based on Fuzzy similarity to interpret result can be described as follows. In the method presented in this work, PCA is utilized as dimension reduction and feature selection tool and similarity classifier based on Łukasiewicz-Structure is used as similarity measurement tool.

At the first step the time response signal captured from sensors are pre-processed to remove noise. Then PCA is applied on the de-noised data to reduce dimension and extract the features. After selecting the adequate number of features, all selected features from pristine structure and also from current status of structure are scaled between [0 – 1] to fulfill the requirement of the methodology as mentioned in the Fuzzy similarity description. Data from pristine structure are used to calculate the representative of the structure called *ideal vector* in healthy condition. This could be done by calculating the mean of learning data in pristine condition or any other statistic representative such as median, etc. Data feature from current status of the structure is compared with ideal vector using similarity classifier measure-

Figure 6. Damage detection algorithm using Similarity Classifier

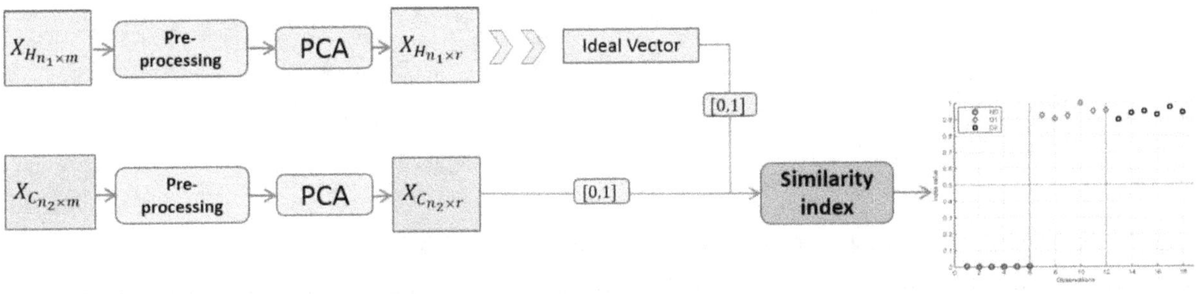

ment. The quantity of similarity for any observation is calculated. This quantity is used as an indicator to distinguish observations that may be affected due to damage. This means that the similarity value between ideal vector and any observation belonging to the healthy status of structure is supposed to be different from the similarity value between ideal vector and any other observation that may be affected by existence of damage in the structure. The described algorithm is briefly depicted in Figure 6.

Case Study

The experiments performed on a skin panel of the torsion box of the wing. This structure is divided into small sections by stringers and ribs. The schematic and snapshot of this case study is shown in Figure 7 and Figure 8. As an acquisition setup, the structure is equipped with piezoelectric transducers (PZT) from PI Ceramics with reference number PIC151 with 10*mm* diameter. The positions of PZT's are distributed over the surface to excite the structure and detect time varying strain response data. The transducers dimensions are 26*mm* in diameter and 0.4*mm* in thickness. A burst signal with three peaks at different frequencies are used as excitation signal. Before applying the signal to the structure, it is amplified to 30 V peak-to-peak using a wideband power amplifier. All data are averaged to obtain one signal for each 10 experiments in order to reduce the noise in the experiments. All data collections are done in room temperature without any temporal changes during the experiments. Due to the limitation to apply real damages on the structure, damages are simulated by placing an appropriate mass in different locations. Adding mass to a structure cause changing acoustic impedance change of propagating

Figure 7. Skin panel schematic

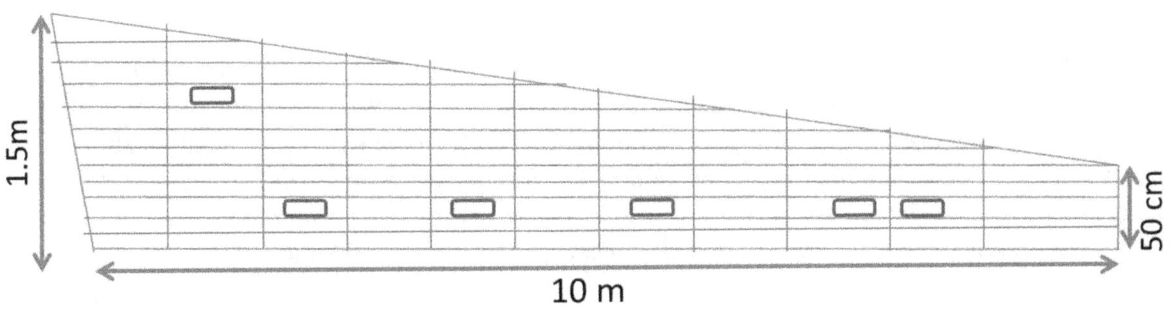

Figure 8. Skin panel snapshot

wave that simulates the damage effect. The experiments are performed on both faces of the structure and they are described as follows.

Case Study 1: The Face without Stringers and Ribs

As a first case study, PZTs are mounted on the plane surface of skin panel where there are neither stringers nor ribs as obstacles. They are located at $9.5cm$ distance from each other in the straight line and $5cm$ distance from the closest edge of structure. Figure 9 shows the sensor location and damage position in this case study. The actuating signal is applied to the first PZT ($S1$) and the response is captured on the second PZT (S2). Damage is located on the route between S1 and S2. The structure was excited by signals of 75KHz to 370KHz with 5KHz steps The whole captured signal is used without performing any mode selection. All captured signals are denoised using discrete wavelet transform with Daubechies wavelets family until third level. In this case study, a number of 50 observations for any status of the structure is captured.

Case Study 2: The Face with Stringers and Ribs

The second case study is performed on the second face of the structure. This side is divided into many sections by stringers (vertical lines) and ribs (horizontal lines). Sensors are located somehow that each pair of PZT (S1 and S2) and (S3 and S4) are located in the same section. Figure 11 shows a scheme of sensor location and damage positions. Two damages are simulated by adding mass as shown in Figure 12. Excitation is performed on different frequencies from 125 KHz to 365KHz with 10 KHz steps. In this case study a number of 30 observations are captured or each status of the structure.

Organizing Data

To validate the proposed methodology, an extensive study is performed. Initially, the similarity index is used to demonstrate its ability to distinguish the different status of the structure between healthy and damaged. This is performed for different exciting frequencies. In the next step, the algorithm is tested on actuator-based data captured from the set of transducers attached on the surface of the structure. Then, the threshold value is studied to find a criterion able to certify the existence of damage. After that, influence of the parameter p is analyzed. This parameter is relevant in tuning the index in Equation (5). In

Figure 9. Schematic

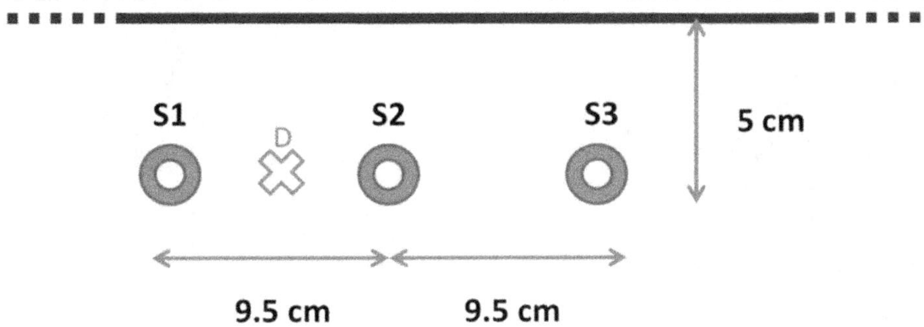

Figure 10. Snapshot

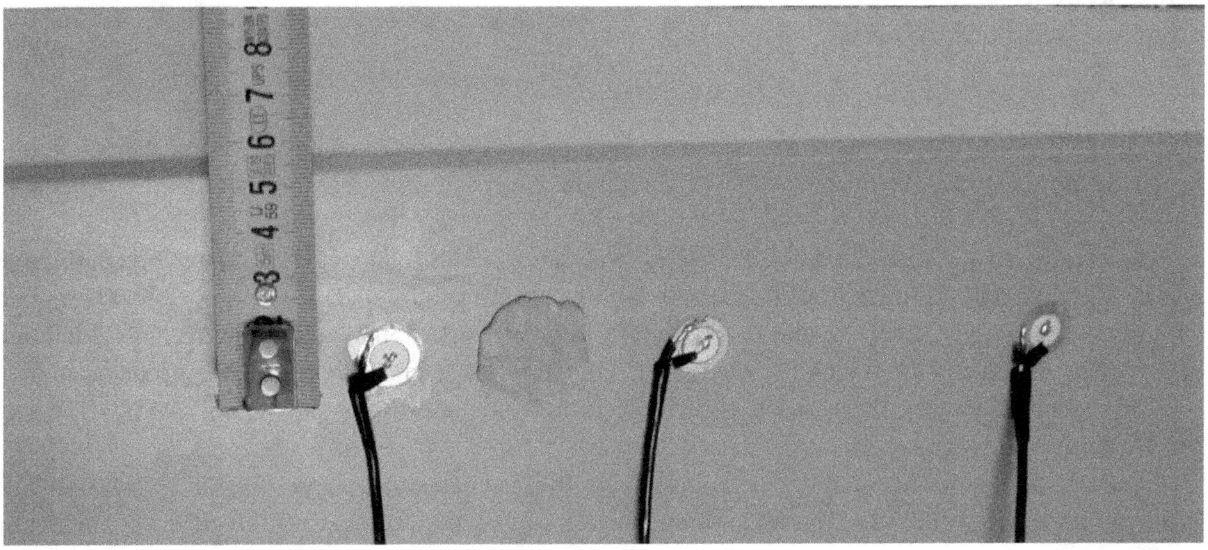

Figure 11. Sensor location and damage position schematic

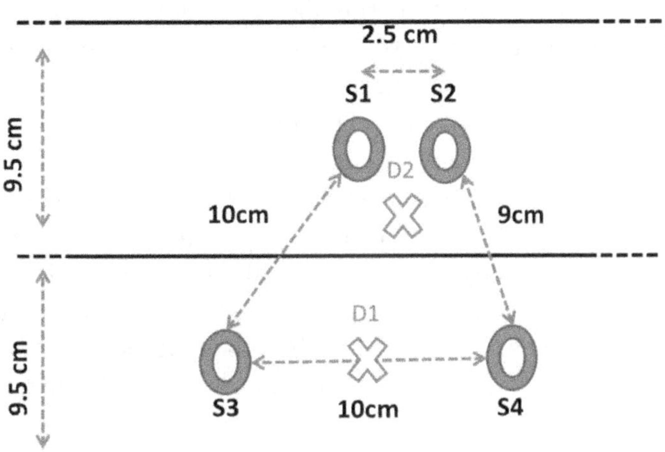

Figure 12a. Damage position, case study 2

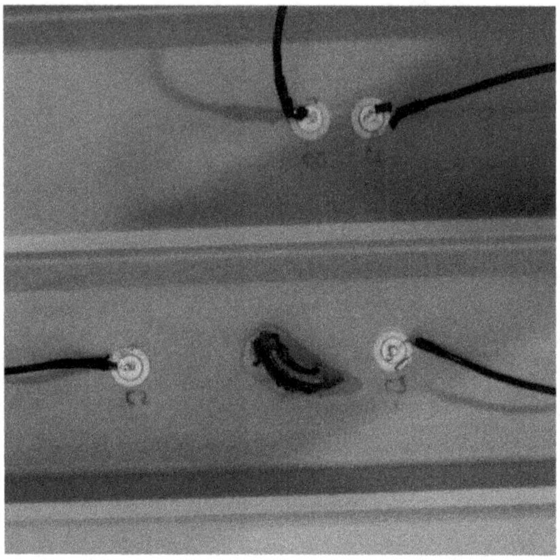

Figure 12b. Damage position, case study 2

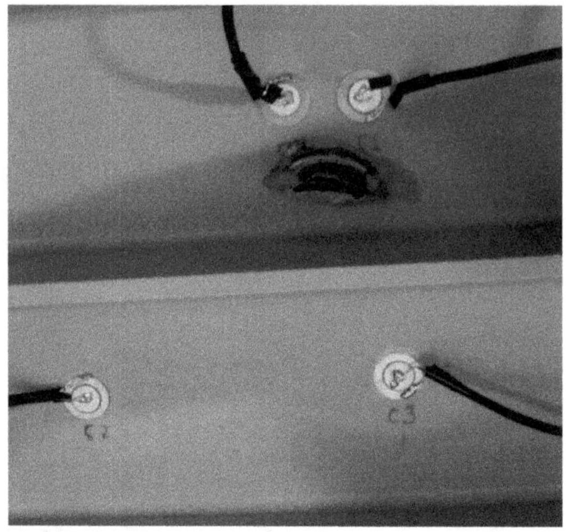

the next step, the proposed index is compared with the Euclidean distance and finally the algorithm proposed in this work is compared with the general damage detection methodology based on classical PCA.

The captured data from the laboratory wing case studies are organized as follows. For both case studies, 80% of data from healthy structure is used as training set and 20% as testing set. Therefore, for the first case study, 40 observations captured from the pristine status are used as a training data set and 10 as a testing dataset, which are captured from both pristine and different status of the structure. For the second case study, the same approach is followed that leads to 24 observations for training and 6observations for testing.

Figure 13 shows this result of the methodology for the first case study for $p = 0.5$ and $p = 0.6$ in 145 KHz. As it can be seen in this figure, similarity index for the first 10 observations that belong to the healthy structure are in the same range, whereas similarity index for the next 10 observations that belong to the damaged status are clearly separated as they stand in a different range. It should be mentioned that the similarity values are reduced from 1 ($|1 - SV|$) and then are normalized between [0, 1] for sake of simplicity for comparison purposes considering the maximum value as 1 that means more distance from the healthy condition. The results show a clear distinguishing and separation between the similarity values for observation from healthy and damaged structure.

The test continues by proving the efficiency of the method at different frequencies and a constant p. As it can be seen in Figure 14, a clear detection is achieved for all performed frequencies. This shows the flexibility of the algorithm to achieve acceptable results in different frequencies.

To have a comprehensive validation, the methodology, is performed for the case study 2. As it is mentioned before, data in this case study are collected (folded) for any individual actuator. Therefore, the test can be performed for any actuator separately. This is done to prove the efficiency of the algorithm for this type of data folding, which is widely used (Gharibnezhad, Mujica, & Rodellar 2011; L. E. Mujica et al. 2008). Figure 15 shows the result. It may clearly be seen that all damages are well distinguished from the healthy structure.

Figure 13a. Damage detection using fuzzy similarity, exciting S1 at 145KHz and capturing from S2 (○-undamaged, ◇ – damaged); P = 0.5

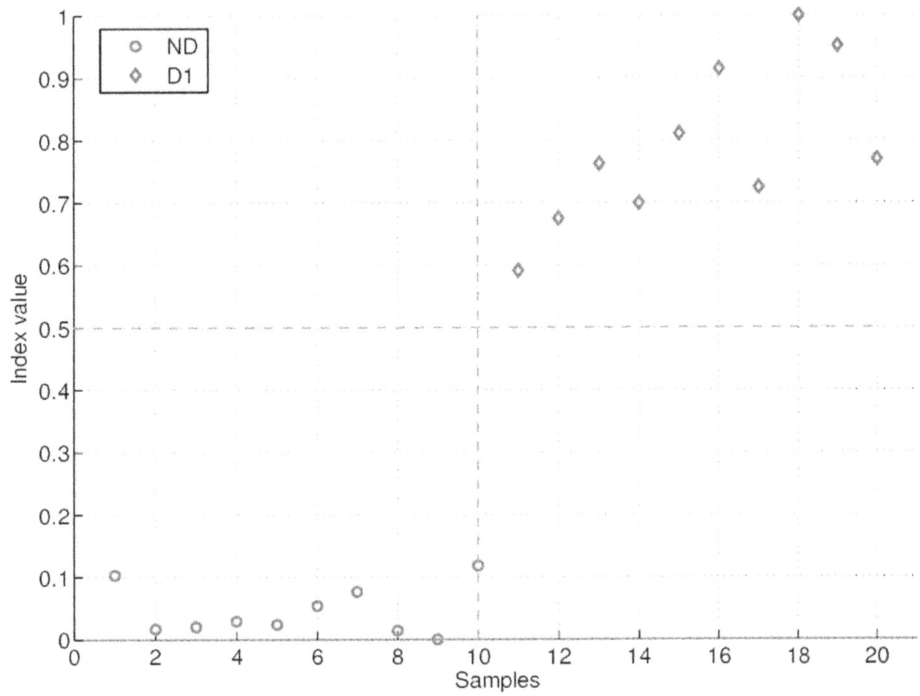

Figure 13b. Damage detection using fuzzy similarity, exciting S1 at 145KHz and capturing from S2 (○-undamaged, ◇ – damaged); P = 0.6

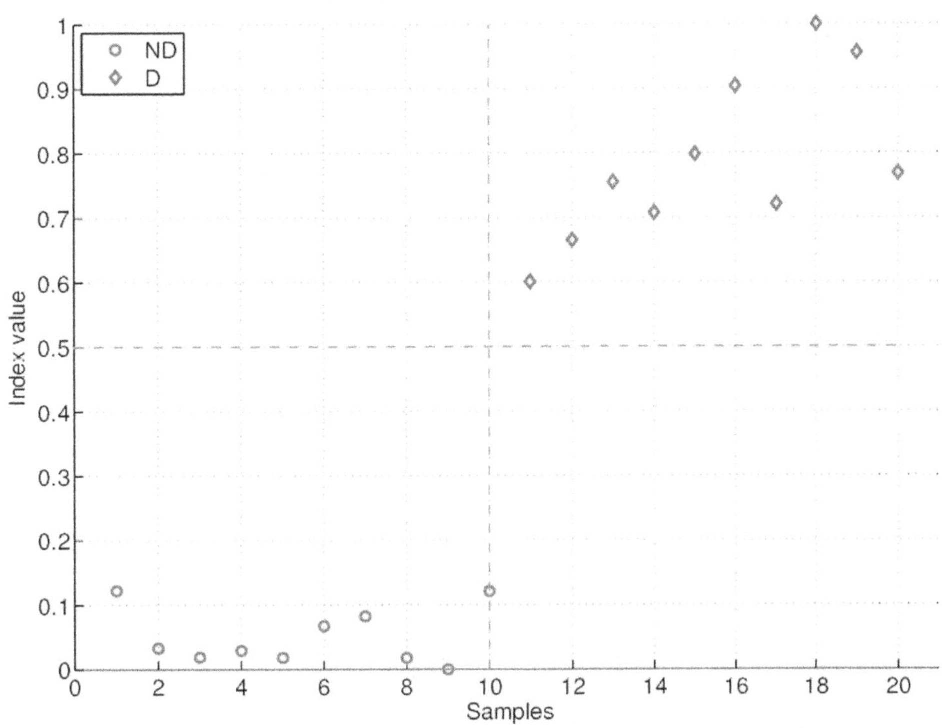

Figure 14a. Damage detection using Fuzzy similarity at different frequencies with p = 0.6 (○-Undamaged, ◇ – Damaged); 75Khz

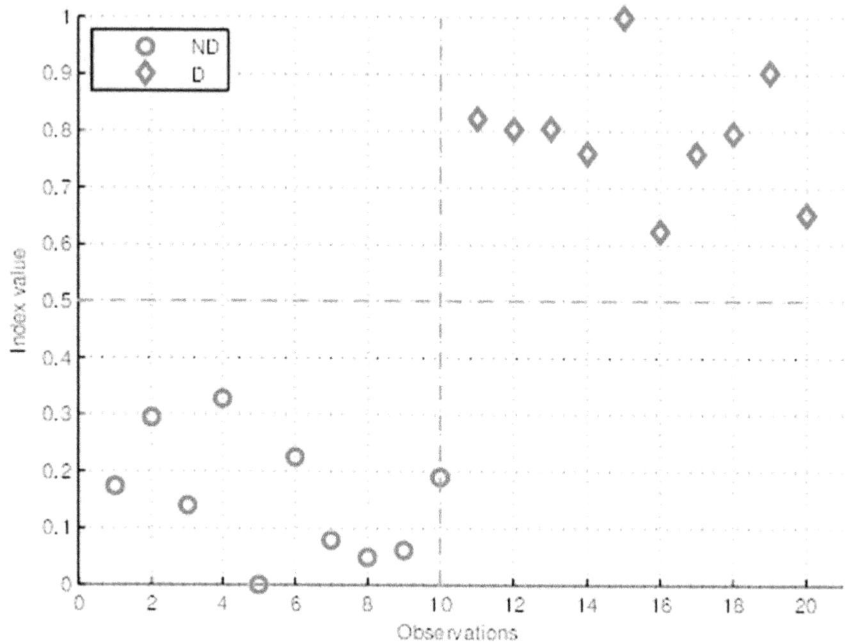

Figure 14b. Damage detection using Fuzzy similarity at different frequencies with p = 0.6 (○-Undamaged, ◇ – Damaged); 185KHz

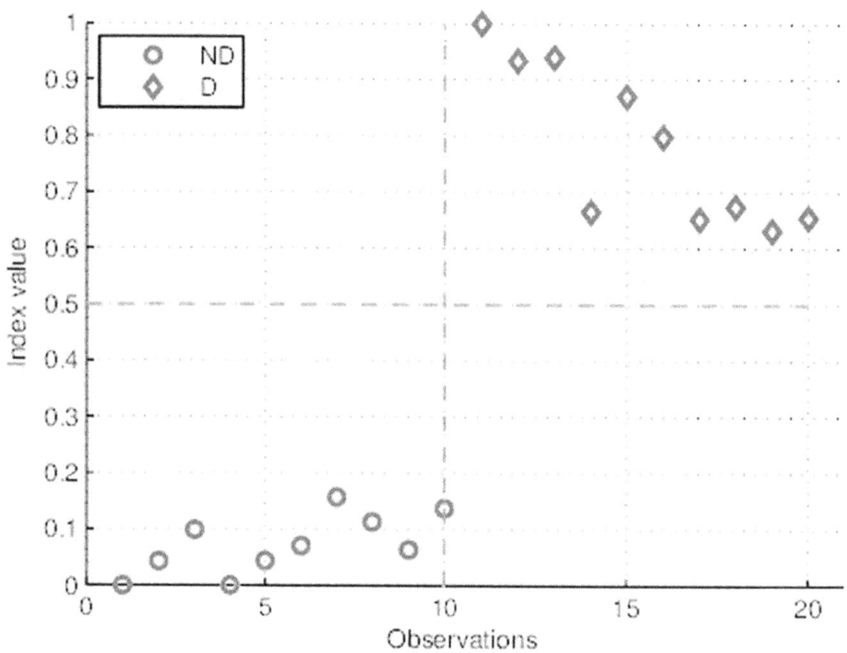

Figure 14c. Damage detection using Fuzzy similarity at different frequencies with p = 0.6 (○-Undamaged, ◇ – Damaged); 300KHz

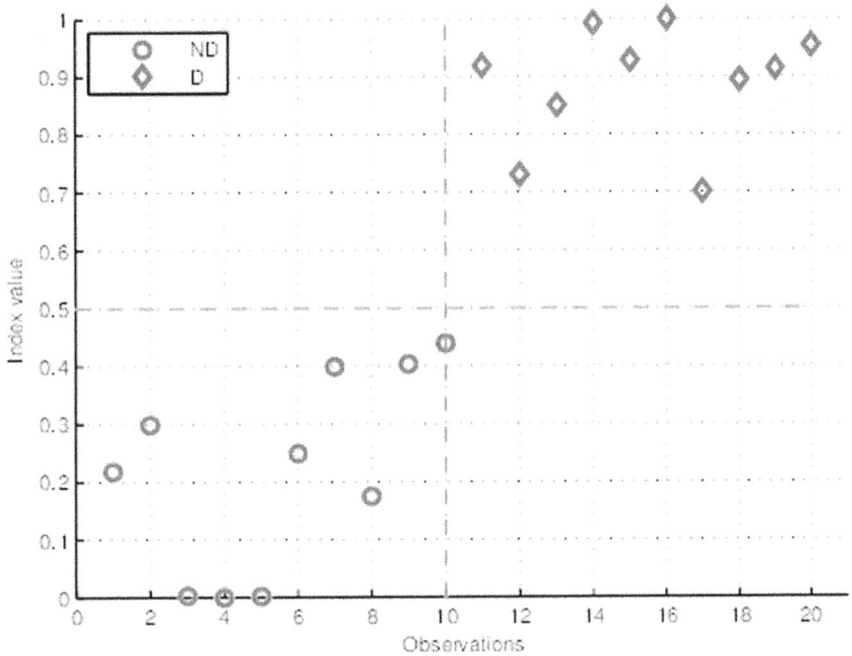

Figure 14d. Damage detection using Fuzzy similarity at different frequencies with p = 0.6 (○-Undamaged, ◇ – Damaged); 360KHz

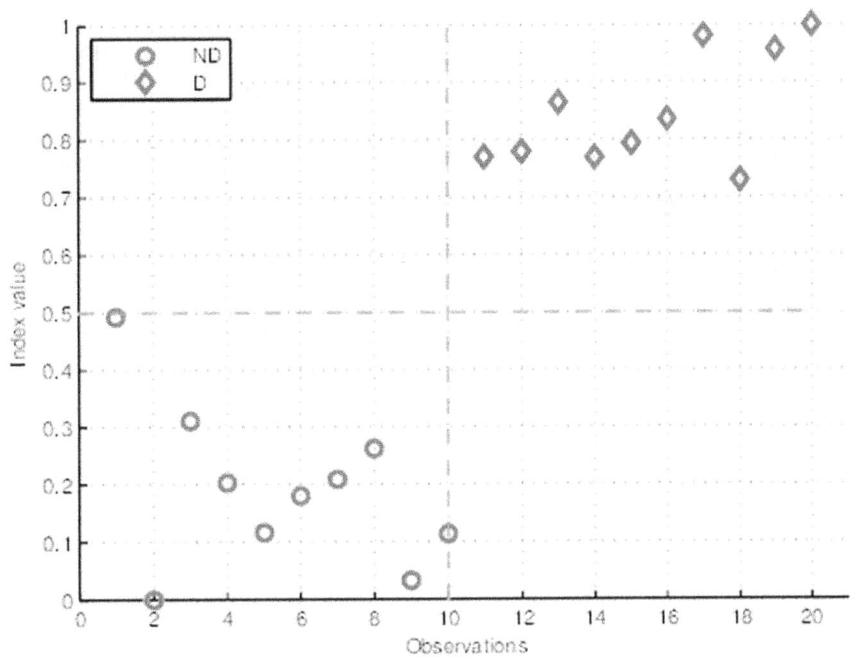

Figure 15a. Damage detection using Fuzzy similarity applied on folded data, at 125 KHz and p = 0.6; Actuator number 1

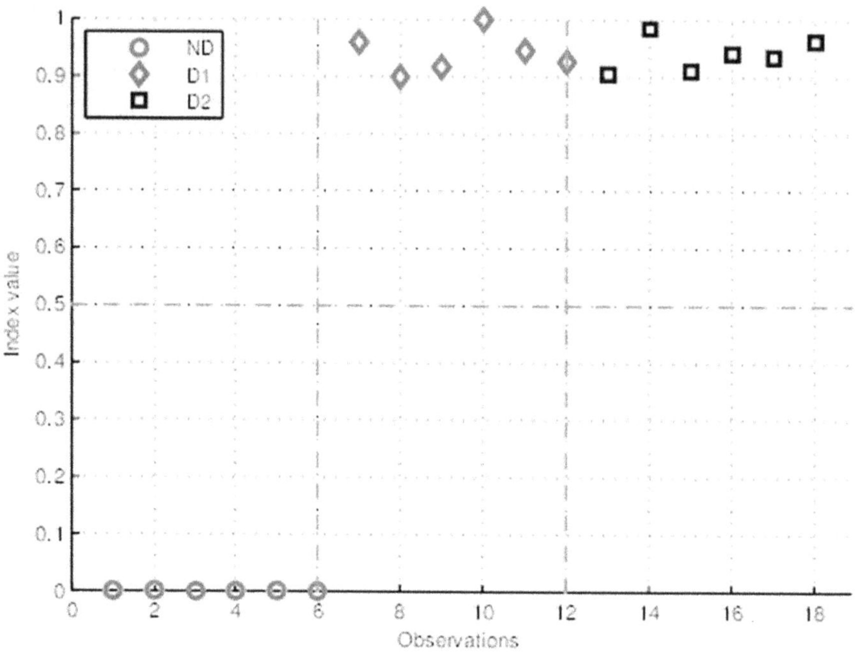

Figure 15b. Damage detection using Fuzzy similarity applied on folded data, at 125 KHz and p = 0.6; Actuator number 2

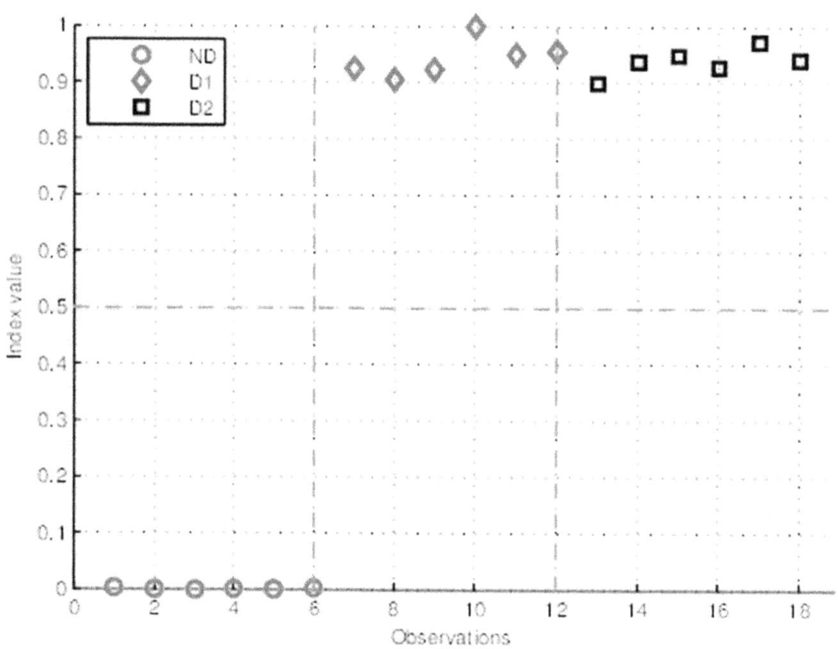

Figure 15c. Damage detection using Fuzzy similarity applied on folded data, at 125 KHz and p = 0.6; Actuator number 3

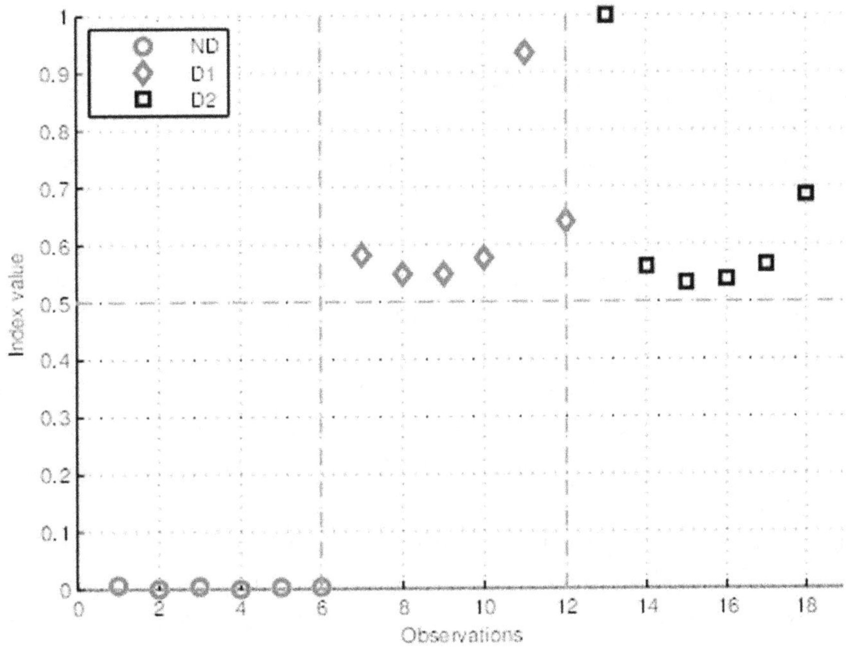

Figure 15d. Damage detection using Fuzzy similarity applied on folded data, at 125 KHz and p = 0.6; Actuator number 4

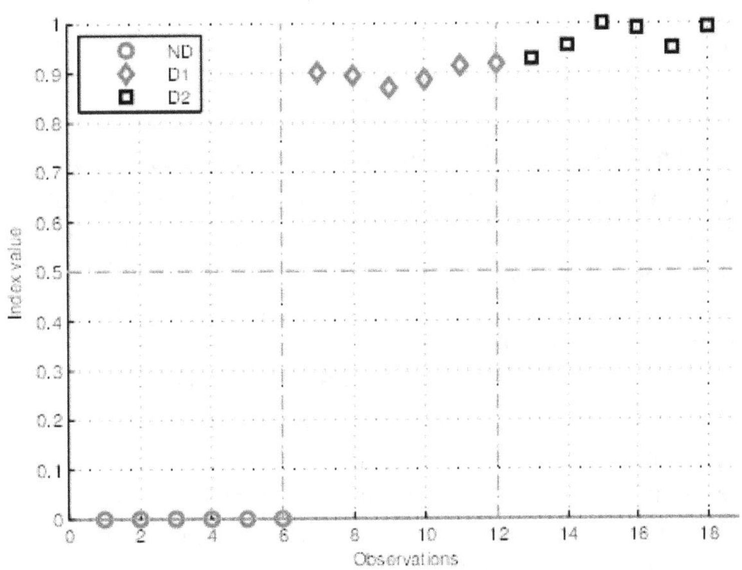

In next step, the methodology is repeated for the first actuator for constant p ($p = 0.5$) for different exciting frequencies. The results re-confirm the efficiency of the algorithm for different frequencies when folded data are used (see Figure 16).

Selecting the Appropriate Value for *P*

As it was described in previous section, the similarity is a function of the parameter p. Although p is an arbitrary value, the results show that the best value for p is around $p = 0.5$. Increasing the value of p, decreases the accuracy of the detection, somehow that the damage index is not capable of separate data from damaged structure and non-damage structure that causes *false-positive* detection. Figure 17 confirms the claim. As it may be seen from the figure, for $p \geq 5$, damage detection is failed. The horizontal line in each figure depicts the calculated threshold level.

To assess the effect of changing p in the sensitivity and accuracy of the damage detection method, the receiver operating characteristic (ROC) curve is plotted regarding to change of p. The ROC is a graphical plot which illustrates the performance of a binary classifier system as its discrimination threshold is varied. It is created by plotting the fraction of true positives out of the actual total positive vs. the fraction of false positives out of total actual negatives at various threshold settings. The closer the curve follows the left-hand border and then the top border of the ROC space, the more accurate the test is. On the other hand, the closer the curve comes to the 45 – degree diagonal of the ROC space, the less accurate the test is. The ROC test is performed when p is changing. As it can be seen in Figure 18, increasing p causes decreasing in sensitivity of the algorithm.

Comparing With Euclidean Distance Novelty Index

Euclidean distance is the most similar novelty index to the index that is presented in this work. The Euclidean metric is the "ordinary" distance between two points x and y. It is the length of the line segment connecting them defined as follows:

$$d(x,y) = \sqrt{\sum_{i=1}^{n} (x_i - y_i)^2} \tag{6}$$

The competency of the new damage index is compared with classical Euclidean distance at different frequencies. The same damage detection methodology is used. It means, after dimension reduction and obtaining the ideal vector, Euclidean distance is used to calculate the similarity of observations with the ideal vector. Figure 19 shows the result of using Euclidean distance in the first case study. As it may be seen in this figure, for the majority of applied frequencies (except in 185 KHz), the Euclidean distance results in false-positive or false-negative decision for some observations as it can be seen from the figure. For instance, in Figure 19a the majority of observation belonging to the damaged status are not detected (false-negative) or in Figure 19-d, some observation that belong to the healthy status are detected as damaged (false-positive).

To further evaluate the proposed methodology, it is compared with the general damage detection strategy based on PCA as both dimension reduction and data interpreting tool (L. Mujica et al. 2010). The result for the first case study shows that this method is not able to distinguish the different status of

Figure 16a. Damage detection using Fuzzy similarity conducted on the second case study for transducer 1 and p = 0.5 at different frequencies; 125KHz

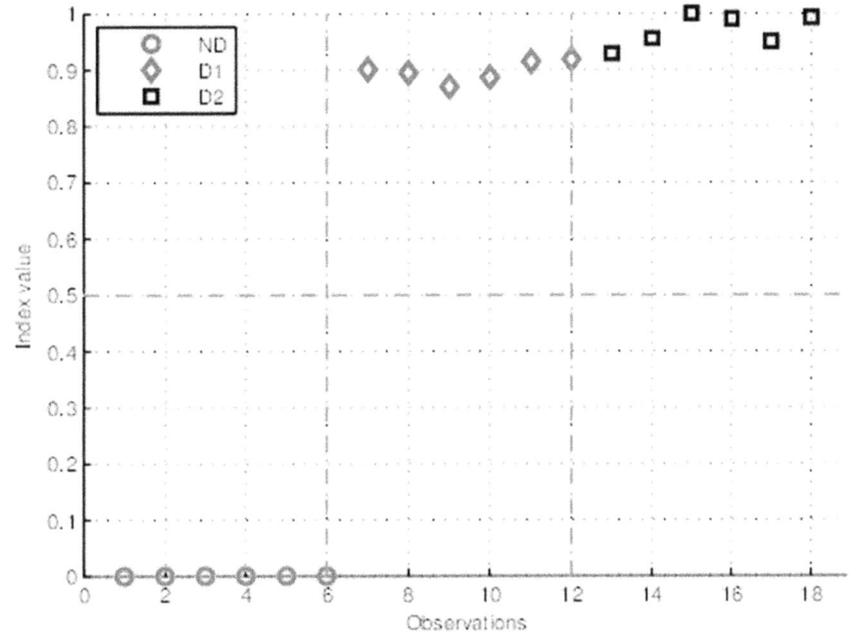

Figure 16b. Damage detection using Fuzzy similarity conducted on the second case study for transducer 1 and p = 0.5 at different frequencies; 175KHz

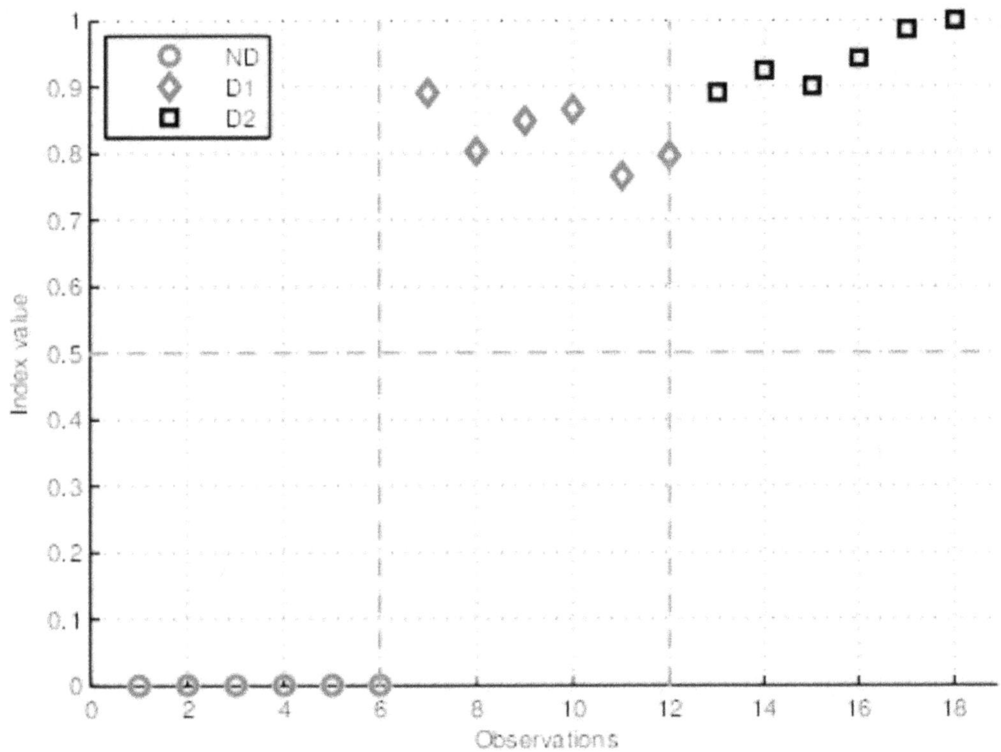

Figure 16c. Damage detection using Fuzzy similarity conducted on the second case study for transducer 1 and p = 0.5 at different frequencies; 275KHz

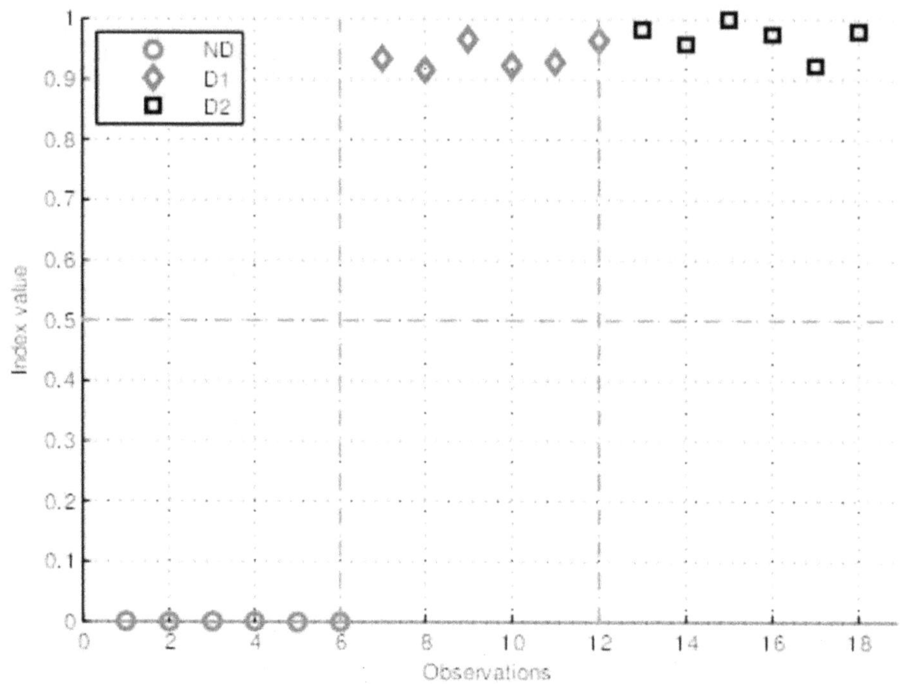

Figure 16d. Damage detection using Fuzzy similarity conducted on the second case study for transducer 1 and p = 0.5 at different frequencies; 365KHz

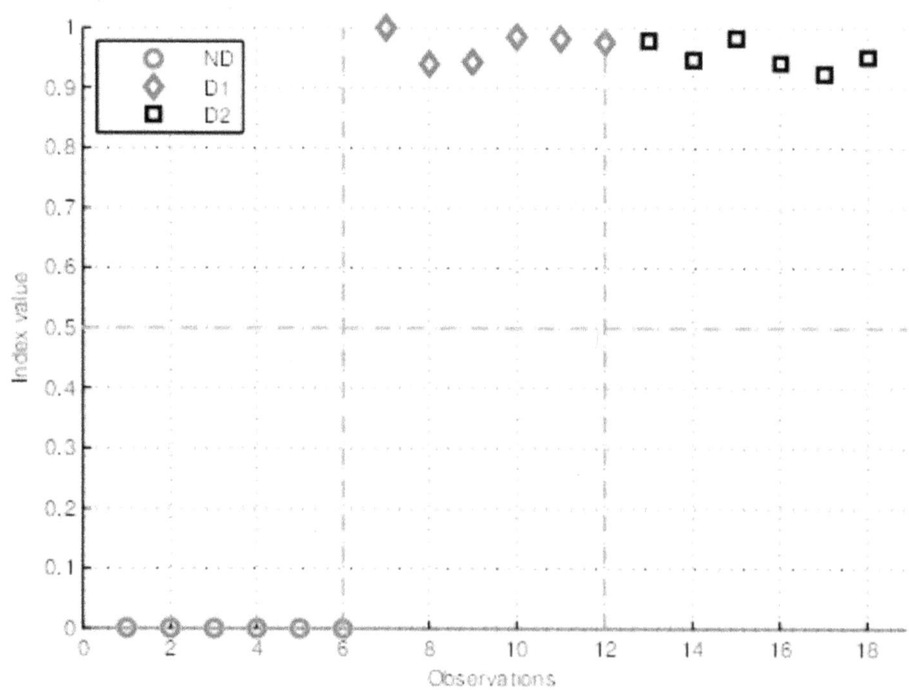

Figure 17a. Damage detection with different values of p conducted on the first case study, UPM SHM Lab; p = 0.3

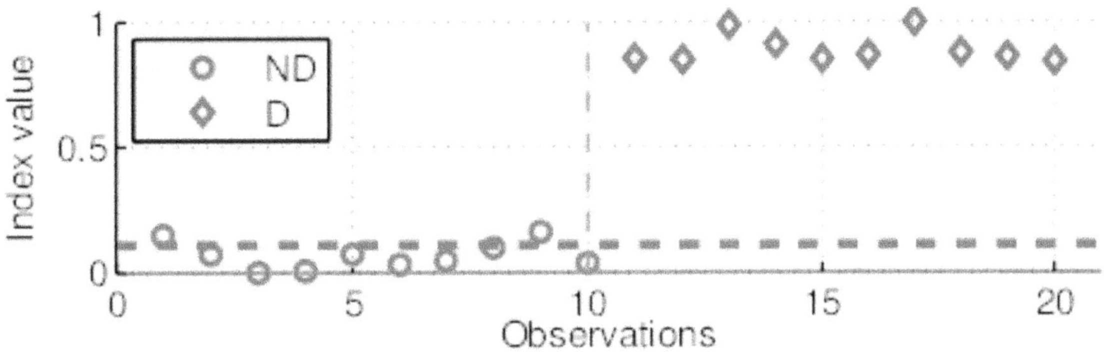

Figure 17b. Damage detection with different values of p conducted on the first case study, UPM SHM Lab; p = 0.5

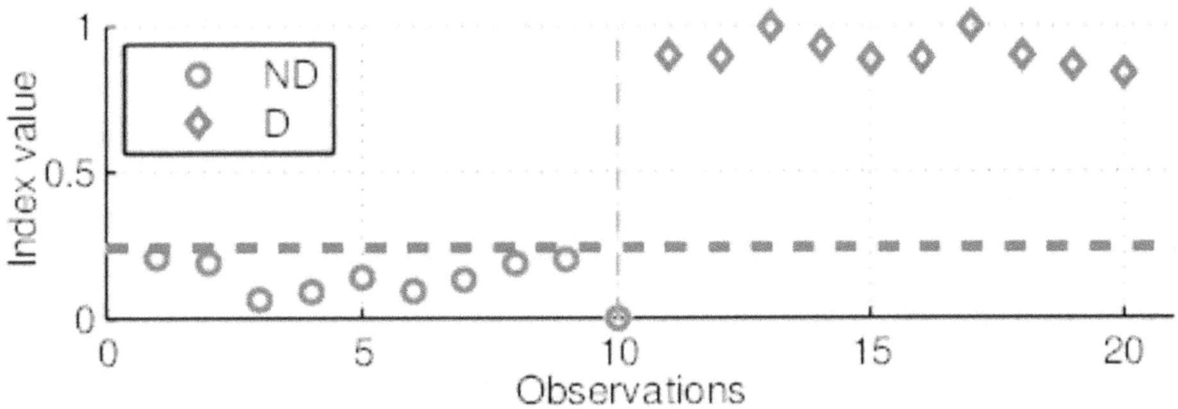

Figure 17c. Damage detection with different values of p conducted on the first case study, UPM SHM Lab; p = 1

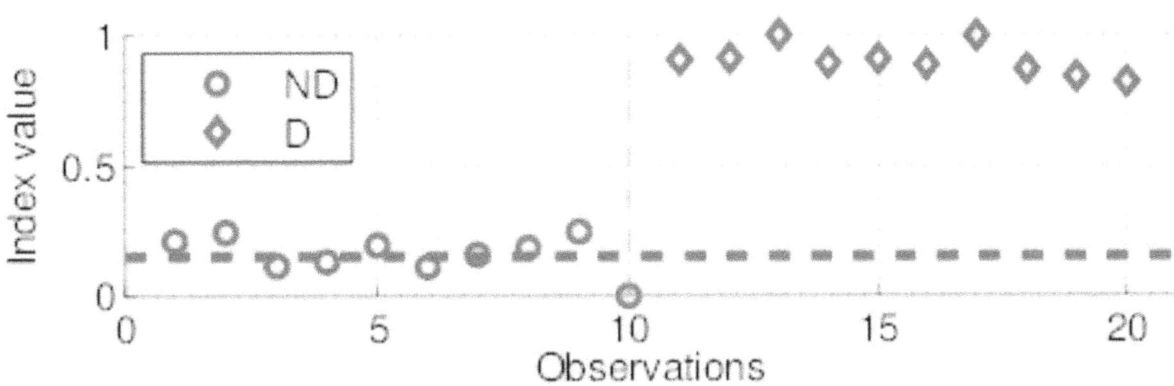

Figure 17d. Damage detection with different values of p conducted on the first case study, UPM SHM Lab; p = 2

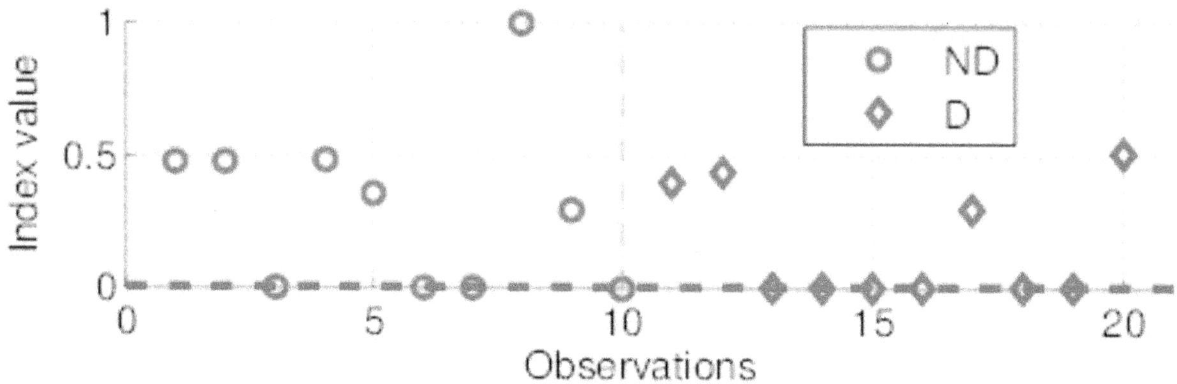

Figure 17e. Damage detection with different values of p conducted on the first case study, UPM SHM Lab; p = 5

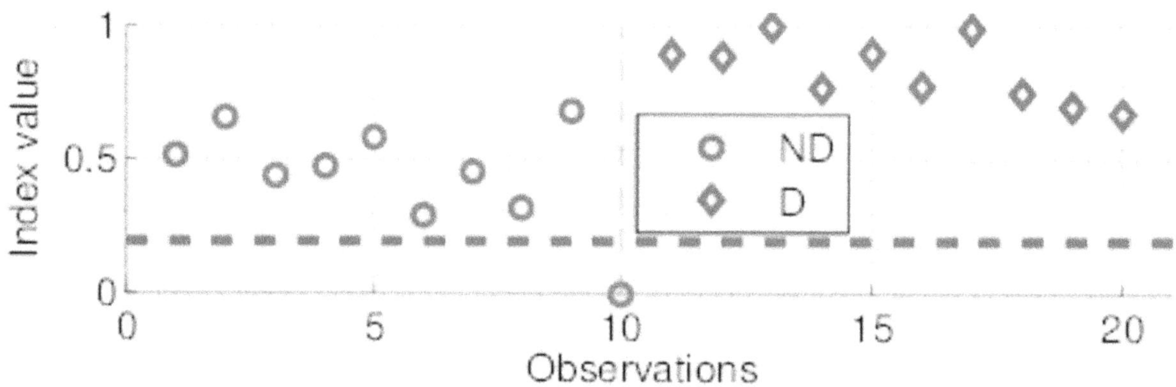

Figure 17f. Damage detection with different values of p conducted on the first case study, UPM SHM Lab; p = 10

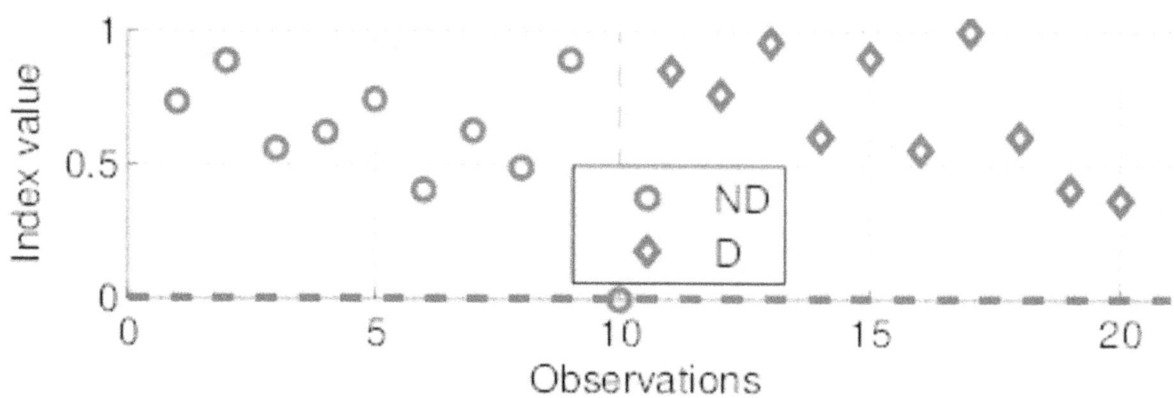

Figure 18. ROC for Fuzzy similarity classifier by changing p

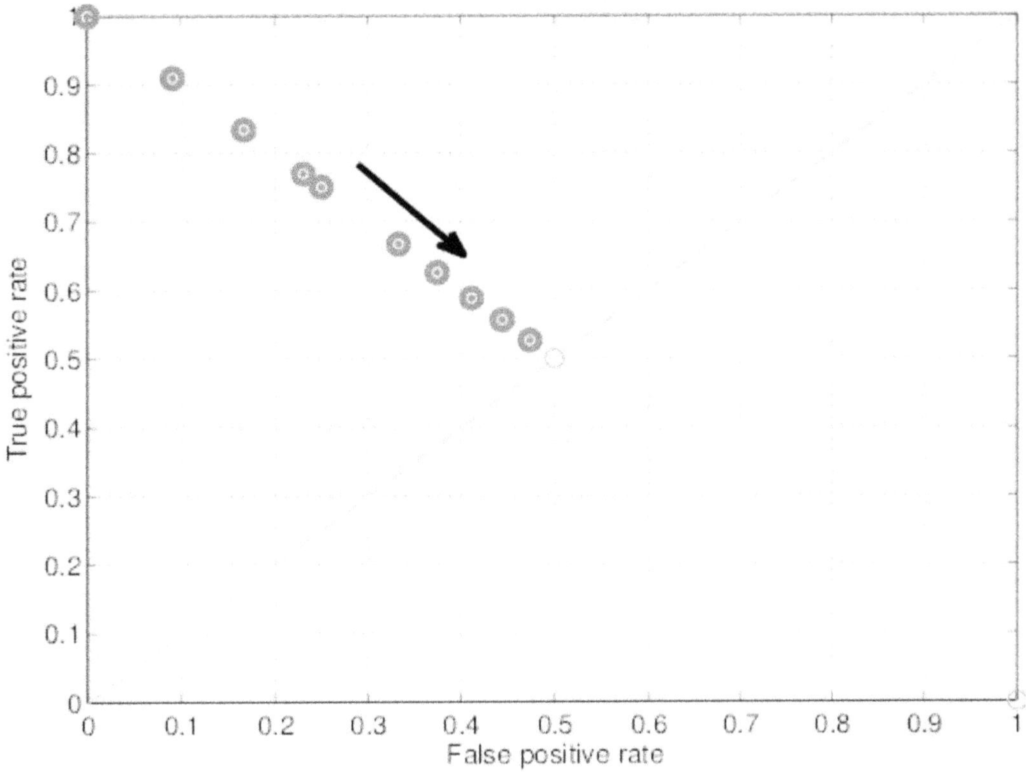

Figure 19a. Euclidean distance, first case study, different frequencies; 75 KHz

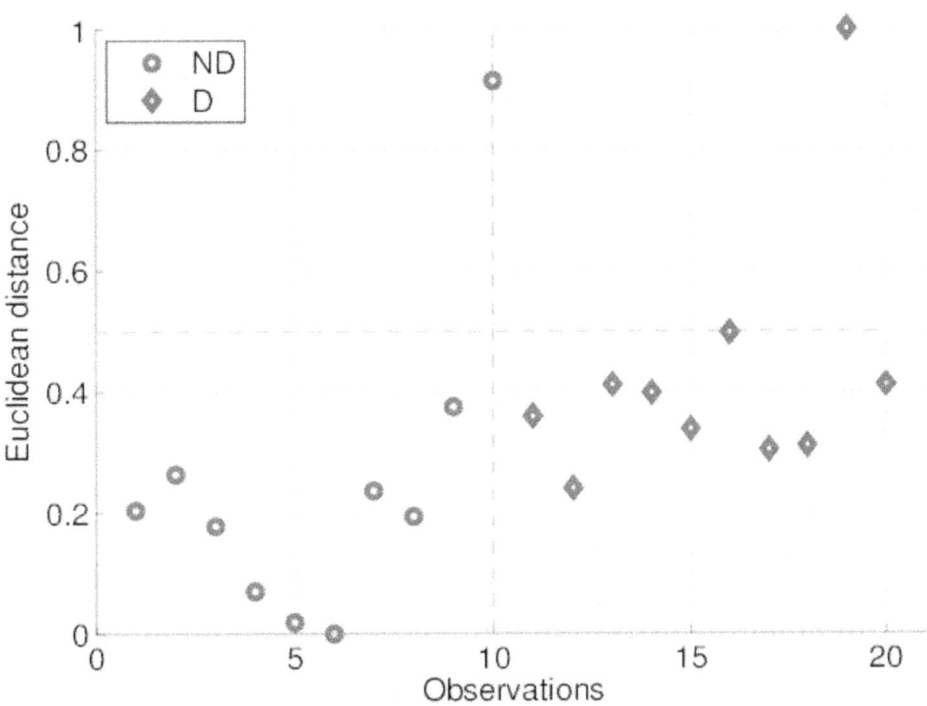

Figure 19b. Euclidean distance, first case study, different frequencies; 185 KHz

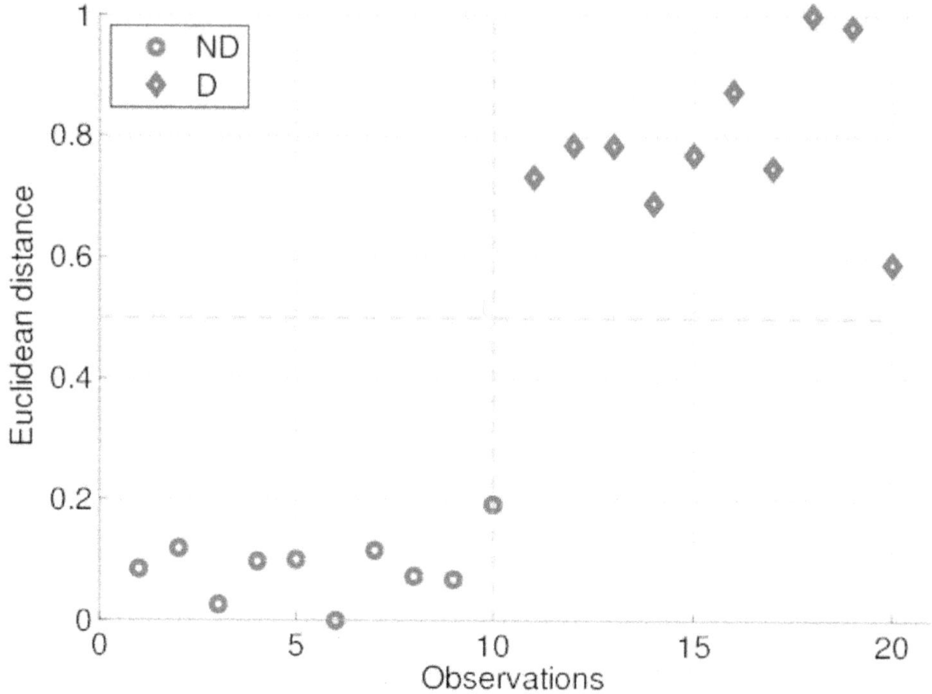

Figure 19c. Euclidean distance, first case study, different frequencies; 200 KHz

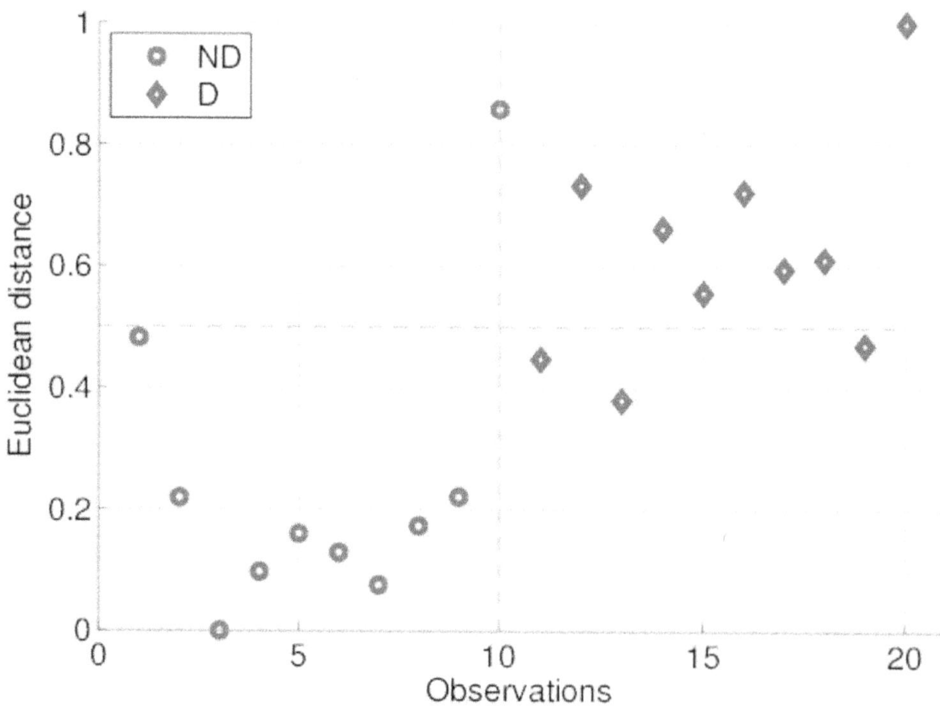

Figure 19d. Euclidean distance, first case study, different frequencies; 360 KHz

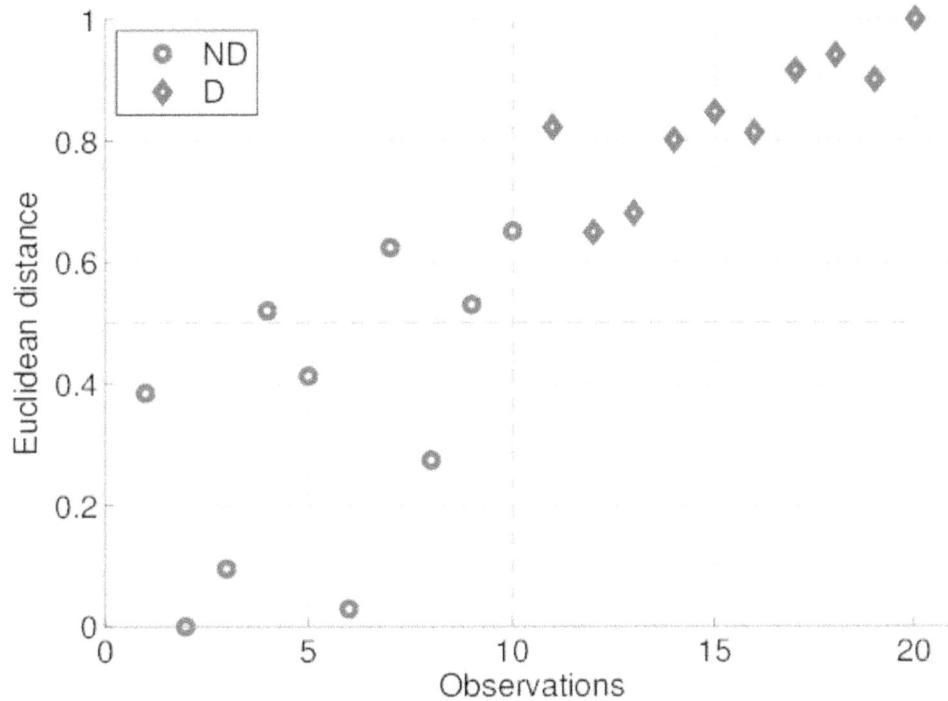

the structure. As it may be seen in Figure 20, there is no clear separation between observations captured from pristine structure and damaged condition.

The analogous result is achieved when the described methodology is applied on the second case study. For instance, as it may be seen in Figure 21, damage detection is not obtained using the data that are folded for actuator 1 or 3. Results show that observations are not separated from each other, which means failure in damage detection.

In this section, Fuzzy similarity is introduced as a new damage index. Its performance is evaluated and confirmed using different case studies. To provide more comprehensive study on this index, it is compared with two other common damage detection methodologies such as Euclidean distance. In addition, the optimal value for p parameter is evaluated and is has been demonstrated that, according to the ROC curves increasing the value of p causes decreasing the sensitivity of the index.

TEMPERATURE EFFECT AND COMPENSATION

Environmental variation is the one of the main obstacles for developing SHM solutions. In fact, these changes can often mask the structural changes caused by damage [Sohn2007]. Among various environmental conditions, temperature changes are of particular interest because temperature is a global environmental condition and its variations substantially alter the recorded waveform (Lu & Michaels 2005).

As many structures exhibit daily and seasonal temperature variations and SHM systems will need to operate across a variety of environmental conditions, understanding and considering the effect of temperature on SHM methodologies is critical. In another words, SHM systems will not be accepted

Figure 20a. Damage detection based on PCA scores conducted on the first case study; 145KHz

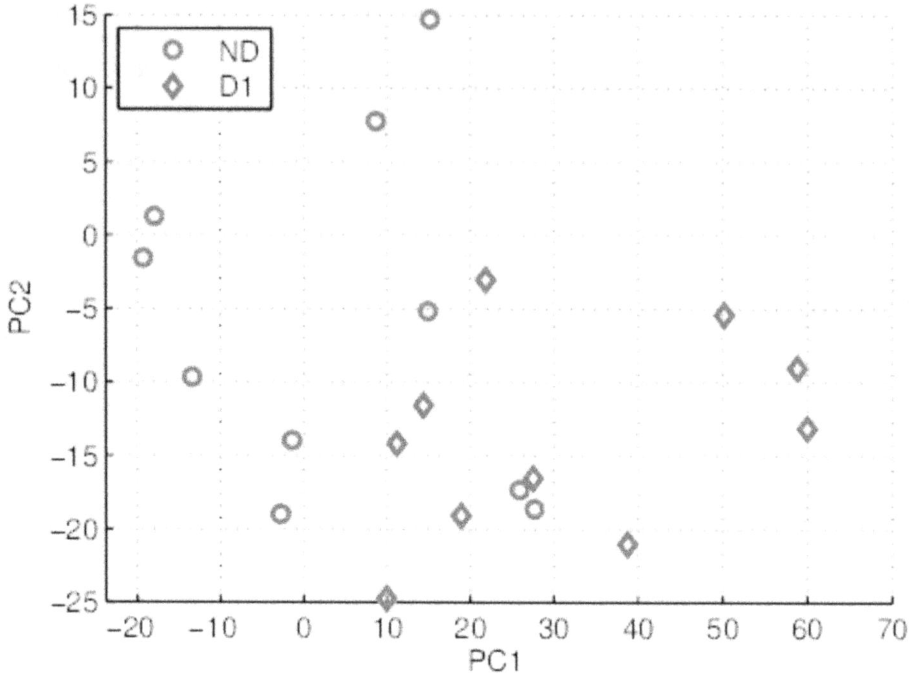

Figure 20b. Damage detection based on PCA scores conducted on the first case study; 185KHz

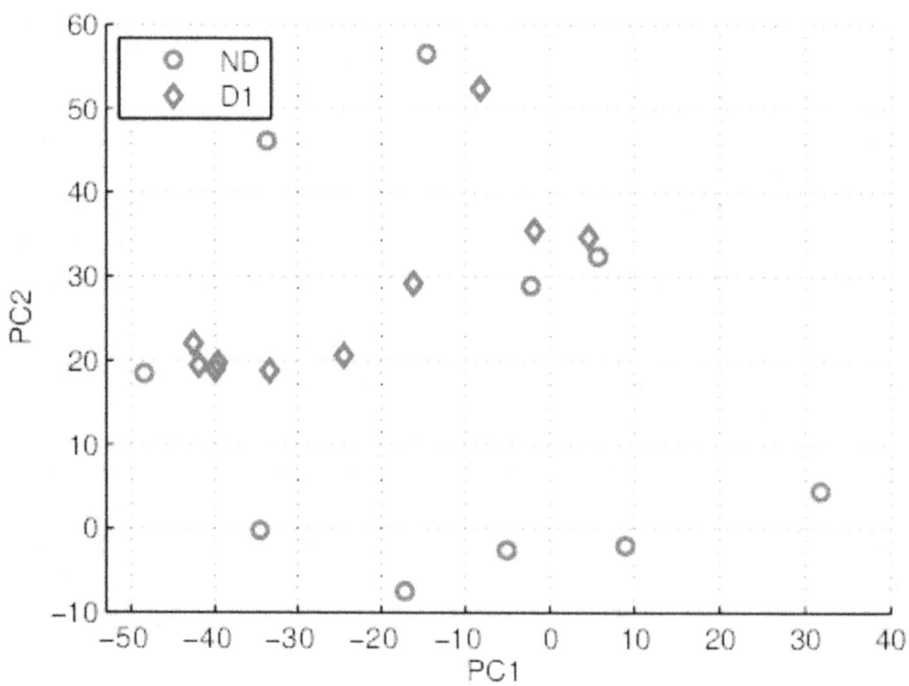

Figure 21a. Damage detection based on PCA scores, second case study, 125 KHz using two actuators; Actuator 1

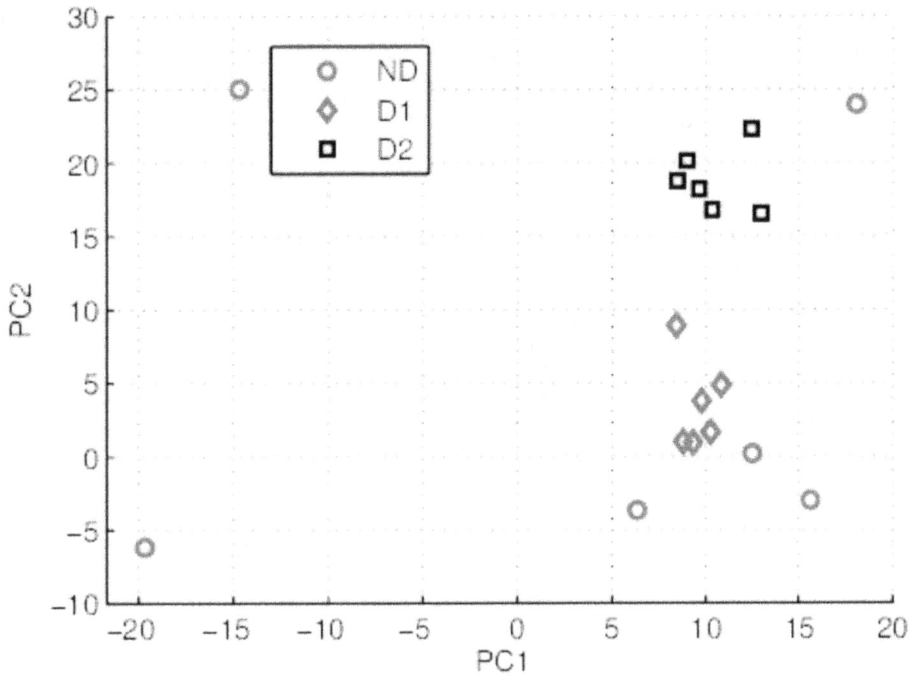

Figure 21b. Damage detection based on PCA scores, second case study, 125 KHz using two actuators; Actuator 3

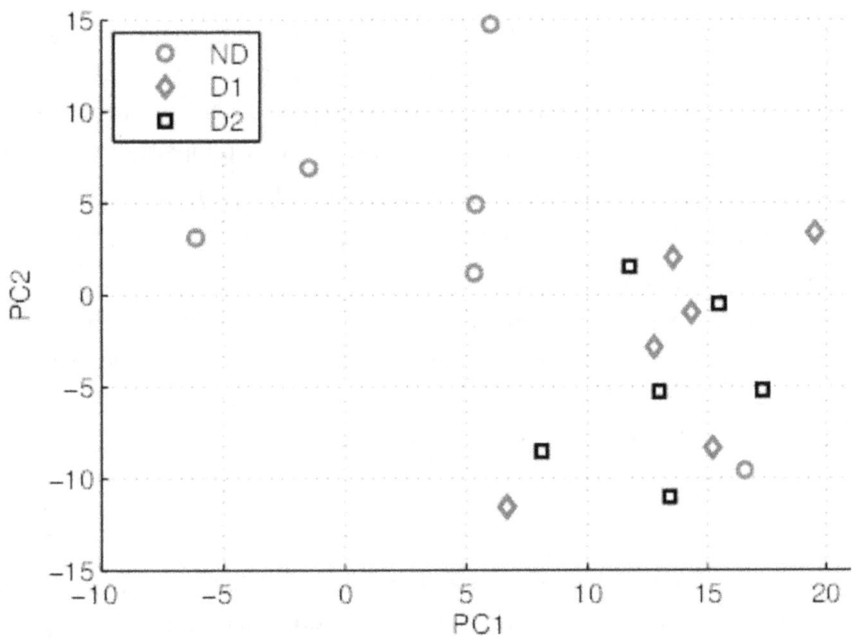

in practical applications unless robust techniques are developed to explicitly account for environmental and operational constraints/conditions of the systems to be monitored (Sohn 2007).

Several investigation were performed to find out the effect of temperature in SHM methodologies and various researchers acknowledged potential adverse effects of varying operational and environmental conditions on vibration-based damage detection (Blaise & Chang 2001; Konstantinidis, Wilcox, & Drinkwater 2007; Lu & Michaels 2005).

The main reason to such effects are categorized as follows:

1. Change in the properties of piezoelectric transducer (majority) such as piezoelectric constants (Raghavan & Cesnik 2008; Williams, Inman, & Wilkie 2004).
2. Alternation in properties of adhesive used to bond transducers to the host structures.
3. Thermal expansion such as change in plate thickness, piezo dimensions and distances traveled by the guided wave in the structure (Raghavan & Cesnik 2008).
4. Change in elastic properties including density and Young module that causes changes in wave velocity (Su & Ye 2009).

When environmental or operational conditions change, the propagating medium and guided wave behavior also change. Therefore, simple baseline comparison methods are unable to distinguish damage from environmental and operational effects (Harley & Moura 2012).

Temperature Compensation

Due to the effect of temperature fluctuation mentioned above, a simple baseline comparison methods are unable to distinguish damage from environmental and operational effects.

Different methodologies are proposed to compensate the effect of temperature such as interpolation and extrapolation of baseline (Inman et al. 2005) baseline signal stretch (BSS) (Lu & Michaels 2005) or optimal baseline selection (OBS) (Michaels 2008).

In this work, first the effect of temperature is analyzed on the performance of the damage index and then the OBS technique is used to compensate the effect of temperature on the damage detection index.

In OBS technique (See Figure 22), to discriminate the effects of damage from those of environmental changes, a "bank" of baseline signals acquired at the various temperatures $\left(T1, T2, \ldots, Tn\right)$ and later the response data (from unknown status) is compared with baseline database to find the closest match $\left(T_i\right)$. The selected baseline is used as the data representing pristine structure in this temperature. Unlike BSS, that technique does not change the time-trace but simply reduces the temperature difference seen. To achieve the highest performance, the baseline database should be formed at number of temperatures spanning the expected service range. In the next step, the best baseline signal is selected using appropriate measurement distance. Michaels (Michaels 2008) suggests three differential features which calculate the difference between the signal and the baseline.

Case Studies

Heating/cooling chamber is used to change the environmental temperature and observe the effect of temperature change on wave propagating and damage detecting techniques. Using this chamber the

Figure 22. Schematic of OBS temperature compensation technique

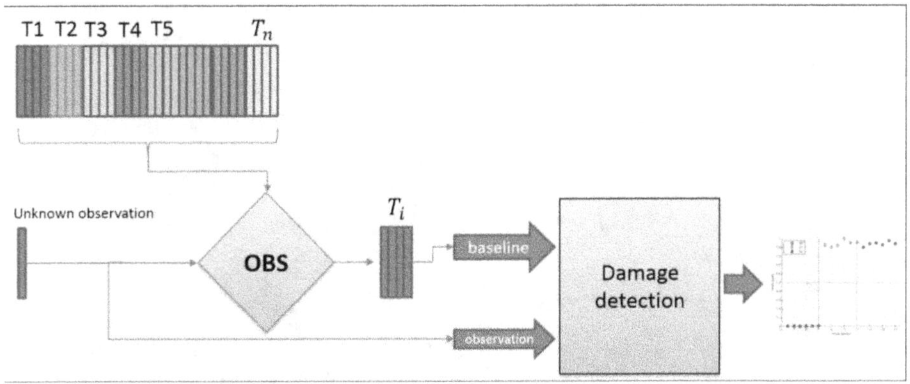

temperature is changed from 20, 25, 30, 40, 46, 55, 60, 65 ºC (heating) $\left(T_1,\ldots T_7\right)$. Figure 23, Figure 24 and Figure 25 shows the chamber and the composite plate inside.

Damages are simulated by adding an appropriate mass in different locations, the same temperature sweeping is repeated on the structure when there is damage on the surface of the structure.

Temperature Effect on Similarity Classifier and Compensation

In this section, the effect of temperature on similarity classifier method is considered. Basically, as this method is applied on the original data and it was previously showed that signal is affected by the temperature, therefore the same phenomenon is expected for this method. To check the claim above, the similarity classifier is applied on the composite plate equipped with PZTs. The analysis is performed on 350 KHz for a route between two PZTs where a damage is simulated by placing a mass. As it could be

Figure 23. Specimen inside the chamber

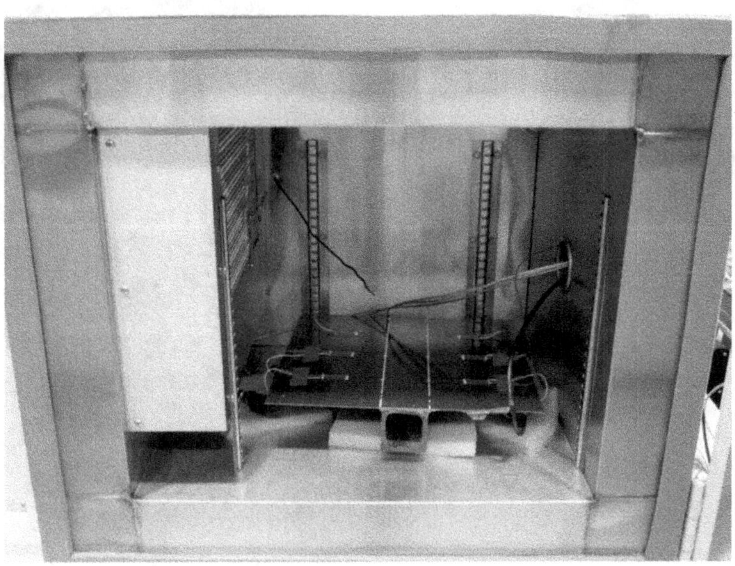

Figure 24. Damage 1 snapshot

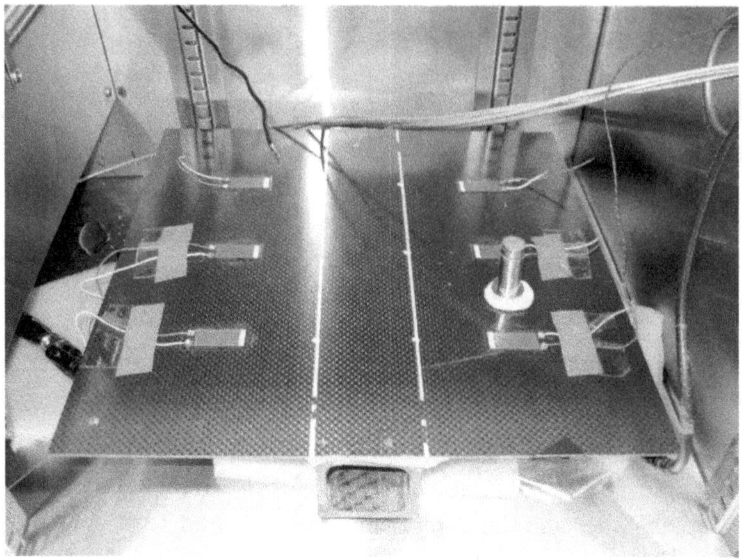

seen in Figure 26, temperature has a clear effect on the index but it does not mask the effect of damage. In other words, the damage is still distinguishable from the other status.

To achieve better analysis, data captured from the structure are analyzed for all temperatures. Figure 27 shows the result of this analysis. As it is depicted, the same result is achieved for other frequencies. For instance, for 400 KHz the damage effect dominate the effect of temperature. The same result is achieved for 300 KHz, 350 KHz and 400 KHz.

Figure 25. Specimen snapshot in chamber

Figure 26. Temperature effect on similarity classifier

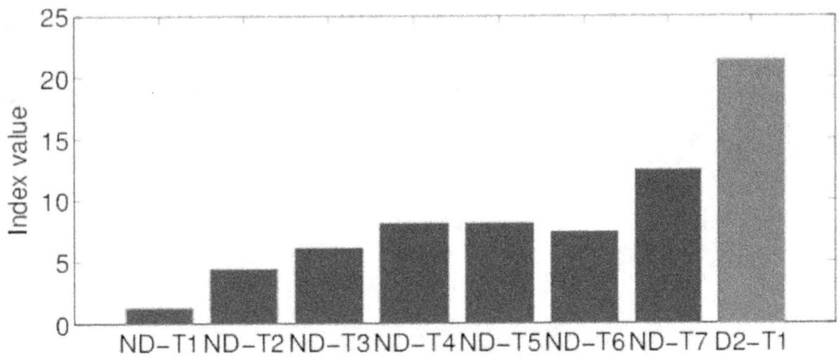

Figure 27a. Temperature effect on similarity classifier; 300KHz

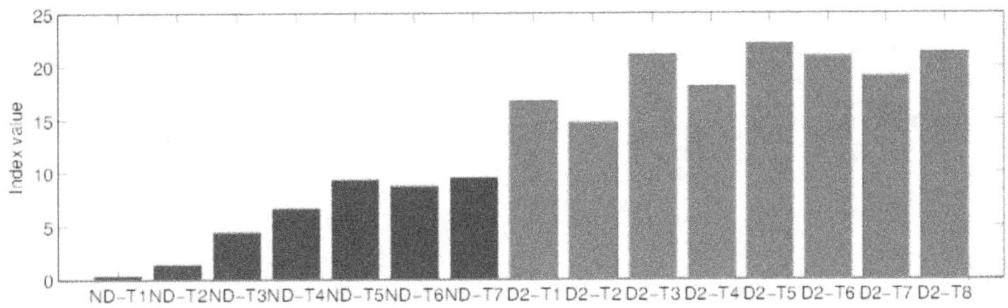

Figure 27b. Temperature effect on similarity classifier; 350KHz

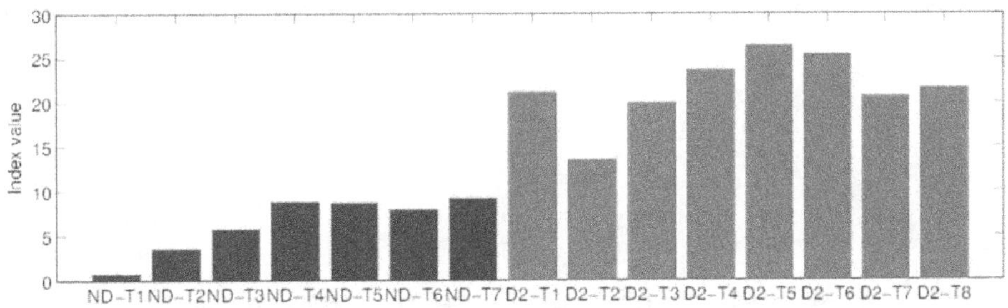

Figure 27c. Temperature effect on similarity classifier; 400KHz

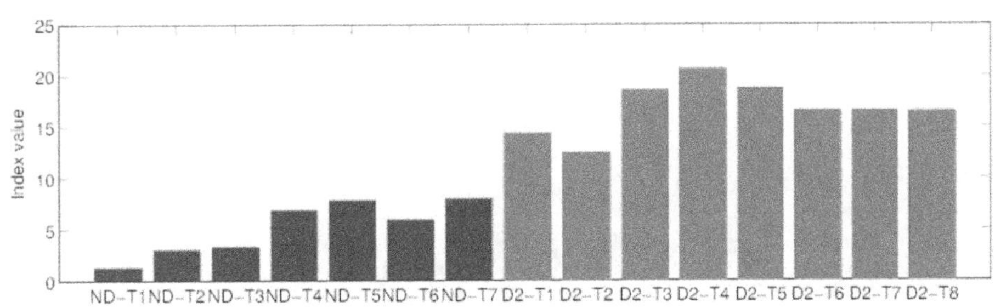

Although similarity classifier shows resistance against the temperature effect, at least in this case study, it is still favorable to apply temperature compensation on the methodology. The result of applying OBS technique is demonstrated for different frequencies (300 KHz, 350 KHz and 400 KHz) in Figure 28. As it can be seen from the figure, applying OBS leads to relieve the effect of temperature and makes the damage effect more separable.

CONCLUSION

This chapter is dedicated to present new techniques for damage detection based on Fuzzy Similarity and Wave Clustering. Three new presented indices can be used as damage detection tools. The efficiency of presented techniques are confirmed applying on different case studies with different complexity. In

Figure 28a. Temperature compensation: similarity classifier, different frequencies; 300KHz

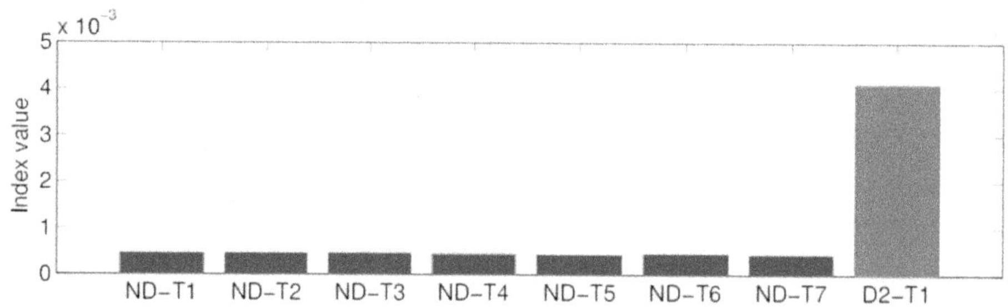

Figure 28b. Temperature compensation: similarity classifier, different frequencies; 350KHz

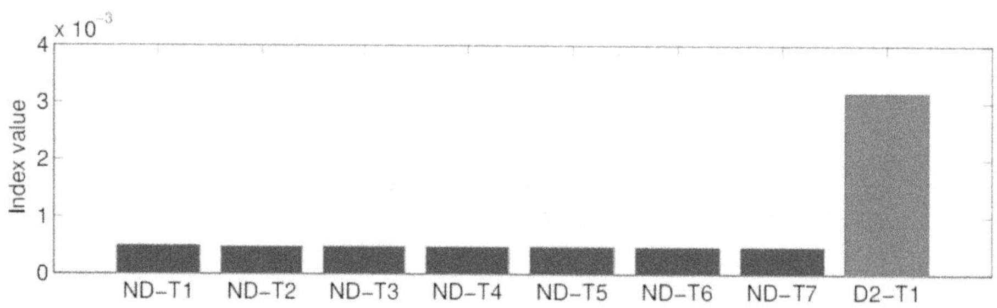

Figure 28c. Temperature compensation: similarity classifier, different frequencies; 400KHz

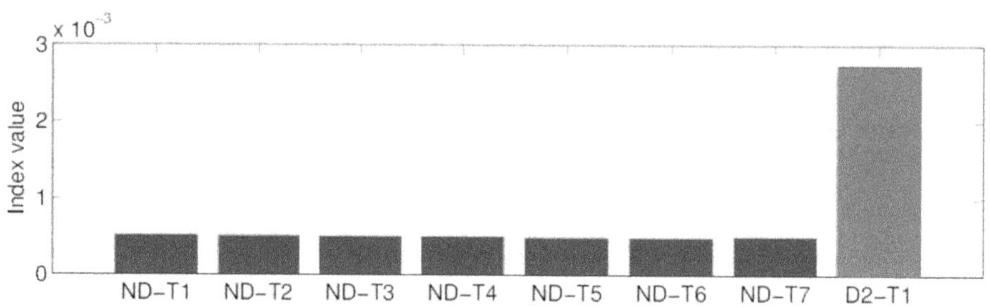

addition, the effect of temperature fluctuation is studied and it is shown how the presented indices are sensitive to temperature change. After that, an appropriate temperature compensation method is applied to mitigate the effect of temperature.

ACKNOWLEDGMENT

This work has been done with support of "Ministerio de Ciencia e Innovación" in Spain through the coordinated projects DPI2008-06564 and DPI2011-28033. Special thanks also to Control, Dynamics and Applications (CoDAlab) and its all members for all valuable days that I spent working in this group.

This work is deeply grateful to three SHM groups that allowed us to conduct the case studies. Therefore, special thanks goes to Professor Dr.-Ing. Claus-Peter Fritzen and his group in "Institut für Mechanik und Regelungstechnik, Universitat Siegen", and all of its members and also to Professor Michael Todd and his group in university of California, San Diego (UCSD) and his group and finally professor Alfredo Guemes from Technical University of Madrid.

REFERENCES

Blaise, E. J., & Chang, F. K. (2001). Built-in Damage Detection System for Sandwich Structures under Cryogenic Temperatures. In *Proceedings of Third International Workshop on Structural Health Monitoring*, (pp. 154–63). Academic Press.

Gharibnezhad, F., Mujica, L. E., & Rodellar, J. (2011). Damage Detection Using Andrew Plots. In *Proceedings of the 8th International Workshop on Structural Health Monitoring*. Stanford University.

Gharibnezhad, F., Mujica, L. E., & Rodellar, J. (2014). Applying Robust Variant of Principal Component Analysis as a Damage Detector in Presence of Outliers. *Mechanical Systems and Signal Processing, 50-51*, 467–479. doi:10.1016/j.ymssp.2014.05.032

Harley, J. B., & Moura, J. M. F. (2012). Scale Transform Signal Processing for Optimal Ultrasonic Temperature Compensation. *Ultrasonics, Ferroelectrics and Frequency Control. IEEE Transactions on, 59*(10), 2226–2236.

Horn, B. (1986). *Robot Vision*. MIT Press.

Inman, D. J., Farrar, C., Lopes, C., & Steffen, V. (2005). *Damage Prognosis: For Aerospace, Civil and Mechanical Systems*. John Wiley & Sons Inc. doi:10.1002/0470869097

Jolliffe, I. T. (2002). *Principal Component Analysis*. John Wiley & Sons, Ltd.

Kaufman, L., & Rousseeuw, P. J. (2005). *Finding Groups in Data: An Introduction to Cluster Analysis*. Wiley.

Konstantinidis, G., Wilcox, P. D., & Drinkwater, B. W. (2007). An Investigation into the Temperature Stability of a Guided Wave Structural Health Monitoring System Using Permanently Attached Sensors. *Sensors Journal, IEEE, 7*(5), 905–912. doi:10.1109/JSEN.2007.894908

Lukasiewicz, J. (1970). *Selected Works*. Cambridge, UK: Cambridge University Press.

Luukka, P. (2009). Classification Based on Fuzzy Robust PCA Algorithms and Similarity Classifier. *Expert Systems with Applications*, *36*(4), 7463–7468. doi:10.1016/j.eswa.2008.09.015

Luukka, P., & Leppälampi, T. (2006). Similarity Classifier with Generalized Mean Applied to Medical Data. *Computers in Biology and Medicine*, *36*(9), 1026–1040. doi:10.1016/j.compbiomed.2005.05.008 PMID:16159657

Luukka, P., Saastamoinen, K., & Kononen, V. (2001). A Classifier Based on the Maximal Fuzzy Similarity in the Generalized Lukasiewicz-Structure. In *Proceedings of Fuzzy Systems, 2001. The 10th IEEE International Conference On* (pp. 195–98). IEEE. doi:10.1109/FUZZ.2001.1007281

Martin, E., Kriegel, H. P., Sander, J., & Xiaowei, X. (1996). A Density-Based Algorithm for Discovering Clusters in Large Spatial Databases with Noise. In *Proceedings of the 2nd International on KDD* (vol. 96, pp. 226–31). Academic Press.

Michaels, J. E. (2008). Detection, Localization and Characterization of Damage in Plates with an in Situ Array of Spatially Distributed Ultrasonic Sensors. *Smart Materials and Structures*, *17*(3), 35035. doi:10.1088/0964-1726/17/3/035035

Mujica, L. E., Rodellar, J., Fernandez, A., & Guemes, A. (2010). Q-Statistic and T2-Statistic PCA-Based Measures for Damage Assessment in Structures. *Structural Health Monitoring*, *10*(5), 539–553. doi:10.1177/1475921710388972

Mujica, L. E., Vehí, J., Ruiz, M., Verleysen, M., Staszewskic, W., & Worden, K. (2008). Multivariate Statistics Process Control for Dimensionality Reduction in Structural Assessment. *Mechanical Systems and Signal Processing*, *22*(1), 155–171. doi:10.1016/j.ymssp.2007.05.001

Raghavan, A., & Cesnik, E. (2008). Effects of Elevated Temperature on Guided-Wave Structural Health Monitoring. *Journal of Intelligent Material Systems and Structures*, *19*(12), 1383–1398. doi:10.1177/1045389X07086691

Rokach, L., & Maimon, O. (2010). Distance Measures for Numeric Attributes. In Data Mining and Knowledge Discovery Handbook, (pp. 322–52). Academic Press.

Sheikholeslami, G., Chatterjee, S., & Zhang, A. (1998). Wavecluster: A Multi-Resolution Clustering Approach for Very Large Spatial Databases. In *Proceedings of 24th VLDB Conference*. New York, NY: VLDB.

Sheikholeslami, G., Chatterjee, S., & Zhang, A. (2000). WaveCluster: A Wavelet-Based Clustering Approach for Spatial Data in Very Large Databases. *The VLDB Journal The International Journal on Very Large Data Bases*, *8*(3-4), 289–304. doi:10.1007/s007780050009

Smith, L. (2002). *A Tutorial on Principal Components Analysis*. Academic Press.

Sohn, H. (2007). Effects of Environmental and Operational Variability on Structural Health Monitoring. *Philosophical Transactions Series A: Mathematical, Physical, and Engineering Sciences, 365*(1851).

Su, Z., & Ye, L. (2009). Heat and Mass Transfer. In Identification of Damage Using Lamb Waves: From Fundamentals to Applications. Springer-Verlag Berlin Heidelberg.

Tibaduiza, D. A. (2013). *Design and Validation of a Structural Health Monitoring System for Aeronautical Structures*. (PhD. Thesis). Technical University of Catalunya.

Urquhart, R. (1982). Graph Theoretical Clustering Based on Limited Neighbourhood Sets. *Pattern Recognition*, *15*(3), 173–187. doi:10.1016/0031-3203(82)90069-3

Williams, R. B., Inman, D. J., & Wilkie, W. K. (2004). Temperature-Dependent Thermoelastic Properties for Macro Fiber Composite Actuators. *Journal of Thermal Stresses*, *27*(10), 903–915. doi:10.1080/01495730490498386

Xu, X., & Martin, E. (1998). A Distribution-Based Clustering Algorithm for Mining in Large Spatial Databases. In *Proceedings of Data Engineering*. IEEE.

Yinghui, L., & Michaels, J. E. (2005). A Methodology for Structural Health Monitoring with Diffuse Ultrasonic Waves in the Presence of Temperature Variations. *Ultrasonics*, *43*(9), 717–731. doi:10.1016/j.ultras.2005.05.001 PMID:15992847

Zhang, T., Ramakrishnan, R., & Livny, M. (1996). An Efficient Data Clustering Method for Very Large Databases. In *Proceeding SIGMOD '96*. ACM.

Chapter 3
Wavelet Transform Modulus Maxima Decay Lines:
Damage Detection in Varying Operating Conditions

Andreas Kyprianou
University of Cyprus, Cyprus

Andreas Tjirkallis
University of Cyprus, Cyprus

ABSTRACT

An important task in structural health monitoring (SHM) is that of damage detection under varying environmental and operational conditions. Structures, under varying environmental conditions, change their mass, elasticity and damping properties whereas changing operational conditions cause changes to excitations. A damage detection methodology implemented in these circumstances faces serious challenges since changes to structural behaviour imparted by environmental or operational conditions could be wrongly attributed to damage. The part of a damage detection decision algorithm that removes environmental and operational effects is called normalization. In this chapter a normalization methodology that is based on the similarity between continuous wavelet transform maxima decay lines is presented. This methodology is implemented on both simulated and experimental data. Simulated data were obtained from a three degree of freedom system. Varying environmental conditions were simulated by temperature dependent stiffness parameters and operating conditions by changing the colour of random excitation. Experimental data were obtained from damaged cantilever beams that were subjected to random excitations of different colour and varying temperatures.

INTRODUCTION

The essential constituent of a structural health monitoring (SHM) methodology is its damage detection procedure. The underlying operating principle of damage detection algorithms is the comparison of the

DOI: 10.4018/978-1-4666-8490-4.ch003

behavior of the structure under monitoring, against its expected (nominal) behavior. If a significant difference between the two behaviors is observed, the damage detection algorithm has to decide whether or should be attributed to damage. Over the years many damage detection methodologies have been developed, that exploit the effects of damage on structural responses. The most demanding challenge faced by these methodologies is that of detecting damage of structures in service; for they are operating under varying environmental and loading conditions.

Various studies have demonstrated that the structural changes induced by varying environmental and operational conditions could be falsely attributed to damage. Therefore, damage detection algorithms should somehow be taking into consideration the effects of changing environmental and operating conditions. The action of removing these effects from the data measured from structures operating under these circumstances was given the name of normalization (Sohn & Farrar, 2001; Sohn, Worden, & Farrar, 2002; Sohn, 2007).

Over the years different normalization techniques have been developed by taking into consideration:

1. Information about the actual operational and environmental conditions, usually available in terms of environmental measurements,
2. The lack of this information, and
3. That such information might or might not be available.

In the context of (1) and (2) damage detection algorithms utilize models, called base line models, which capture the behavior of the undamaged structure under varying environmental and operational conditions. In (3) damage detection features that are insensitive to environmental or operational conditions are sought, and hence base line models are not required. A review of the normalization methodologies developed up to 2007 can be found in Sohn (2007) and the references therein.

The normalization methodologies developed after 2007 in the context of (1) relied on statistical analysis. Deramaeker et al. (2008) implemented statistical factor analysis of the eigen-properties. They used stochastic subspace identification methodology and the Fourier transforms of modal filters in order to compute the eigen-properties. Limongeli (2010) used spline interpolation to obtain frequency response functions (FRF) at locations on structures where measurements are not available. She proposed as a damage detection feature the difference of interpolation errors between a reference structure and the monitored in-service structure as a damage detection feature. In order to avoid the environmental effects on this feature she employed a statistical classification criterion. This was realised by formulating a statistical hypothesis test on the probability distribution of the error that interpolating FRFs yielded.

Hypothesis testing on the predictive ability of a model generated by embedding delay vectors was proposed in Figuiredo et al. (2010) for the situations where information about environmental conditions is not available. Chang and Sohn (2009) presented an approach that was based on unsupervised support vector machines to obtain non-linear principal components that capture the effects of unmeasured environmental conditions on the structure. This procedure was also used with impedance measurements (Lim, Kim, & Sohn, 2011). An alternative way that has been motivated by the biological immune system, for damage detection of structures operating in varying environmental conditions where information about the actual operating conditions is not available was proposed in Surace and Worden (2010).

A methodology that exploits the interaction of Lamb waves with both cracks and strategically positioned piezoelectric sensors on plates to detect damage without using any base line model was proposed in Kim and Sohn (2007), An and Sohn (2010), Sohn et al. (2011). Another way of removing the environmental effects that uses the concept of cointegration was developed in Cross et al. (2011).

This chapter presents a new methodology of detecting structural damage under varying environmental and/or operating conditions. The underlying concept of the new methodology is the wavelet transform modulus maxima decay line, denoted by WTMM, and its properties. The WTMM decay line is the curve obtained by connecting together the local maxima of the modules of continuous wavelet transform of a given time series. Since, in a certain sense, the wavelet transform is the multiscale derivative of the time series it analyses, the local maxima of its derivatives carry information about the singular (non-differentiable) behavior of the time series. The actual nature of the singular behavior can be unveiled by the evolution of the wavelet transform modulus maxima across the scales, i.e the WTMM decay line (Mallat & Huang, 1992; Mallat 2009). In this chapter, the view that the singular behavior of the time series of structural responses is determined by the environmental and operating conditions is taken. For example, the singular nature of the response of a structure excited by wind is different from that excited by white noise. The power spectrum of the wind is different to that of white noise and in addition if turbulent flow prevails the excitation can become non-stationary. Furthermore, temperature changes could impart trends in the response that render the response time series singular. The nature of singularities in the time series obtained for the situations discussed above are different and these differences are reflected in the characteristics of the corresponding WTMM maxima decay lines.

The proposed damage detection methodology is based on the similarity of the WTMM decay lines. Similar WTMM decay lines within each time range are determined using the normalized cross correlation, NCC, technique. Two WTMM decay lines are similar if their NCC value is higher than a Similarity Threshold Value (STV), which is specified by the application. A damage identification methodology is devised based on the number of WTMM similar decay lines between time ranges. This number is compared to a specified constant Damage Possibility Similarity (DPS) number, which is the number of similar WTMM decay lines between successive time ranges indicating that damage possibly exists in the structure, and a specified constant Damage Identification Similarity (DIS) number, which indicates that damage certainly exists.

When decision about damage existence cannot be made directly from the similarity between WTMM decay lines a technique based on the Difference of the Pointwise Summation of similar WTMM decay lines, abbreviated by DSSD is proposed. It actually computes the distance between the WTMM decay line with the maximum length within a time range with a reference one. This principle is applied separately to lower and higher scales to account for the differences that damage and varying environmental and operational conditions induce in the multiscale features of the structural response.

The rest of this chapter is organized as follows. The next section briefly discusses the fundamental theory underlying the proposed methodology; namely, the theory of the continuous wavelet transform and its modulus maxima lines and the theory of similarity that was used to compute the similarity between two WTMM decay lines. Then in a separate section the proposed damage detection methodology is presented followed by the sections of the case studies and conclusions.

BACKGROUND OF FUNDAMENTAL THEORY

This section discusses briefly the fundamental theory underpinning the damage detection methodology development.

Continuous Wavelet Transform and Its Modulus Maxima Lines

The mathematical theory of continuous wavelet transform, CWT, is well developed and can be found in many textbooks such as Kaiser (1994) and Mallat (2009). This section gives the definition of the CWT and explains how its modulus maxima lines are obtained.

The CWT of a signal, $f(t)$, is defined by,

$$Wf\left(s,u\right) = \frac{1}{\sqrt{s}} \int_{-\infty}^{\infty} f(t) \psi^{*}\left(\frac{t-u}{s}\right) dt \tag{1}$$

where $\psi\left(t\right)$ is the mother wavelet function, which is non-zero only over a finite time interval and its integral over that interval is zero. These two conditions render $\psi\left(t\right)$ a small wave, hence its name. The mother wavelet can be dilated or compressed through the scale parameter s, for $s>1$ $\psi\left(t\right)$ is dilated whereas for $s<1$ is compressed. In order to enable analysis for all the time period $f(t)$ is defined the dilated or compressed wavelet should be translated across the time period that $f(t)$ is defined. This achieved by the use of u which is known as the translation parameter. The symbol * denotes complex conjugation. Equation (1) can be interpreted as the similarity at a translation u between compressed or dilated versions of $\psi\left(t\right)$ and $f\left(t\right)$.

Therefore, the CWT transforms the one dimensional signal $f(t)$ into the two dimensional function $Wf\left(s,u\right)$. The graph of $Wf\left(s,u\right)$, or of its modulus $\left|Wf\left(s,u\right)\right|$ can be visualized as a landscape over the translation by scale plane, $u \times s$. This transform, by varying s at a specific position $u=t_1$, captures the multiscale nature of $f(t)$ at the specific instant t_1. The scale s measures the size of the neighborhood of t_1 in which $\psi\left(\frac{t-t_1}{s}\right)$ is non-zero.

A local extremum of $Wf\left(s,u\right)$ occurs at a point $\left(u_0, s_0\right)$ of the $u \times s$ plane where the derivative $\frac{\partial\left(Wf\left(s,u\right)\right)}{\partial u}\Big|_{\left(u_0,s_0\right)}$ has a zero crossing (Mallat et al., 1992). A wavelet modulus maxima line is any connected curve in the $u \times s$ plane along which all points on the graph of $\left|Wf\left(s,u\right)\right|$ are local maxima.

Similarity between WTMM Decay Lines

In general, two WTMM decay lines are similar if their shape is the same. It has been demonstrated that the similarity of WTMM decay lines can be used to discriminate realizations of the same nonstationary stochastic processes, different realizations of different stochastic processes and the nonstationary stochastic processes obtained when the values of the parameters of ARIMA model change (Tjirkalis, 2013). In order to quantify the similarity between different maxima decay lines Tjirkalis (2013) proposed the use of the Normalized Cross Correlation. It defines the similarity between the decay lines DL_n and DL_m of the response time series n and m, as the maximum of the normalized cross correlation given by,

$$\max NCC(DL_n, DL_m) = \frac{\max\limits_{1 \le k \le N} (CR_{DL_n, DL_m}(k))}{\left\| DL_n \right\|_2 \left\| DL_m \right\|_2} \tag{2}$$

where N is the number of elements in DL_n and DL_m, $CR_{DL_n, DL_m}(.)$ the statistical correlation between DL_n and DL_m given by,

$$CR_{DL_n, DL_m}(k) = \frac{1}{N} \sum_{i=1}^{N} DL_n(i) DL_n\left(i + k\right) \tag{3}$$

and $\left\| . \right\|_2$ is the Euclidean norm of its argument.

An indicator that could identify non identical similar decay lines is also used in the proposed damage detection methodology below. For example, Figure 1 depicts the time response and the modulus of its continuous wavelet transform (CWT), computed by Equation 1, of mass 1 of the simulated three degree of freedom system that is used as a case study for the proposed damage detection methodology. Figure 2(a) depicts the maxima lines extracted from the CWT plot of Figure 1. The decay of these maxima lines is shown in the plot of Figure 2(b). The similarity between the decay lines is computed using Equation (2). In some cases the situation that is shown in Figure 3 arises, namely existence of similar decay lines that are not identical, left plot of Figure 3, and of decay lines that are partially similar, right plot of Figure 3.

A method called the difference of the Pointwise Summation of Similar WTMMM decay lines, abbreviated to DSSD is proposed in order to identify whether two similar decay lines are identical or not. This is achieved by quantifying the vertical distance between two similar decay lines using the following equation,

$$DSSD_n = \sum_{d=1}^{k} \left| y\left(DL_{t_n}\right)_d \right| - \sum_{d=1}^{j} \left| y\left(DL_{t_{n-1}}\right)_d \right| \tag{4}$$

where $y\left(DL_{t_n}\right)$ and $y\left(DL_{t_{n-1}}\right)$ are the values along the WTMM decay lines of the time series n and n-1, and k and j are the total number of values of the corresponding WTMM decay lines. It is clear that Equation 4 gives the difference in summation of the absolute values of all points along a WTMM decay line of a time range n with the corresponding, similar decay line of the previous time range n-1. If the mean DSSD value is 0 then it is concluded that the compared WTMM decay lines are identical, otherwise they are not. Equation 4 is used in the proposed damage methodology in order to distinguish changes in the time response signals due to changes in forcing conditions or/and due to different force realisations, from changes due to damage.

DAMAGE DETECTION

The proposed damage detection methodology relies on the principle of the WTMM decay line similarity and the DSSD technique, which they were both described above. The methodology uses the time

Figure 1. The time history (top plot) and continuous wavelet transform (bottom plot) of the response of the simulated three degree of freedom system

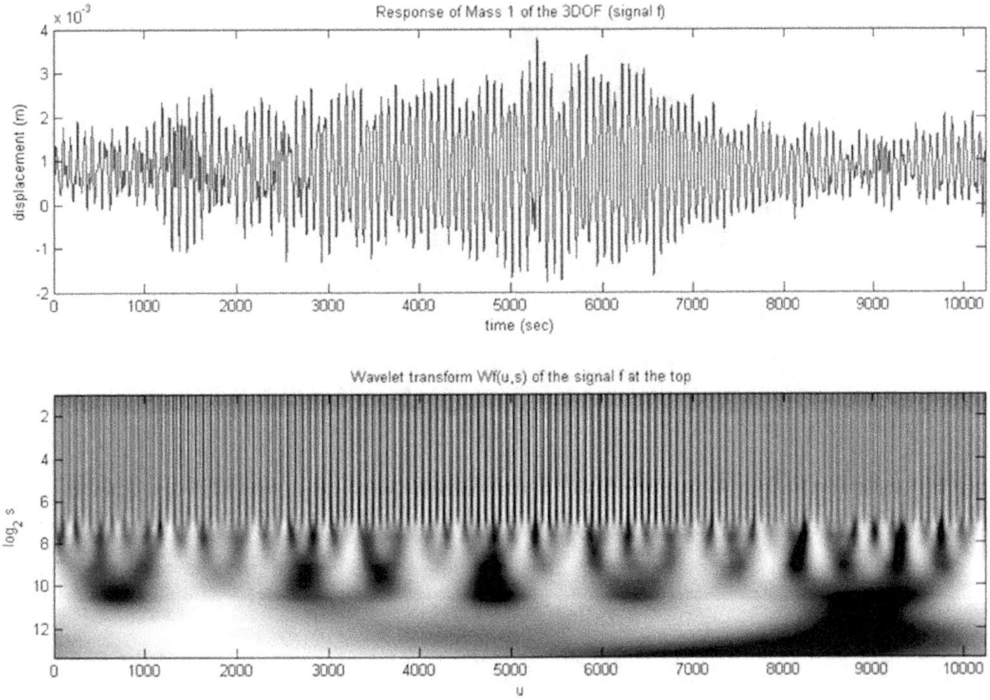

series of the response of vibrating structures. The time series used should be normalized by setting their maximum value to unity. In order to ensure that the maxima lines used capture multiscale features that characterised the responses, only the decay lines that have a decay line length DLL above an application

Figure 2a. The WTMM lines of the CWT shown in Figure 1; the maxima decay lines of the wavelet transform of Figure 1

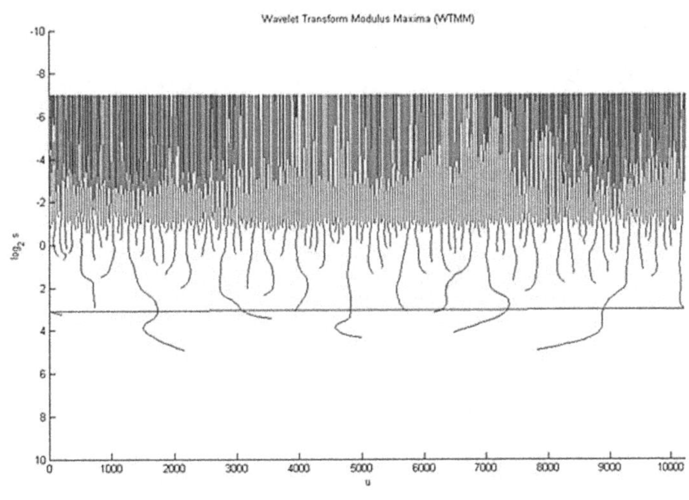

Figure 2b. The WTMM decay lines of the WTMM lines shown in (a)

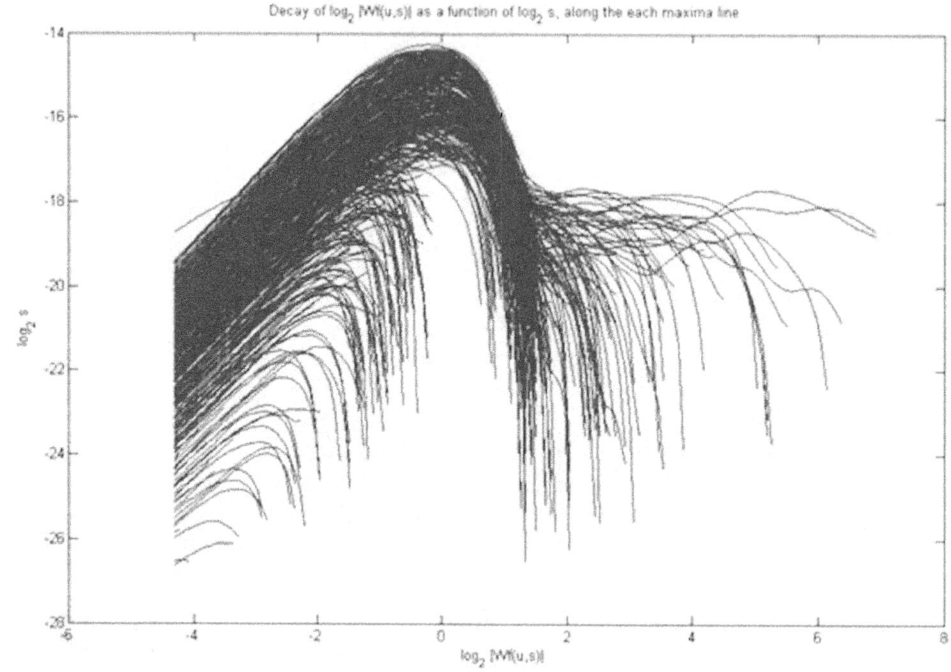

Figure 3. Left: two identical similar decay lines; right: two partially similar decay lines

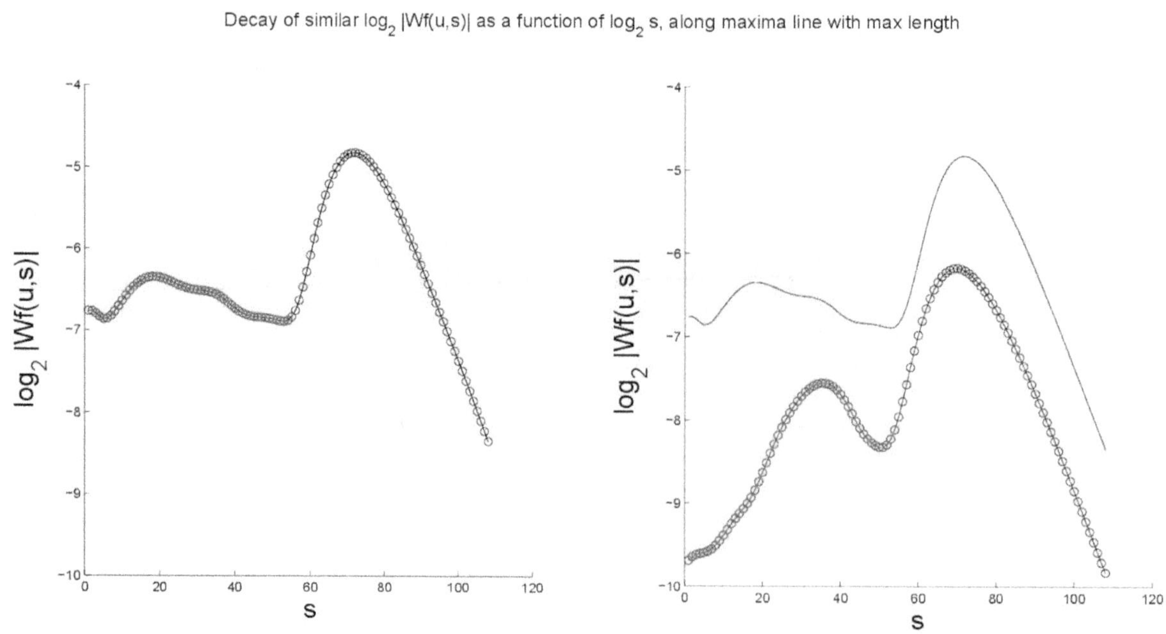

specific threshold value are selected for further analysis. This has the additional benefit of removing the effect of noise, since noise generates short decay lines (Peng, Chu, & Che, 2005).

First, the methodology computes the CWT using Equation 1 and then extracts the wavelet transform modulus maxima, WTMM, of the normalized time series of the structural responses. Then the decay lines whose length is bigger than an application related threshold are selected. Subsequently, the similarity between the WTMM decay lines of successive time ranges is computed using the NCC principle, Equation 2. This gives a similarity value between pairs of decay lines that is stored in an array designated as "Similarity Array". Two decay lines are taken to be similar when this value exceeds a certain Similarity Threshold Value, STV, which also depends on the application.

The number of decay lines between the successive time ranges that have a similarity value greater or equal to STV is denoted by SDS, Decay Line Similarity. If this number is greater than a threshold value, the Damage Possibility Similarity, DPS, number then the structure does not contain any damage, and any changes in the time series responses are attributed to operational or/and environmental variability. On the other hand if this number is less than a set value, called the Damage Indication Similarity, DIS, number, damage has been introduced in the structure after the end of the previous time range.

At the case, which the SDS number is greater than the DIS but it is less than the DPS, it cannot be deduced whether the changes in the time series are either due to varying environmental and/or operating conditions or due to imparted damage. In order to distinguish between these cases the DSSD principle between the WTMM decay line with the maximum length, i.e. the one that extends to all the scales, in the current and the previous time ranges with a reference WTMM decay line are compared with a reference WTMM decay line. The reference decay line is the decay line with maximum length that is determined during the first time range observed when the damage detection monitoring begins.

Figure 4 shows the WTMM decay lines of the responses of mass 1 of the simulated system when the pink force realisation was changed, left plot and when the colour of the forcing was changed from pink to white, right plot. By computing the DSSD numbers for each halve of the plot the changes that are observed on the WTTM decay lines can be attributed to changes in forcing conditions and not to the presence of damage.

CASE STUDIES

This section presents the results of implementing the damage detection methodology described above to a simulated three degree of freedom mass spring system and to an experimental cantilever beam.

Simulated Three Degree of Freedom Mass Spring System

The damage detection methodology is applied to the three degree of freedom mass, spring and damper system shown in Figure 5. The values of all the mass and damping elements were taken to be 1Kg and 0.2Ns/m respectively. These values were kept constant throughout the damage detection analysis. This system was excited by force realizations of both white and pink noise on mass 1. The realizations were changing during the implementation of the damage detection methodology. The variation of environmental conditions was changed by varying the temperature whose effect was simulated by increasing linearly the values of all spring constants with respect to the running time. Damages were simulated by reducing the value of the spring constant of anyone of the stiffness elements of the system. The system

Figure 4. Left: comparison between decay lines obtained by changing the time series realisation of the excitation; right: WTMM decay lines obtained by changing the colour of the excitation from pink to white

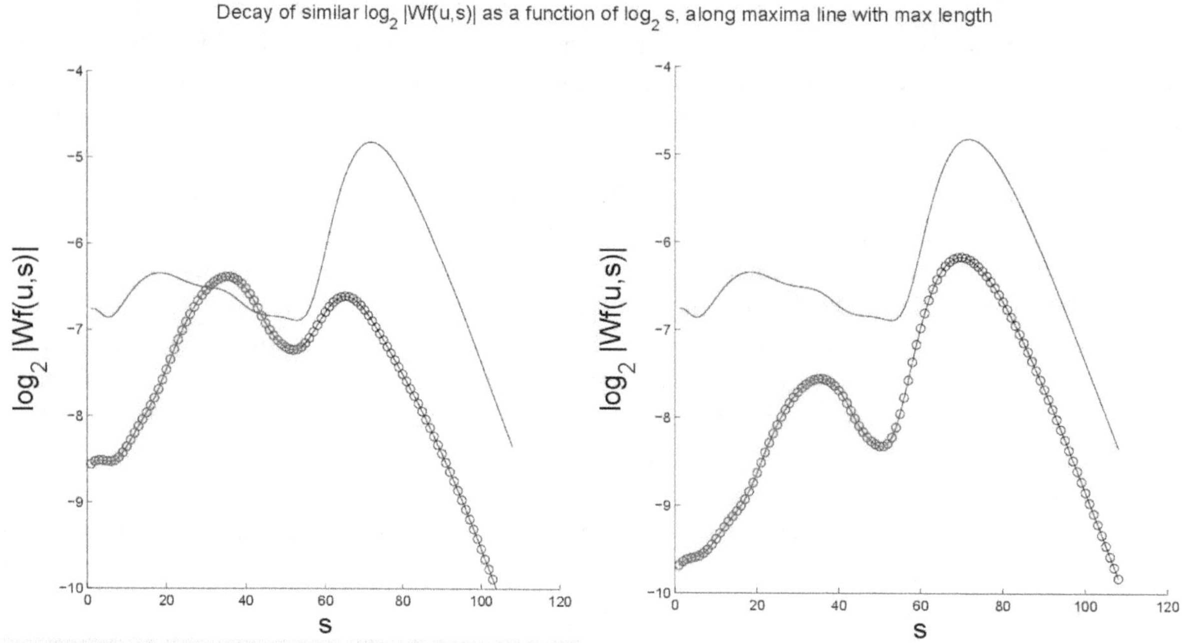

was considered undamaged when all its spring constant values were equal to 450 N/m. The system was simulated under 11 different cases covering a number of different environmental and operational conditions, and health states. The pink and white noise force time series were generated in Labview® and the system was simulated in Simulink®.

The cases listed in Table 1 were studied for both the simulated and experimental systems. The time series of the response of mass 1 together with its continuous wavelet transform and WTMM decay for the cases 1 and 7 are shown in Figures 6 and 7 respectively. The number of similar decay lines, SDS between two successive tine ranges, one designated as a current time range and the other as previous time range was computed and tabulated in Table 2.

Figure 5. The three degree of freedom mass, spring, and damper system

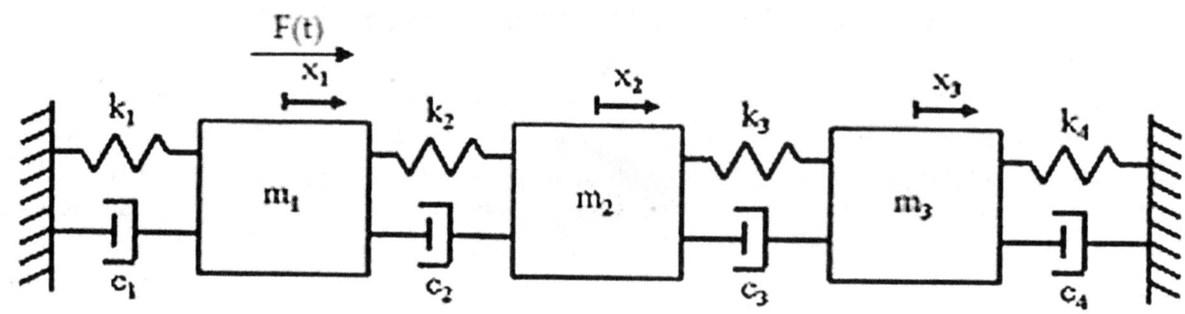

Table 1. The cases studied for both simulated and experimental systems

Case	Description
1	Undamaged, Constant Temperature, Pink Noise Excitation
2	Undamaged, Constant Temperature, Pink Noise Excitation – different realisation
3	Undamaged, Constant Temperature, White Noise Excitation
4	Undamaged, Varying Temperature, Pink Noise Excitation
5	Undamaged, Varying Temperature, Pink Noise Excitation-different realisation
6	Undamaged, Varying Temperature, White Noise Excitation
7	Damaged, Constant Temperature, Pink Noise Excitation
8	Damaged, Constant Temperature, Pink Noise Excitation-different realisation
9	Damaged, Constant Temperature, White Noise Excitation
10	Damaged, Varying Temperature, Pink Noise Excitation
11	Damaged, Varying Temperature, White Noise Excitation

The mean value of the DSSD for the first half of the WTMM decay line for the current time range of case 3 to the previous time range of case 1 is 2.1 whereas for the second half of the WTMM decay line is 0.3. Similarly, the mean value of the DSSD for the first half of the WTMM decay line computed between the current time range of case 6 to the previous time range of case 1 is 2.30 whereas for the second half of the WTMM decay line this value drops to 0.85. These situations are depicted in Figures 8(a) and (b) respectively. This means that the SDS number for these cases is due to changes in the force realization and/or forcing conditions only.

Figure 6. Time series, CWT, and WTMM decay lines of the response of mass 1 for the Case 1

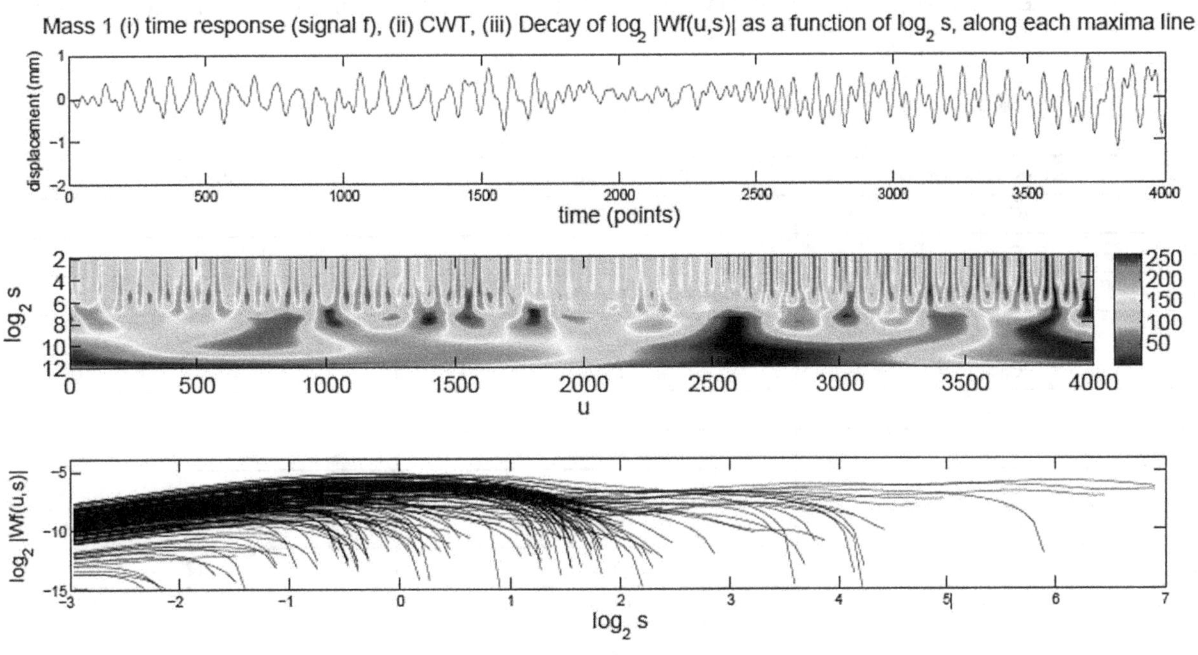

Figure 7. Time series, CWT, and WTMM decay lines of the response of mass 1 for the Case 7

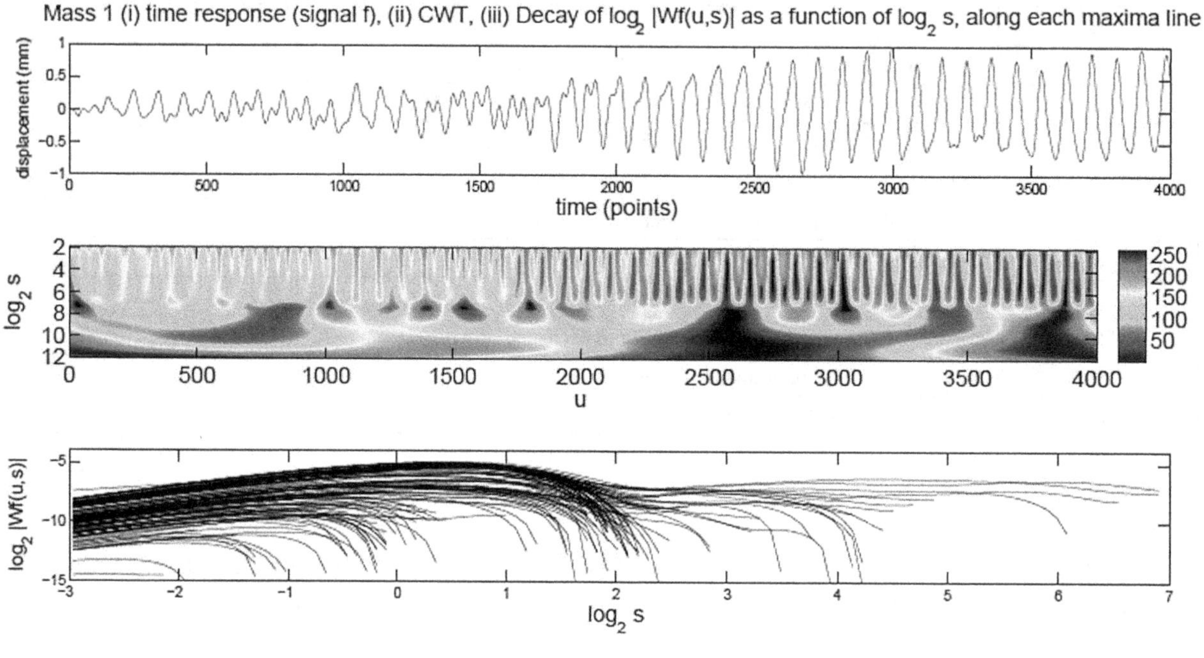

In contrast the mean values of DSSD of the first half of the WTMM decay line between the current time range of case 10 and the previous of case 1 is 0.21 and for the second half is 0.29, Figure 8(c). Since the value of mean DSSD does not increase indicates that the SDS number for this case is due to damage.

The conclusion from these results is that by the variation in the forcing condition simulated by changing the colour of knowledge affects the lower scales of the response. This is in accordance with the relevant theory that links the singularity properties of realisations of stochastic processes with the lower scale behavior of the WTMM decay line (Mallat. 2009, Ch 4). On the other hand the introduction of damage

Table 2. The SDS number for the cases of Table 1

Previous Time Range	Current Time Range	SDS Number
Case 1	Case 1	98
Case 1	Case 2	21
Case 1	Case 3	16
Case 1	Case 4	43
Case 1	Case 5	22
Case 1	Case 6	15
Case 1	Case 7	12
Case 1	Case 8	4
Case 1	Case 9	8
Case 1	Case 10	17
Case 1	Case 11	12

Figure 8a. Case 3: DSSD1st=2.1 and DDD2nd=0.3

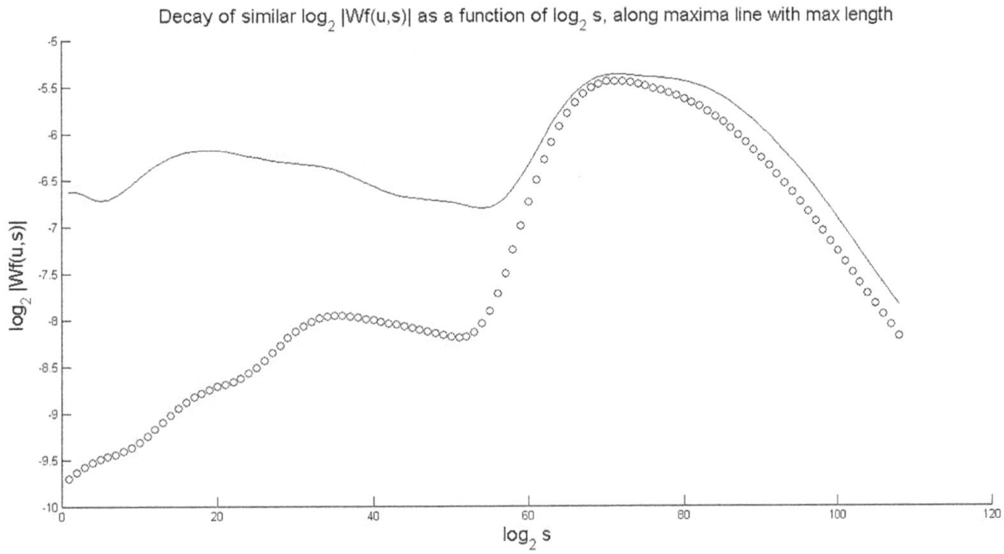

Figure 8b. Case 6: DSSD1st=2.3 and DDD2nd=0.85

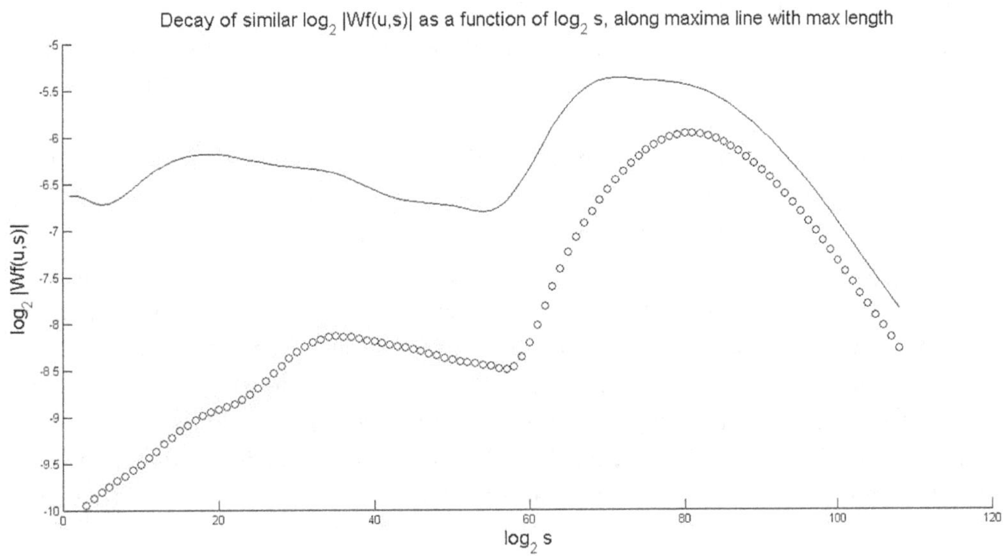

on average does the effect across scales is similar. The computation of the continuous wavelet transform and the decay lines was done using the Wavelab toolbox of Matlab®.

Experimental Cantilever Beam

A cantilever beam with the characteristics listed in Table 3 was set up as it shown in the photograph of Figure 9. A permanent magnet shaker, LDS V201, connected to the cantilever via a stringer rod was

Figure 8c. Case 10: DSSD1st=0.20 and DDD2nd=0.29

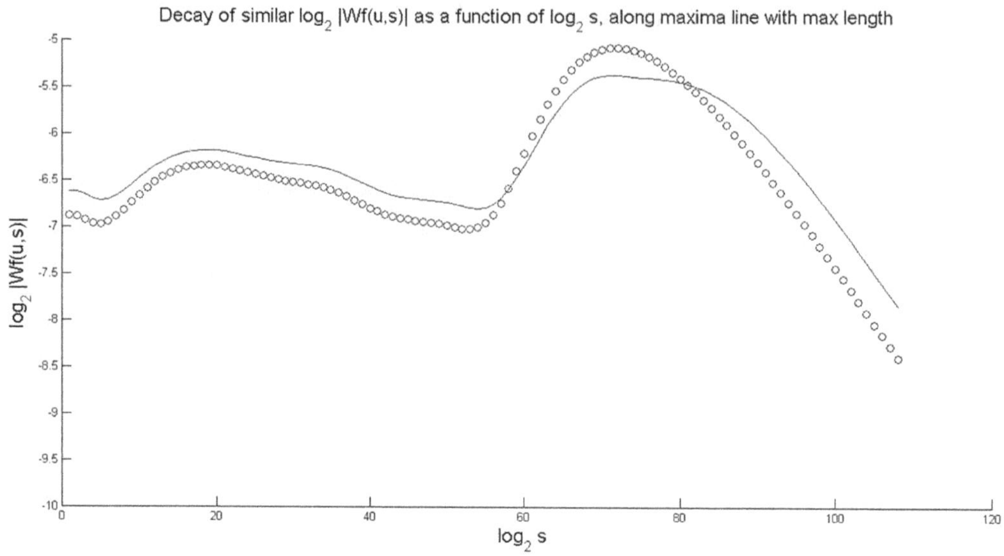

used to excite the beam by pink and white noise force. The different realizations of pink and white noise excitations were generated in Labview®.

A crack of depth d, width b and opening e, is introduced at a distance of 452 mm from the fixed end of the beam, Figure 10. The width, opening and depth of the crack were set to 30 mm, 2 mm and 0.75 mm respectively.

The time response on more than 15 points along the length of beam was measured simultaneously using a high speed camera, namely, Mega Speed MS70K MDG2 High Speed Camera. The maximum resolution of this camera is 504x504 pixels, its frame rates reach up to 100000 frames per second and its shutter speed is in the range of 1 µs to 30 ms. The lighting conditions during the experiments were set in such a way as to ensure that the light intensity was adequate and constant across the length of the beam. The camera was located at a position that ensured that the cantilever was situated in the middle of its lens. Edge detection was used to in order to produce the state of the vibrating structure at a specific point in time. The edges were identified by using a wavelet transform edge detector (Gentile & Messina, 2003).

Figures 11, 12 and 13 show the time series, the CWT and the maxima decay lines orresponding to the response at 405 mm from the fixed end of the cantilever, for Cases 1, 2 and 11 of Table 1 respectively.

Table 3. The properties of the cantilever beam

Property	Value
Length (l)	592 mm
Width (b)	30 mm
Thickness (t)	1.5 mm
Density (ρ)	7.87x10^{-6} Kg/mm^3
Young Modulues (E)	200x10^6mN/mm^2
Second Moment of Area (A)	8.44 mm^4

Figure 9. Photograph of the experimental set-up

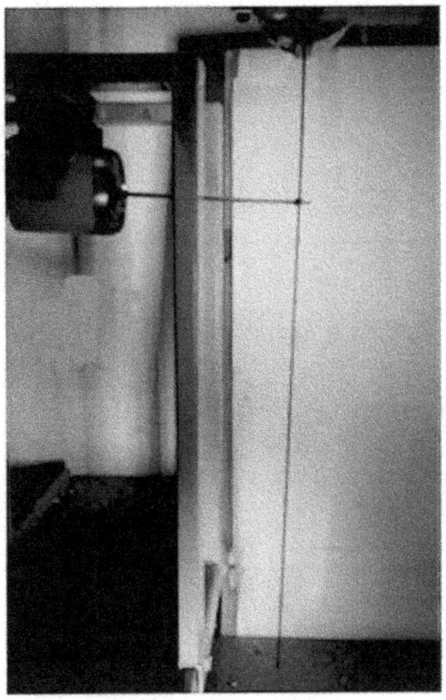

(a) Experimental Setup

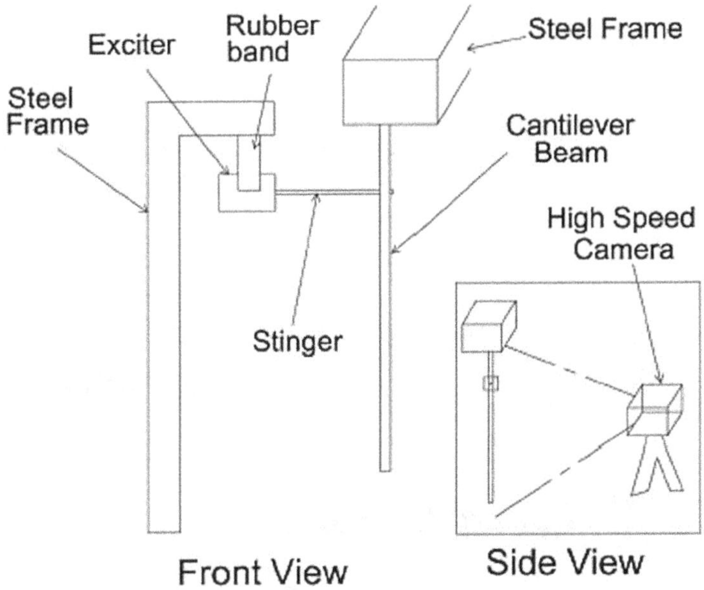

(b) Experimental Setup Schematic

Figure 10. The geometry of the crack

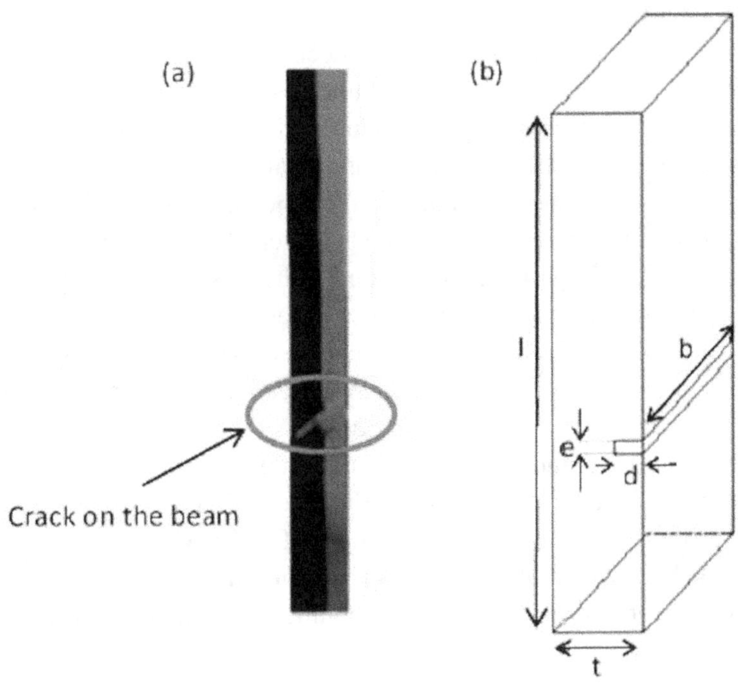

Figure 11. Time series, CWT, and WTMM decay lines of the response at 405 mm from the fixed end of the cantilever for Case 1

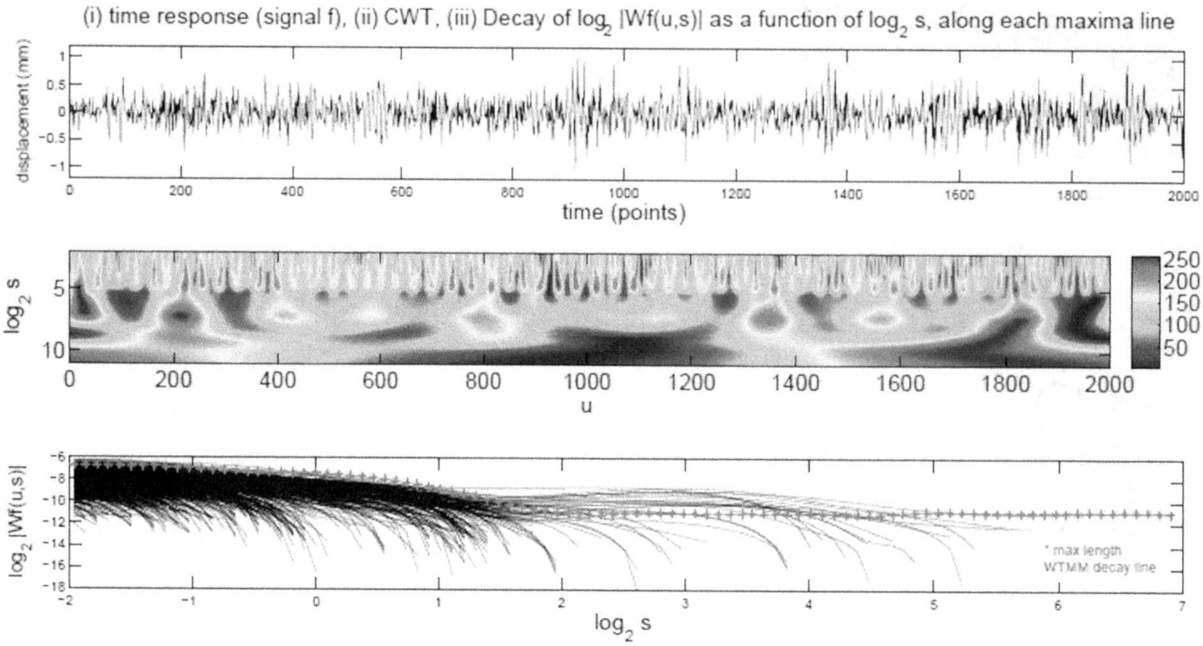

Table 4 lists the SDS numbers, the number of similar WTMM decay lines between successive time ranges, designated as 'Previous Time Range' and 'Current Time Range', between cases 2 to 11 and case 1. For this application the DPS and DIS numbers were set to 22 and 14 respectively and these numbers were computed at the six different locations across the beam also given in Table 4.

The data in Table 4 is used to make a decision about the state of the structure according to the flow chart shown in Figure 14. At the beginning of the decision making process, represented by the top rhombus of the flow chart, if the computed DPS number is less than the SDS number then it can be concluded that there is no damage in the system. This is clearly what happens with cases 1,2, 4 and 6, taBle 4, where SDS numbers at all beam positions are greater than the DPS number that was set at 44. If the SDS number is lower than the DPS number then the condition DPS < DIS represented by the lower rhombus in the flowchart of Figure 14 should be checked. DIS for this study was set to 14. If it is satisfied then damage has been imparted to the system. Otherwise the technique that is based on DSSD should be used. This condition is clearly satisfied at the cases 7 and 11 and for the majority measurement locations of the case 10. Cases 3, 5, 8 and 9 neither can be classified as damaged nor undamaged since the effects from environmental and operational conditions cannot be distinguished from those caused by damage. Therefore, according to the proposed methodology the DSSD based technique should be used.

The DSSD results for cases 3 and 9 are shown in Figures 15 and 16 respectively. The mean DSSD value for the first half of the WTMM decay line for the case 3, Figure 15, is 1.47 whereas for the second is 0.67. This drop of DSSD values for both scale ranges of WTMM line indicated that the changes are due to the change of the forcing condition. Since by changing the colour of noise the nature of the time domain singularities, a lower scale feature, changes significantly. In case 9, Figure 16, the mean DSSD

Figure 12. Time series, CWT, and WTMM decay lines of the response at 405 mm from the fixed end of the cantilever for Case 2

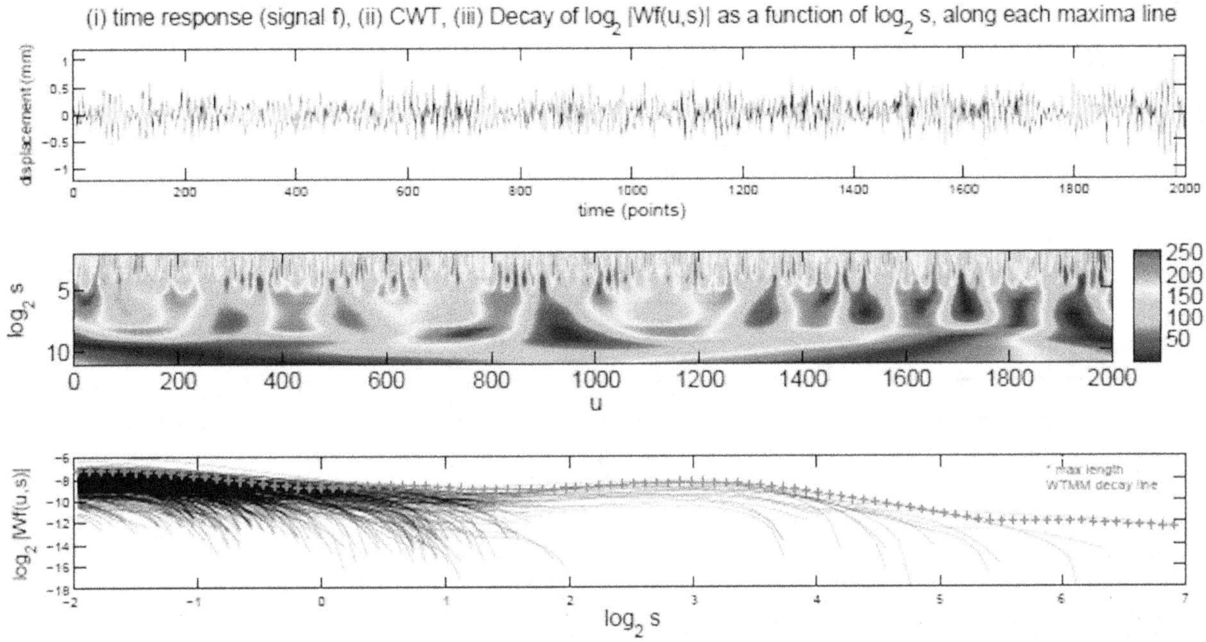

Figure 13. Time series, CWT, and WTMM decay lines of the response at 405 mm from the fixed end of the cantilever for Case 11

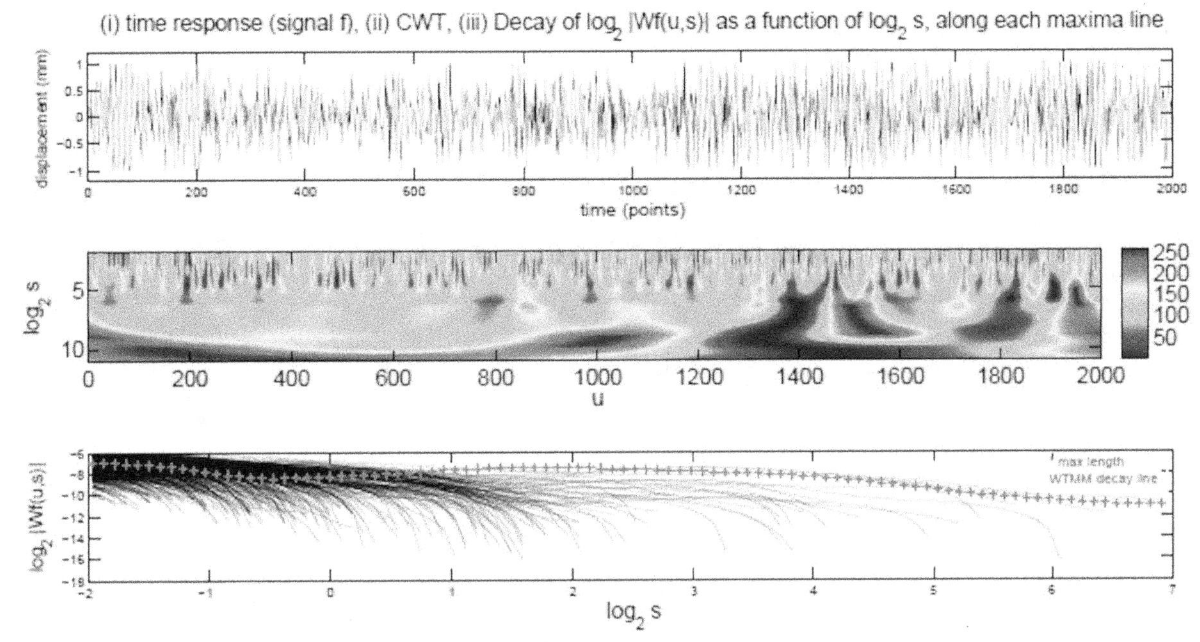

Table 4. Number of similar WTMM decay lines between successive time ranges for all the cases

		Locations across the Length of the Beam in Pixels from the Fixed End					
		240	270	300	330	360	390
Previous Time Range	**Current Time Range**	**SDS Numbers**					
Case 1	Case 1	44	44	44	44	44	44
Case 1	Case 2	28	28	27	27	27	27
Case 1	Case 3	**20**	**20**	**20**	**17**	**17**	**14**
Case 1	Case 4	28	28	28	28	26	24
Case 1	Case 5	**20**	24	23	22	**22**	**19**
Case 1	Case 6	24	23	24	24	24	23
Case 1	Case 7	14	12	11	11	10	9
Case 1	Case 8	**20**	**20**	12	12	12	12
Case 1	Case 9	22	**20**	**18**	**18**	**18**	**17**
Case 1	Case 10	**19**	12	12	11	12	**14**
Case 1	Case 11	12	5	8	9	11	12

value for the first half of the WTMM decay line is 0.96 and for the second one 1.08. This indicates that the corresponding SDS value in Table 4 is due to damage.

CONCLUSION AND DISCUSSION

The number of similar decay lines of a reference time series changes as its characteristics change. For example the number of similar decay lines between time series corresponding to different realizations of

Figure 14. Decision making for damage detection

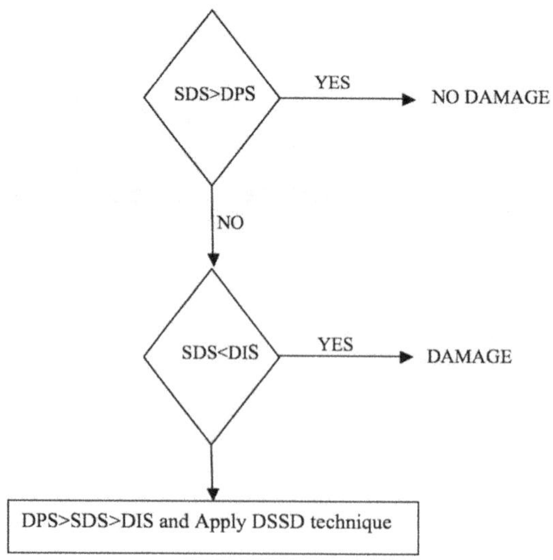

Figure 15. Case 3: DSSD1st=1.47 and DSSD2nd=0.67

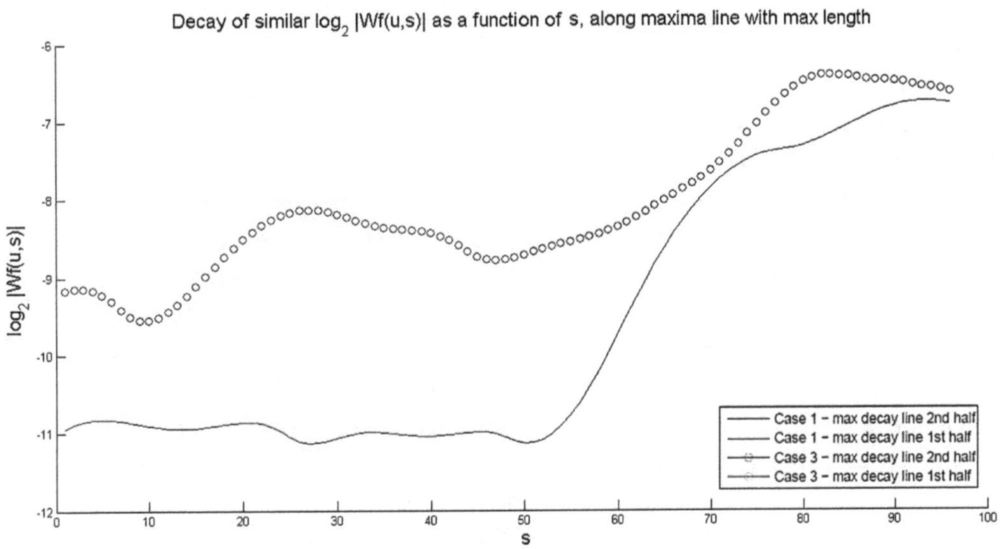

a stationary stochastic process is higher than that between realizations of stationary and non-stationary stochastic processes. A damage detection algorithm that rejects the environmental and operational effects on a vibrating system was proposed that takes advantage the similarity properties of the WTMM decay lines of the system's responses. This algorithm uses the responses only, not the excitations, of vibrating structures. In first place the algorithm computes the number of similar WTMM between the time series of various response measurements. If this number is not conclusive with respect to the health state of

Figure 16. Case 9: DSSD1st=0.96 and DSSD2nd=1.08

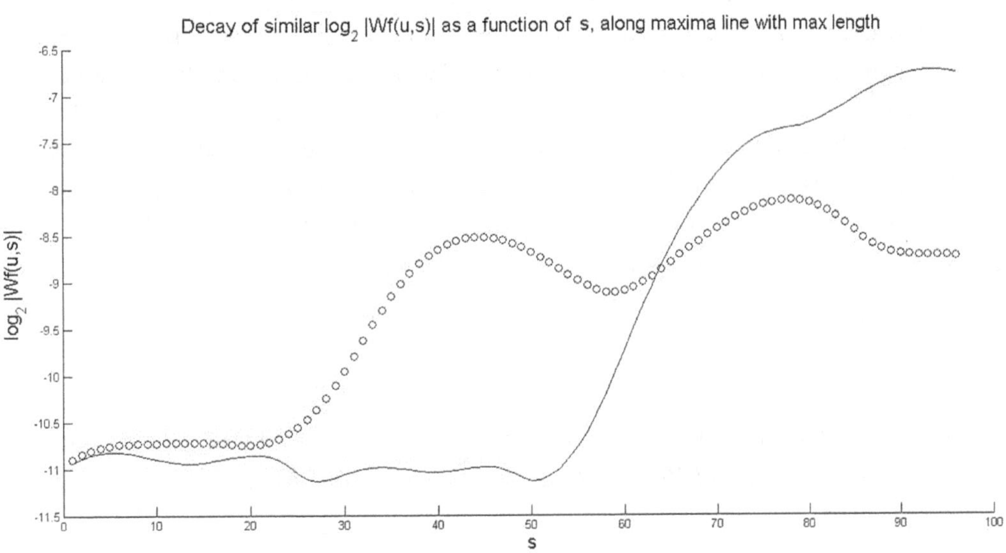

the system a measure of the similarity is used to enable decision. The time required for all the sequence to be computed and reach the final conclusion was two minutes.

The experimental cases 3 and 9 corresponding to damaged and undamaged system with measurements taken at constant temperature when the system is subjected to white noise were the most inconclusive with respect to damage identification when the SDS number was used only. As it can be seen from Table 3 the SDS numbers for these cases almost remained unaltered. The reference time series, Case 1 in Table 1, was the response of the undamaged system subjected to pink noise excitation whereas cases 3 and 9 were response to white noise at constant temperature. The stochastic processes corresponding to pink noise power spectrum are non-stationary whereas white noise power spectra correspond to stationary processes. This demonstrates that the change of the nature of the operating condition i.e. from stationary to non-stationary the number of similar decay lines could not separate the damage from the undamaged case. In this case the DSSD principle that measures not only how similar but also how identical are two WTMM decay lines was used. It is important to notice that under varying temperature and white noise excitation by solely using the SDS number could identify the state of the structure. This happened because the time temperature variation induced non-stationary properties to the system responses.

Structural damage detection under varying environmental and operating conditions is a challenging problem with as many possibilities for a false decision as the various environmental or operational conditions a structure can be found in during its service. This makes of paramount importance of having effective and robust damage detection algorithm. It emerges from the work presented in this chapter that in general WTMM decay lines and their properties can be used as the main constituent of such algorithms. A general hypothesis for future research that tends to stem from the results is that for damage detection algorithms that analyse time series in regards to the effectiveness, and possibly the robustness of the algorithm the features contain in the reference time series are important. For example, it is known that responses obtained under varying environmental conditions stand a good chance of being non-stationary. It looks that instead of creating a detailed model of how a structure actually behaves under these conditions and then utilize the information obtained from such a model to devise a damage detection algorithm one could excite the structure by non-stationary process and use the response as reference. The results of this chapter by using the pink force indicate just this. In the more specific context of the work presented in this chapter further work is required in identifying the effects of locally concentrated environmental and operating conditions, and whether the proposed methodology is still capable for damage detection under these conditions. Further research can also be done, in extending the proposed methodology into monitoring the healthy state of a system continuously for a number of days, while an initially induced crack is gradually propagating. This will enable the testing of the effectiveness of the proposed methodology in the continuous monitoring of crack propagation in real life operations. One more possibility is to extend the idea of WTMM decay line similarity to identification of damages directly from acquired images.

REFERENCES

An, Y., & Sohn, H. (2010). Instantaneous crack detection under varying temperature and static loading conditions. *Structural Control and Health Monitoring*, *17*(7), 730–741. doi:10.1002/stc.394

Chang, O., & Sohn, H. (2009). Damage diagonosis under environmental and operational variations using unsupervised support vector machine. *Journal of Sound and Vibration*, *325*(1-2), 224–239. doi:10.1016/j.jsv.2009.03.014

Cross, E., & Worden, K. (2012). Cointegration and why it works for SHM. *Journal of Physics: Conference Series*, 382.

Cross, E., Worden, K., & Chen, Q. (2011). Cointegration: A novel approach for the removal of environmental trends in structural health monitoring data. *Proceedings of the Royal Society of London. Series A*, *467*(2133), 2712–2732. doi:10.1098/rspa.2011.0023

Deraemaeker, A., Reynders, E., De Roeck, G., & Kullaa, J. (2008). Vibration based structural health monitoring using output only measurements under changing environment. *Mechanical Systems and Signal Processing*, *22*(1), 34–56. doi:10.1016/j.ymssp.2007.07.004

Figuiredo, E., Todd, M., Farrar, R., & Flynn, E. (2010). Autoregressive modeling with state-space embedding vectors for damage detection Ander operacional variability. *International Journal of Engineering Science*, *48*(10), 822–834. doi:10.1016/j.ijengsci.2010.05.005

Gentile, A., & Messina, A. (2003). On the continuous wavelet transform applied to discrete vibrational data for detecting open cracks in damaged beams. *International Journal of Solids and Structures*, *40*(2), 295–315. doi:10.1016/S0020-7683(02)00548-6

Kaiser, G. (1994). *Friendly Guide to Wavelets*. Boston, MA: Birkhäuser.

Kim, S., & Sohn, H. (2007). Instantaneous reference free crack detection based on polarization characteristics of piezoelectric materials. *Smart Materials and Structures*, *16*(6), 2375–2387. doi:10.1088/0964-1726/16/6/042

Kyprianou, A., & Staszewski, W. (1999). On the cross wavelet: Interpretation and application to Duffing oscillator. *Journal of Sound and Vibration*, 228(1), Lim, J., Kim, M., & Sohn, H. (2011) Impedance based damage detectionunder varying temperature and loading conditions. *NDT & E International*, *44*, 740–750.

Limongelli, M. (2010). Frequency response function interpolation for damage detection under changing environment. *Mechanical Systems and Signal Processing*, *24*(8), 2898–2913. doi:10.1016/j.ymssp.2010.03.004

Mallat, S. (2009). *A Wavelet Tour of Signal Processing*. Burlington, MA: Elsevier.

Mallat, S., & Hwang, L. (1992). Singularity detection and processing with wavelets. *IEEE Transactions on Information Theory*, *38*(2), 617–642. doi:10.1109/18.119727

Nikias, C., & Mendel, J. (1993). Signal processing with higher order spectra. *IEEE Signal Processing Magazine*, *10*(3), 10–37. doi:10.1109/79.221324

Peng, K., Chu, L., & Tse, P. (2005). Singularity analysis of the vibration signals by means of wavelet modulus maxima method. *Mechanical Systems and Signal Processing*, *21*(2), 780–794. doi:10.1016/j.ymssp.2005.12.005

Sohh, H., Dutta, D., & An, Y. (2011). Temperature independent damage detction in plates using redundant signal measurements. *Journal of Nondestructive Evaluation*, *30*(2), 106–116. doi:10.1007/s10921-011-0096-6

Sohn, H. (2007). Effects of environmental and operational variability on structural health monitoring. *Philosophical Transactions of the Royal Society*, *A365*, 539–556. PMID:17255051

Sohn, H., & Farrar, R. (2001). Damage diagnosis using time series analysis of vibration signals. *Smart Materials and Structures*, *10*(3), 446–451. doi:10.1088/0964-1726/10/3/304

Sohn, H., Worden, K., & Farrar, R. (2002). Novelty detection in a changing environment: Regression and interpolation approaches. *Journal of Sound and Vibration*, *285*(4), 741–761.

Surace, C., & Worden, K. (2010). Novelty detection in changing environment: A negative selection technique. *Mechanical Systems and Signal Processing*, *24*(4), 1114–1128. doi:10.1016/j.ymssp.2009.09.009

Tjirkalis, A. (2013). *Continuous Structural Health Monitoring Under Varying Environmental and Operational Conditions using Wavelet Transform Modulus Maxima Decay Lines Similarity*. PhD Thesis.

Tsa, M., & Chiang, H. (2002). Rotation-invariant pattern matching using wavelet decomposition. *Pattern Recognition Letters*, *23*(1-3), 191–201. doi:10.1016/S0167-8655(01)00099-X

KEY TERMS AND DEFINITIONS

Continuous Wavelet Transform: Mathematical transformation that converts time domain data to scale-translation domain data.

Continuous Wavelet Transform Modulus Maxima: Local maxima of the continuous wavelet transform modulus.

Continuous Wavelet Transform Modulus Maxima Decay Lines: The curves traced in the scale-translation plane by joining together the local maxima of the continuous wavelet transform modulus.

Damage Detection Algorithm: Algorithm that analyses structural responses and decides whether there is a damage in the structure.

Data Normalization: In the context of damage detection the process or method of removing from the data the effects of the operational and environmental conditions.

Excitation Colour: Spectral characteristics of random excitations. White noise corresponds to flat spectrum and the spectrum of the pink noise decays inversely proportional with respect to frequency.

Normalized Cross Correlation: A statistical expression that measures the similarity between two data sets.

Chapter 4
Development of a System for Detecting Weld Failures

Jairo Alejandro Rodríguez
Universidad Santo Tomás, Colombia

Edwin F. Forero
Universidad Santo Tomás, Colombia

ABSTRACT

Considering the possibility that from the area of electronics can be provided feasible solutions in the field of non-destructive testing, this chapter present a prototype and methodology that allows energize an ultrasound transducer. This system is used to evaluate for detecting weld failures at the junctions of metallic parts. Subsequently, in order to validate the design quality of that source, a computer system that allows control of a card developed ultrasound. Finally, it is implemented the ultrasonic imaging by time of flight diffraction technique, in order to obtain an objective comparison methodology to both systems.

INTRODUCTION

Ultrasonic methods of non-destructive testing have been widely used in welding inspection; however, varied defectology, search and interpretation, is often highly dependent on subjective factors such as visual fatigue, mood and physical fatigue of the evaluator, which affects the quality of the inspection (Polikar, Upda & Taylos, 1998).

Now; venture into the field of non-destructive testing, or ultrasound, involves having access to appropriate instrumentation and often expensive, so that its development has leaded to the realization of projects with the participation of students and teachers interesting challenges. In that vein, it was designed and built an essential component in ultrasonic instrumentation; that is, the excitation system (also called button). The purpose: to design and build an ultrasonic system that allows the capture of acoustic information from welded joints to identify any discontinuities present in them. With the capture of echoes generated by any defects, develop and implement algorithms for the ultrasonic imaging and diagnosing faults that threaten the integrity in industrial components such as welded joints. The development of ultrasound

DOI: 10.4018/978-1-4666-8490-4.ch004

Figure 1. Representation of the sound spectrum

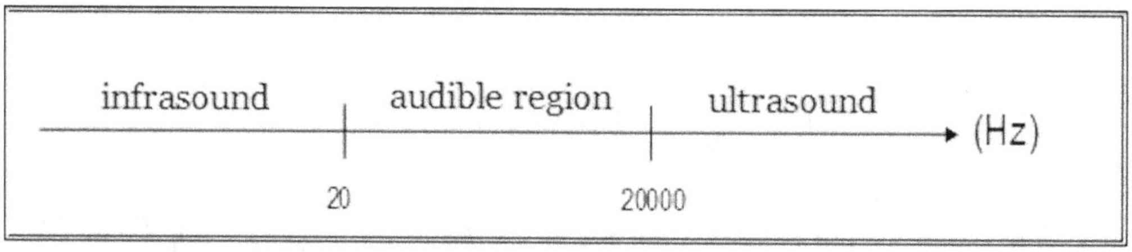

systems and ultrasound imaging algorithms can become a source of research narrows the links between industry and academia, while conducive to more simplicity and reliability of the work of the inspectors.

FUNDAMENTALS OF ULTRASOUND

The sound generated beyond 20,000 Hz is called ultrasound, but the frequency range used in nondestructive testing ranges from 100000 Hz to 50 MHz, and like any sound wave meets the physical properties of sound, but differs in the short wavelength, because it can interact with small particles, such as discontinuities eventually in welded joints regions. Considering that the wavelength is inversely proportional to frequency, it follows that to achieve short wavelengths it is necessary that the disturbance in a particular environment often vibrate every second.

$$\lambda = \frac{c}{f} \tag{1}$$

It is known that the human ear, on average, is capable of capturing vibrations between 20 and 20000 Hz, as it cannot perceive the range below 20 Hz and above 20,000 Hz, infrasound and ultrasound regions, respectively. See Figure 1.

Among the physical properties that obey ultrasound, are the reflection, scattering, refraction and diffraction, all of them are very useful in non-destructive testing with ultrasound. As is known, the light is also a wave disturbance, but there is a notable difference between light waves and sound waves. While light can propagate in a vacuum, sound cannot do it; ie, the sound needs a medium to be spread, whether it is fluid or solid.

The sound propagates in fluid media as a longitudinal disturbance; this means that the traveling direction is parallel to vibration the medium through which it propagates. However, in solid media, the sound is propagated as a longitudinal wave and as transverse cutting wave, which means that the molecules vibrate perpendicularly to the direction of sound propagation. Also, exist Rayleigh waves; these have an elliptical vibration surface, and Lamb, which are disturbances whose wavelength is greater than the thickness of the solid through which they propagate (Charlesworth, J. P. 2001).

Considering a longitudinal transmission of a sound wave when moving by a certain medium, which occurs are spatial variations in time due to the compression and expansion of the molecules making up said means to measure the pressure wave propagates. See Figure 2.

Figure 2. Illustration of the concept of longitudinal wave, and the physical meaning places for which the wave amplitude takes the maximum and minimum values

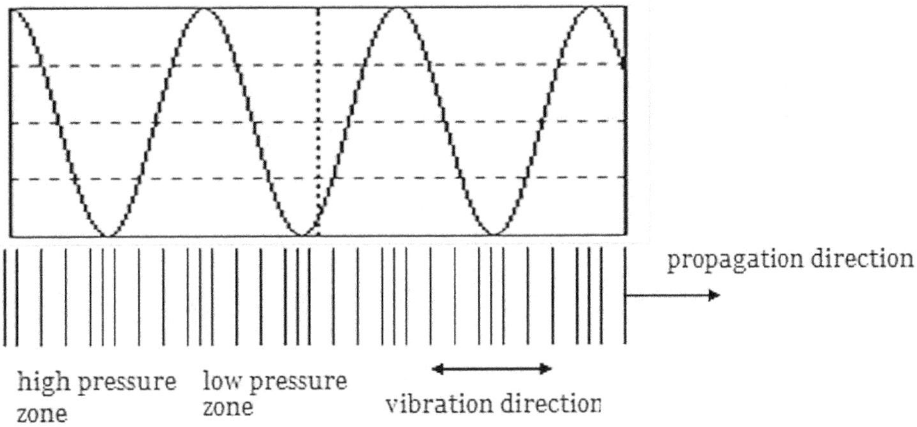

The governing equation of the variations shown in Figure 2, is given by,

$$\frac{\partial^2 p}{\partial x^2} = \frac{1}{c^2}\frac{\partial^2 p}{\partial t^2} \tag{2}$$

The expression (2) is called wave equation in one dimension, where p (x, t) represents the pressure at a location within the environment in which the disturbance to propagate a certain moment, and c the speed spreads the variation of pressure through said medium, and the solution is of the form,

$$p(x,t) = p_0 e^{j(kx - wt)} \tag{3}$$

In (3), p_0 represents the maximum value of the pressure, $k = \frac{2\pi}{\lambda}$, the wave number, $w = 2\pi f = \frac{2\pi}{T}$, the angular frequency, where λ and T mean wavelength and period vibration, respectively, and f is the cyclic frequency. The following graph of (1.2) allows to clarify these concepts.

An examination of Figure 3, support the conclusion that the wavelength is the distance between two identical values of the pressure in various locations, for a certain time, and measures the time interval between two identical pressure values for different times in a given location.

GENERATION OF ULTRASOUND

An essential for the practice of non-destructive testing using ultrasonic component, the transducer is used to generate it. In general, the term refers to a transducer device that is capable of converting one form of energy into another form of energy, so that an ultrasonic transducer (also called a probe) is a device capable of converting electrical energy into acoustic and vice versa; mechanical energy (sound) into electrical. These characteristics make a centerpiece transducer for transmitting and receiving ultra-

Figure 3. Graphical representation of equation (2); the image on the left gives the value of the pressure at a certain moment, and the picture on the right, shows the value of the pressure for a given location.

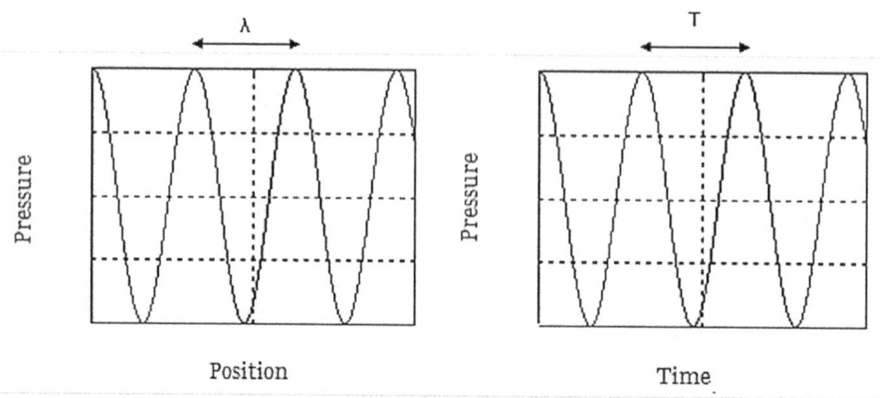

sonic energy, in addition to functioning as an intermediary between the material under inspection and electronics involved.

Certain crystals such as quartz and barium titanate, have an electromechanical property, called piezoelectricity. When a piece of such material is subjected to mechanical stress arises within the electric polarization, and when the piece is now applied to an oscillating electric field, the material will deform accordingly. See Figure 4.

According to Figure 4, the piezoelectric direct (left) and reverse (right) effect shall be given by the expressions (4) and (5),

$$P_j = d_{ij} S_i \tag{4}$$

$$W_j = d_{ij} E_i \tag{5}$$

Figure 4. Illustration piezoelectric effect forward and reverse

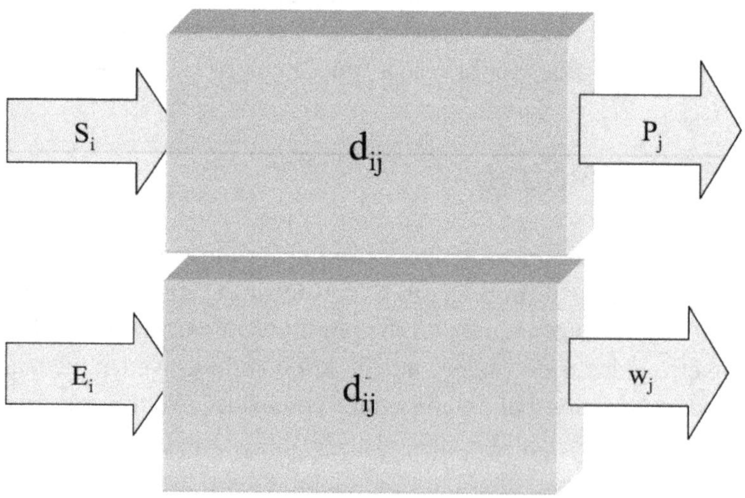

In the term dij ((4) and (5)) represents the piezoelectric coefficient of the material, Pj, electrical polarization, Si, mechanical stress, Wj, mechanical deformation and Ei the electric field. The subscripts indicate that the glass does not necessarily answer is obtained in the same direction in which the stimuli are applied to it. (Kasap, 2000).

Proper processing of the piezoelectric crystals, makes possible the construction of piezoelectric transducers which generate ultrasonic; however, this is not enough, because to ensure the incorporation of energy efficient the specimen to be examined, it is necessary to add in one side of the glass as a material officiate acoustic adapter; also, if you want to have control over excessive vibration of the glass, as when the stick hits a saucer, it is essential to add in the back of the glass, a material that plays a role of cushion to lessen these vibrations. The grid involves placing electrodes on both sides of the glass and finally encapsulates all this in a cover, usually metal. See Figure 5.

Figure 5 is a representation of a synthetic ultrasound system provided with an inlet (excitation pulse), and the impulse response of the transducer that is, a typical ultrasonic waveform output radiated this. It is important to add that it is sometimes necessary to condition a kind of acoustic transducer wedge shoe called to allow an angle of incidence of the ultrasonic wave in order to provide some guidance of refraction within the specimen under study.

ULTRASOUND SYSTEM

A typical ultrasound equipment used for measurements in non-destructive testing, consists of an excitation source, one or more transducers, a unit who receives the ultrasonic waves reflected at the border and from inside the specimen under study, a processing system of the reflected waves and a method for displaying the results.

Figure 5. Schematic diagram showing the parts of a piezoelectric ultrasound transducer

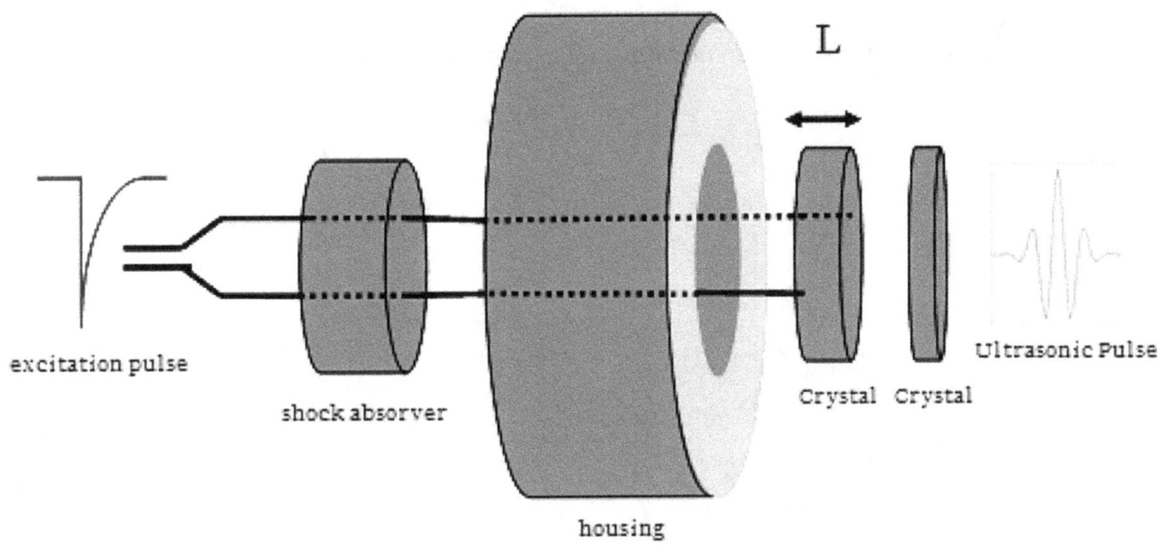

The excitation source includes an oscillator that activates a system responsible for controlling the switching that allows the generation of high voltage pulses, at a rate such that the reflected waves, or echoes do not overlap with those excitation pulses.

The signal containing echoes normally contains additive noise which must be eliminated, for which the system responsible for processing information includes filters, analog or digital, but not substantially alters the phase of said signals. Regarding the methodology for displaying the results, usually on a screen, you may be able to present information in the form of images in one or two dimensions. (Bentley, 1993).

ELECTRONIC TRIP CIRCUIT

The shape and amplitude of the ultrasonic pulse transmitted from the transducer are dependent on the characteristics of the waveform at the output of the power generated by this stimulus. In that vein, to achieve high accuracy in the axial direction, that is, to distinguish echoes generated by discontinuities in the specimen under study, situated one behind the other but close to each other, the key is the resolution along the depth to be scanned, which can be enhanced by an appropriate design of the electronic system which stimulates the transducer with the suitable impulse. To achieve this purpose, it is necessary a high voltage pulse on the transducer with times of the order of nanoseconds. In essence, what is sought is a quickly download over the transducer via a high voltage stored in a capacitor through a switch circuit.

As shown in Figure 6, the circuit designed to meet the stage to the transducer, comprises three distinct subsystems: frequency generator and pulse repetition circuit driver MOSFET high voltage circuit, and Shooting conditioner circuit. (Ramos, Sanz, Montero, & Riera, 1986).

The first section consists of a microcontroller 16F628, which plays a role in stimulating the high voltage circuit. The output signal of said composite device comprises pulse bursts of 5.0 volts and 500 microseconds in width which are repeated every 10 milliseconds.

The output signal from the microcontroller to the second feeding circuit section, comprising two 74HC07 buffer that increase the trigger voltage of 12.0 volts, and that by being connected in parallel, to

Figure 6. Diagram of the excitation circuit implemented for the purposes document

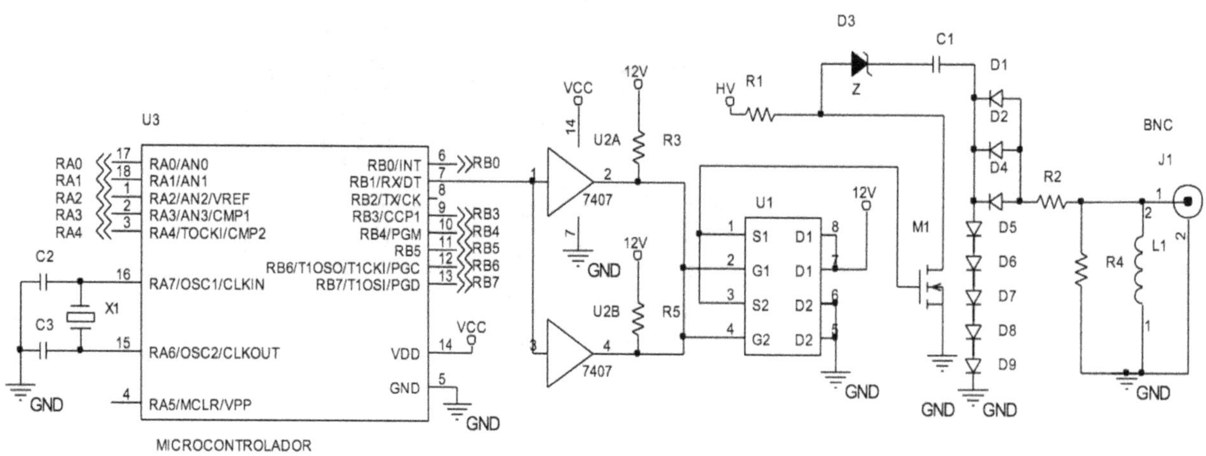

achieve adequate levels of current to ensure a correct triggering IRF7509 MOSFET, for proper switching power MOSFET IRF830.

The third section consists of a conditioning pulse circuit, which is responsible for coupling the excitation signal (from MOSFET IRF830) with the transducer, in terms of making compatible the impedances of both in order to achieve maximum power transfer.

For development of the above drive circuit is used as the reference waveform emitted by an ultrasound system Krautkramer USD10, and further inquired about the spectral content of both. These results are summarized in the images contained in the following figure.

Characteristic waveform, called in slang sprig of ultrasound images seen in the left of Figure 7. On the lower left image, which represents the signal achieved a sprig exciter output developed, it is important to note the curly little when compared to the waveform Krautkramer, which greatly reduces the possibility of generating spurious echoes.

Figure 7. Graphing parallel between the waveforms in the time domain and frequency between a sales team and developed for this work

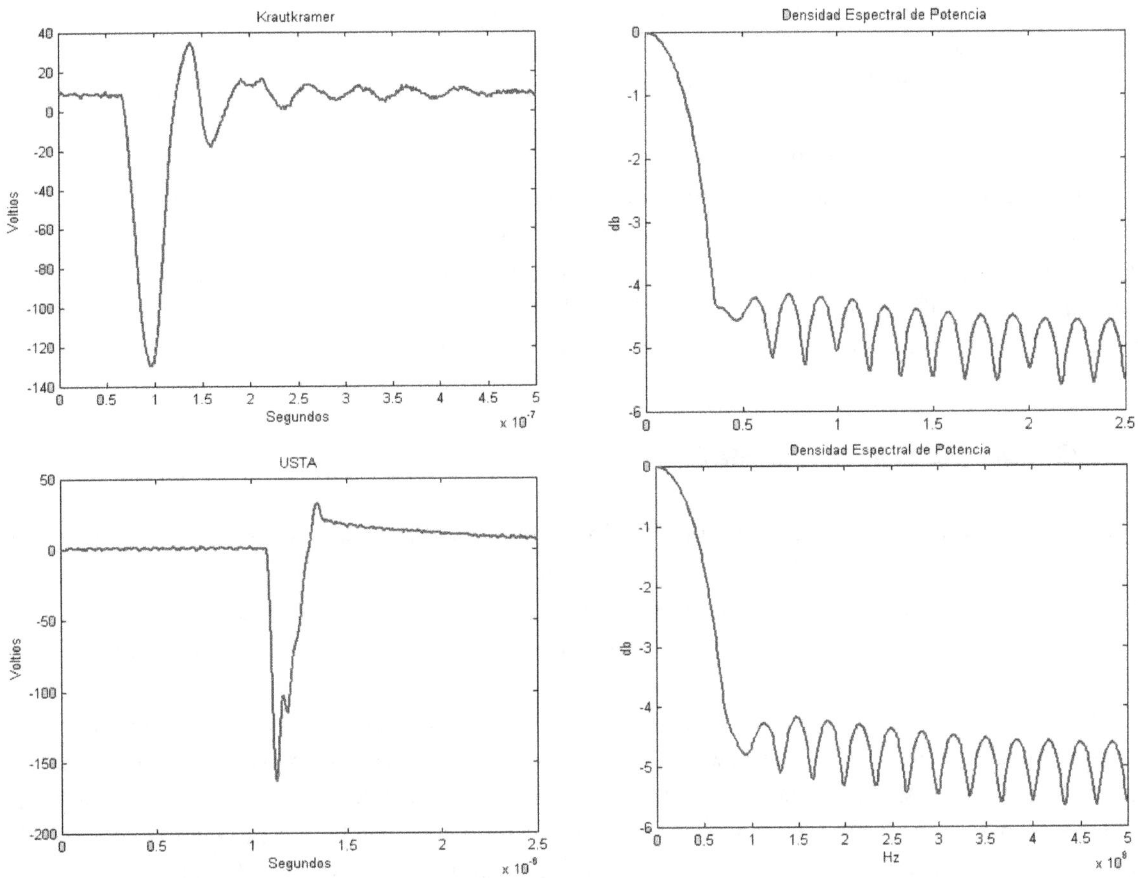

Figure 8. The left image shows a weld joining two metal sheets. On the right, the beveled shape of the ribbon is illustrated, containing some defects.

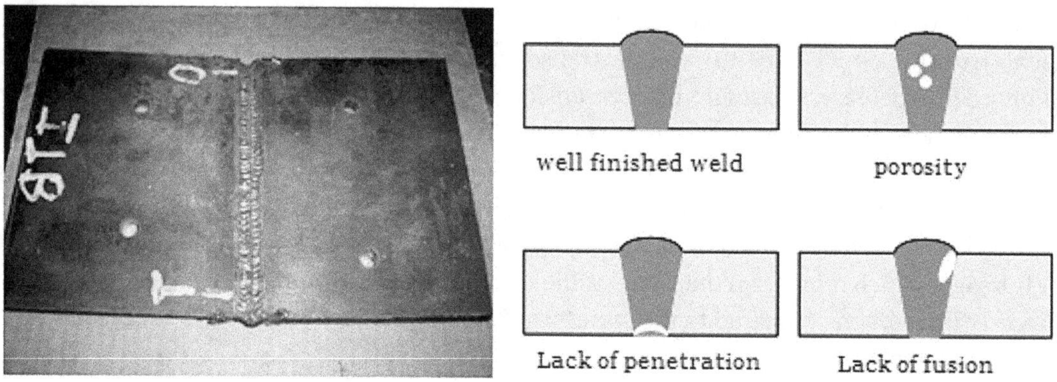

BREAKS IN WELDING

An application of the ultrasound system built what is the detection of defects in welding, for which they designed and built to help welded specimens system validation.

The welding defects are a major cause of the weakening structures, and are generated by different circumstances at the time of welding, producing discontinuities are unacceptable from the perspective of codes or standards established officially.

In welded joints, as in the left image in Figure 8, you can have defects or discontinuities due to reasons such as inadequate selection of base materials and filler, or inappropriate designs of welding procedures, or poor designs joints, among others.

Discontinuities often appear in welded joints are the following: lack of penetration at the root of the weld; lack of fusion, or discontinuity between the weld metal and the base metal bezel on the face; slag inclusion or non-metallic solids trapped in the weld metal between it and the base metal; porosity, or gas trapped during solidification of the weld metal; cracks, or linear rupture in the weld metal; undercut, or groove by melting in the root adjacent base metal. Some of these are illustrated in Figure 8.

INSPECTION METHODOLOGIES

The pulse-echo using shear waves to counter the low thickness of the specimen (12.0 mm) Longitudinal Waves: In order to unravel the information about any defects into a welding sample three strategies were implemented using a transmitter and another transducer as a receiver, and the time of flight diffraction, to exploit the interactions between the defects and the ultrasonic signal in order to generate X-rays similar to those maps. However, it is very relevant to mention that the correct assessment that provides inspection depends heavily on the reliability of the ultrasound system so that detail the following characteristics.

Ultrasonic equipment to characterize two standards, which generally do not differ from each other (Garcia, Katchadjian & Desimone, 2007), the European standard EN 12668-1 and ASME American 1324 may be taken into account (for systems between 100kHz and 25MHz) ultrasonic equipment which divide into the following parts: power supply, driver or push, receiving system, linearity.

Figure 9. Various alterations pulse waveform excitation

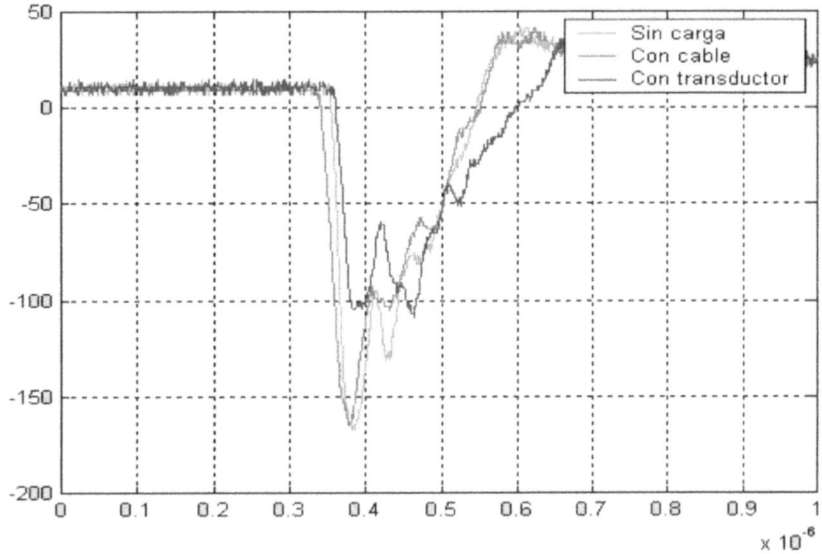

The first item is assured because the device connection is made to the online university laboratory, and this is properly designed to ensure the proper functioning of various electronic and electrical systems.

As the exciter waveform (second item), it was found that in a possible future development line constituting the adaptive design a reconfigurable circuit impedances, as each different demand electromechanical transducer characteristics. Associated with the pulse shape, measure the decay time, which is the length of the pulse to exceed 10% of its maximum amplitude.

The changes in the pulse waveform produced by the driver are illustrated in Figure 9, but under different conditions:

Evidenced for example, by connecting the transducer is no impedance mismatch, which results in the transfer of energy to the transducer [10]. As the fall time, and following the criteria of the rule, which requires it to be unloaded, has a value of 15 nanoseconds.

The other parameter is the pulse duration, which is 10% of the maximum amplitude on the falling edge and the maximum value on the rising edge. The values obtained lead to a duration of 209 nanoseconds, has a marked effect on the duration of the pulse generated by the transducer, because if it becomes too short, the pulse can have low amplitude, but on the contrary, is very wide, can deform and impoverish the axial resolution.

The maximum peak amplitude is 167V, which is slightly below the stipulated standards; that is, between 200 and 400V.

Regarding the repetition frequency of the pulses, which affects the correct display for a given range, the measurements yielded fairly reliable data.

With regard to the frequency spectrum of the pulse, the power spectral density for the Welch method, Figure 10 shows that for-3dB is worth approximately 20MHz.

The third item refers to the measurements in the receiver; that is, vertical linearity, horizontal linearity test was performed by the method shown in Figure 11

Figure 10. Spectrum of wave frequencies to the output pad

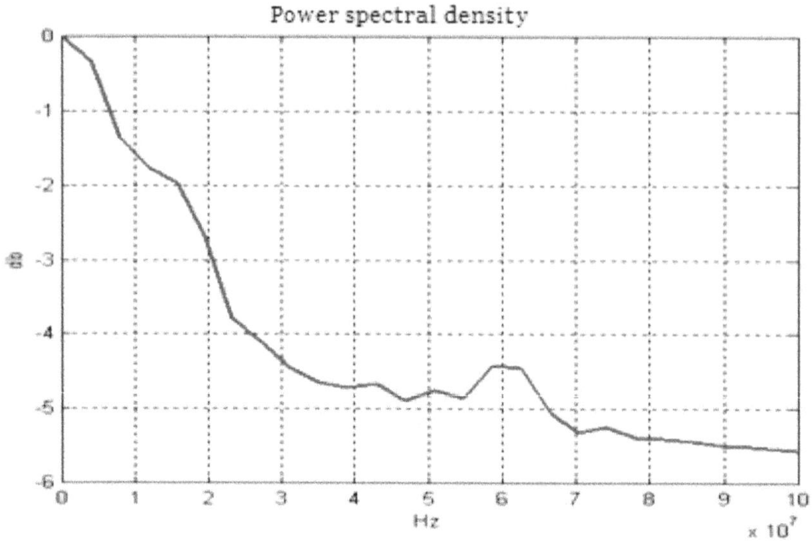

Figure 11. Mounting for verification of linearity

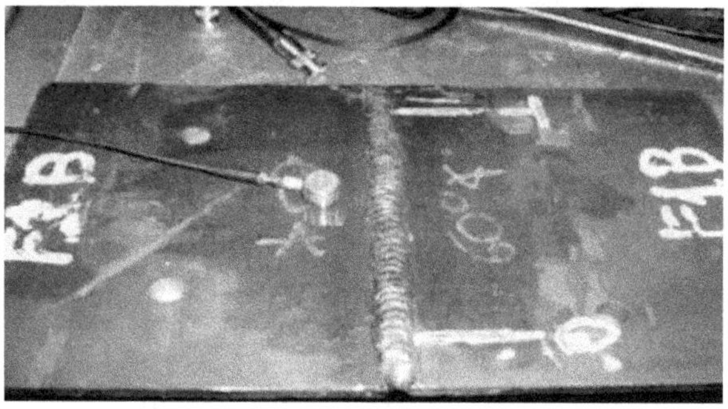

Using the technique illustrated in Figure 11, called pulse-echo transducer official both as transmitter and receiver of the ultrasonic waves. Tripping of the switch causes the transducer to generate an ultrasonic pulse that is reflected repeatedly between the two interfaces of the sample, resulting in a pattern of echoes such as those shown in Figure 12.

Figure 12 shows three consecutive echoes, and tangent to the maxima of the same lines. We can see a vertical linearity quite satisfactory, since otherwise, no maximum fall in line. This linearity is very important, because this ensures that the reading on the display is proportional to the input amplitude.

As for the horizontal linearity measurements affecting flight time of the wave, the intervals between echo and echo duration should be very similar. In the case of this study, and from the figure, we successive times of 16.9, 12.6, 8.30 and 3.98 microseconds, which leads to a range of 4.3 microseconds between consecutive echoes.

Surpassed the control step described above equipment, an angle transducer device capable of generating shear waves with frequency of 5MHz with an incidence angle of 65 ° was chosen and scanned on a specimen was carried out, as shown in Figure 13.

Figure 12. Pattern of echoes from acoustic impedance changes of the upper and lower borders of the specimen

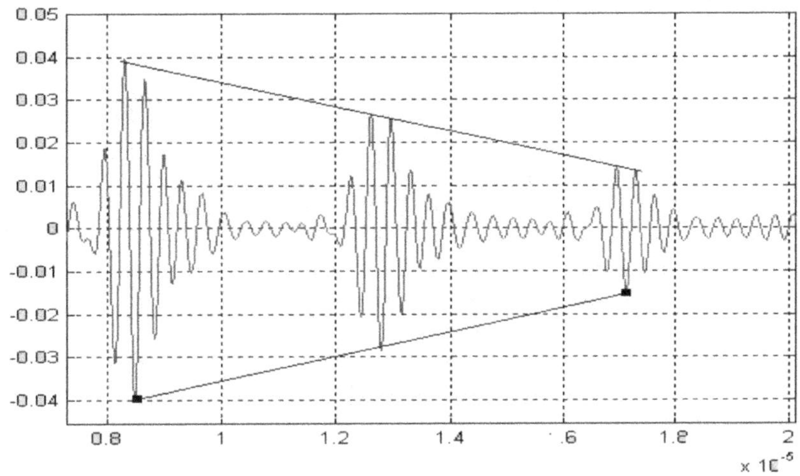

The photograph shows the transducer generating shear waves with which scanning are performed at the weld joining the two metal foils. (Rodríguez, Vitola, Sandoval & Forero, 2007, 2,008, 2009). In the graph to the right one of the displacement transducer is illustrated, with the objective to carry out exploration in search of defects. The other displacement occurs along the cord; ie perpendicular to the plane of the image direction.

The third technique, known as Time of Flight Diffraction, TOFD (Charlesworth, 2001, Baby, Balassubramanian, Pardikar Palaniappan & Subbaratnam, 2007 Bossuat, Walaszeck & Falvenot, 2006 & Abolhasany Riahi, 2006) requires the use of two transducers ; officiating one transmitter T and the other as a receiver R. These transducers are designed to deliver an angular beam width, so that, when the transmitter sends a wave, it passes through four paths.

The first is parallel to the surface of the specimen where the transducers are called direct or surface wave; the following called bottom echo due to reflection of the wave on the bottom border of the specimen due to the change of acoustic impedance; and the other, as a result of the interaction of the wave front with the edges of a discontinuity in the solder, resulting in a wave phenomenon known as diffraction; that is, the property that has a wavelength of around obstacles. See Figure 14.

Figure 13. Experimental arrangement for exploration along the bezel of a weld bead

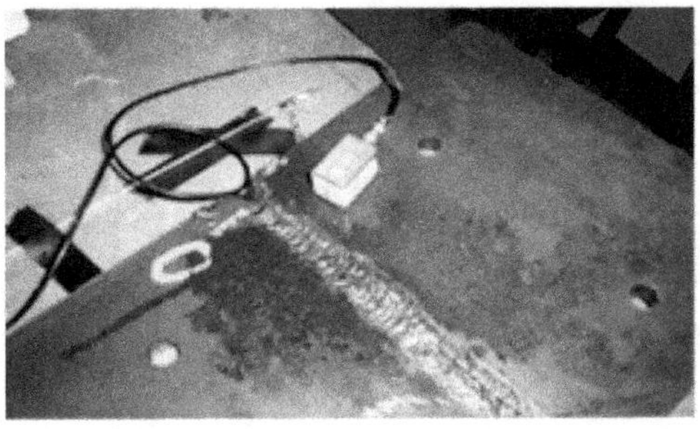

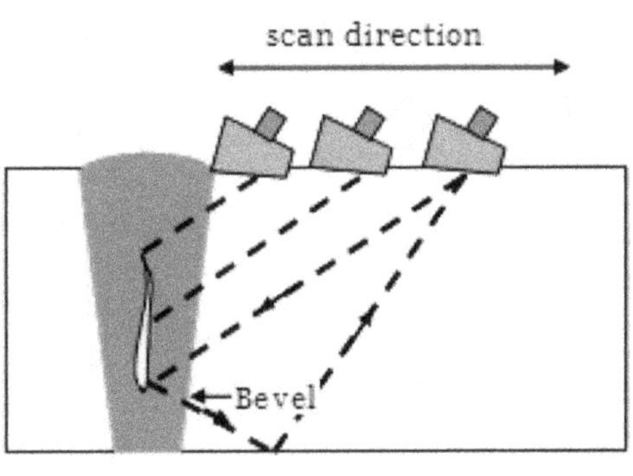

Figure 14. Experimental arrangement for inspection using TOFD

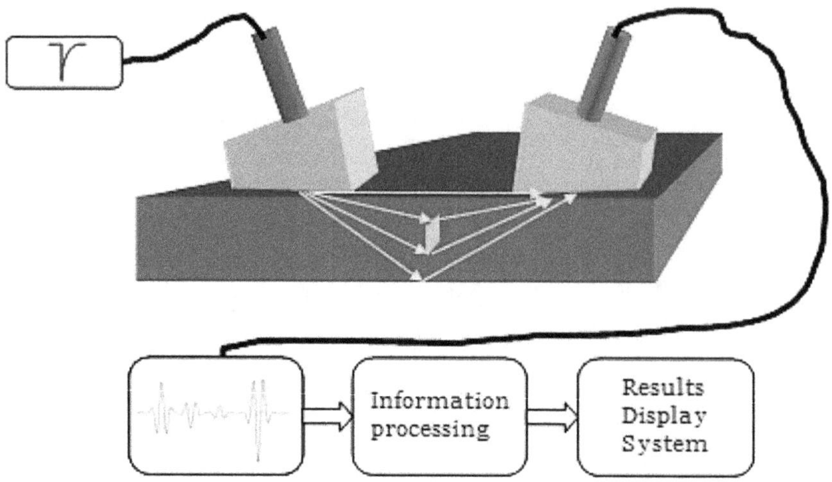

The TOFD transducers are manufactured so that the emitted sound field is broad in order to ensure the widest possible exploration of the area of interest of the part under study. Once stimulated transmitting transducer T, and if there were no defect, the signal to reach the receiving transducer R is formed by a wave which runs slightly below the surface, and a wave produced by the reflection of the acoustic field in the bottom wall of the part under study. Right image in Figure 15.

In the same figure, the image on the right as we infer that the four signals invest different transit times, which arrive at the receiver with different delays and phase shift. The first pulse represents the direct wave, the second wave is diffracted by the upper edge of a discontinuity, the third one is at the lower edge and the fourth is the reflection of the wave at the bottom wall of the sample. Note also, the phase change of them all, which provides strategies for correct procedure of exploration.

Returning to Figure 14, in this experimental setup required to perform nondestructive tests illustrated END using the TOFD technique, in particular, the arrangement of transducers for sending and receiving information. The weld joining two pieces of metal, and which optionally contains embedded defects are also observed.

The process begins by performing a sweep in parallel to the cord, which is achieved by moving the pair of transducers with the smallest possible step. Moving each transducer to a certain step, sending

Figure 15. Waveforms, A-Scan, TOFD technique, with and without default respectively

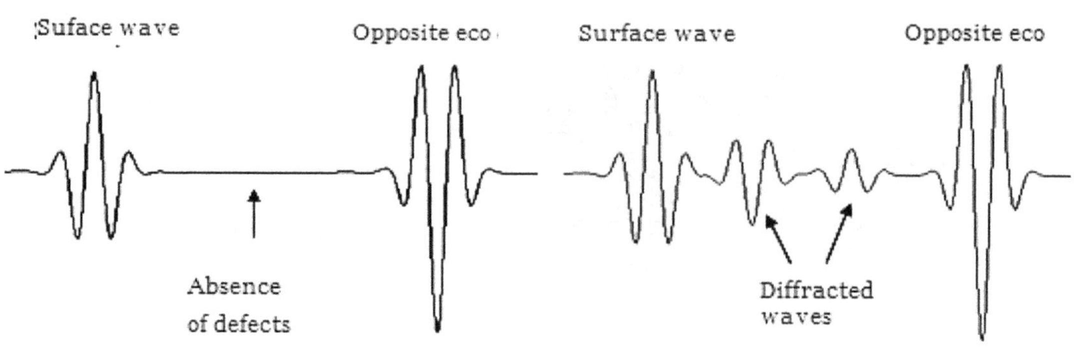

Figure 16. Example of a TOFD image

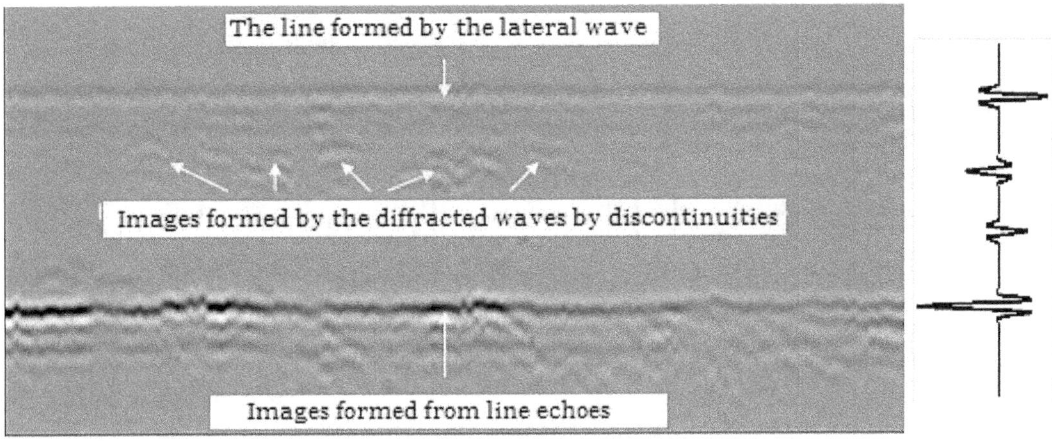

signals to the receiver, known as A-scan, as shown in Figure 15, which contain information about the presence or absence of discontinuities, then that information is stored.

After the scanning of all the weld bead will have stored signals sufficient, A-scan, to superimpose them one after another to form a two-dimensional map, called D-Scan, which contains information on the cross sections of weld in line with each step forward of the pair of transducers. More specifically, an image, D-Scan, is obtained by grouping a series of sweeps, A-Scan, but currently the top view of the grouping, and tones assigned proportional to the relative amplitude of the signal at each position gray.

With the information contained in the A-Scan signals, the map D-Scan, built based on these, displays an image of the inside of the weld bead, similar to an x-ray. (Subbaratnam, Venkatraman & Raj, 2006).

Added a wave A-scan to visualize the relationship between the two.

As mentioned at the beginning of this paper, in order to validate the developed system, ultrasound OPCARD card manufactured by Polish company OPTEL was chosen. (Rodriguez and Segura, 2011, 2012 and 2013). This card is connected to a computer through a PCI port, and for the capture of signals application owned by the manufacturer is used, which is responsible for managing the card and setting, acquire the information and display the waveform of the captured signal. The captured information is stored using the same application via a button or a function key, being housed in a comma separated file. Now; the necessary applications for the research demands more than one sample, so that the procedure is repeated, but with the disadvantage that it cannot be stored in a single file rather than a sample, A-Scan; ie if twenty of these samples is needed, should generate twenty files. The unification of this information in a spreadsheet is done outside the application supplied by the manufacturer, which increases the duration of the process and makes the process of acquiring information and to consolidate file demanded too long.

To achieve the solution to the problem mentioned statement of versatility and also provide the handling of said card, in terms of accelerating the process of acquiring information, the implementation of a program that helps improve information capture times arose.

As the manufacturer of the card provides a library with (Application Programming Interface) required for setup and management of the card, these APIs are the starting point for the implementation of the program model. The proposed model is illustrated in Figure 17.

The proposed in Figure 17, model consists of the following modules:

Figure 17. Initial model program to control OpCard card

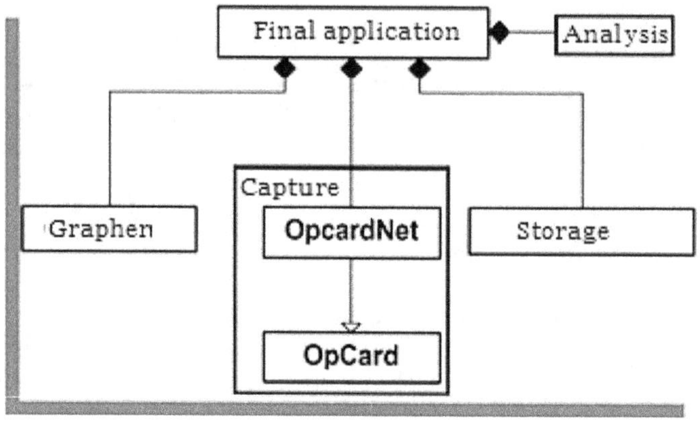

- The graphing module, which is responsible for displaying the information, but independently of the origin of it.
- The storage modulus, which is intended to keep the task of non-volatile manner information for further analysis in a file compatible with MatLab, a database or any other means, depending on the need.
- The capture module, whose function is to ensure transparency between the hardware used (Opcard card) and other modules.
- The analysis module is reserved for information processing.

This model is intended that the final application, capture, storage and analysis of consolidated information faster mode is performed. The model is implemented on the platform. NET using the C # language under development environment Visual Studio Express. This platform is a framework for virtual machine similar to the Java software.

Now; the methodology described is limited, because one must resort to own MatLab functions, so that a strategy based on a model developed on the raised platform Net, but that removes these limitations.; that is, the use of the application submitted by the manufacturer of the card for data capture, and MatLab for analysis and visualization.

Then the realization of the proposed new model is described, taking into account that the implementation of this was carried out on the platform. Net, using the programming language C #. (Boer, Bonsague & Van Buuren, 2007).

The process of data capture is performed by two components: the first is physical, the OpCard card, which allows you to perform non-invasive inspection by ultrasound, and the second, a component of the software delivered by the manufacturer to control, using the APIs (Application Program Interface). As the above APIs are for use in Win32 platform designed component performs the link between these and the. NET through a dynamic link library (DLL) that enables control OpCard card.

The captured information is stored in a file that contains numeric data signal (output of the digital to analog conversion by the card) in plain text, complete column space and row terminated with a new-line character. Each column represents a sample of the ultrasound signal digitized, and all these forms a record. In turn each record or line in the file represents a point-making information that is not more

Figure 18. Program end pattern control OpCard card

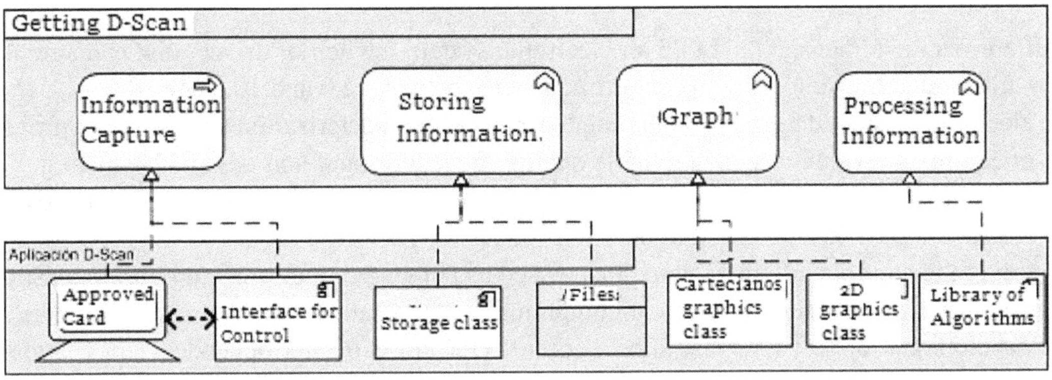

than the digitized A-scan signal, which by definition is the entire file with multiple records that have the information of a D- Scan.

This function graph consists of two software components; Cartesian graphs class and two-dimensional graphics.

The first is responsible for displaying the information that is delivered in a grid and adjust the range displayed to optimal settings for the data shown, and also allows viewing both single A-Scan as several at a time, selecting for each a different color. The second class of graphs is able to take information from one or more A-Scan and view it as a 2D image, transforming the signal amplitudes of A-Scan in grayscale and joining them into a single image.

The function intended for the task of processing information is composed of a class containing the algorithms necessary for processing the signals. The following figure shows the steps described above are illustrated.

Counting now with the control of OpCard card, then it is proceeded to examine the same specimen using the two systems. For the experiment a pair of transducers, transmitter and receiver, center frequency of 10.0 MHz, shoe 70 ° was used and arranged as shown in Figure 14. The test result is shown below.

As shown in Figure 19, the two images contain information, allowing the inference that the driver built meets the appropriate design, to perform research tasks algorithms for signal processing of ultrasonic NDT.

Figure 19. Comparison of D-Scan images obtained using the OpCard (left) card and the developed system (right)

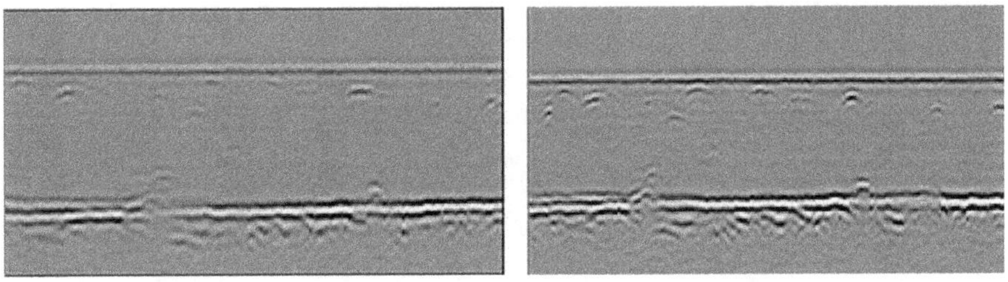

CONCLUSION

With little investment managed to build an electronic system (switch or driver) that can stimulate, efficiently, ultrasound transducers with center frequencies between 2.0 and 10.0 MHz

The electronic system design meets the requirements of characterization to ensure the proper functioning of equipment which results in reliable quality procedures and non-destructive testing.

The combination of various inspection methods, such as radiography and D-Scan images are very useful for providing additional information.

Regarding the control card ultrasound device used as compared to developed button, constituted an effective cooperation between the fields of computing and generating and capturing ultrasonic signals.

The development of software that allows control OpCard card, not dependent applications using computational tools such as MatLab or similar.

The development of the button, and the software control OpCard card, will have the necessary instrumentation that will test the design of algorithms that facilitate the extraction of information contained in ultrasonic signals.

REFERENCES

Baby, S., Balassubramanian, T., Pardikar, R. J., Palaniappan, M., & Subbaratnam, R. (2007). *Time of Flight Diffraction Technique for Accurate Sizing of Surface Breaking Cracks*. Institute of Materials and Minig.

Bentley, J. P. (1993). *Sistemas de Medición. Compañía Editorial Continental*. S.A. Mexico.

Boer F., Bonsague, M., & Van Buuren R. (2007). *Concepts for Architectural Description*. Telematica Instituut / Archimate Consortium.

Bossuat, B., Walaszeck, & Falvenot, J. (2006). *Non Destructive Evaluation by Time of Flight Diffraction*. Method Mechanical Applications, ECNDT.

Charlesworth, J. P. (2001). *Engineering Applications of Ultrasonic Time of Flight Diffraction*. Research Studies Press LTD.

Cohen, L. (1995). *Time-Frequency Analisys*. Prantice Hall.

Edwin, F., & Rodríguez, J. A. (2008). Procesamiento Digital de Señales TOFD. *Journal, STSIVA*, *1*, 93–97.

Gang, T., & Chi, D. (2007). Nobel Approach to Enhancement of Ultrasonic, TOFD B-Scan, image for measurement of weld crack. *Science and Technology of Welding and Joining*, *12*(1), 87–92. doi:10.1179/174329307X159847

García, A., Desimone, C., & Katchadjian, P. (2007). *Caracterización de equipos de END. In IV Conferencia Panamericana de END*. Buenos Aíres: AAENDE.

Kaiser, G. A. (1994). *Friendly Guide to Wavelets*. Birkhäuser.

Kasap, S. O. (2000). *Principles of Electrical Engineering Materials and Devices*. McGraw-Hill.

Matz, V., Kreidl, M., & Smíd, R. (2004). *Signal to Noise Ratio Improvement Based of the Discrete Wavelet*. Academic Press.

Polikar, R., Upda, L., Upda, S., & Taylor, T. (1998). Frequency Invariant Classification of Ultrasonic Weld Inspection Signals. *IEEE Transactions on Ultrasonics, Ferroelectrics, and Frequency Control*, *45*(3), 614–624. doi:10.1109/58.677606 PMID:18244213

Proakis, M. (2007). *Tratamiento Digital de Señales*. Prentice Hall.

Ramos, P., Sanz, F., & Montero, E. (1986). *Riera*. Mundo Electrónico.

Riahi, M., & Abolhasany, M. (2006). Substitution of the Time-of-Flight Diffraction Technique for Nondestructive Testing of Welds and Thick Layers of Steel. *Russian Journal of Nondestructive Testing*, *42*(12), 794–801. doi:10.1134/S1061830906120047

Rodríguez, J. A., Sandoval, S., & Vitola, J. (2007). Fundamentos Teórico-Prácticos de Ultrasonido. *Revista Tecnura*, *10*(20), 4–18.

Rodríguez, J. A., Vitola, J., & Sandoval, S. (2009). Diseño y Construcción de un Sistema de Ultrasonido para la detección de Discontinuidades en Soldadura. *Revista Colombiana de Física*, *41*(1), 159–161.

Rodríguez, J. A., Vitola, J., Sandoval, S., & Forero, E. (2009). Diseño y Construcción de un Sistema para Examen no Destructivo de Fallas y Defectos en Metales Utilizando Señales Ultrasónicas. *Revista EIA*, *12*, 9–21.

Segura, D., & Rodríguez, J. A. (2013). Algoritmo que Reduce Distorsión de Imágenes Ultrasónicas en Ensayos no Destructivos. *CIMM*, 65.

Segura D., A., & Rodríguez J., A. (2011). Interfase para el Control de Tarjeta de Ultrasonido. *CIMM*, 74.

Segura, D. A., & Rodríguez, J. A. (2012). Sistema de Formación de Imágenes TOFD para Ensayos no Destructivos. *Revista EIA*, *17*, 139–149.

Subbaratnam, R., Venkatraman, B., & Raj, B. B. (2006). Time of Flight Difracttion- an Alternative to Radiography Examination of Thick Walled Stainless Weldmentes. In *Proceedings of A-PCNDT -Asia-Pacific Conference on NDT*. A-PCNDT.

Chapter 5
Structural Damage Assessment using an Artificial Immune System

Maribel Anaya Vejar
Universitat Politècnica de Catalunya, Spain & Universidad Santo Tomás, Colombia

Diego Alexander Tibaduiza Burgos
Universidad Santo Tomás, Colombia

Francesc Pozo
Universitat Politècnica de Catalunya, Spain

ABSTRACT

Structural damage assessment methodologies allow providing knowledge about the current state of the structure. This information is important because allows to avoid possible accidents and perform maintenance tasks in the structure. This chapter proposes the use of an artificial immune system to detect and classify damages in structures by using data from a multi-actuator piezoelectric system that is working in several actuation phases. In a first step of the methodology, principal component analysis (PCA) is used to build a baseline model by using the collected data. In a second step, the same experiments under similar conditions are performed with the structure in different states (damaged or not). These data are projected into the different baseline models for each actuator, in order to obtain the damages indices and build the antigens. The antigens are compared with the antibodies by using an affinity function and the result of this process allows detecting and classifying damages.

INTRODUCTION

Structural health monitoring (SHM) is an important discipline which seeks to assess the proper performance of a structure. To achieve this objective, SHM makes use of sensors permanently installed in the structure for inspecting and defining its current state based on the analysis of structural responses. As a result, the collected structural responses are analyzed and compared with baseline patterns in order

DOI: 10.4018/978-1-4666-8490-4.ch005

to detect abnormal characteristics and define the structural integrity. The obtained information can be used to define whether the structure can operate and under which conditions.

In general, the damage identification can be performed by two main approaches; the first consists in obtaining a reliable physics-based model of the structure, while the second is based on data-driven approaches which normally tackle the problem as one of pattern recognition.

One advantage of the use of data driven approaches is the reliability in the analysis since the indication of damage could be directly determined with the comparison between a baseline and the data collected. However, to ensure the reliability of the analysis performed to the signals collected in several experiments, it is necessary to ensure the proper functioning of the sensors, actuators and hardware used to inspect the structure. Among the big quantity of damages that can be presented in the normal service of a structure, the following categories can be distinguished (Farrar, Doebling, & Nix, 2001), (Doebling, Farrar, & Prime, 1998):

- Gradual damage such as fatigue, corrosion, and aging.
- Sudden and predictable damage like aircraft landing and planned explosions in confinement vessels.
- Sudden and unpredictable damage originating from foreign-object-impact, earthquakes and wind loads.

At the same time, these different kinds of damage can also be classified depending on its severity in three big groups (Tibaduiza, 2013):

- **Light Damage:** This corresponds to the initial stage of damage, which can be relatively easily-repairable and is not dangerous for the normal operation of the structure.
- **Moderate Damage:** In comparison with the previous one, this damage requires major repairs and need to be evaluated more carefully in order to define if the structure can operate in normal conditions.
- **Severe Damage:** This type of damage unlike previous damages requires big reparations or the replacement of the structure.

For a structure in service, the variability in its dynamic properties can be a result of time-varying in environmental and operational conditions. Some of the environmental conditions to consider are humidity, wind loads, temperature, and pressure, among others. Operational conditions include loading conditions, operational speed and mass loading (Sohn, 2007). The damage identification techniques need in this way to consider several variables and most of the cases the damage identification procedure depends on the critical damage admissible in the structure.

The damage diagnosis is often grouped in four levels (Rytter, 1993), starting by the damage detection. In this level the objective is to know whether there are some changes in the structure and if these changes are due to damage. Second level considers the damage localization. Third level is used to define the type of damage and its size. The last level is defined for calculating the Remaining lifetime of the structure. Recently, an extra level is considered which includes intelligent structures with auto-healing (Figure 1).

Computational algorithms inspired by the nature have become in a new trend in data-driven. These algorithms mimic the behavior and properties of several natural systems and have been used to solve complex problems in engineering showing a high effectively. Among these computational intelligences,

Figure 1. Level in SHM

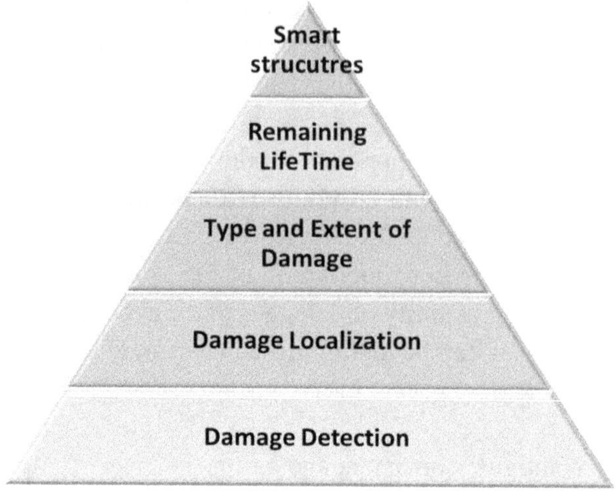

a relatively new approach called artificial immune system (AIS) has presented a good results in intrusion detection (Elsadig, Adbullah, & Samir, 2010), data analysis and optimization (Binitha & Siva, 2012) among others.

The use of AIS is relatively new, and its application in structural health monitoring has a reduced number of works. Costa et al. (2003) developed three algorithms in order to detect fault in a squirrel-cage induction motor. These algorithms are based on immunologic principles. The first one is called T-module and is able to distinguish between self and non-self condition. The second algorithm is called B-module, this algorithm analyze the occurrence of both condition (self and non-self). D-module is the last algorithm and its task is report the unusual condition to a central unit, as a candidate anomaly. In the work of Costa et al the feature vector of the normal operation is built from frequency spectrum of the signal motor. Zhang et al. (2008) developed an algorithm to solve the problem of sensors distribution in a structure in order to detect impacts. This algorithm is based on an immunologic principle called the clonal selection. The authors tested a composite plate instrumented with seventeen piezoceramics transducers. Chen and Zang (2009) used artificial immune system as pattern recognition in order to classify damage in structures. In this work the features used to distinguish each damage pattern are obtained from the structure dynamic response by using an auto regressive algorithm and least square method. The approach proposed by the authors is validated using a benchmark structure proposed by the international association for structural control- American society of civil engineers (AISC-ASCE). Chen (2010) applied an agent-based artificial immune system in order to develop a system for adaptive damage detection. The immune cells (B-cells) are represented by a group of agents. These cells go over a distributed sensor network installed in the structure in order to carry out the damage diagnosis by analyzing the structural dynamic response data. Chilengue et al. (2011) presented a pattern recognition algorithm based on artificial immune system in order to identify faults in the stator and rotor circuits of an induction machine. In this work the authors apply the Clarke transformation to the stator current to obtain the feature of the pattern and build the feature vector.

More recently, Zhou et al. (2012) developed a classifier of damage in structures based on the immune principle of clonal selection. Using evolution algorithms and the immune learning, a high quality memory cell is generated in the classifier to recognize several damage patterns in structures. Xiao.

(2012) developed a structural health monitoring and fault diagnosis system based on artificial immune system. In Xiao's approach, the antigen represents the structural state (healthy or damaged), whereas the antibody represent database information that identify a damage pattern. In this work, the features space is built by natural frequencies and modal shapes obtained by simulation of the structure in free vibration and seismic response. Yue et al. (2013) perform a structural detection and classification algorithm based on artificial immune pattern recognition and diagonal distance. This algorithm is tested by using a benchmark proposed by the IASC-ASCE. As result of this work, the authors show that is possible produce a high quality memory cells and obtain a high detection percentage. Also in this year, Li et al. (2013) developed a hybrid methodology for structural health monitoring based on immune algorithms (IA) and symbolic time series analysis (STSA). In this work a negative selection principle is used to detect damage, the clonal selection principle is used to localize and quantify the damage and the data symbolization is used to deal the noise in the acceleration data. The methodology is tested in a five-story shear frame structure. Parra et al. (2014) presented a method to identify and characterize structural fault in an aircraft; this method is based on the principle of negative selection of the natural immune system. The proposed method is tested in an aluminum beam and show excellent results, robustness and accuracy.

In this work, the damage detection and classification are tackled by using bio-inspired data-driven approaches. These are applied to analyze the signals collected by several experiments from a piezoelectric system which is permanently attached to the structures.

Theoretical Background

This section presents a brief introduction to the different concepts and methods used in the methodology presented on this paper. More detailed information can be consulted in the references included in this section.

Human Immune Systems

The human immune system is a complex and robust defense mechanism composed by a large network of specialized cells, tissues and organs. The system further includes an elevated number of sensors and a high processing capability. This system has demonstrated its effectiveness in the detection of foreign elements by protecting the organism against disease. Among its skills are: to distinguish between its own cells (self) and foreign cells (non-self), to recognize different invaders (called antigen) in order to ensure the protection of the body and to learn from specific antigen and adapt to them in order to improve the immune response to this kind of invader.

To protect the organism, this system has different defense levels such as external barriers, innate immune system and adaptive immune system. The first concerns the major line of defense into the human body and includes elements such as the skin, the mucus secreted by the membranes, the tears, saliva and urine which present different physiological conditions that are harmful to the antigens as the temperature and the pH level, among others. The response of these barriers is equal for any foreign invader (Delves, Martin, Burton, & Roitt, 2011). The second owes its name to the fact that the organism born has the ability to recognize certain antigens and destroy them immediately on first encounter. This barrier acts when the first barrier is broken and this immunity is unchanging (Cruz, 2004). The last barrier is the final defense level and responds to the stimulus of foreign cells or antigens that evade both the external barriers and the innate immune defense (Cruz, 2004). This stimulus provides the capability of learning into the immune system. This way, in a future detection of this antigen, the system knows how to proceed. The immune systems also includes different type of cells, some of these are cells born in the bone marrow and usually called white blood cells (de Castro & Von Zuben, 1999). Among the white blood cells it is possible to highlight the T-cells and the B-cells. The T-cells are called so because their maturation takes place in the thymus, they have high mobility and can be found in the blood and the lymph (Trejo, 2005). The B-cells produce and secrete a special protein called antibody, which recognizes and binds the antigen. Each B-cell is the responsible to produce a specific antibody. This protein is used for signaling other cells to remove them from the body. The T-cells and the B-cells proliferates when they recognize antigens and, some of them, become memory cells that remain in the immune system in order to eliminate the same antigen in the future in a more effective manner (Trejo, 2005), (de Castro & Von Zuben, 1999). Three immunological principles are used in artificial immune system:

- **Immune Network Theory:** This theory explains how the immune memory is built by means of the dynamic behavior of the immune system cells. These cells can recognize by themselves, detect invaders, as well as interconnect between them allowing the stabilization of the network (Cruz, 2004).
- **The Negative Selection:** The negative selection is a process that allows to identify and remove cells that react to the own body cells giving to the system a tolerance to own cells. This ensures a proper functioning of the immune system since it is able to distinguish between foreign molecules and self-molecules, avoiding autoimmune diseases (de Castro & Von Zuben, 1999).
- **The Clonal Selection:** This is a process of adaptive immune responses in which the cells of the system are adapted to identify an invader element (Trejo, 2005). Antibodies capable of recognizing an antigen can proliferate and the rest are eliminated. The new cells are clones of their parents and they are subjected to an adaptation process by mutation. From the new antibody set, the cells that have the greatest affinity with the primary antigen are selected as memory cells and the remainders are eliminated.

Artificial Immune Systems

Artificial immune system (AIS) is an adaptive and bio-inspired computational system based on the processes, performance of the human immune system and its properties such as diversity, error tolerance, dynamic learning, adaptation, distributed computation and self-monitoring. Nowadays, this computational system is used in several research areas such as optimization (Eroglu, Erden, & Erden, 2011), pattern recognition (Zhang, Worden, & Staszewski, 2008) among others (Xiao, 2012). Table 1 presents the analogy between the natural and artificial immune systems.

In the implementation of an artificial immune system, it is necessary to bear in mind three important aspects: First of all, it is necessary to define the role of the antigen and the antibody in the context of the application. Generally, an antigen is the solution to be achieved, for instance when it is necessary check the data to know if it is an intrusion, or the element to classify, while the antibodies are the remaining data or those that have been previously identified (Aickelin & Dasgupta, 2005). Both antigens (Ag) and antibodies (Ab) are represented or coded in the same way; this representation is generally given by a string of binary or real numbers. A second important aspect is to define the mechanism by which the degree of correspondence between Ag and Ab molecules is measured. This measure is related to the

Table 1. Analogy between the biological immune system and artificial immune system

Biological Immune System (BIS)	Artificial Immune System (AIS) in Structural Health Monitoring (SHM)
Antibody	A detector of a specific pattern.
Antigen	Structural health or damage conditions.
The matured antibodies	Database or information system for damage detection.
Recognition of antigens	Identification of health and damage condition.
Process of mutation	Training procedure.
Immune memory	Memory cells

(Xiao, 2012).

distance between them (de Castro & Timmis, 2002). Finally, there exists the adaptation process of the molecules in the artificial immune system. This adaptation allows including the dynamic of the system, for instance, the antibodies excitation, cloning all the excited antibodies and the interconnection between them. All these elements are adapted from the three immunologic principles previously introduced.

PRINCIPAL COMPONENT ANALYSIS

PCA is a classical method used in statistical multivariate analysis, feature extraction and dimensionality reduction. The general idea in the use of PCA is to find a smaller set of variables with less redundancy but with a minimal loss of information (Hyvarinen, Kahunen, & Oja, 2001). The application of PCA starts with a matrix X that contains information from *m* sensors and *n* experimental trials (Tibaduiza, 2013). Since the data organized in this matrix come from experimental trials and could have different magnitudes and scales, it is necessary to apply a preprocessing step to scale the data using the mean of all measurements of the sensor at the same time and the standard deviation of all measurements of the sensor. Using this preprocessed matrix it is possible to calculate the covariance matrix as follows:

$$C_x = \frac{1}{n-1} X^T X \tag{1}$$

C_X is a square symmetric m x m matrix that measures the degree of linear relationship within the data set between all possible pairs of variables (sensors). The subspaces in PCA are defined by the eigenvectors and eigenvalues of the covariance matrix as follows:

$$C_x \tilde{p} = \tilde{P} \Lambda . \tag{2}$$

where the eigenvectors of C_x are the columns of P and Λ is a diagonal matrix whose diagonal elements are the eigenvalues and the off-diagonal terms are zero. The columns of matrix $\tilde{P}$ are sorted according to the eigenvalues by descending order and they are called principal components of the data set. Choosing only a reduced number $r < n$ of principal components, those corresponding to the first eigenvalues, the reduced transformation matrix could be defined as a model for the structure. In this way, the new matrix P ($\tilde{P}$ reduced) is called the PCA model. Geometrically, the transformed data matrix T (score matrix) represents the projection of the original data over the direction of the principal components P:

$$T = XP \tag{3}$$

Damage Detection Indices Based on PCA

Several damage detection indices based on PCA has been proposed and applied with excellent results in pattern recognition applications. In particular, two damage indices are commonly used: Q-index (or square prediction error SPE-index) and the Hotelling's T^2 statistic (or D-statistic). The first one is based on analyzing the residual data matrix **X** to represent the variability of the data projection in the residual subspace (Zhou, Tang, Zang, & Zhou, 2012). The second index is based on analyzing the score matrix **T** to check the variability of the projected data in the new space of the principal components.

The Q-index and the T^2 statistic can be calculated according to Equations (4) and (5), respectively:

$$Q_i = e_i e_i^T = x_i (I - PP^T) x_i^T \tag{4}$$

$$T_i^2 = \sum_{j=1}^{r} \frac{t_{sij}^2}{\lambda_j} = t_{si} \Lambda^{-1} t_{si}^T = x_i P \Lambda^{-1} P^T x_i^T \tag{5}$$

More details about these indices can be found in [3].

Fuzzy Clustering

The purpose of data clustering is to identify partitions or groups of data from a large data set, in order to produce a precise representation of a system's behavior (Babuska, 2012). In the traditional clustering approaches the groups or clusters are generated and in each of them each data belongs only to a one cluster. Fuzzy cluster is an unsupervised technique normally used to organize or classify data into groups. The algorithm builds these groups by grouping data with similar information or features. In this technique a data is simultaneously associated to many clusters or groups by using a membership function. A huge value to the membership indicates a high confidence in the assignment of the data to the cluster (Rokach & Maimon, 2005). This technique has been widely used in areas as pattern recognition, image processing, classification and fuzzy model and identification (Karim, Yun, & Siva Krishna Madani, 2010).

The data that can be applied to the clustering technique are numerical (quantitative), categorical (qualitative) or a mixture of both. These data are usually observations of a process wherein each observation is an n-dimensional row vector that group n measurement variables (6). A set of N observations is denoted by a $n \times N$ matrix (7) wherein the columns are named patterns and the rows are named features whereas that the matrix **X** is referred as data matrix (Balasko, Abonyi, & Feil, 2005). (6)

$$\boldsymbol{X} = \begin{bmatrix} x_{11} & x_{12} & \cdots & x_{1N} \\ x_{21} & x_{22} & \cdots & x_{2N} \\ \vdots & \vdots & \ddots & \vdots \\ x_{n1} & x_{n2} & \cdots & x_{nN} \end{bmatrix} \tag{7}$$

The definition of cluster depends on the context of the application. Generally the cluster is a group of data that are most similar to one another that the data of other group. The *similarity* is a mathematical measure that is often defined by a distance norm. The geometrical shapes, sizes and densities of the clusters can be spherical, hollow or elongated. Furthermore, the cluster can be a linear or nonlinear subspace; besides, this cluster can be overlapping each other, well separated or continuously connected (Matlab, 2014), (Karim, Yun, & Siva, 2010).

The clustering methods can be classified according to the characteristics of the subsets (fuzzy or crisp (hard)). In the fuzzy clustering method it is possible that an object or data with different degrees of membership belong to several clusters. However, the hard clustering method just allows a data to belong to a cluster (Babuska, 2012).

There are several algorithms in fuzzy clustering. In this work the most popular technique, called fuzzy c-means algorithm (FCM), was used. This technique was developed and introduced by Dunn in 1973 and improved in 1981 by Jim Bezdek. In the pattern recognition field, this technique is frequently used (Yang, Hwang, & Chen, 2004), (Döring, Lesot, & Kruse, 2006). In this algorithm, a p-dimensional data set is grouped into a specific number c of clusters depending on a membership degree wherein each cluster is represented by its own center ci. The measure of membership between a data point and a cluster is given by the Euclidean distance (Matlab, 2014), (Balasko et al., 2005).

Damage Classification Methodology

The developed methodology consists of three steps. First, a training step is developed by using data from the healthy structure; then, a testing step is applied to the data from the structure in an unknown state in order to define the state of the structure (healthy or damaged). In both steps, the data from the structure are collected and organized in an unfolded matrix (Tibaduiza, Mujica, Güemes, & Rodellar, 2010) that contains the information of the signal collected by the sensors in different parts of the structure by each actuation phase. Finally, once the data set is defined as data from the structure with damage, these are organized in a matrix and a fuzzy clustering algorithm is applied in order to classify them.

To apply the methodology, as a preliminary step, normalization is applied to each matrix in both steps. Previous works by the authors (Tibaduiza, 2013), (Tibaduiza et al., 2010) have shown that group scaling presents a better performance compared with other kind of normalizations, because it considers changes between sensors and does not process them independently. More information about the normalizations can be found in (Tibaduiza, 2013).

Training Step

This step only considers the feature vector from the healthy state of the structure. To perform the training of the healthy state, it is first necessary to select a number k of random experiments from the healthy data set. These experiments are divided and grouped in training antibodies (AB_train) and antigens (AG_train). It is worth noting that the experiments used in both groups need to be different to include diversity in the system. Once these groups are defined, a comparison between the antibody training set and the antigen training set is performed. This comparison is applied by evaluating the affinity between both data sets. The affinity between AB_train and AG_train is defined in the same way as in (Chen & Zang, 2011) by the expression in the next equation:

$$aff(AB\_train_i, AG\_train_j) = 1 - \frac{1}{2} d(AB\_train_i - AG\_train_j) \quad (8)$$

where *aff* corresponds to the affinity and ***d*** to the Euclidean distance between the two data sets (AB_train and AG_train) which is defined by each vector of the matrix as follows:

$$d(AB\_train_i, AG\_train_j) = \sqrt{\sum_{k=1}^{N} (AB\_train_{ik} - AG\_train_{jk})^2} \quad (9)$$

Note that both AB_train and AG_train are $n \times N$ matrices with *n* rows or experiments and *N* features. Depending on the affinity value obtained in the previous step, the antibody must evolve and a new antibody set is obtained as a result. This evolution consists on the mutation of the antibody and it is performed as in (Chen & Zang, 2011). After that, this new set of data, that is the result of the evolution, is compared again with the antigen training data set by means of the affinity. If this new affinity value is larger than or equal to an affinity threshold T_h, then the antibody initializes the set of memory cell with the healthy state (MCH_SET); otherwise, it is eliminated. This dataset is saved in the memory of the AIS algorithm and corresponds to the training step.

Testing Step

The damage detection process is depicted in Figure 2. This process is based on the affinity value between the memory cell with the healthy state (MCH_SET) that acts as an antibody, and the unknown data state that act as an antigen (data test). When the affinity is greater than or equal to the affinity threshold (T_h), the result is a healthy state. Otherwise, the result is defined as a damage state.

Figure 2. Damage detection process

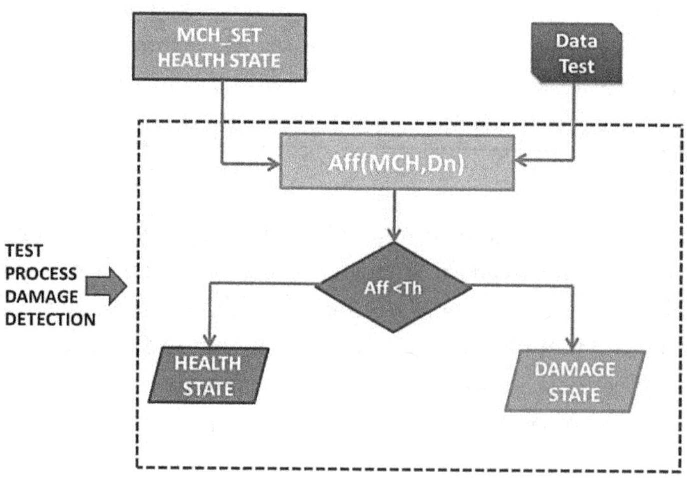

Classification Process

After the detection of damages, the abnormal data are used as input to the fuzzy clustering algorithm. To perform the classification, a fuzzy clustering algorithm, called FC-means, is used. At each actuator-phase, an antibody is selected and the affinity value and the Euclidean distance are obtained by comparing this antibody with each experiment from the structure with damages, as it is shown in Figure 3. Using these measures, a matrix input data by each actuation phase is built, where the first column corresponds to the Euclidean distance and the second is the affinity value as the pattern of the data input. The number of clusters used is 3, the weighting exponent used is 4 –this parameter determines the fuzziness of the clusters– and the tolerance of the clustering method is 10^{-6}.

Experimental Set Up

The experimental setup consists of an aircraft skin panel instrumented with a piezoelectric system that works in several actuation phases. This skin panel is divided in small sections by means of stringers and ribs as shown in Figure 4. For testing the approach, two sections of this structure were used. The dimensions of each section and the damage description are depicted in Figure 5. These sections were

Figure 3. Damage classification process

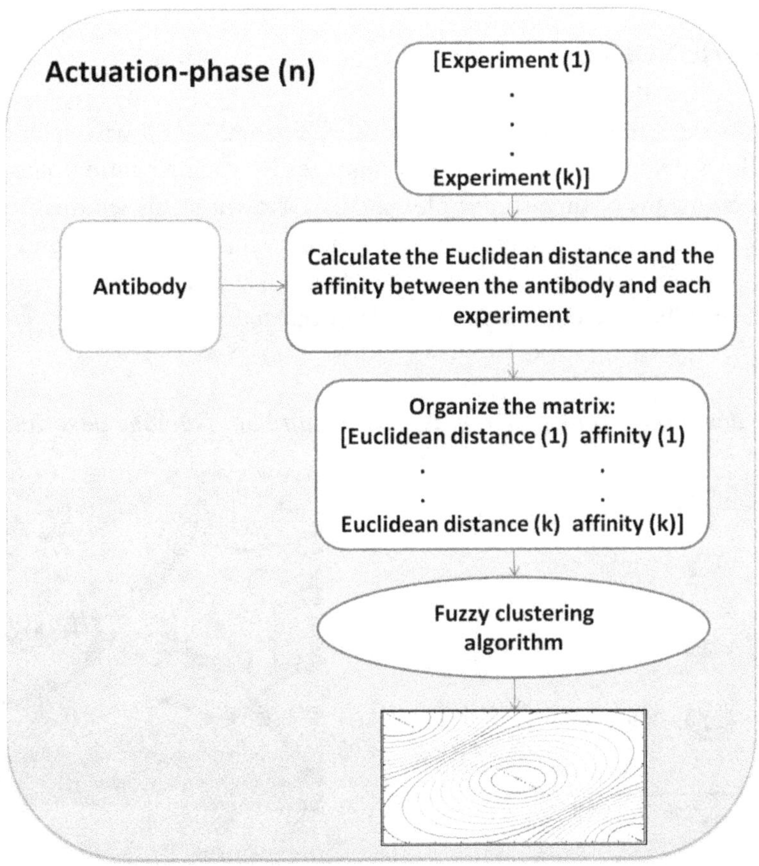

Figure 4. Aircraft skin panel

instrumented with 6 PZT transducers as in Figure 5(b). As excitation input, a BURST signal with 205 KHz as central frequency and nine peaks was used. Four different states of the structure were analyzed: the healthy structure and the structure with three different damages. Damages were simulated by adding a mass at three different positions (Figure 5a), two of them on the skin and the other on the stringer. 100 experiments were performed and recorded: 25 with the undamaged structure and 25 more per damage. The aim of this artificial damage is to introduce reversible changes in the structure.

To ensure a good signal to noise ratio each signal was averaged 10 times. To apply and collect the signals to the PZT transducers, a chassis PXI 1033 from National Instruments ® was used. Due to the complexity and the size of this structure, a wideband power amplifier model 7602M of Krohn-Hite Corporation is used to amplify the signal applied to the actuators.

EXPERIMENTAL RESULTS

In a previous paper by the authors (Anaya, Tibaduiza, & Pozo, 2014) it was demonstrated that the use of affinity values allows detecting the presence of damages by each actuation phase. Figure 6 presents the results obtained by means of the affinity plot. As it is shown in this figure, the three damages can be distinguished because presents different range of affinity values. Since it is previously known which experiment contains information about the different structural states, these experiments were labeled in the Figure 5. Similar results were obtained in the rest of the actuation phases.

Figure 5a. Damage description and placement

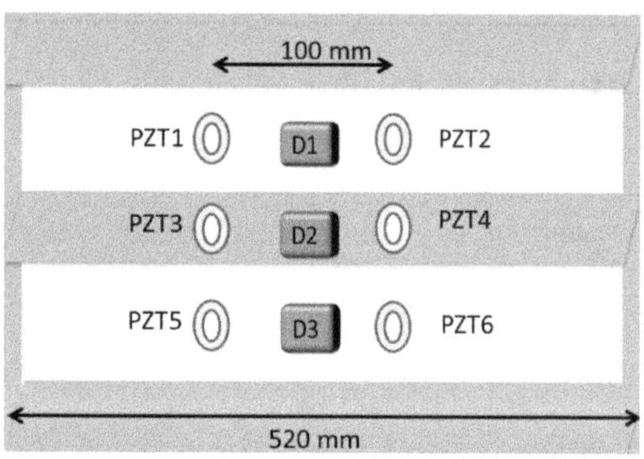

Figure 5b. Damage description and placement

Figure 6. Damage detection using the affinity value

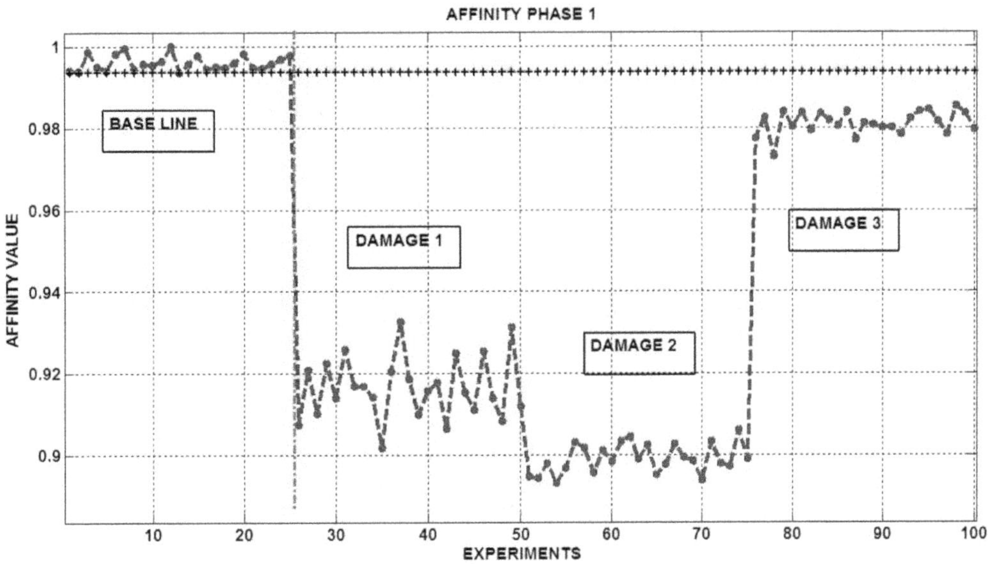

Based on the results, the classification is performed by applying the fuzzy clustering algorithm as it was shown in Figure 3. The c-means method was used in each actuation phases to classify the three different damages simulated to the structure. Figures 7, 8, and 9 show the result of the damage classification. As it can be seen, three cluster groups are presented since in the fuzzy clustering algorithm only the data from the three simulated damages was included. The data belong to an only cluster or group; this means that the classification step has a good performance. The results in these figures are evidence that each damage presents different features that can be distinguished by the algorithm.

Figure 7. Damage classification using data from the actuation phase 1

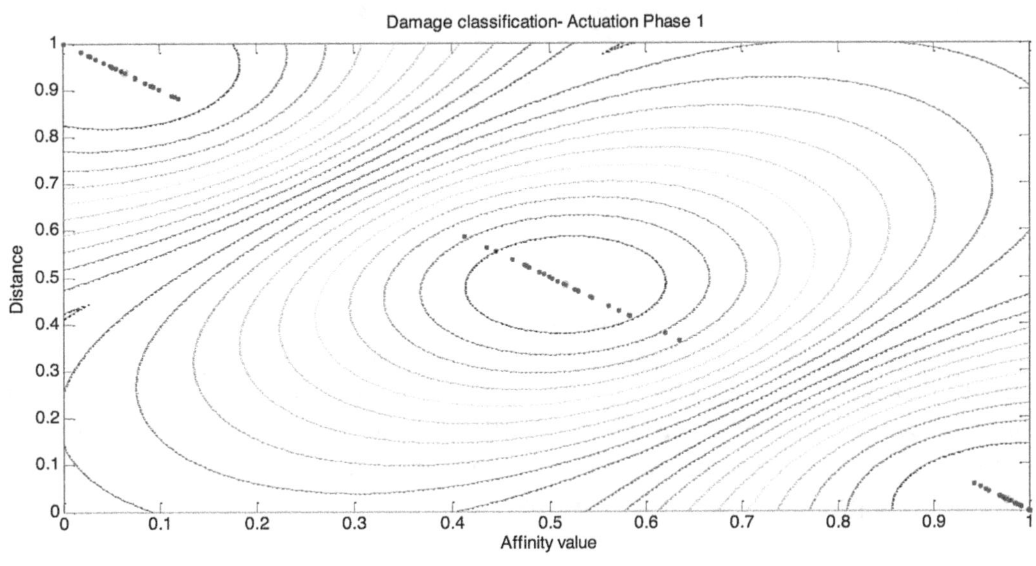

Figure 8. Damage classification using data from the actuation phase 3

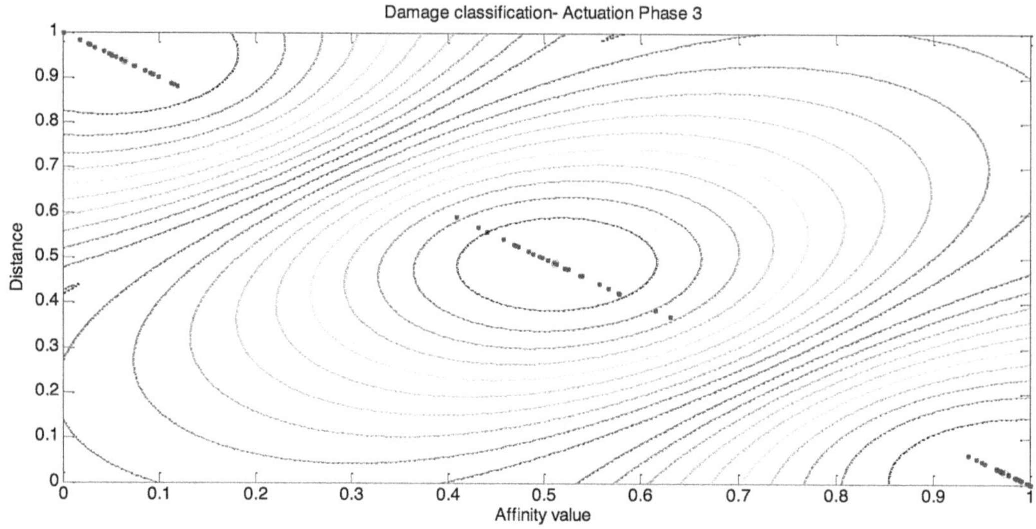

Figure 9. Damage classification using data from the actuation phase 6

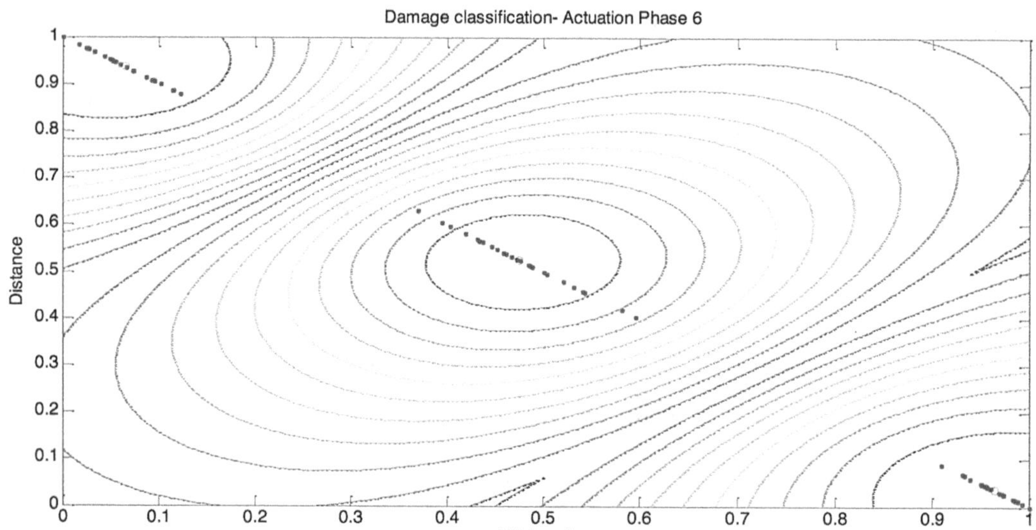

CONCLUSION

An application of fuzzy clustering based on artificial immune system to classify damage in structures is presented in this work. In the proposed methodology, the damage is first detected. Then, the experiment that represents damage is used as the input to the fuzzy cluster algorithm and the pattern used are the Euclidean distance and the affinity value obtained with the comparison between a memory cell and the experiments. The presented approach has been used to classify damage in an aircraft skin panel. The results show that the parameters chosen for the fuzzy clustering are suitable for the structure damage classification. The approach proposed in this work performs with excellent results to classify damages using artificial immune systems and fuzzy clustering, allowing to discern between the different damages.

Although the methodology presented in this chapter it was applied to the analysis of data from different PZTs which were attached to an aircraft skin panel, the methodology can be applied to any metallic or composite structure instrumented with PZTs or other sensor such as accelerometers. This is because it can be used in the same way when the collected data are organized by actuation phases, regardless of the number of sensors used.

ACKNOWLEDGMENT

This work is supported by Universidad Santo Tomás through Grants FODEIN 2014, project code 047, and FODEIN 2015, project code 8410083003-036. The authors would like to express their gratitude to the Center of Composite Materials and Smart Structures at the Universidad Politécnica de Madrid (Spain) and to Professor Alfredo Güemes to allow us to test the structure in a research visit performed by Diego Tibaduiza Burgos at their Center. Furthermore, the authors would like to thank Mr. Fahit Gharibnezhad and Dr. Luis E. Mujica at the UPC for their support during the execution of the experiments.

REFERENCES

Aickelin, U., & Dasgupta, D. (2005). Artificial Immune System. In E. Burke & G. Kendall (Eds.), *Search Metodologies: Introductory Tutorials in Optimization and Decision Support Techniques* (pp. 375–399). New York, NY: Springer. doi:10.1007/0-387-28356-0_13

Anaya, M., Tibaduiza, D., & Pozo, F. (2014). Data driven methodology based on artificial immune systems for damage detection. In *Proceedings of the 7th European Workshop on Structural Health Monitoring*. Nantes, France: Academic Press.

Babuska, R. (2012). *Fuzzy Modeling for control*. Netherlands: Springer.

Balasko, B., Abonyi, J., & Feil, B. (Eds.). (2005). *Fuzzy Clustering and Data Analysis Toolbox (for Use with Matlab)*. Academic Press.

Binitha, S., & Siva Sathya, S. (2012). A survey of bio-inspired optimization algorithms. *International Journal of Soft Computing and Engineering, 2*, 137-151.

Chen, B. (2010). Agent-based artificial immune system approach for adaptative damage detection in monitoring network. *Journal of Network and Computer Applications*, *33*(6), 633–645. doi:10.1016/j.jnca.2010.03.011

Chen, B., & Zang, C. (2009). Artificial immune pattern recognition for structure damage classification. *Computers & Structures*, *87*(21-22), 1394–1407. doi:10.1016/j.compstruc.2009.08.012

Chen, B., & Zang, C. (2011). Emergent damage pattern recognition using immune network theory. *Smart Structures and Systems*, *8*(1), 69–92. doi:10.12989/sss.2011.8.1.069

Chilengue, Z., Dente, J., & Costa Branco, P. (2011). An artificial immune system approach for fault detection in the stator and rotor circuits of induction machines. *Electric Power Systems Research, 81*(1), 158–169. doi:10.1016/j.epsr.2010.08.003

Costa Branco, P. J., Dente, J. A., & Vilela Mendes, R. (2003). Using immunology principles for fault detection. *Industrial electronics. IEEE Transactions on, 50*, 362–373.

Cruz Cortes, N. (2004). *Sistema Inmune Artificial para Solucionar Problemas de Optimización.* (PhD thesis). Cinvestav.

de Castro, L. N., & Timmis, J. (2002). Artificial Immune Systems: A Novel Approach to Pattern Recognition. In J. Corchado, L. Alonso, & C. Fyfe (Eds.), *Artificial Neural Networks in Pattern Recognition* (pp. 67–84). Paisley: University of Paisley.

de Castro, L. N., & Von Zuben, F. J. (1999). Artificial immune systems: Part I basic theory and applications. Campina, Brazil: Technical report. School of Computing and Electrical Engineering. State University of Campinas. Brazil, No. DCA-RT 01/99.

Delves, P. J., Martin, S. J., Burton, D. R., & Roitt, I. M. (2011). *Roitt's essential immunology.* Wiley-Blackwell.

Doebling, S., Farrar, C., & Prime, M. (1998). A summary review of vibration-based damage identification methods. *The Shock and Vibration Digest, 30*(2), 91–105. doi:10.1177/058310249803000201

Döring, C., Lesot, M. J., & Kruse, R. (2006). Data Analysis with Fuzzy Clustering Methods. *Computational Statistics & Data Analysis, 51*(1), 192–214. doi:10.1016/j.csda.2006.04.030

Elsadig, M., Adbullah, A., & Samir, B. (2010). Intrusion prevention and self-healing algorithms inspired by danger theory. In *Proceedings of 2nd international conference on computer and automation engineering (ICCAE).* Singapore, Singapore: IEEE Press.

Eroglu, A., Erden, Z., & Erden, A. (2011). Bioinspired conceptual design (bicd) approach for hybrid bioinspired robot design process. In *Proceedings of Mechatronics (ICM), 2011 IEEE International Conference on* (pp. 905-910). IEEE.

Farrar, C. R., Doebling, S. W., & Nix, D. A. (2001). Vibration-based structural damage identification. *Philosophical Transactions: Mathematical, Physical & Engineering Sciences, 359*(1778), 131–149.

Hyvarinen, A., Kahunen, J., & Oja, E. (2001). *Independent Component Analysis.* John Wiley & Sons. doi:10.1002/0471221317

Karim, M. E., Yun, F., & Siva Krishna Madani, S. P. (2010). *Fuzzy Clustering Analysis.* (Master Thesis). Blekinge Institute of Technology.

Li, R., Mita, A., & Zhou, J. (2013). Hybrid methodology for structural health monitoring based on immune algorithms and symbolic time series analysis. *Journal of Intelligent Learning System and Applications*, 48-56.

Matlab. (2014). *Fuzzy Logic Toolbox User`s Guide*. MathWorks.

Parra dos Anjos Lima, F., Chavarette, F., dos Santos e Souza, A., Silva Frutuoso de Souza, S., & Martins, M. L. (2013). Artificial immune systems with negative selection applied to health monitoring of aeronautical structures. *Advanced Materials Research, 871*, 283-289.

Rokach, L., & Maimon, O. (2005). Clustering Methods. In O. Maimon, & L. Rokach (Eds.), Data Mining and Knowledge Discovery Handbook (pp. 321-352). Springer.

Rytter, A. (1993). *Vibration Based Inspection of Civil Engineering Structures.* (PhD Thesis). Aalborg University, Aalborg, Denmark.

Sohn, H. (2007). Effects of environmental and operational variability on structural health monitoring. *Phil. Trans. R. Soc. A*, *365*(1851), 539–560. doi:10.1098/rsta.2006.1935 PMID:17255051

Tibaduiza, D. (2013). *Design and validation of a structural health monitoring system for aeronautical structures.* (PhD Thesis). Universitat Politècnica de Catalunya.

Tibaduiza Burgos, D., Mujica, L., Guemes, A., & Rodellar, J. (2010). Active piezoelectric system using PCA. *Proceedings of the Fifth European Workshop on Structural Health Monitoring*, (pp. 164-169). Sorrento.

Trejo Perez, D. (2005). *Optimización global en espacios restringidos mediante un sistema inmune artificial.* (Master Thesis). Cinvestav.

Xiao, W. (2012). *Structural health monitoring and fault diagnosis based on artificial immune system.* (PhD Thesis). Worcester Polytechnic Institute.

Yang, M. S., Hwang, P. Y., & Chen, D. (2004). Clustering Algorithms for Mixed Feature Variables. *Fuzzy Sets and Systems*, *141*(2), 301–317. doi:10.1016/S0165-0114(03)00072-1

Yue, Z., Shi, T., Xuesong, J., Dongwei, Z., & Chuanzhi, Z. (2013). Structural damage detection and classification algorithm based on artificial immune pattern recognition. *Journal of Shenyang Jianzhu University*, 2.

Zhang, J., Worden, K., & Staszewski, W. (2008). Sensor optimization using an immune system metaphor. In *Proceedings of 26th International Modal Analysis Conference (IMAC)* (pp. 1-10). IMAC.

Zhou, Y., Tang, S., Zang, C., & Zhou, R. (2012). An artificial immune pattern recognition approach for damage classification in structures. In D. Zeng (Ed.), *Advances in information technology and industry applications, LNEE 136* (pp. 11–17). Berlin: Springer. doi:10.1007/978-3-642-26001-8_2

KEY TERMS AND DEFINITIONS

Actuation Phase: Methodology used to inspect a structure by using piezoelectric transducers as sensors and actuators.

Affinity: Measure of the degree of correspondence between antigen and antibody in the artificial immune system, this measure is calculated by using the distance between them.

Antibody: Detector of a specific pattern in an artificial immune system. In the algorithm, an antibody is a pattern which represents the healthy state of the structure.

Antigens: Pattern that represents structural health or damage condition. This pattern can stimulate the generation of memory in the artificial immune system.

Artificial Immune System: Computational and adaptive system inspired by immune system. This system is mainly used to solve different kind of problems in the area of the computation security and engineering.

Clonal Selection: Principle immune that establishes that only those cells that be able to recognize the antigen are selected to proliferate.

Damage Detection Indices: Statistical indices based on PCA proposed and applied as a pattern recognition technique.

Fuzzy Clustering: Unsupervised technique used to classify data into groups. The classification is performed grouping data with similar features.

Negative Selection: Immune cells ability to identify the foreign as well as own cell in the body but react only to foreign cells.

Principal Component Analysis (PCA): Technique used in multivariate statistical analysis in order to reduce the dimensionality of a data set, feature extraction, among others. The main objective is find a small number of variables with less redundancy.

Chapter 6
Structure Impact Localization Using Emerging Artificial Intelligence Algorithms

Qingsong Xu
University of Macau, China

ABSTRACT

Extreme learning machine (ELM) is a learning algorithm for single-hidden layer feedforward neural networks. In theory, this algorithm is able to provide good generalization capability at extremely fast learning speed. Comparative studies of benchmark function approximation problems revealed that ELM can learn thousands of times faster than conventional neural network (NN) and can produce good generalization performance in most cases. Unfortunately, the research on damage localization using ELM is limited in the literature. In this chapter, the ELM is extended to the domain of damage localization of plate structures. Its effectiveness in comparison with typical neural networks such as back-propagation neural network (BPNN) and least squares support vector machine (LSSVM) is illustrated through experimental studies. Comparative investigations in terms of learning time and localization accuracy are carried out in detail. It is shown that ELM paves a new way in the domain of plate structure health monitoring. Both advantages and disadvantages of using ELM are discussed.

INTRODUCTION

Plate structures have been applied extensively in aerospace, civil, and mechanical engineering, etc. The reason lies in that pate structures provide some advantages including light weight, high strength, and design flexibility. In practice, maintenance is needed to be performed by workers to guarantee the structure integrity for the safety reason, even though the structures are designed to tolerate some damage. It is known that one origin of structural damage is the impact exerted on the structures. Therefore, it is necessary to monitor the location and magnitude of the impact force (Wang & Deng, 1999). Such a monitoring process will be beneficial to the subsequent maintenance work of the structures (Haywood, Coverley, & Staszewski, 2005).

DOI: 10.4018/978-1-4666-8490-4.ch006

In the past few years, different techniques have been proposed to facilitate the damage localization of plate structures (Tibaduiza, Mujica, & Rodellar, 2013; Tibaduiza, Torres-Arredondo, & Mujica, 2013). For instance, Park, Sohn, & Olson (2012) developed an impact localization technique using scanning laser Doppler vibrometer (SLDV) and surface-mounted PZT transducers. The technique is able to locate the impact events simply by comparing an actual impact response with IRFs obtained from a grid of training points. The training data are collected by individually exciting PZT transducers and scanning the corresponding responses over the target area using the SLDV. Shin, Yang, & Lee (2013) proposed a group delay based location template matching (G-LTM) method for impact source localization on a plate. The method utilized a concept of the measure of similarity which is introduced by examining the phase structure of the fictitious frequency response function. Aiming at an on-line application, Yuan, Liu, & Qiu (2012) proposed a digital impact localization method targeting at finding a localized search area for further inspections for online monitoring of structural health, which is realized using a PZT sensor array. Qiu, Yuan, & Liu (2013) reported a digital sequence based impact area localization method combined with a de-noising method of the digital sequences dedicated to the task of on-line impact monitoring of large-scale composite structures in their whole service lifetime. By fusing the merits of triangulation method and the inverse analysis method, Liang, Yuan, & Liu (2013) presented a distributed coordination algorithm for improving the preciseness of real-time impact location of a composite structure. Ciampa & Meo (2010) reported a new in situ structural health monitoring method for locating an acoustic emission source in complex composite structures based on the differences of the stress waves gathered by six surface-attached acoustic emission piezoelectric sensors. Dehghan Niri, Farhidzadeh, & Salamone (2012) proposed an adaptive multisensor fusion algorithm for acoustic emission source location in isotropic plate-like structures in noisy environment. Based on the differences of the stress waves measured by six surface-attached acoustic emission piezoelectric sensors, Ciampa, Meo, & Barbieri (2012) proposed an in situ structural health monitoring method able to locate the impact source and to determine the flexural Lamb mode A_0 velocity in composite structures with unknown lay-up and cross section. Jeong & Cho (2013) presented a structural health monitoring technique for locating impact position in a composite plate, which employed a single sensor and spatial focusing properties of time reversal and inverse filtering. Baravelli, Senesi, & Ruzzene (2013) fabricated a wavenumber frequency-steerable acoustic transducer and applied it to record the plate response. The source location was then determined through a time-frequency analysis of the received signal by exploiting the frequency selective response of the transducer which directly maps the dominant component of the received signal to the direction of arrival of the incoming wave. Perelli, De Marchi, & Marzani (2014) presented a novel impact localization algorithm based on the frequency warping unitary operator applied to E-spline wavelet multiresolution analysis to identify the location of acoustic emission (AE) sources due to impacts. The method applies a dispersion compensation procedure on the signals acquired by passive sensors, thus overcoming the difficulties associated with arrival time detection based on classical thresholding procedures.

In addition, Liu, Qiu, & Chang (2010) reported an approach for acousto-ultrasonic source (passive damage or impact) localization based on metal core piezoelectric fiber rosettes. The localization procedure is based upon the computation of the principal direction of the wave strain. Coelho, Hiche, & Chattopadhyay (2010) developed a novel methodology for impact localization and prediction of loading as a function of time. The localization method proposed is based on the maximum strain amplitude measured by Fiber Bragg grating (FBG) sensors during impact and the relative placement of all sensors. A machine learning estimation approach based on support vector machine was applied to both simulated and experimental tests on a composite wing. Kirikera, Balogun, & Krishnaswamy (2011) described a passive structural health monitoring system for locating foreign-object impact using a network of FBG sensors which monitor high frequency dynamic strains. Using permanently attached sensors, Ward, Croxford, & Paget (2012) proposed the detection and location of impacts to a carbon fibre reinforced plastic panel with stringers. Instead of using deterministic knowledge of the propagation of Lamb waves in the structure, a statistical measure of the signal is used to determine the arrival time of elastic waves propagating in the structure as a result of the impact.

Generally, these methods can be classified into active and passive approaches. In view of the efficiency of the passive method, which is implemented using sensors attached to structures, it is the focus of this chapter. Data-driven techniques based on artificial intelligence (AI) have been widely employed in the literature (Mujica, Vehi, & Staszewskic, 2008). In particular, neural networks (NNs) have been popularly adopted in previous works to establish a regression model (Jones, Sirkis, & Friebele, 1997; Worden & Staszewski, 2000). However, it has been recognized that there is no universal method to determine an optimal NN structure in terms of the number of hidden layers and number of neurons in each layer (LeClerc, Worden, & Staszewski, 2007). Furthermore, NN exhibits some shortcomings including overfitting and sinking into local optima.

Alternatively, support vector machine (SVM) is a promising tool to localize the impact on structures accurately (Suykens, 2001; Wong, Xu, & Vong, 2012). As compared with NN, SVM owns the major advantages of global optimization and higher generalization capability. Moreover, least squares support vector machine (LSSVM) simplifies the traditional SVM by introducing equality constraints instead of inequality constraints (Xu & Wong, 2011). The potentials of SVM and LSSVM in impact localization of plate structures have been demonstrated in recent work (Fu & Xu, 2013; Xu, 2014).

Concerning an AI-based data-driven algorithm, the most practical issue is the learning time. Generally, a typical NN requires considerable learning time to generate an accurate regression model, which blocks its real applications in rapid implementation scenarios. To overcome this issue, extreme learning machine (ELM) has been recently proposed by Huang, Zhu, and Siew (2006) to realize an extremely fast learning. ELM is a learning algorithm for single-hidden layer feedforward neural network. In theory, this algorithm is able to provide good generalization capability at extremely fast learning speed. Comparative studies of benchmark function approximation problems reveal that ELM can learn thousands of times faster than conventional NN and can produce good generalization performance in most cases.

After its proposal, ELM has been widely applied to solve different engineering problems. However, the research on damage localization using ELM is rare to find in the literature. In this chapter, the ELM is extended to the domain of damage localization of plate structures. Its effectiveness in comparison with typical AI algorithms, i.e., back-propagation neural network (BPNN) and LSSVM, are illustrated through experimental studies. Comparative investigations in terms of learning time and localization accuracy have been carried out in detail. Results show that ELM paves a new way in the domain of plate structure health monitoring. Moreover, it can be expanded into other fields as well.

ELM MODELING

Given N arbitrary samples $\left(x_i, t_i\right)$ with $x_i \in R^n$ and $t_i \in R^m$, a standard single-layer feedforward neural network with L hidden neurons and activation function $g(x)$ can be mathematically described as:

$$\sum_{i=1}^{L} \beta_i g\left(w_i x_j + b_i\right) = d_j,\ j = 1,\ldots,N \tag{1}$$

where w_i is the weight vector connecting inputs and the ith hidden neurons, β_i is the weight vector connecting the ith hidden neurons and output neurons, b_i is the threshold of the ith hidden node, and d_j is the ELM output for the jth data point.

Ideally, this network can approximate these N samples with zero error $\sum_{j=1}^{L}\left\|d_j - t_j\right\| = 0$, that is, there exists β_i, w_i, and b_i to satisfy the condition:

$$\sum_{i=1}^{L} \beta_i g\left(w_i x_j + b_i\right) = t_j, j = 1,\ldots,N \tag{2}$$

The above N equations can be expressed into a compact form:

$$H\beta = T \tag{3}$$

in which

$$H = \begin{bmatrix} h(x_1) \\ \vdots \\ h(x_N) \end{bmatrix} = \begin{bmatrix} g(w_1 x_1 + b_1) & \cdots & g(w_L x_1 + b_L) \\ \vdots & \cdots & \vdots \\ g(w_1 x_N + b_1) & \cdots & g(w_L x_N + b_L) \end{bmatrix}_{N\times L} \tag{4}$$

$$\beta = \begin{bmatrix} \beta_1^T \\ \vdots \\ \beta_L^T \end{bmatrix}_{L\times m} \quad and\ T = \begin{bmatrix} t_1^T \\ \vdots \\ t_L^T \end{bmatrix}_{N\times m} \tag{5}$$

Within the ELM, the hidden node parameters (w_i, b_i) remain fixed once generated randomly. In order to train the network, it is simply equivalent to find a least-square solution $\hat{\beta}$ of the linear system (3). That is,

$$\left\|H\hat{\beta} - T\right\| = \min_{\beta}\left\|H\beta - T\right\| \tag{6}$$

The smallest norm least-square solution to (3) can be obtained as:

$$\hat{\beta} = H^{\dagger}T \tag{7}$$

where $H^{\dagger}$ is the Moore-Penrose generalized inverse of the matrix H.

Therefore, once given a training set

$$\left\{\left(x_i, t_i\right) \mid x_i \in R^n, t_i \in R^m, i = 1,\ldots,N\right\},$$

the training process of the basic ELM algorithm is very straightforward.

However, in the case that when the hidden layer feature mapping $g(x)$ is unknown to user, a kernel matrix of ELM can be defined as follows.

$$\Omega_{ELM} = HH^T : \Omega_{ELM\, i,j} = h\left(x_i\right)h\left(x_j\right) = K\left(x_i, x_j\right) \tag{8}$$

Then, the output function of ELM becomes

$$f(x) = h(x)H^T\left(\frac{I}{\lambda} + HH^T\right)^{-1} T = \begin{bmatrix} K\left(x, x_1\right) \\ \vdots \\ K\left(x, x_N\right) \end{bmatrix}^T \left(\frac{I}{\lambda} + \Omega_{ELM}\right)^{-1} T \tag{9}$$

where a positive term $\frac{I}{\lambda}$ is added to the diagonal HH^T of the Moore-Penrose generalized inverse H, to improve the stability of the learning (Huang, Wang, & Lan, 2011). In this kernel-based ELM, the hidden layer feature mapping $h(x)$ needs not to be known to user. Instead, the kernel $K(u,v) = \exp\left(-\gamma \left\| u - v \right\|^2\right)$ is given to user, where γ is the kernel parameter. Additionally, the number of hidden nodes L needs not be specified by the user.

Thus, once given a training set $\left\{\left(x_i, t_i\right) \middle| x_i \in R^n, t_i \in R^m, i = 1, \ldots, N\right\}$, the kernel ELM algorithm is further simplified as a single learning step implementation as follows (Huang, Wang, & Lan, 2011).

Step 1: Calculate the output function.

$$f(x) = \begin{bmatrix} K\left(x, x_1\right) \\ \vdots \\ K\left(x, x_N\right) \end{bmatrix}^T \left(\frac{I}{\lambda} + \Omega_{ELM}\right)^{-1} T \tag{10}$$

Owing to its simplicity, the kernel ELM algorithm is employed in this chapter for the impact localization of a plate structure.

IMPACT LOCALIZATION STRATEGY

When a structure plate is impacted using kinetic energy, propagating waves are induced, which can be detected by the surface-mounted piezoelectric sensors. The essence of impact localization is to detect these induced propagating wave signals and to process the signals to determine the location of the impact.

Data Acquisition

In this work, the impact experiments were carried out on an aluminum alloy (Al-6061) plate. The plate structure with sensors is sketched in Figure 1. The plate has a dimension of 490 mm x 390 mm x 2.5

Figure 1. Schematic of the plate structure with four surface-mounted PZT sensors

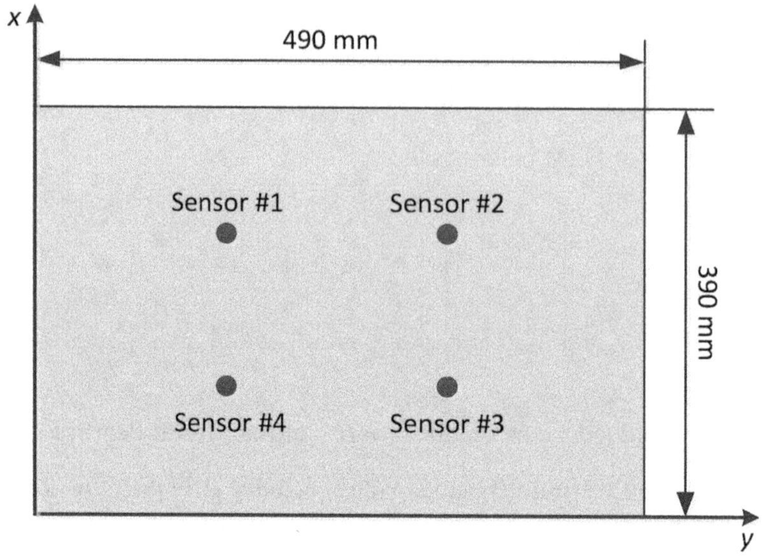

mm and is simply clamped on a table using four fixing screws. Four piezoelectric (PZT) sensors (material: Navy Type II) of 6.35 mm diameter and 0.191 mm thickness are bonded on the surface of the plate using conductive adhesive. As illustrated in Figure 1, the four transducers are mounted on the *xy* plane with the coordinates of (170 mm, 290 mm), (320 mm, 290 mm), (320 mm, 100 mm), and (170 mm, 100 mm), respectively. The impacts are applied on the plate surface using an instrumented impulse force hammer (from PCB Piezotronics, Inc.). The magnitudes of the impact forces are kept small in order not to cause real damage to the structure.

When an impact is exerted by the hammer, the analog voltage output signals of the piezoelectric sensors and the force output signal of the hammer are acquired by an NI PCI-4472 board, which provides eight input channels with 24-bit resolution. The data acquisitions of the four piezoelectric sensor channels are conducted using a sampling rate of 100 kHz, and 10,000 data are recorded for each of the four channels. The data acquisition starts when the amplitude of the force sensor output exceeds a predefined threshold.

Data Processing

However, the overall amount of the acquired strain data is too large to be used directly. Hence, feature extraction is necessary for practical applications. It has been shown that different features can be extracted from the original data. It has been shown that the time after impact of maximum response and the magnitude of maximum response are the best among the known features (Worden & Staszewski, 2000). Hence, these two features are employed in this work.

To extract these two features from each sensor output, the original sensor signals are processed to obtain their envelopes. In particular, the Hilbert transform is used to create an analytic signal. Then, the envelope of the signal is obtained. More details about the algorithm can be found in the literature (Quek, Tua, & Wang, 2003).

To generate better localization accuracy, it is suggested to normalize the input data into the range between −1 and 1 (Huang, Zhu, & Siew, 2006). However, in this research, preliminary testing reveals that the normalization to zero mean and unit variance produces a better prediction result for the impact localization problem. Hence, each of the eight signals s is normalized through the following transformation:

$$N(s) = \frac{s - \mu}{\sigma} \tag{11}$$

where μ and σ represent the mean and variance of the signal s, respectively.

Due to relatively small number of sensors and features, the signals are processed as conducted above. More details about the signal processing procedure can be found in the literature (Quek, Tua, & Wang, 2003). It is worth mentioning that if the dimension of the data set is too high, a dimensionality reduction operation is more convenient for practical applications (Mujica, Vehi, & Ruiz, 2008).

EXPERIMENTAL INVESTIGATIONS

Experimental Setup

A photo of the experimental setup is shown in Figure 2, and the setup has been described in previous section. Through experimental impact testing, two sets of impact data were acquired with the aforementioned experimental setup. The first set includes a regular grid of 63 impacts, as shown in Figure 3. It is used for the training of the ELM model. Another set of 90 random impacts, as shown in Figure 4, is employing for the model testing.

In order to quantify the localization accuracy, an error function is employed to provide a measure of the accuracy. Specifically, to visualize the results, the averaged errors of x and y coordinates are calculated, and the ratio between the product of them and the plate area is considered as the error evaluation index.

$$E = \frac{\sum_{i=1}^{N} \left(x - x_i\right) \sum_{i=1}^{N} \left(y - y_i\right)}{N^2 A} \tag{12}$$

Figure 2. Experimental setup of a plate structure for impact testing

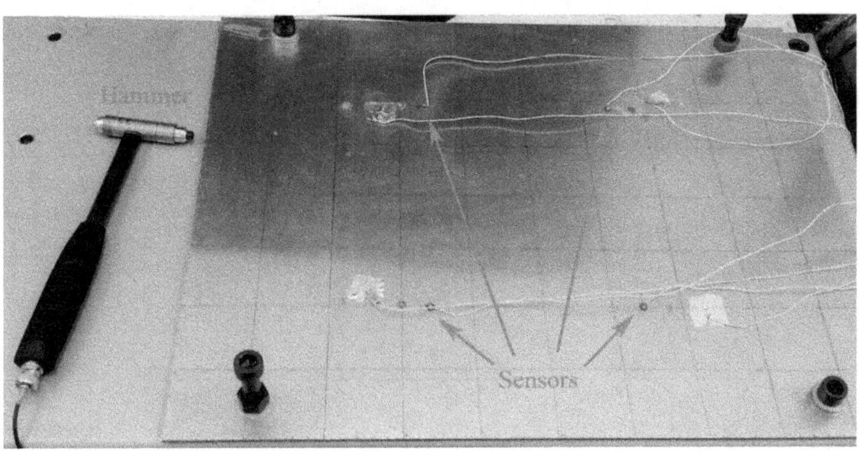

Figure 3. Impact locations of the training set in xy plane

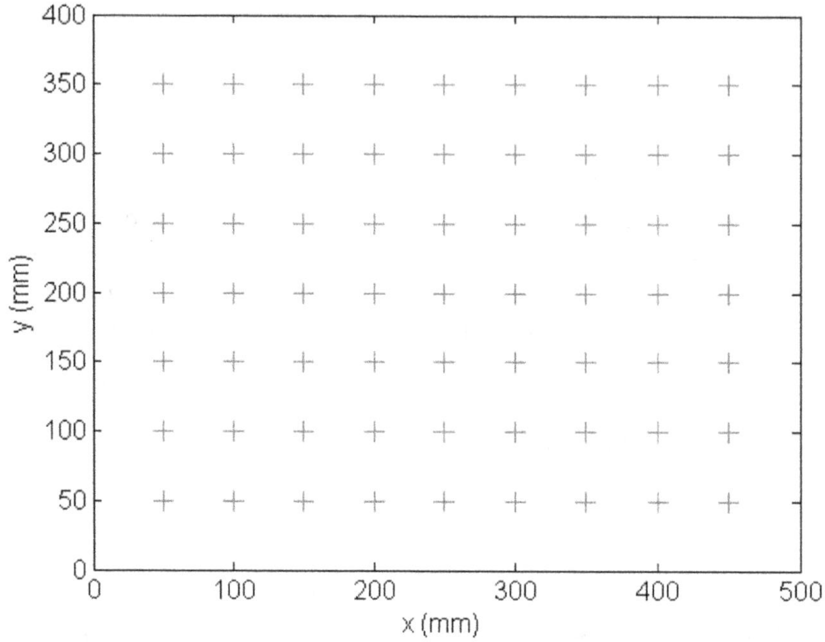

Figure 4. Impact locations of the testing set in xy plane

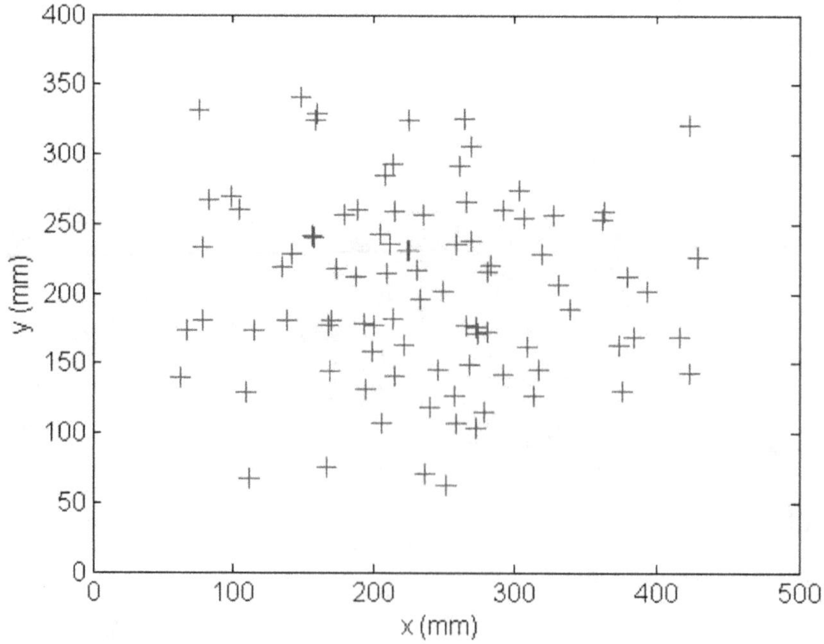

where A denotes the area of the plate structure, and E is the ratio between the error area and the plate area. The lower the index is, the better the detection accuracy for the impact location will be. However, in order to better describe the situation when the error along one axis is small, other evaluation index may be needed. An investigation of the evaluation indices will be conducted in the future work.

BPNN Model Results

For a comparative study, a BPNN is constructed and trained due to its popularity. It is well known that BPNN owns universal approximation capabilities. It has been shown that a three-layer perceptron NN with one hidden layer is capable of representing any continuous function (Negnevitsky, 2005). Therefore, the BPNN of three-layer structure is employed in the current research. Additionally, the popular Levenberg–Marquardt training algorithm is adopted. In consideration of the input and output variables, one BPNN model can be developed to estimate the two coordinates of the impact location using eight neurons in the input layer and two neurons in the output layer, respectively. Even so, to reduce the complexity of the network and to illustrate a fair comparison with LSSVM and ELM, two BPNN models are established for the two outputs. Besides, the transfer functions for the hidden and output layers are selected as the tangent-sigmoid and linear functions, respectively.

Concerning the selection of optimal number of neurons for the hidden layer, there is little guidance available. Hence, a preliminary search has been performed to determine an optimal number of neurons in hidden layer. By varying the number of hidden neurons between 1 and 30, results show that 15 hidden neurons produce better performance in this study. The first 63 sets of impact data are used to train each of the two BPNN models. During the training procedure, the data sets are divided into three subsets of 60%, 20%, and 20%, which are used for the purposes of training, validation, and testing, respectively. Two stop criteria are set for the training process, that is, the training is stopped when the average squared error between network outputs and target outputs is lower than 10^{-6} or when the number of iterations exceeds 2000. Moreover, in order to remove the overfitting and improve the network generalization ability, an early stopping technique is applied by monitoring the validation error. That is, the training is stopped when the validation error increases for a number of five consecutive iterations, and the weights and biases at the minimum of the validation error are returned.

It is known that the result of NN can heavily rely on the initial values assigned to the weights (Haywood, Coverley, & Staszewski, 2005). To examine the effects of different initial conditions on the NN performance, the two BPNN models are trained by multiple times. Moreover, in order to obtain more accurate results, the NN models have been carried out 30 times to conduct a large-sample statistical analysis. Before each training process, the weight values of the NN are initialized randomly. The result of the best trained BPNN is shown in Figure 5, and the training errors are shown in Figure 6. The resulted mean errors of the coordinates x and y are 41.6 and 36.9 mm, respectively, which correspond to a percentage area error of 0.80%.

Once the BPNN is trained, the remaining 90 sets of impact data are used for generalization testing, and the testing results are depicted in Figure 7. The testing errors are illustrated in Figure 8. It is observed that the testing mean errors are 55.9 and 48.3 mm for the coordinates x and y, respectively, that is, 1.41% percentage area error.

Figure 5. BPNN training results of the impact locations

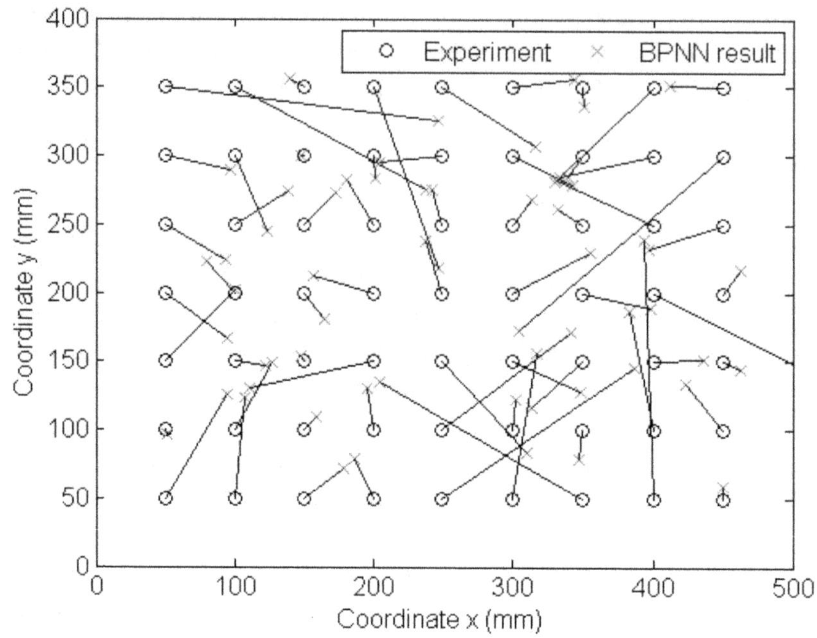

Figure 6. BPNN training errors of x and y coordinates

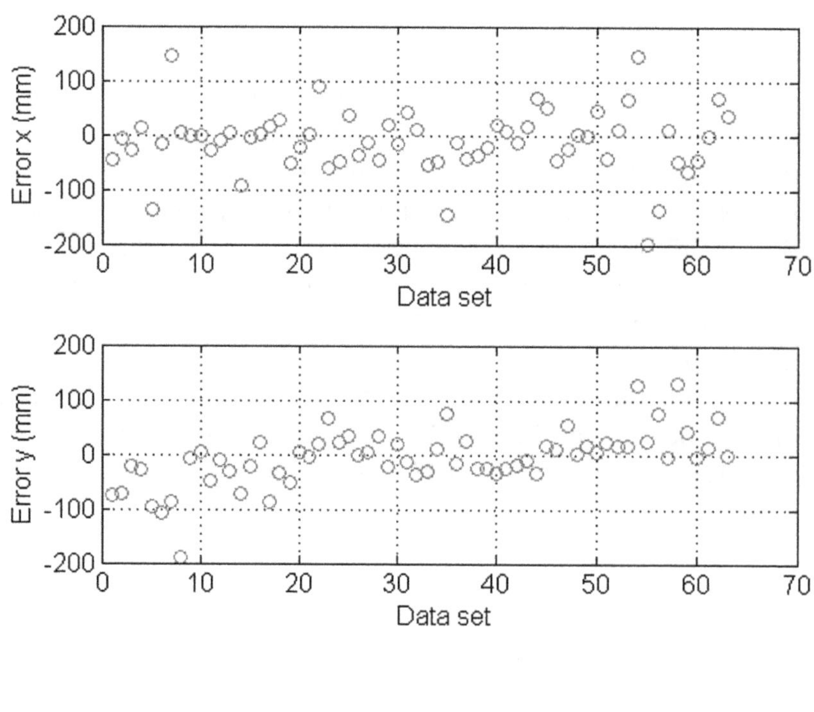

LSSVM Model Results

Concerning the LSSVM, the aforementioned 63 and 90 data sets are used for the purposes of training and testing, respectively. In view of the output ability of LSSVM (i.e., a single output), two LSSVM models are trained to detect the location coordinates *x* and *y*, respectively. More details about the LSSVM model setup can be found in the literature (Xu, 2014).

To illustrate the influences of the initial conditions on LSSVM performance, the training procedures of the three LSSVM models are conducted by 30 times. Results show that majority of the training results give *E* around 0.183%, indicating the effects of the different initial conditions. For the 30 times of LSSVM training, the average training time is less than 0.15 s.

Since the best performance of the algorithms is more concerned, only the results of best trained LSSVM models are presented. The best trained LSSVM models produce the results of the locations as shown in Figure 9. The training errors of the two coordinates are shown in Figure 10. It is found that the LSSVM delivers the mean errors of 24.8 and 0.22 mm for *x* and *y* coordinates, respectively, which are equivalent to a very small percentage area error. It is found that the *y*-axis error is smaller than the *x*-axis error. The reason is that the training data are obtained by arranging the grid points row by row (see Figure 3). By this way, the *y*-axis position is an increased step function, which is simpler than the *x*-axis position function. Hence, a smaller error is obtained for the training of *y*-axis position.

Concerning the best trained LSSVM models, their generalization testing results are shown in Figure 11, and the testing errors are plotted in Figure 12. It is observed that the mean error of coordinates x and y are 39.4 and 35.7 mm, respectively, which correspond to 0.74% percentage area error with respect to the overall plate size.

Figure 7. BPNN testing results of the impact locations

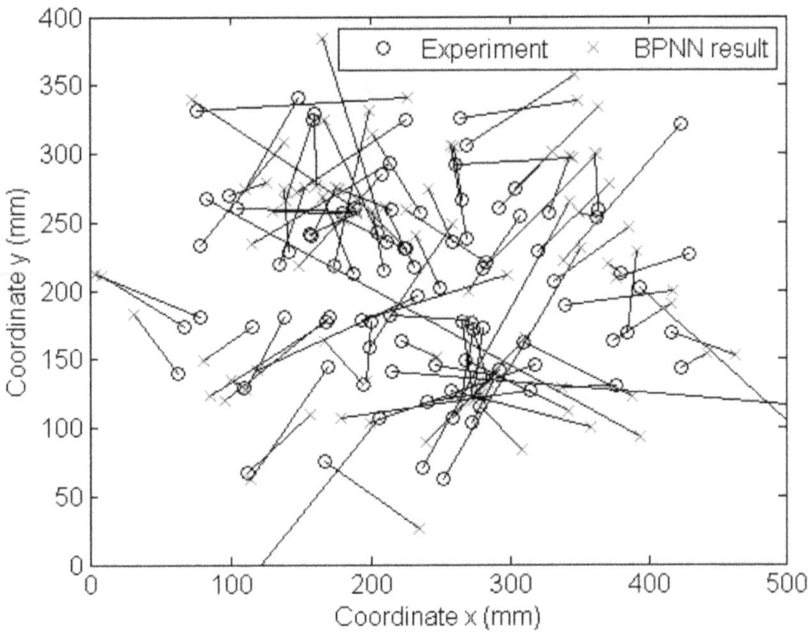

Figure 8. BPNN testing results of x and y coordinates

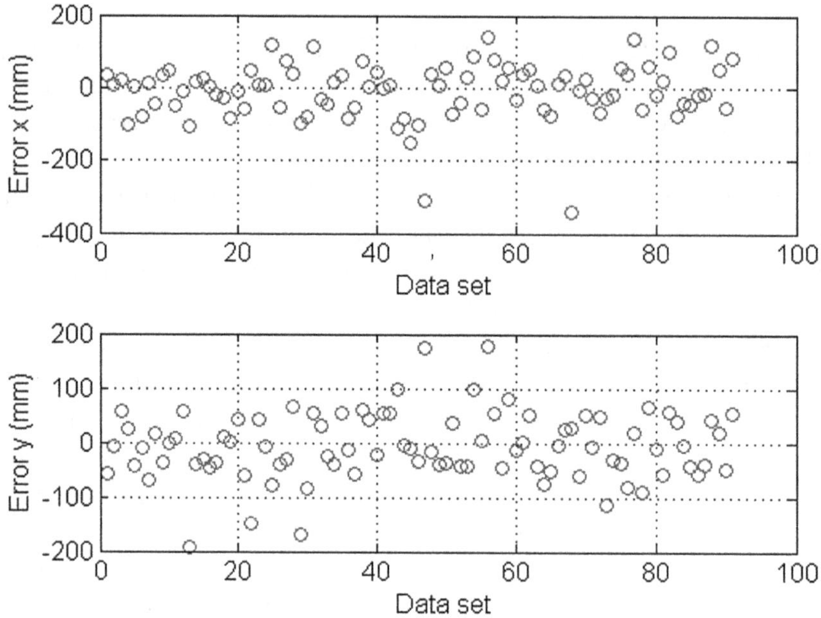

Figure 9. LSSVM training results of the impact locations

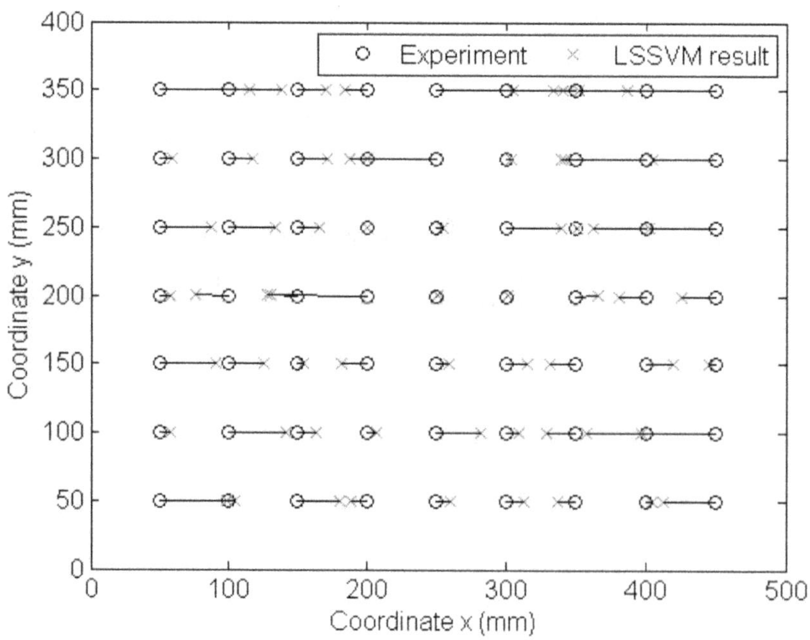

Figure 10. LSSVM training errors of x and y coordinates

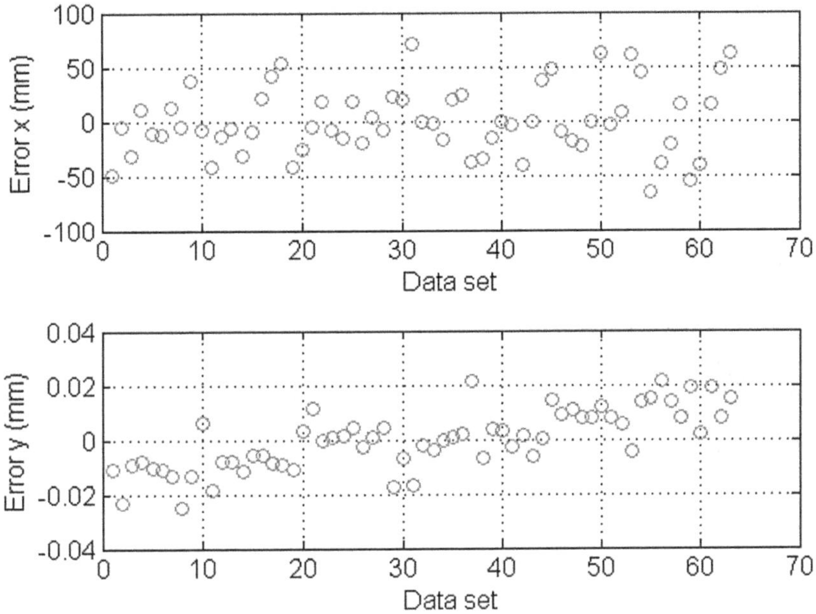

Figure 11. LSSVM testing results of the impact locations

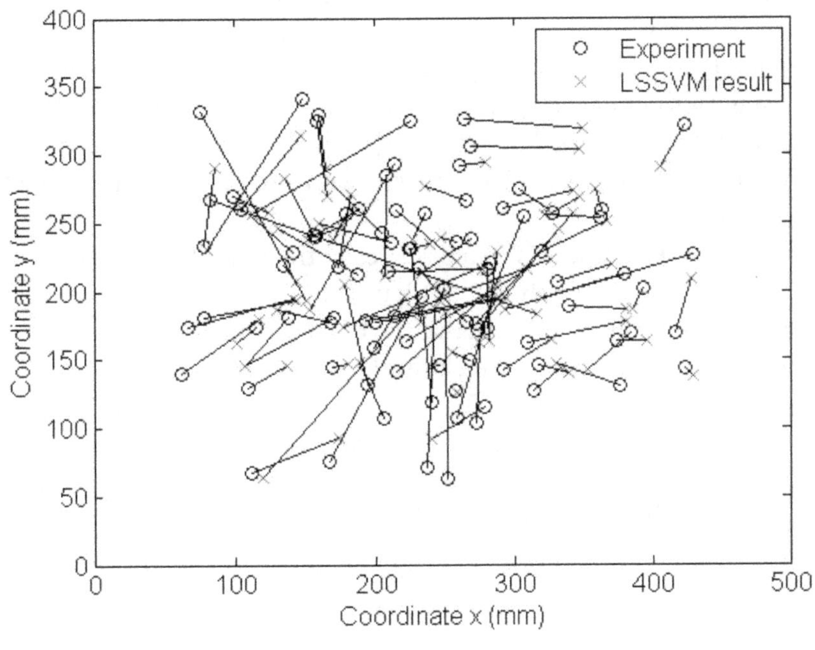

Figure 12. LSSVM testing results of x and y coordinates

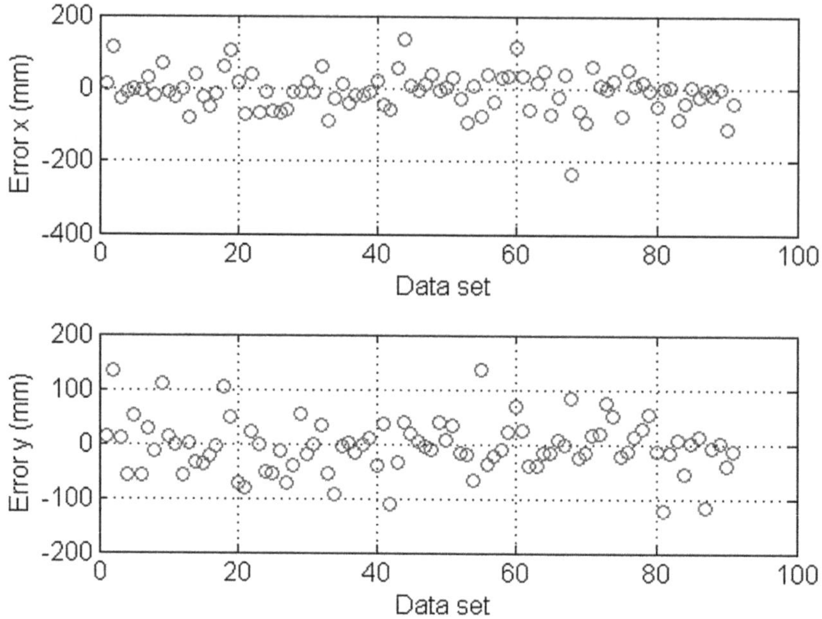

ELM Model Results

As for the ELM model, the radial basis function (RBF) kernel is selected owing to its popularity. Two kernel ELM models are developed for the detection of x and y coordinates for the impact. By selecting the regularization coefficient $\lambda = 1$, preliminary results indicate that the training and testing performances of ELM models are dependent on the RBF parameters.

Specifically, as the RBF parameter increases, the training and testing accuracies for the x and y coordinates are shown in Figures 13 and 14, respectively. It is seen that the training errors increase as the rising of the RBF parameter. Whereas the testing errors arrive at the minimum values when the RBF parameters are 65 and 20 for x and y coordinates, respectively. Thus, to produce better testing results, the RBF parameters of the two ELM models are selected as 65 and 20 for the prediction of x and y coordinates, respectively.

By running each model 30 times, it is found that each time leads to the same result. The training results of the two ELM models are illustrated in Figure 15. In addition, the training errors of the two coordinates are shown in Figure 16. It is found that the ELM produces the mean errors of 62.7 and 61.4 mm for x and y coordinates, respectively, which results in a training error index of 2.02%. Using the two trained ELM models, the testing results for the two coordinates are obtained, as shown in Figure 17. The corresponding testing errors of the x and y coordinate predictions are shown in Figure 18. The ELM models produce the mean errors of 37.0 and 40.0 mm for the x and y coordinates, respectively, which is equivalent to the testing error index of 0.77%.

Figure 13. ELM training and testing accuracy of coordinate x versus RBF parameter

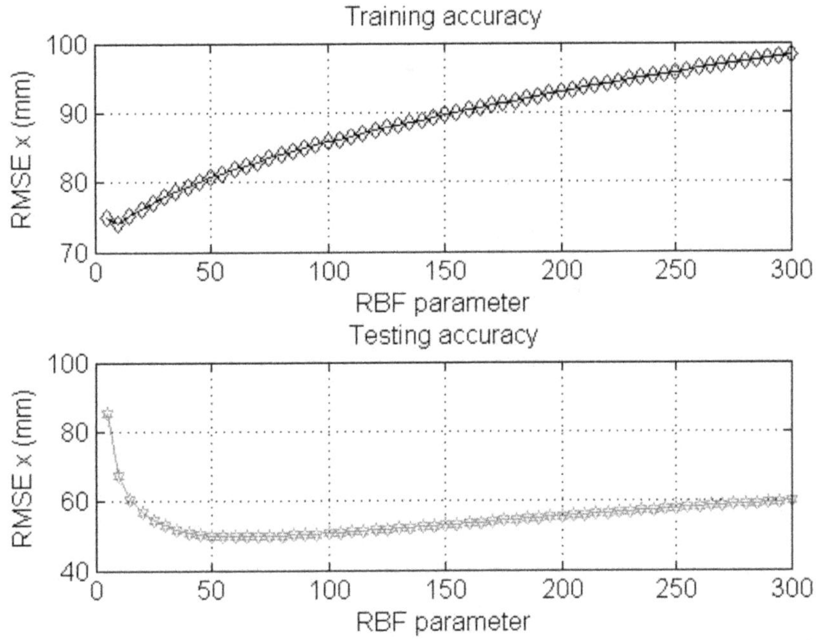

Figure 14. ELM training and testing accuracy of coordinate y versus RBF parameter

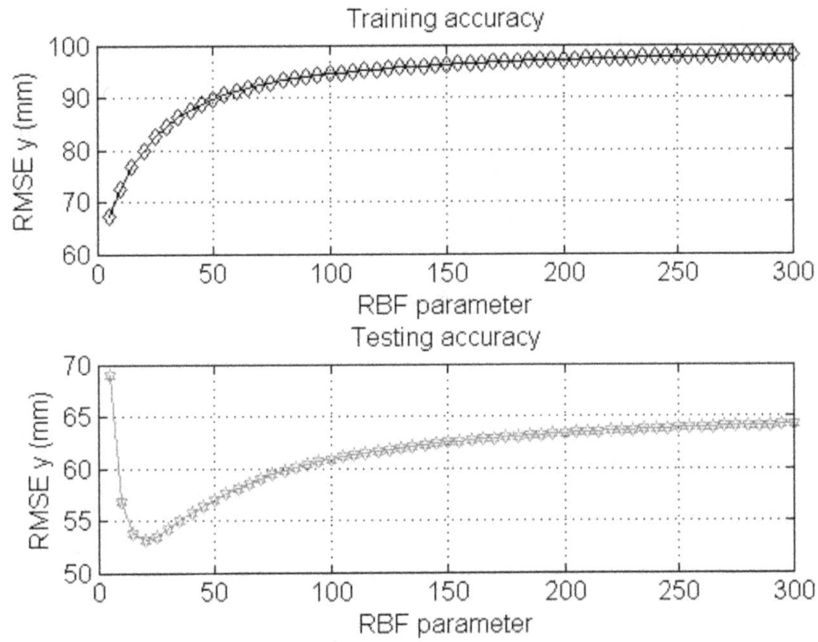

Figure 15. ELM training results of the impact locations

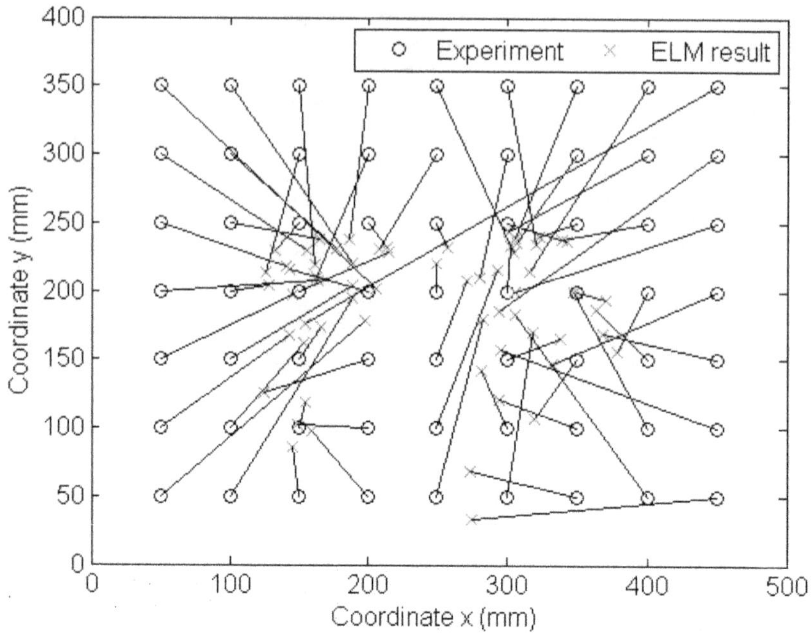

Figure 16. ELM training errors of x and y coordinates

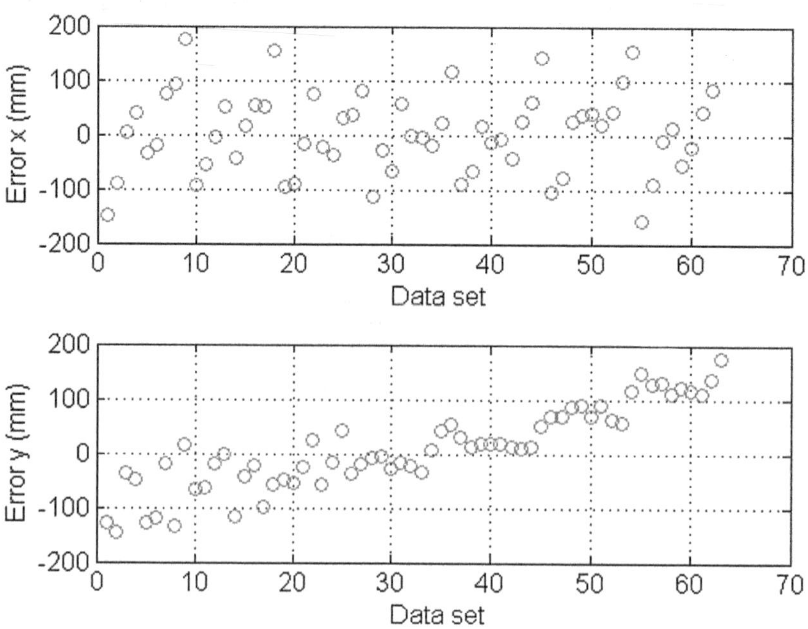

Figure 17. ELM testing results of the impact locations

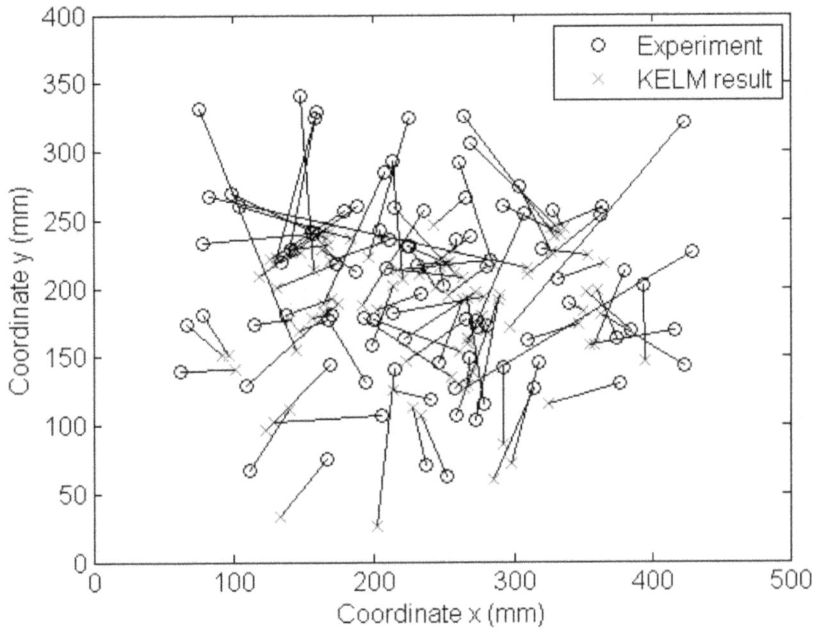

Discussion

For a clear comparison of the detection performance of the BPNN, LSSVM, and ELM models, the results are summarized in Table 1. The training time means the time required by each model for the model training for a single coordinate prediction. The testing time is the needed execution time by each model to predict a single coordinate of the impact.

By comparing the training and testing time, it is evident that the ELM requires the minimum time for both training and testing processes. The BPNN needs the maximum time for the model training. In addition, BPNN calls for the longest execution time for the impact localization.

In addition, given the training samples, it is deduced from Table 1 that the LSSVM is able to produce the minimum training error, which is equal to zero approximately. Instead, the ELM creates the maximum training error. Therefore, the LSSVM provides an excellent function approximation capability.

Moreover, from the testing accuracy of the three models, it is found that the LSSVM is able to deliver the most accurate localization result (0.7358%). It is further seen that the BPNN gives the lowest accuracy of the impact localization although it produces the lower training error than LSSVM. By comparing the testing accuracy of ELM and LSSVM, it is seen that the ELM localization error is only 5.3% higher than that of LSSVM.

In summary, the ELM produces the sub-optimal localization accuracy among the three types of intelligent models. It requires the shortest training and testing execution time. Therefore, the ELM is more suitable for the situations where the execution time is the most critical requirement, whereas the LSSVM is more appropriate to the scenarios when the localization accuracy is the most desirable performance.

Figure 18. ELM testing errors of x and y coordinates

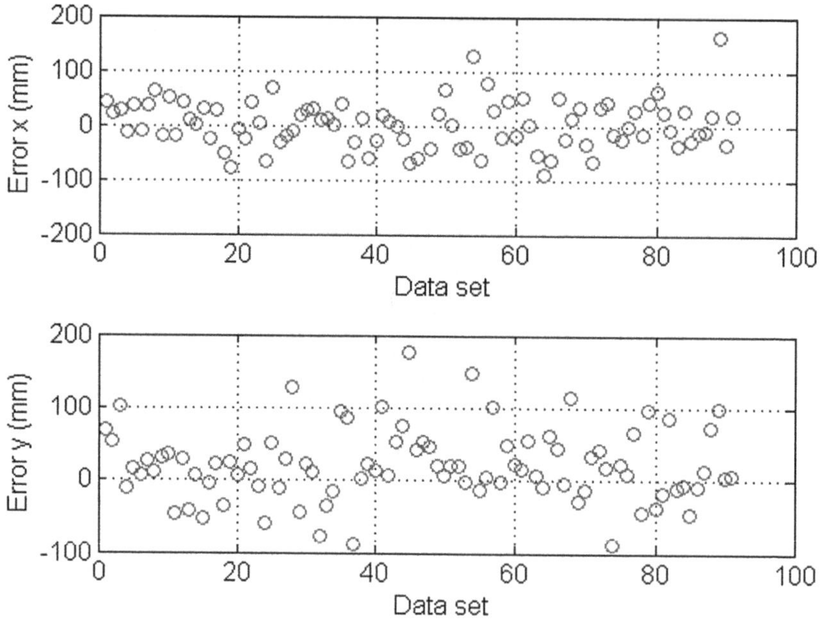

Table 1. Comparison of BPNN, LSSVM, and ELM localization performance

Performance	BPNN	LSSVM	ELM
Training time (s)	0.4109	0.1438	0.0134
Training error (%)	0.8029	0.0000	2.0154
Testing time (s)	0.0280	0.0098	0.0007
Testing error (%)	1.4137	0.7358	0.7746

CONCLUSION

This chapter presents a comparison study of BPNN, LSSVM, and ELM dedicated to impact localization of a plate structure with surface-mounted PZT sensors. By extracting useful features from each sensor output using Hilbert transform, regression models are established to predict the location of impact as exposed by a hammer. The kernel-based ELM algorithm is employed for a comparison with the BPNN and LSSVM approaches.

Experimental results reveal that the LSSVM is able to provide a nice function approximation capability. The ELM requires the minimum execution time for both training and testing processes. The execution time of ELM is only 1/40 and 1/14 of that of BPNN and LSSVM, respectively. In addition, LSSVM is capable of producing the most accurate localization result. ELM delivers the sub-optimal localization accuracy, which is 5.3% worse than that of LSSVM. Therefore, the ELM is more suitable for the scenarios where the execution time is the most critical requirement, whereas the LSSVM is more appropriate to the situations when the localization accuracy is most desired.

In the future, more advanced ELM algorithm will be explored and developed to further improve the localization accuracy while maintaining the shortest execution time. Moreover, its application in real impact localization of aircraft structures will be carried out in the future work.

ACKNOWLEDGMENT

The work was support by the Macao Science and Technology Development Fund under Grant Nos.: 070/2012/A3 and 052/2014/A1 and the Research Committee of the University of Macau under Grant Nos.: MYRG083(L1-Y2)-FST12-XQS and MYRG078(Y1-L2)-FST13-XQS.

REFERENCES

Baravelli, E., Senesi, M., Ruzzene, M., & De Marchi, L. (2013). Fabrication and Characterization of a Wavenumber-Spiral Frequency-Steerable Acoustic Transducer for Source Localization in Plate Structures. *IEEE Transactions on Instrumentation and Measurement*, *62*(8), 2197–2204. doi:10.1109/TIM.2013.2255992

Ciampa, F., & Meo, M. (2010). A new algorithm for acoustic emission localization and flexural group velocity determination in anisotropic structures. *Composites. Part A, Applied Science and Manufacturing*, *41*(12), 1777–1786. doi:10.1016/j.compositesa.2010.08.013

Ciampa, F., Meo, M., & Barbieri, E. (2012). Impact localization in composite structures of arbitrary cross section. *Structural Health Monitoring*, *11*(6), 643–655. doi:10.1177/1475921712451951

Coelho, C. K., Hiche, C., & Chattopadhyay, A. (2010). Impact Localization and Force Estimation on a Composite Wing using Fiber Bragg Gratings Sensors. In *Proceedings of 51st AIAA/ASME/ASCE/AHS/ASC Structures, Structural Dynamics, and Materials Conference*. AIAA.

Dehghan Niri, E., Farhidzadeh, A., & Salamone, S. (2012). Adaptive multisensor data fusion for acoustic emission source localization in noisy environment. *Structural Health Monitoring*, *12*(1), 59–77. doi:10.1177/1475921712462937

Fu, H., & Xu, Q. (2013). Locating impact on structural plate using principal component analysis and support vector machines. *Mathematical Problems in Engineering*, article ID 352149. doi: .10.1155/2013/352149

Haywood, J., Coverley, P. T., Staszewski, W. J., & Worden, K. (2005). An automatic impactmonitor for a composite panel employing smart sensor technology. *Smart Materials and Structures*, *14*(1), 265–271. doi:10.1088/0964-1726/14/1/027

Huang, G.-B., Wang, D. H., & Lan, Y. (2011). Extreme learning machine: A survey. *International Journal of Machine Learning and Cybernetics*, *2*(2), 107–122. doi:10.1007/s13042-011-0019-y

Huang, G.-B., Zhu, Q.-Y., & Siew, C.-K. (2006). Extreme learning machine: Theory and applications. *Neurocomputing*, *70*(1-3), 489–501. doi:10.1016/j.neucom.2005.12.126

Jeong, H., & Cho, S. (2013). Impact source location of composites using a single sensor and time reversal technique. In *AIP Conf. Proc.* (vol. 1511, pp. 246-253). doi:10.1063/1.4789055

Jones, R. T., Sirkis, J. S., & Friebele, E. J. (1997). Detection of impact location and magnitude for isotropic plates using neural networks. *Journal of Intelligent Material Systems and Structures*, *7*(1), 90–99. doi:10.1177/1045389X9700800109

Kirikera, G. R., Balogun, O., & Krishnaswamy, S. (2011). Adaptive Fiber Bragg Grating Sensor Network for Structural Health Monitoring: Applications to Impact Monitoring. *Structural Health Monitoring*, *10*(1), 5–12. doi:10.1177/1475921710365437

LeClerc, J. R., Worden, K., Staszewski, W. J., & Haywood, J. (2007). Impact detection in an aircraft composite panel—a neural network approach. *Journal of Sound and Vibration*, *299*(3), 672–682. doi:10.1016/j.jsv.2006.07.019

Liang, D., Yuan, S. F., & Liu, M. L. (2013). Distributed coordination algorithm for impact location of preciseness and real-time on composite structures. *Measurement*, *46*(1), 527–536. doi:10.1016/j.measurement.2012.08.011

Liu, J., Qiu, J., Chang, W., Ji, H., & Zhu, K. (2010). Metal core piezoelectric ceramic fiber rosettes for acoustoultrasonic source localization in plate structures. *International Journal of Applied Electromagnetics and Mechanics*, *33*, 865–873.

Mujica, L. E., Vehi, J., Ruiz, M., Verleysen, M., Staszewski, W., & Worden, K. (2008). Multivariate statistics process control for dimensionality reduction in structural assessment. *Mechanical Systems and Signal Processing*, *22*(1), 155–171. doi:10.1016/j.ymssp.2007.05.001

Mujica, L. E., Vehi, J., Staszewskic, W., & Worden, K. (2008). Impact damage detection in aircraft composites using knowledge-based reasoning. *Structural Health Monitoring*, *7*(3), 215–230. doi:10.1177/1475921708090560

Negnevitsky, M. (2005). *Artificial intelligence: A guide to intelligent systems* (2nd ed.). London: Addison-Wesley.

Park, B., Sohn, H., Olson, S. E., DeSimio, M. P., Brown, K. S., & Derriso, M. M. (2012). Impact localization in complex structures using laser-based time reversal. *Structural Health Monitoring*, *11*(5), 577–588. doi:10.1177/1475921712449508

Perelli, A., De Marchi, L., Marzani, A., & Speciale, N. (2014). Frequency warped cross-wavelet multiresolution analysis of guided waves for impact localization. *Signal Processing*, *96*, 51–62. doi:10.1016/j.sigpro.2013.05.008

Qiu, L., Yuan, S., Liu, P., & Qian, W. (2013, July). Design of an All-Digital Impact Monitoring System for Large-Scale Composite Structures. *IEEE Transactions on Instrumentation and Measurement*, *62*(7), 1990–2002. doi:10.1109/TIM.2013.2247491

Quek, S. T., Tua, P. S., & Wang, Q. (2003). Detecting anomalies in beams and plate based on the Hilbert-Huang transform of real signals. *Smart Materials and Structures*, *12*(3), 447–460. doi:10.1088/0964-1726/12/3/316

Shin, K., Yang, H., Lee, S.-K., & Lee, Y.-S. (2013). Group delay based location template matching method for the identification of the impact location on a plate. *Journal of Sound and Vibration*, *332*(8), 2111–2117. doi:10.1016/j.jsv.2012.12.007

Suykens, J. A. K. (2001). Support vector machines: A nonlinear modeling and control perspective. *European Journal of Control*, *7*(2-3), 311–327. doi:10.3166/ejc.7.311-327

Tibaduiza, D. A., Mujica, L. E., & Rodellar, J. (2013). Damage classification in structural health monitoring using principal component analysis and self-organizing maps. *Structural Control and Health Monitoring*, *20*(10), 1303–1316. doi:10.1002/stc.1540

Tibaduiza, D. A., Torres-Arredondo, M.-A., Mujica, L. E., Rodellar, J., & Fritzen, C.-P. (2013). A study of two unsupervised data driven statistical methodologies for detecting and classifying damages in structural health monitoring. *Mechanical Systems and Signal Processing*, *41*(1-2), 467–484. doi:10.1016/j.ymssp.2013.05.020

Wang, Q., & Deng, X. (1999). Damage detection with spatial wavelets. *International Journal of Solids and Structures*, *36*(23), 3443–3468. doi:10.1016/S0020-7683(98)00152-8

Ward, J., Croxford, A., & Paget, C. (2012). Passive impact localisation for the structural health monitoring of new airframe materials. 11th Anglo-French Physical Acoustics Conference. *Journal of Physics: Conference Series*, *457*, 012010. doi:10.1088/1742-6596/457/1/012010

Wong, P. K., Xu, Q., Vong, C. M., & Wong, H. C. (2012). Rate-dependent hysteresis modeling and control of a piezostage using online support vector machine and relevance vector machine. *IEEE Transactions on Industrial Electronics*, *59*(4), 1988–2001. doi:10.1109/TIE.2011.2166235

Worden, K., & Staszewski, W. J. (2000). Impact location and quantification on a composite panel using neural networks and a genetic algorithm. *Strain*, *36*(2), 61–70. doi:10.1111/j.1475-1305.2000.tb01175.x

Xu, Q. (2014). Impact detection and location for a plate structure using least squares support vector machines. *Structural Health Monitoring*, *13*(1), 5–18. doi:10.1177/1475921713495083

Xu, Q., & Wong, P. K. (2011). Hysteresis modeling and compensation of a piezostage using least squares support vector machines. *Mechatronics*, *21*(7), 1239–1251. doi:10.1016/j.mechatronics.2011.08.006

Yuan, S., Liu, P., & Qiu, L. (2012). A miniaturized composite impact monitor and its evaluation research. *Sensors and Actuators. A, Physical*, *184*, 182–192. doi:10.1016/j.sna.2012.07.011

Chapter 7
Case-Based Reasoning for Stiffness Changes Detection in Structures:
Numerical Validation by using Finite Element Model

Rodolfo Villamizar Mejia
Universidad Industrial de Santander, Colombia

Jhonatan Camacho Navarro
Universidad Industrial de Santander, Colombia

Wilmer Alexis Sandoval Caceres
Universidad Industrial de Santander, Colombia

ABSTRACT

This chapter presents an expert monitoring algorithm approach to detect, locate and quantify stiffness variations in structures. The algorithm is based on pattern recognition and artificial intelligence techniques that emulate knowledge based on human reasoning. The expert system (ES) uses time-frequency information about dynamics of structure, which is processed by using discrete wavelet transform (DWT), self-organizing maps (SOM), case-based reasoning (CBR) and principal component analysis (PCA). In addition, two applications are considered in order to evaluate the effectiveness of vibration analysis methodology and CBR in damage detection. The first application (Camacho 2010) uses the environmental excitation to detect and quantify damage in a Mechanical UBC ASCE Benchmark. The second one (Sandoval 2010) uses a predesigned signal to detect geometric damages on a gas pipeline. In both cases, a finite element model (FEM) is used to simulate different damages scenarios, which correspond to stiffness variations in different location.

DOI: 10.4018/978-1-4666-8490-4.ch007

INTRODUCTION

Condition monitoring systems (CMS) can be described as the process of implementing a structural damage identification algorithm, which uses structural dynamical measurements in order to determine a diagnosis about the current structural health (Mujica, Vehí, & Rodellar 2005). Some of the benefits acquired by implementing fault detection systems include the avoidance of premature breakdowns, reduction of maintenance costs, continuous remote diagnosis and economic benefits (Carden & Fanning, 2004) in terms of an operational life extension, among others. Many CMS approaches based on vibration analysis have been reported in literature (Dyke & Caicedo, 2000), where a high tolerance in the diagnosis is one of the main objectives. However, few methodologies with the capability for damage quantification have been reported.

In this chapter, the capability for damage detection, location and quantification, is demonstrated by using a case-based reasoning methodology (Camacho, & Villamizar, 2010). Vibration measurements of structures are processed by applying wavelet transform analysis and cross correlation functions (Sandoval, 2010). Principal Component Analysis is used to obtain features sensible to damage condition and a Self-Organizing Map is implemented in order to organize and classify the assessment of the structural current state (Mujica, 2006). The methodology is validated by using two numerical cases: a metallic framework structure (Camacho, 2010) and a pipe (Sandoval, 2010), which have been modeled by using the finite element technique (Inman 2001). The methodology can be extended to other applications as for example aerospace, aircraft industries and civil engineering (Mujica, 2006).

BACKGROUND

The methodology described in this chapter emphasized the advantages of using vibrational based damage detection algorithms. Vibration based condition monitoring refers to the use of in situ non-destructive sensing and analysis of system characteristics –in the time, frequency or modal domains –for the purpose of detecting changes, which may indicate damage or degradation (Carden & Fanning, 2004). Vibration based damage assessment has been reported as successful for a great variety of applications, which include structures as spacecraft components (Mujica, 2006), bridges (Villamizar, 2005), aircrafts, beams (Begambre, 2004), offshore structures and railroads, among others (Yolken & Matzkanin, 2008). The damage diagnosis is achieved by configuring in a proper way three major components (Li et. al., 2014): Sensor system, a data processing system and a health evaluation system. The reported literature can be divided into three categories (Sinou, 2009): modal testing, vibration-based methods with signal-based and model-based methods, and non-traditional methods.

Modal testing refers to methods where damage assessment is based on identifying changes in estimated modal parameters such as frequencies and modes shapes (Hatch, 2001; Fan & Qiao, 2011). As example, the effectiveness of damage identification methods based on natural excitation Technique (NeXT) using ambient vibration it is demonstrated (Caicedo, 2011).

In signal-based methods, different statistical indexes of the time structure response to a specific excitation (vibration - forced) are used to distinguish between damage and undamaged states. Some of the applications involve experimental testing of structural fault detection algorithms based on features from principal component analysis and piezoelectric active systems, which are validated on pipeline structures, laminate plates, aircraft sections and composite materials, among others (Mujica et al., 2010).

Figure 1. Case based reasoning for stiffness recognition
Adapted from (Mujica, 2005).

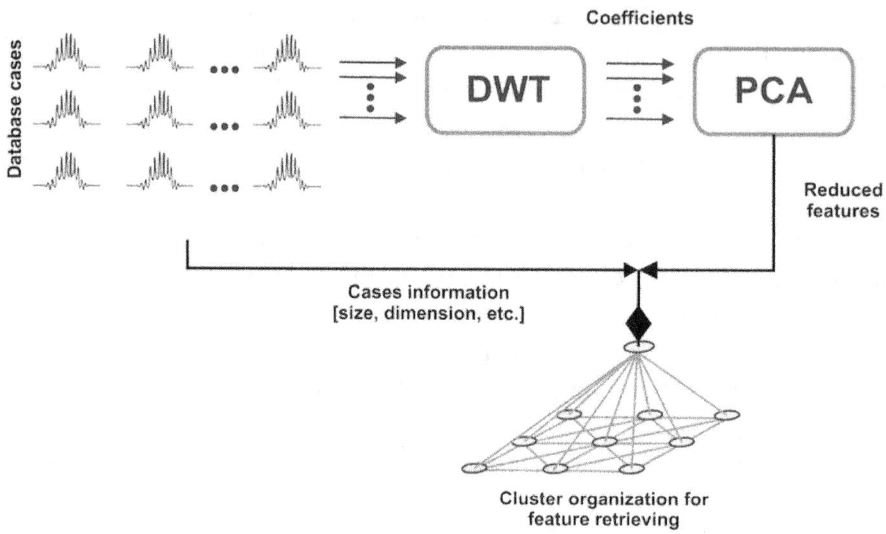

Model-based methods propose to correlate experimental damaged/undamaged measurements with analytical or numerical models in order to identify the current state of the structure. For instance, a finite element (FE) model updating procedure is used to identify damages of a 5-story framed structure (Yu et. al., 2014).

The non-traditional methods are based on applying pattern recognition algorithms such a genetic algorithms (Chou & Ghaboussi, 2001), neural networks (Sandoval, 2010; Camacho, 2010; Lin & Qun, 2013), and/or digital signal processing techniques such a wavelet analysis (Taha et. al, 2006) or Wigner-Ville transforms (Chen et. al., 2014) in order to detect damages in structures. As example, artificial neural networks have been used as solution for structural design, material behavior modeling, structural damage assessment, and structural analysis (Berrais, 1999).

Since non-traditional methods outperform algorithms based on modal parameters (Yan et al., 2007) for environmental conditions treatment, sensitivity to low scale damages and multiple scenario damage classification, in this chapter it is described a methodology for damage detection, location and quantification based on non-traditional techniques. Figure 1 shows the general scheme to identify damages using case based reasoning (CBR) (Begambre, 2004; Mujica, 2005; Camacho, &Villamizar 2010).

The steps in order to identify critical changes in a structure are the following:

1. **Case Database Construction:** It is a memory array which contains full dynamic simulations of the model structure. They describe damage patterns and facilitate the search of similar damages. When a new record is sensed it is possible to classify its severity using information previously storage. The similarity measure is commonly Euclidean distance. In Figure 1, the SOM evaluate this similarity.
2. Apply Discrete Wavelet Transform to time signals of acceleration sensors located in some elements of the structure. The resulting detail and approximation wavelet coefficients are the feature vector.

3. Reduce the feature vector using Principal Component Analysis in order to extract the most relevant information.
4. Obtain a diagnosis by analogy using Case-Based Reasoning. In this sense, a SOM neural network is used.

One of the most important part of the algoritm is the mathematical model, which is used to simulated the dynamical response of the building and the pipeline system. Both systems can be characterized by the Equation 1 using a mass-damper-spring system. Where, m corresponds to mass, c is damping, k represent the stiffness, $\ddot{x}(t)$ is the acceleration, $\dot{x}(t)$ is the velocity, $x(t)$ is displacement and f is the force applied to the system (De Carvalho et al., 2003).

$$m\ddot{x}(t) + c\dot{x}(t) + kx(t) = f(t) \tag{1}$$

Taking into account that the matrices m, c and k describe completely the dynamics of the system, the mathematical model can be characterized by the Equations 2 and 3, using space-state representation.

Where, the state equation is represented by $\dot{x}$, u corresponds to the input vector, x is called state vector, A is the state matrix, B belongs to the input matrix, y is the output vector, C is the output matrix, and D belongs to the direct transition matrix (Hatch 2001).

$$\dot{x} = Ax + Bu \tag{2}$$

$$y = Cx + Du \tag{3}$$

In order to represent the mathematical model into a space-state system, it is necessary to define the variables of state x_1, x_2 and x_3 as is showed on the Equation 13, which $\ddot{x}(t)$ is given by Equation 1 divided by m.

$$\begin{aligned} x_1 &= x(t) \\ x_2 &= \dot{x}_1 = \dot{x}(t) \\ x_3 &= \dot{x}_2 = \ddot{x}_1 = -m^{-1}kx_1(t) - m^{-1}cx_2(t) + m^{-1}f(t) \end{aligned} \tag{4}$$

Other important component of the methodology used is related with the signal processing. Consequently, some mathematic technics as DWT and PCA are used to get feature of the dynamic response in different locations/elements of the structure as shows in the Figure 2.

The dynamic responses' features from different scenarios of damage are used to train and evaluate the ES based on a group of SOM neural networks, which organize and retrieve the information according with the similarity of the different scenarios of damage.

Finally, the damage detection is accomplished by estimating two parameters called dimension or size and severity, as shows in Equation 5 and 6. Dimension (Dim) of damage is defined as the number of elements with simultaneous stiffness reduction. Size of damage is defined as the percent of stiffness reduction in one element of the structure and it corresponds to a numerical damage measure for quantification.

Figure 2. Feature matrix for CBR stiffness recognition
Adapted from (Mujica et al., 2005).

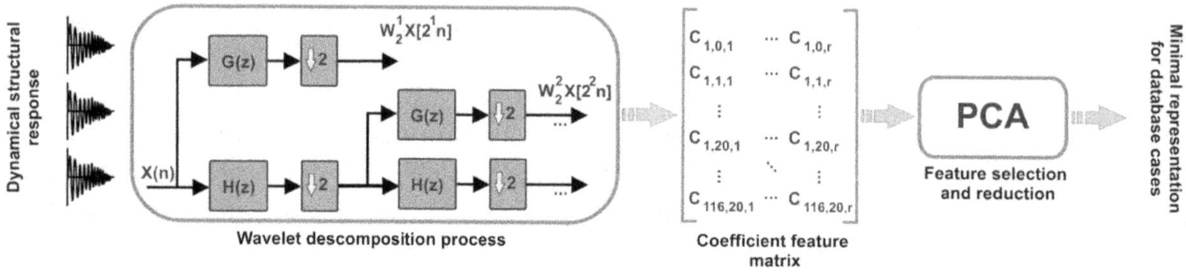

$$Dimension = \sum_{j=1}^{n} \dim(j) \frac{d(1)/d(j)}{\sum_{i=1}^{n} d(1)/d(i)} \tag{5}$$

$$Severity_E = \sum_{j=1}^{m} dam(j) \frac{d(1)\,/\,d(j)}{\sum_{i=1}^{m} d(1)\,/\,d(i)} \tag{6}$$

In Equation 5 and 6, n and m are the number of scenarios of damage or cases retrieved that have similar features, d(1) corresponds to the closer case by Euclidean distance to the evaluation case, d(j) represent the Euclidean distance of each one of the retrieved cases respect to the evaluation case. Additionally, $dim\left(j\right)$ and $dam\left(j\right)$ are the dimension and damage of each case retrieved.

METALLIC FRAMEWORK CASE

Figure 3 shows the first structure, a two-bay by two-bay steel framework, considered to validate the methodology for damage detection, location and quantification. It corresponds to a metallic framework hosted in the University of British Columbia (UBC) (Dyke, & Caicedo. 2000).

FINITE ELEMENT MODEL DESCRIPTION

An analytical 3D Finite Element Model for the UBC structure is available for this Benchmark problem, which have been developed by the task group of the IASC-ASCE (Dyke, Bernal & Ventura, 2003). The numerical model was constructed with the assumption that each of the four floors is rigid. The structure was modeled with 116 Euler-Bernoulli elements as is shown in Figure 4. In this model, the vertical DOF is restrained allowing the floors rotate with respect to x, y and z axes. Floors have four steel plates each one. The nominal masses on floor one is 800Kg, the second and third floor has 600 Kg slabs and 750Kg for fourth floor. It is assumed that the braces have zero moment of inertia. The model has a total of 120 DOF.

Figure 3. Benchmark structure for CBR methodology validation; left: four-floor scaled structure, right: elements properties
Retrieved from (Dyke, & Caicedo 2000).

Property		Columns	Floor Beams	Braces
section type		B100×9	S75×11	L25×25×3
cross-sectional area	A [m^2]	1.133×10^{-3}	1.43×10^{-3}	0.141×10^{-3}
moment of inertia (strong direction)	I_y [m^4]	1.97×10^{-6}	1.22×10^{-6}	0
moment of inertia (weak direction)	I_z [m^4]	$.664\times10^{-6}$	$.249\times10^{-6}$	0
St. Venant torsion constant	J [m^4]	8.01×10^{-9}	38.2×10^{-9}	0
Young's Modulus	E [Pa]	2×10^{11}	2×10^{11}	2×10^{11}
Mass per unit length	ρ [kg/m]	8.89	11.0	1.11

Figure 4. Diagram of analytical model for the metallic framework
Retrieved from (Dyke, Bernal & Ventura, 2003).

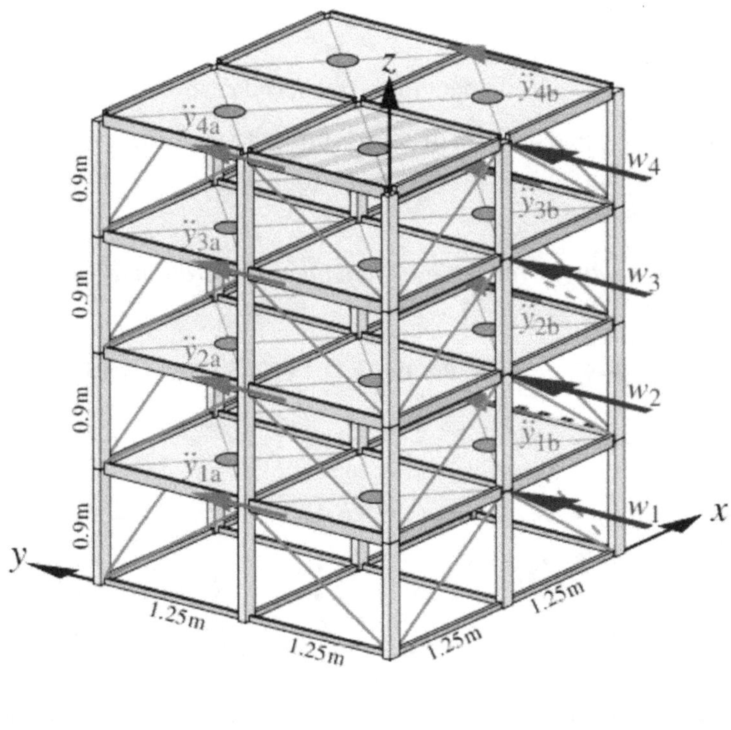

DAMAGE IDENTIFICACION METHODOLOGY

The first step in order to apply case based reasoning for damage identification is the generation of cases representing percentage stiffness variations (respect to nominal state) in one or more elements of the structure. Thus, solving the structural finite element model represented in Equation (1) by using numerical methods, acceleration records can be obtained when the structure is excited by means of environmental excitation forces (Gaussian noise).

The next step consists in estimate cross-correlation function between acceleration responses respect to a reference measurement. When different damage cases are simulated, a Data-set from cross correlation functions is organized as is shown in Figure 5.

The third step is to implement a feature extraction procedure by means of discrete wavelet transform (DWT). Thus, features correspond to the coefficients obtained in the wavelet decomposition process applied to cross-correlation signals as is shown in Figure 2. Also, a feature reduction and selection procedure is implemented through principal component analysis to DWT features, in order to obtain a minimal representation for the case database (Figure 5).

The reduced features are then used to train a self-organizing map (SOM) (T, Kohonen., 2001). Information concerning to damage severity, dimension, and location of damage cases are stored in clusters as is shown in Figure 6.

When a new case from validation data-set is processed, SOM network response is a retrieved cluster with the most similar cases (Mujica, et al 2005). A probability factor is estimated by means of damage occurrence and similarity measure (distance to the closer cluster):

$$F_{El} = \sum_{k=1}^{R} \frac{1}{d\left(k\right)} \tag{7}$$

Equation (7) is computed for each Element in the retrieved cases $\left(F_{El}\right)$. R correspond to the number of cases containing element F_{El} and d(k) are distance similarity measures. A normalized probability bar plot, similar to Figure 7, is generated.

Figure 5. Data set-for stiffness changes recognition; left: time acceleration records, right: cross-correlation signals of accelerations response

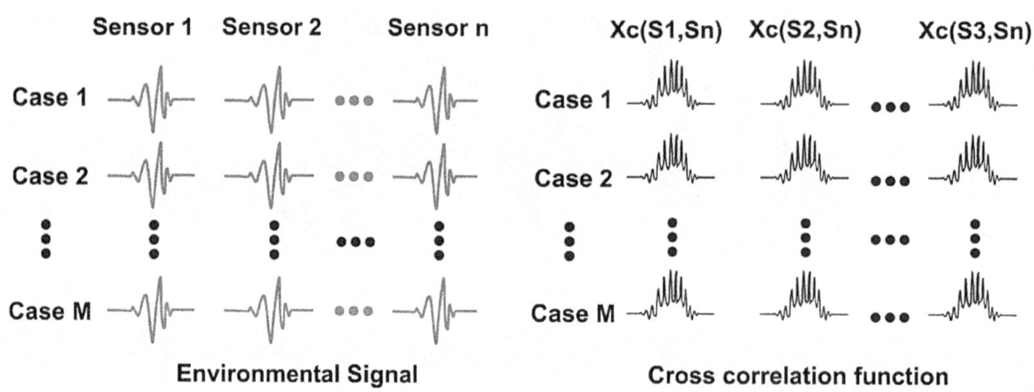

Figure 6. Database information organized in clusters
Adapted from (T, Kohonen, 2001).

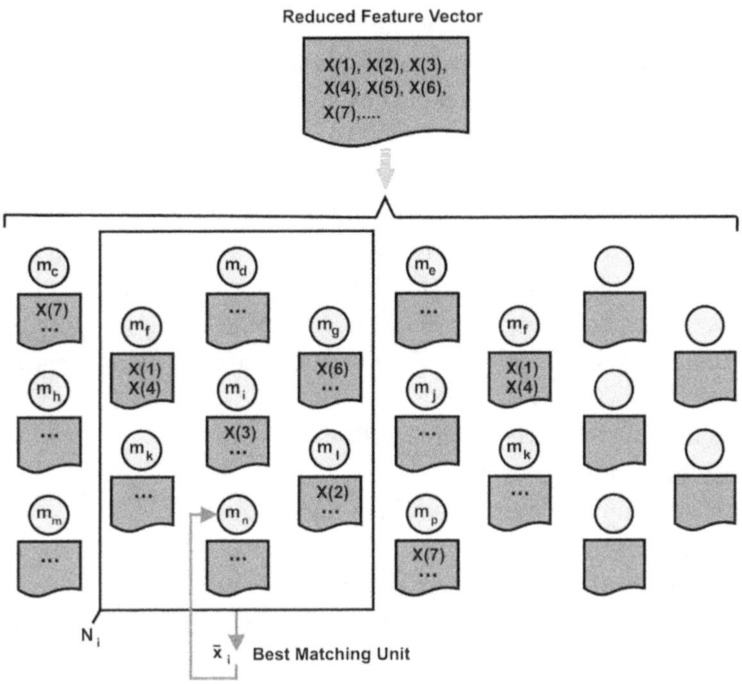

By using information in Figure 7, it is possible to identify what is the element in the structure with major probability of damage taking into account similarity measure for previous storage damages. In this sense, the validation case has similar information (Dimension, size, type and location) to the SOM retrieved cases.

Figure 7. Factors and damage probability

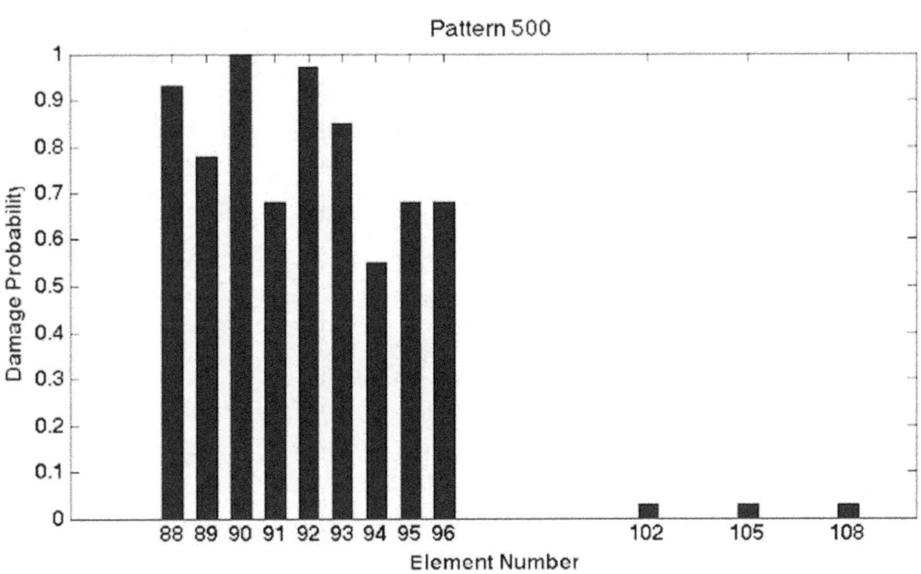

DAMAGE CASES DATABASE

In order to generate damage scenarios from the analytical model, a 100 Hz Gaussian noise signal was simulated as input by using the finite element model. Figure 4 specify the direction of the induced force (W_i) which allow to obtain 16 acceleration records as is shown in Figure 8. White noise is added to accelerations in order to simulate measurement noise. The sample time is 125 [Hz] for 90 [s] record length.

Damages are simulated as a percent stiffness reduction on each of the 116 framework elements. Element types in the structure are braces (diagonal), columns (vertical) or beams (horizontal), which are identified by means of number tags to facilitate results interpretation. Thus, on each floor exist 9 columns, 12 beams and 8 braces. Maximum 4 elements (high severity) are considered with simultaneous damage. A 100% stiffness reduction corresponds to a removed element. A maximum of three simultaneously removed elements (full damage) are considered. The following cases were taken into account as scenarios:

- Undamaged structure, which have nominal stiffness.
- All braces in first floor were removed.
- All braces in first and third floor are removed simultaneously.
- One brace in first floor is removed.
- One brace in first and third floor are removed simultaneously.
- One brace are removed in first and third floor simultaneously. Also, a beam in the first floor is damaged.
- One element of any type on any floor is damaged.

Figure 8. Acceleration records obtained from analytical model for the metallic framework
Retrieved from (Dyke, Bernal & Ventura, 2003).

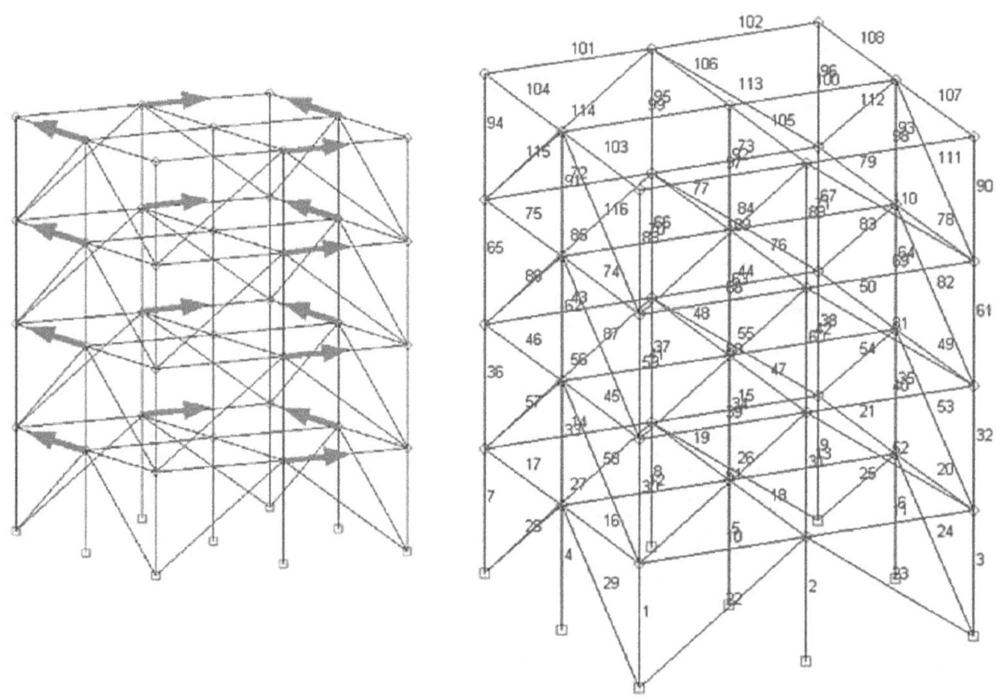

- Several same type elements are damaged. Damages are produced in the same floor.
- Simultaneous damaged element of any type located in the same floor.

Stiffness reduction for damaged scenarios is in the range [10% - 50%] with 10% step. For training purposes, the distribution of 6814 damaged scenarios is the following: 100 undamaged cases, 300 for one damaged element, 5860 with simultaneous damaged elements and 548 cases when the element is removed. In order to validate the generalization of the algorithm, for validation purposes 1870 scenarios were generated and distributed as: one damaged element (280), simultaneous damages (1360), Full-Damage (204) and undamaged structure (20).

ERROR INDEXES FOR DAMAGE IDENTIFICATION EVALUATION

In order to evaluate the effectiveness of the damage detection algorithm, the following indexes are calculated:

Severity and Dimension errors. The subscripts E and R indicate estimated and real values respectively.

$$\%\left(error_{Dimension}\right) = \left|\frac{Dimension_E - Dimension_R}{Dimension_R}\right| \times 100 \tag{8}$$

$$\%\left(error_{Severity}\right) = \frac{\sum_{i=1}^{n}\left|\frac{Severity_E\left(i\right) - Severity_R\left(i\right)}{Severity_R\left(i\right)}\right|}{n \times 100} \tag{9}$$

Type and Floor errors

$$Dam_{EI} = \max_{k \in n} Dimension_k ; n = 1,2,3,4 \tag{10}$$

$$Severity_{EI} = \frac{\sum_{k \in n}\left(Severity_E\right)_k}{k} ; n = 1,2,3,4 \tag{11}$$

Equation 10 is a count index to estimate the number of elements belonging to a respective floor, and Equation 11 is the mean severity error for elements in the same floor. Type error is estimated similar to dimension error.

RESULTS DISCUSSION FOR METALLIC FRAMEWORK CASE

Tests were conducted for daebuchies, symlet and coiflet wavelet families, because they have been reported as successful wavelet mothers used for pattern recognition tasks. Table 1 shows the errors for type element (errTy) and floor (errFl). The first kind of error refers if the element is a beam, a column or a brace and errFl indicates if the element belong to the first, second, third or fourth floor in the UBC structure. Additionally, the best results of the table are shown at the Figure 9.

In Table 1, the suffix Corr indicate that Principal Component Analysis was conducted by using the Correlation matrix, whilst the suffix Cov is the Covariance matrix.

According to the results, following comments are found:

- Stiffness change detection using the proposed algorithm is achieved it with rates above 70%, properly locating the type element damage (> 90%) and their location on the respective floor (> 80%).
- In general, is better to use covariance matrix for principal component analysis. It is because it allows to take into account magnitude of the coefficient wavelet for feature selection. Also, it requires less components to represent minimal information of the cases.
- The mother wavelet used in stiffness detection does not represent significant variations in the results. This is because the used wavelet families have similar frequencies properties, and stiffness changes are characterized by frequency information rather than its temporal information about abrupt changes.
- It is necessary to generate more undamaged cases for the training stage. It could be improve results regarding to floor detection.

Table 1. Stiffness recognition results

	Training Set				Validation Set			
	errFl (%)		errTy (%)		errFl (%)		errTy (%)	
	Mean	Std.	Mean	Std.	Mean	Std.	Mean	Std.
coif3_Corr	18.02	0.92	7.54	0.86	21.51	1	8.49	0.6
coif3_Cov	14.8	1.1	6.23	0.7	16.97	2.19	6.96	0.56
coif4_Corr	18.28	1.62	7.33	0.6	22.26	2.75	8.1	0.79
coif4_Cov	15.59	1.34	5.87	0.68	18.18	1.25	6.46	1.1
coif5_Corr	15.91	1.91	6.7	0.53	18.97	1.54	6.85	0.49
coif5_Cov	15.92	0.75	6.7	0.22	18.05	0.5	6.98	1.27
db4_Corr	17.35	1.38	7.92	0.19	21.61	1.61	8.33	0.82
db4_Cov	15.9	1.1	6.22	0.18	17.57	0.28	6.71	0.41
db5_Corr	18.95	0.79	7.89	1.29	22.23	1.04	8.39	1.71
db5_Cov	16.57	3.19	6.72	0.89	19.0	1.82	7.29	1.15
db6_Corr	16.99	2.26	7.83	1.33	20.27	1.57	8.42	1.6
db6_Cov	15.97	1.15	6.09	0.38	18.96	1.71	6.58	0.33
sym4_Corr	18.41	3.11	7.62	0.64	21.63	4.11	8.08	1.0
sym4_Cov	15.67	1.1	6.2	0.79	17.97	2.21	6.88	0.29
sym5_Corr	17.13	1.09	7.63	0.03	20.49	1.91	8.31	0.5
sym5_Cov	16.22	0.79	6.93	0.56	19.44	1.14	7.46	0.37
sym6_Corr	16.74	1.23	7.62	0.37	20.58	1.42	8.2	0.56
sym6_Cov	15.89	0.77	6.09	0.58	18.89	1.8	6.58	0.44

Figure 9. Severity mean error for best result in Table 1 (Coif3_Cov)

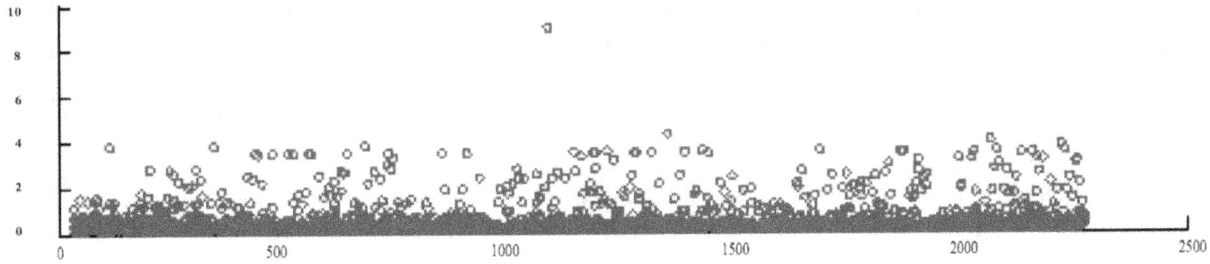

- It is suspected that reasons for which error floor is greater than type error is because more information is stored in database cases about this item. It is a similar performance for differences between training and validation errors. However, additional work is required in this sense.
- When the results detailed in this chapter is compared to other methodologies for damage identification using the benchmark as reported by (Caicedo, 2011), it is possible to conclude that same damage scenarios are well recognized by using CBR methodology. However, additional information is obtained for damaged types and damaged floors. Also, the sensitivity to stiffness changes are better, as is shown in the severity mean error figure; where the quantification deviation is lower than 10%.
- Taking into account the results presented in this paper, the authors propose future works related with family wavelet selection. A proper alternative would be building a wavelet mother for each engineering application. The result should be a methodology to extract the most relevant time-frequency information in the wavelet mother construction.

PIPELINE CASE

Expensive but accurate In-line inspection (ILI) equipment is commonly used for the technical examination of pipeline networks (Tiratsoo 1999). However, scientists around the word have been researching cheaper, more reliable and more feasible structural assessment methods, which are based on artificial intelligence (AI) and analysis vibration (Mujica 2006). Therefore, stiffness damage location and severity on pipelines can be assessed by using vibration-based damage identification (BVDI) and an expert system (ES) founded on case-based reasoning (CBR).

In this application, a structural health monitoring (SHM) system has been used to assess stiffness damage on pipelines, which is explained in four sections "problem description and background, mathematical model, ES configuration and its results".

Problem Description and Background

In Latin America, companies which use pipelines to transport their products have issues with the pipeline damage assessment because it is limited by the inspection cost, accessibility to the ILI tools per year and the extreme environment conditions which affect the pipes (Jones, & Holtz 1973).

The major preoccupation of the petrochemical industry (i.e., water, oil, gas, etc.) is to reduce the risk of catastrophic failures in pipelines to avoid accidents that could negatively affect the environment, mankind's way of life, normal operation of the pipeline transportation system, and the economy of the business sector (Ii'kaev et al., 2005). For that reason, the pipeline systems are periodically tested throughout their lives by using sophisticated ILI tools (i.e., intelligent pigs.), which are based on the traditional Non-Destructive Testing (NDT) techniques (Nestleroth 2005).

The principal problem related to the structural integrity assessment by using the conventional NDT techniques is the high cost to make an inspection to an extensive infrastructure (i.e., ILI tools). This issue is shared by researchers from different engineering disciplines (e.g. civil, mechanical, aerospace, etc.), who have been researching cheaper alternative damage identification methods (Mujica 2006) that can be applied for structural assessment (Khool et al., 2004)..

Another issue is linked with accessibility to the ILI tools. Companies which provide the pipeline inspection services, usually have their ILI tools testing pipeline networks around the world with a busy schedule. Consequently, the petrochemical companies which are interested in inspecting their pipelines have to rent the inspection services several months prior the pipeline network examination.

Other obstacle concerning pipeline integrity is related to the extreme environment conditions (Wijewickreme 2013). In Latin America, some petrochemical companies have part of their infrastructure located in places which the land movement and unstable soil conditions affect the pipeline's integrity constantly (Contreras et al., 2013). Also, companies cannot afford several inspections over the years with ILI tool. For that reason, they need a cheaper and more reliable method to inspect specific pipe sections.

Mathematical Model

The principal goal of the mathematical model implemented was to be able to simulate the structural response in frequency of the pipeline to the different stiffness scenarios of damage (e.g. stiffness reduction and its location) by using a selected excitation signal. The pipeline dynamical model was implemented in Matlab by using modal beam theory, mass-spring-damper system (MSD), finite element method (FEM) and space-state analysis. The model developed was validated by using specialized commercial structural analysis software "ANSYS".

Beam Theory

The first part of the mathematical model has been based on Equation 12 "Euler-Bernoulli beam model" (Begambre 2004), where $A(x)$ ⊉ belongs to the beam cross section area, ρ is the density of the beam material, E describes the Young's modulus or elastic coefficient, I is the second momentum of inertia for the cross section, $\omega(x,t)$ corresponds to beam depletion, and $f(x,t)$ is the external force applied on the beam. Additionally, the modal analysis of the beam is given by the Equation 13 "a general multi degree of freedom MSD" (De Carvalho et al., 2003), which matrices m, c, k and b corresponds to mass, damping, stiffness and applied force; $f(t)$, x, $\dot{x}$, $\ddot{x}$, are force, displacement, velocity and acceleration vectors respectively

$$\rho A(x)\frac{\partial^2\omega(x,t)}{\partial t^2}+\frac{\partial^2}{\partial x^2}\left(EI(x)\frac{\partial^2\omega(x,t)}{\partial x^2}\right)=f(x,t), \tag{12}$$

$$m\ddot{x}(t) + c\dot{x}(t) + kx(t) = bf(t), \tag{13}$$

Furthermore, the Equation 13 can be described by the Equations 14and 15 "using Laplace transformation technique and finding an equivalent frequency domain model", which ω_n is the undamped natural frequency of the system, and $\omega_n^{\,2}$ corresponds to km^{-1}.

$$mS^2x(s) + cSx(s) + kx(s) = bf(s), \tag{14}$$

$$\frac{x(s)}{bf(s)} = \frac{\frac{1}{m}}{s^2 + \frac{c}{m}s + \frac{k}{m}} \tag{15}$$

Pipeline Modeled with FEM

The pipeline model defined in the table 2 (Mujica 2006) was implemented in Fortran and Matlab by using FEM, Euler-Bernoulli beam model restrictions "according with the boundary configuration" and MSD model described in Equation 12.

In addition, Figures 10 and 11 describe each discretized element (e) of the pipeline model, which has associate two degrees of freedom (DOF) "one is known as transversal displacement (v) and the other as rotation $(\varnothing)$ (Begambre 2004), in each one of its nodes (N).

Consequently, the variables of Equation 14 can be rewritten by substituting the values of the Equation 16, as follows:

$$\ddot{x}(t) = \begin{bmatrix} \ddot{v} \\ \ddot{\varnothing} \end{bmatrix}, \dot{x}(t) = \begin{bmatrix} \dot{v} \\ \dot{\varnothing} \end{bmatrix}, x(t) = \begin{bmatrix} v \\ \varnothing \end{bmatrix}, f(t) = \begin{bmatrix} f_1 \\ f_2 \end{bmatrix} \; and \; b = \begin{bmatrix} 1 & 0 \\ 0 & 1 \end{bmatrix} \tag{16}$$

The values of the matrices m and k to each element are given by Equation 17 (Inman 2001), were calculated by using FEM and they are defined to a beam element with concentrated load. The global matrices m and k can be reduced in size by applying Kidder (1973) and Guyan (1975) FEM theories.

$$m_{element} = \frac{\rho A l}{2} \begin{array}{c} \begin{matrix} v_1 & \varnothing_1 & v_2 & \varnothing_2 \end{matrix} \\ \begin{bmatrix} 1 & 0 & 0 & 0 \\ 0 & 0 & 0 & 0 \\ 0 & 0 & 1 & 0 \\ 0 & 0 & 0 & 0 \end{bmatrix} \end{array} \begin{matrix} v_1 \\ \varnothing_1 \\ v_2 \\ \varnothing_2 \end{matrix}, \; and \; k_{element} = \frac{EI}{l^3} \begin{array}{c} \begin{matrix} v_1 & \varnothing_1 & v_2 & \varnothing_2 \end{matrix} \\ \begin{bmatrix} 12 & 6l & -12 & 6l \\ 6l & 4l^2 & -6l & 2l^2 \\ -12 & -6l & 12 & -6l \\ 6l & 2l^2 & -6l & 4l^2 \end{bmatrix} \end{array} \begin{matrix} v_1 \\ \varnothing_1 \\ v_2 \\ \varnothing_2 \end{matrix} \tag{17}$$

Table 2. Pipeline characteristic

Characteristic Name	Value
Number of elements	32
External radius	$0.042[m]$
Inner radius	$0.040[m]$
E	$210 \times 10^{9}\left[\frac{N}{m^{2}}\right]$
Poisson's ratio (V)	0.28
Length	$4.28[m]$
A	$1.2874 \times 10^{-4}[m^{2}]$
I	$2.7066 \times 10^{-8}[m^{4}]$
ρ	$7800\left[\frac{kg}{m^{3}}\right]$
Model configuration	Simply supported at both ends

Figure 10. DOF associated with beam element

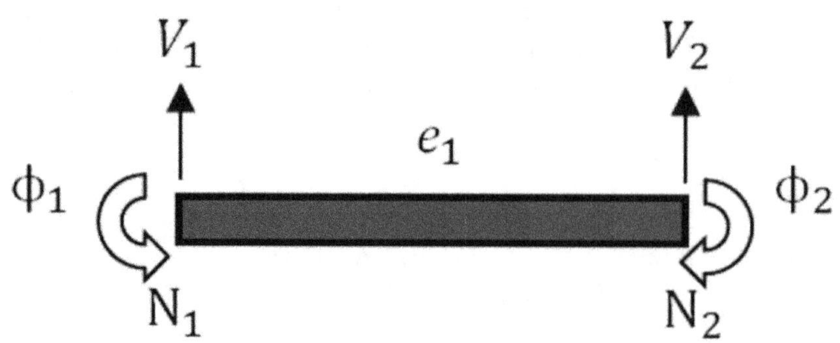

Figure 11. Pipeline model by using multiple MSD systems

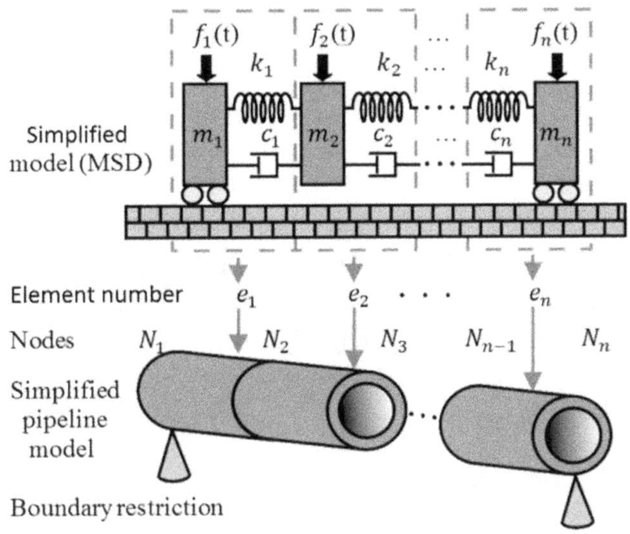

After getting the reduced matrices, the damped matriz described in Equation 18 has been characterized by the eigenvalues (λ) and eigenvectors (Φ) related with ω_n^2; where, f are the frequencies, ζ the damping ratios, and $2\pi f$ correspond to $\sqrt{\lambda}$ (Villamizar 2005).

$$c = m\Phi diag\left(2\zeta\left(2\pi f\right)\right)\Phi^T \tag{18}$$

Dynamic Response of the Pipeline Model

Taking into account that the matrices m, c, and k describe completely the dynamics of the pipeline system, the mathematical model can be characterized by the Equations 2 and 3, using space-state representation.

Therefore, Equation 4 can be expressed in a matrix form as Equation 16. The matrices A and B are described on Equation 20, which can be obtained directly by inspection using Equations 2 and 19.

$$\begin{bmatrix} \dot{x}_1 \\ \dot{x}_2 \end{bmatrix} = \begin{bmatrix} 0 & 1 \\ -m^{-1}k & -m^{-1}c \end{bmatrix} \begin{bmatrix} x_1 \\ x_2 \end{bmatrix} + \begin{bmatrix} 0 \\ m^{-1} \end{bmatrix} \begin{bmatrix} 1 & 0 \\ 0 & 1 \end{bmatrix} \begin{bmatrix} f_1 \\ f_2 \end{bmatrix} \tag{19}$$

$$A = \begin{bmatrix} 0 & 1 \\ -m^{-1}k & -m^{-1}c \end{bmatrix}, B = \left[m^{-1}\right] \tag{20}$$

Similarly, the output vector of accelerations is calculated using Equation 21, which y is given by $\ddot{x}\left(t\right)$. Consequently, The matrices C and D can be written in the form of Equation 22.

$$y = \begin{bmatrix} -m^{-1}k & -m^{-1}c \end{bmatrix} \begin{bmatrix} x_1 \\ x_2 \end{bmatrix} + \begin{bmatrix} m^{-1} \end{bmatrix} \begin{bmatrix} 0 & 1 \end{bmatrix} \begin{bmatrix} f_1 \\ f_2 \end{bmatrix} \tag{21}$$

$$C = \begin{bmatrix} -m^{-1}k & -m^{-1}c \end{bmatrix}, D = \begin{bmatrix} m^{-1} \end{bmatrix} \tag{22}$$

Since, the matrices A, B, C and D have been calculated, the pipeline model can be represented as a space-state system in terms of Equations 23 and 24, from which I represents the unit matrix, and the vectors are given in terms of v and $\varnothing$ (Sandoval 2010).

$$\begin{bmatrix} \ddot{v} \\ \ddot{\varnothing} \end{bmatrix} = \begin{bmatrix} 0 & I \\ -m^{-1}k & -m^{-1}c \end{bmatrix} \begin{bmatrix} \dot{v} \\ \dot{\varnothing} \end{bmatrix} + \begin{bmatrix} 0 \\ m^{-1} \end{bmatrix} \begin{bmatrix} I & 0 \\ 0 & I \end{bmatrix} \begin{bmatrix} f_1 \\ f_2 \end{bmatrix} \tag{23}$$

$$\begin{bmatrix} \ddot{v} \\ \ddot{\varnothing} \end{bmatrix} = \begin{bmatrix} -m^{-1}k & -m^{-1}c \end{bmatrix} \begin{bmatrix} \dot{v} \\ \dot{\varnothing} \end{bmatrix} + \begin{bmatrix} m^{-1} \end{bmatrix} \begin{bmatrix} 0 & I \end{bmatrix} \begin{bmatrix} f_1 \\ f_2 \end{bmatrix} \tag{24}$$

In brief, the pipeline model was implemented in Simulink using the structure showed in Figure 12. The block diagram was used to simulate the dynamic response of the pipeline "acceleration at each node of the finite elements" to the different scenarios of damage related with stiffness variations by using Equation 17.

Figure 12. Pipeline model by using Space-state analysis

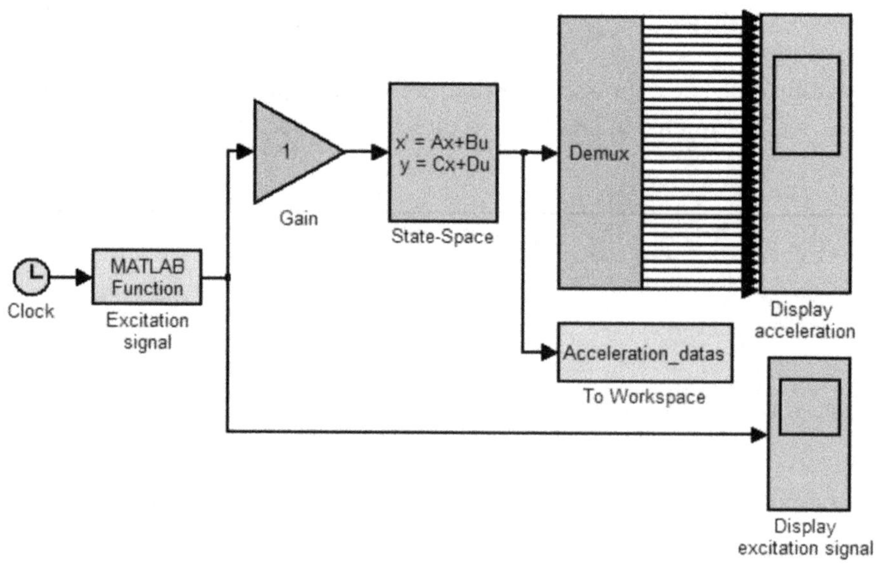

Finally, the excitation signal $\left(Exc\right)$ is defined by equation 25 and it corresponds to a windowed sine burst (Mujica, et al. 2005).

$$Exc\left(x\right) = n * \arcsin\left|\sin\left(\frac{1}{n}2\pi ft\right)\right|\sin\left(2\pi ft\right) \tag{25}$$

where, the parameters f corresponds to the excitation frequency of the beam "a specific vibration mode to be excited or eigenfrequency", and n represents the amount of a sino half-cycles contained in the window (Mujica 2006).

Validation of Pipeline Model

Mathematical model testing can be accomplished efficiently by characterizing the modes of vibration (modal frequency and modal shape) of the pipeline model, and contrasting its results with a specialized structural analysis tool. During the validation process, several stiffness scenarios of damage were simulated with Fortran and Matlab, and its results (measure of amplitude and frequency) were compared with Ansys. Figure 13 illustrates an example of a structure scenario of damage, which the stiffness of elements 2 and 4 were reduced 10%, Dam is the localization of damage, PD represent the percentage of damage, NDE corresponds to the number of damaged elements and NE is the number of finite elements.

Four steps were proposed in order to validate properly the FEM model:

1. Modal frequency response according to variations of location of the damage
2. Modal frequency response depending on variation of *PD*.
   a. Verify components of matrices k and m.
3. Damage position verification.

The first step is used in order to verify variations in frequency of the pipeline model by using the same PD in different locations as is shown in Figure 14. Next, the changes of frequency are verified using a specific element location with different PD variations. The third step is dedicated to contrast each simulated component of the matrices k and m (using a structure with 8 elements) with its cor-

Figure 13. Pipeline scenario of damage

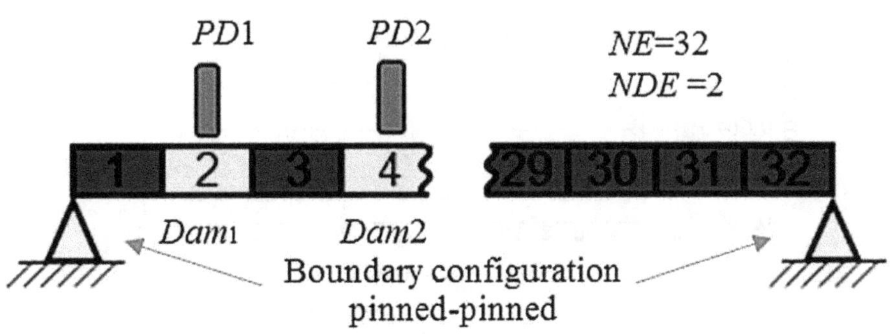

responding theoretical value. Finally, in order to verify that the damage simulated is located in the desired position, a theoretical method curvature damage factor (CDF) is used (Wahab & De Roeck 1999).

In Figure 15, Amplitude belongs to CDF, and the damage position is calculated using the first three pipeline vibration modes and Equation 26, where n represents the number of vibration modes, C_{Ck} corresponds to curvature mode shape without damage, and C_{Dk} is curvature mode shape with damage (Fayyadh & Razak 2011).

$$CDF = \frac{1}{n}\sum_{k=1}^{n}\left|C_{Ck} - C_{Dk}\right| \tag{26}$$

Figure 14. FEM's modal frequency error test Ansys and Fortran

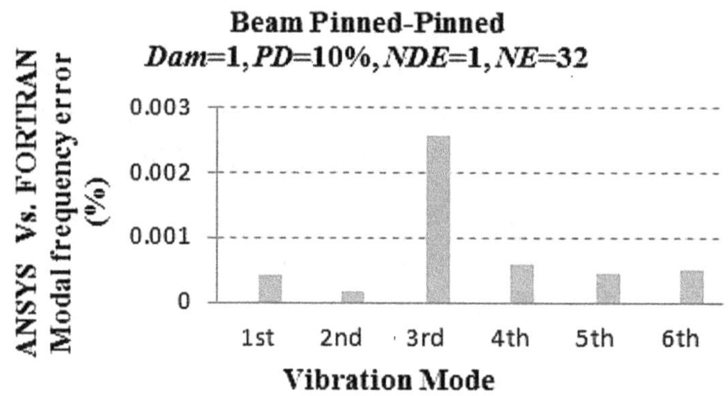

Figure 15. Damage locations by using CDF

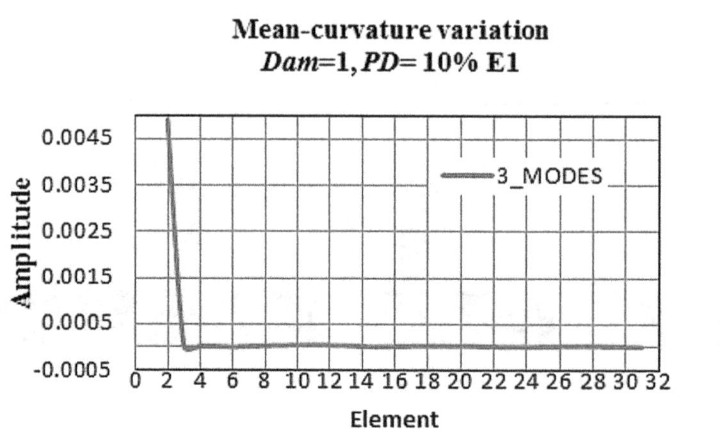

Expert System (ES)

The most important component of the ES is the group of data denominated case library, which contains information or knowledge to train and evaluate the performance of the ES. Each damage scenario is associated with a dynamic response (in acceleration at each element of the structure) when the system is excited by a known signal.

The ES is composed of 6 self-organizing maps (SOM), one for each sensor (S) and each one has its own knowledge base as is illustrated in figure 16. Three different steps were used to implement the ES: generating damage scenarios, processing digital signals and implementing SOMs.

Generation of Damage Scenarios

To recreate the different damage scenarios, three different variables have been used: localization $\left(Dam\right)$, damage severity $\left(PD\right)$ and number of elements affected by stiffness reduction $\left(NDE\right)$, where the value of Dam has a range $\left[1-32\right]$, PD corresponds to a number multiple of 5 in a range $\left[0-50\right]$, the maximum value of NDE is two elements (contiguous or not-contiguous), and S represents each sensor where the propagated signal is collected as described in Figure 16.

In summary, 14,209 different scenarios where simulated using a known excitation signal. The element of each scenario of damage has been associated with a group of 29 acceleration signals $\left(a\right)$ that correspond to each pipeline element (Plus the two additional positions, in which $a = 0$ because of the boundaries configuration). Also, each signal has 2,048 samples of length.

Figure 16. Expert system description

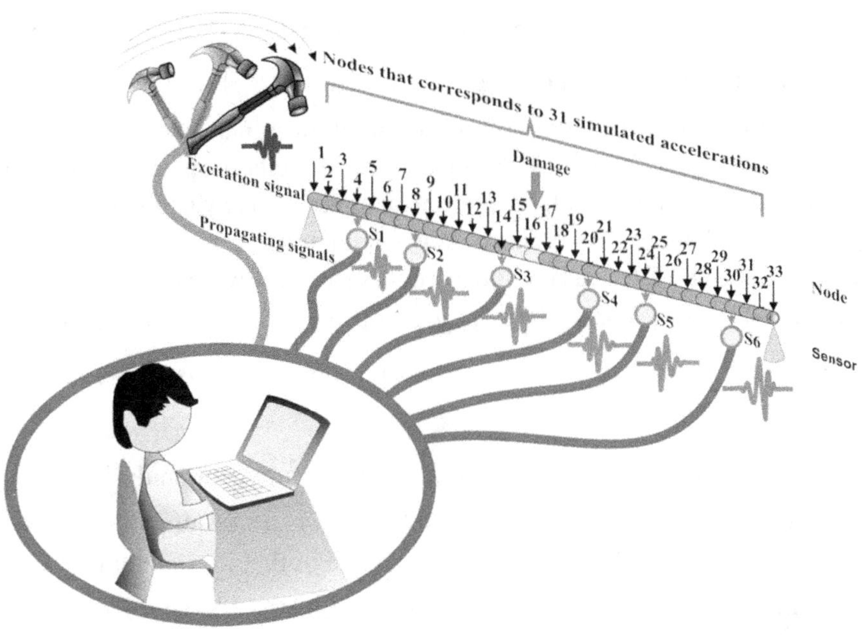

Figure 17. ES data distribution

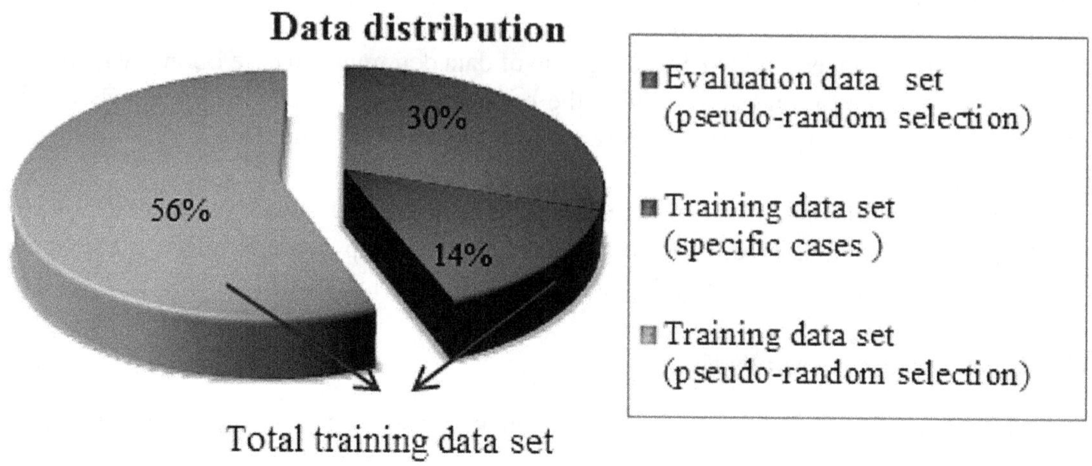

A graphical description of the data sorting used to train and evaluate the ES is shown in Figure 17, where the critical cases are just those cases which the stiffness has been reduced in a maximum or minimum value (e.g. PD is 50% or 5%) and its symmetrical cases (e.g. if $Dam = 1$ and $PD = 50\%$ then the symmetrical case is $Dam = 31$ and $PD = 50\%$); the pseudo-random distribution corresponds to the set of data that are not considered critical data to the training.

Digital Signal Processing

The signal processing was accomplished in the steps called pre-processing, data characterization and data dimension reduction.

During the signal pre-processing, three different kinds of signals were used: the original signal $\left(Os\right)$, the difference $\left(Diff\right)$ between the signals (specifically between the excitation signal and the acceleration on a specific element of the pipeline), and the cross correlation $\left(S_{xf}\right)$.

After the pre-processing step, the length of the acceleration signals can be condensed in size by using a multi-resolution analysis (MRA) based on a discrete wavelet transformer (DWT), so rather than considering the entire acceleration signal, an alternative approach is by using a selected group of signal descriptors (e.g. mean, median, etc.) in order to reduce computational cost when the ES is used.

Consequently, after use DWT ten different descriptors (feature) were used for data characterization; root mean square $\left(Rms\right)$, wavelet energy $\left(We\right)$, square of wavelet energy Swe, standard deviation $\left(\sigma\right)$, variance $\left(\sigma^2\right)$, mean $\left(\mu\right)$, mode $\left(Md\right)$ maximum $\left(Max\right)$, minimum $\left(Min\right)$ and correlation coefficient $\left(C.coef.\right)$.

Finally, data dimension was reduced by using principal component analysis $\left(PCA\right)$, in which the group of characteristics was organized by variance relevance, and the number of principal components $\left(PC\right)$ were selected based on Jollife (1972, 1973) criteria. For that reason, all the PC explained over 98% of the total variation of the data set, were selected to build the ES database.

Figure 18. Data dimension reduction

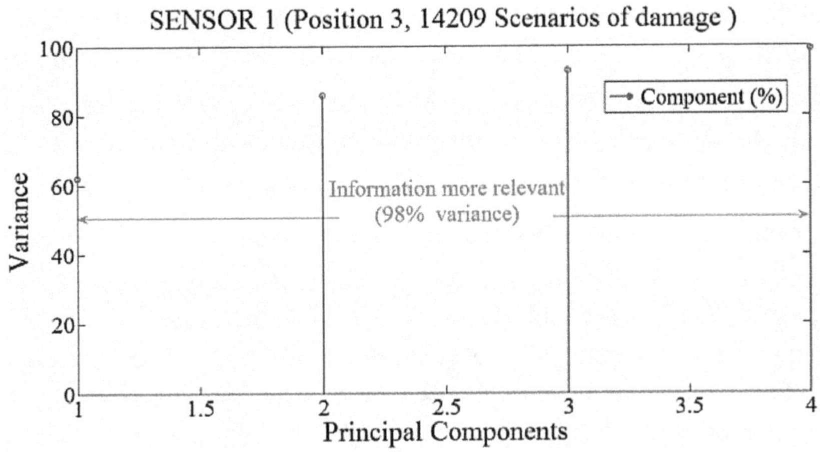

SOM Implementation

The principal purpose of the SOM is to classify or identify (Kohonen, 2001) features extracted from the simulated damages $\left(data\ set\right)$ on the pipeline model. It has two modes of operation called training and evaluation. Thus, the $data\ set$ has to be divided in two groups (as illustrated in Figure 17): one to train the SOM and the other to validate their model.

During the training mode, the group named $training\ data\ set$ (Patterns of 9,945 scenarios of simulated damage as shown in Figure 17) has been used to build the knowledge base $\left(casebase\right)$ according to the parameters shown in Table 3. In this mode, the SOM organize the 10 patterns extracted $\left(PE\right)$ of each scenario of damage simulated (by similitude in dynamical response) into a bi-dimensional map.

Table 3. SOM characteristics

Characteristic Name	Value
Training algorithm	$Batch$
Init	$Lininit$
Shape	$Sheet$
Normalization	$Logistic$
Size	Big
Training	$Long$
Lattice	$Hexa$

Later, each neuron of the trained SOM is associated to an index in order to organize the $casebase$ by cluster. Each SOM cluster contains relevant information (PD, Dam and PE) related to the cases with the similar characteristics. However, the trained SOM is tested with each one of the $training\ data\ set$ in order to detect neurons that are not activated during the test. Further, the inactive neurons were associated with the closer active neuron (using the minimum Euclidean distance).

The second group of 5,684 cases $\left(evaluation\ data\ set\right)$ as shown in Figure 17, is used to validate the performance of the trained SOM, where patterns of each $evaluation\ data\ set$ are compared with the patterns of the $casebase$. Finally, the most similar cases to the evaluated case are retrieved from the $casebase$ to infer about the damage localization and identification (severity).

During the evaluation process, three thresholds of selection ($U1$, $U2$ and $U3$) are defined to choose the most relevant information about the retrieve cases (based with the minimal Euclidean distance). The first threshold $\left(U1\right)$ is applied to select the most relevant SOMs of the $casebase$. The second one $\left(U2\right)$ is used to select the most relevant BMUS of each SOM. The last one $\left(U3\right)$ is use to determine which are the most similar cases of the BMU (related with the evaluated case). Figure 19 shows an example how the threshold $U3$ is use to choose the most relevant information to the expert system; in similar way are used $U1$, $U2$.

Figure 19. Process of damage identification

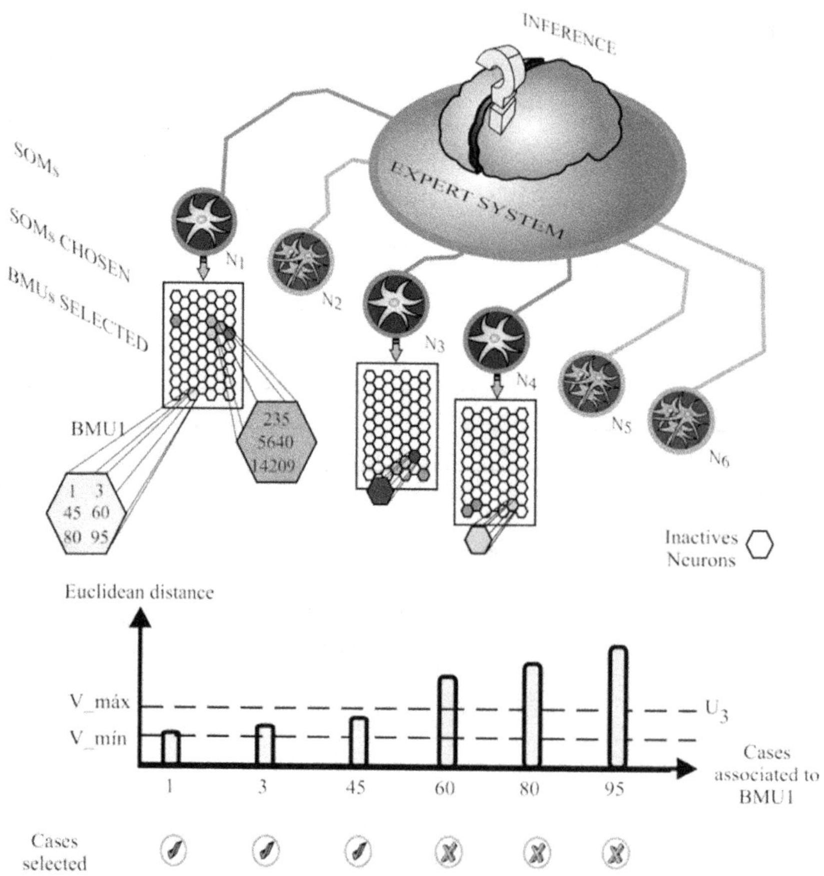

Finally, the performance of the ES is improved by modifying the amount of information analyzed during the validation process. Throughout the calibration of the ES, values of $U1$, $U2$ and $U3$ are changed in order to increase the accuracy of the damage identification system.

Pipeline Damage Detection System Issues

There are some interesting issues related to the right selection of the different parameters that affect the ES performance, such as the type of pre-processing (e.g. OS or $Diff$ or S_{xf}), method used during signal processing (e.g. DWT or PCA or a combination of both), signal segmentation (Sandoval 2010), descriptors of the signal (e.g. Rms, We, σ, μ, etc.), SOM characteristics (e.g. shape, size etc.), and the thresholds of selection ($U1$, $U2$ and $U3$). At this point, it is necessary to say that the right parameters to implement the ES cannot be selected until the expert system is completely operative. Consequently, the ES calibration has to be performed for using the successful response (damage identification) of the system during the validation process.

Several tests have been done in order to find the best ES configuration. The tests were organized in three different types.

The first group of ES was implemented using PCA, different combination of descriptors, after apply DWT, and different number of features, after apply PCA. In this group the maximum accuracy of the ES identifying damages was 21.85% (Sandoval 2010). Next, a second type of test was done, where the processing of the signal was realized with DTW with different wavelet families $\left(Wf\right)$ as Daubechies $\left(Db\right)$, Symlets $\left(Sym\right)$ and Coiflets $\left(Coif\right)$, different levels of decomposition, and different coefficient $\left(Wcoef\right)$. From these tests, the accuracy of the ES identifying damages improved up to 68.08% (Sandoval 2010), after including a signal segmentation process $\left(SSP\right)$.

Then, a set of ES based on the combination of DWT and PCA was implemented, and the feasibility of the new ES configuration in damage identification increased up to 82.01% as is observed in Table 4.

Finally, the parameters $U1$, $U2$ and $U3$ were adjusted in order to increase the performance of the ES based on PCA and DTW. Its highest degree of accuracy is illustrated in Tables 4 and 5.

In Table 5, Wavelet coefficients of approximation and detail correspond to the variables called Det and App.

Furthermore, the priority of ES developed is localizing faults more than evaluating severity. However, meanwhile the accuracy of the system is tested; some cases can be identified in a symmetric localization instead of the right position based on the patterns of similarity in these cases. Figure 20 shows an example of damage detection using the ES in evaluation mode.

In this particular case of application where the size of $database$ is too big to be simulated and stored its results into the physical memory (hyper matrix which contain 31 signals per case, 2,048 samples per each signal, and a total of 14,209 cases) it is recommended to organize the information in small packages, until the $casebase$ is built.

Another suggestion is related to the reduction of size of the original signals in order to reduce computational cost and increase the efficiency of the ES. Certainly, different mathematical methods and tools a (e.g. PCA, DWT, etc.) can extract the most relevant features of the dynamical response to build the $casebase$.

Table 4. SOM based on PCA and DWT

SYS	Processing Based on			Pre-Processing			Descriptors								SSP		Wcoef		Wf	Thr	PC	Accuracy (%)
	WT	PCA	WT+PCA	OS	S_{xf}	Diff	Md	σ^2	Rms	Ew	Rew	Max	Min	C. Coef	Yes	Not	Det+App	Det				
`1			X	X			X	X	X	X	X	X	X	X		X	X		db6	0.7	5	15.76
2			X	X			X	X	X	X	X	X	X	X	X		X		db4	0.5	10	53,59
3			X	X							X				X			X	db4	0.5	9	41.46
4			X		X		X	X	X	X	X	X	X	X		X	X		db4	0.7	4	13.02
5			X		X		X	X	X	X	X	X	X	X	X			X	db4	0.5	10	6,72
6			X		X		X	X	X	X	X	X	X	X	X			X	db6	0.5	10	19,72
7			X		X						X				X			X	db6	0.5	10	37.76
8			X			X	X	X	X	X	X	X	X	X		X	X		db6	0.7	7	44,59
9			X			X					X				X			X	db4	0.7	10	76.44
10			X			X					X				X		X		db4	0.5	10	81,94
11			X			X					X				X			X	db4	0.5	10	82.01

Also, it is important to consider that symmetrical scenarios of damage have similar dynamic response, so the ES can be failing the damage localization. Therefore, during the training data selection it is appropriate to choose the dynamic response for critical scenarios of damage and its symmetric cases.

Table 5. Calibration of parameters $U1$, $U2$ and $U3$

Test	Accuracy (%)	Bmus	THR	U1	U2	U3
1	76.15	10	0.7	1.5	1.07	1.5
2	76.91	2	0.7	1.5	1.07	1.5
3	80.75	10	0.5	1.5	1.07	1.5
4	81.54	2	0.5	1.5	1.07	1.5
5	80.39	4	0.5	1.5	1.07	1.7
6	76.98	2	0.7	1.5	1.07	1.7
7	80.93	4	0.5	1.5	1.07	1.1
8	80.82	10	0.5	1.5	1.07	1.1
9	81.09	10	0.5	1.5	1.1	1.1
10	77.11	10	0.5	1.5	1.5	1.1
11	82.01	4	0.5	1.5	1.5	1.1
12	77.62	4	0.7	1.5	1.07	1.5

Figure 20. ES operating in evaluation mode

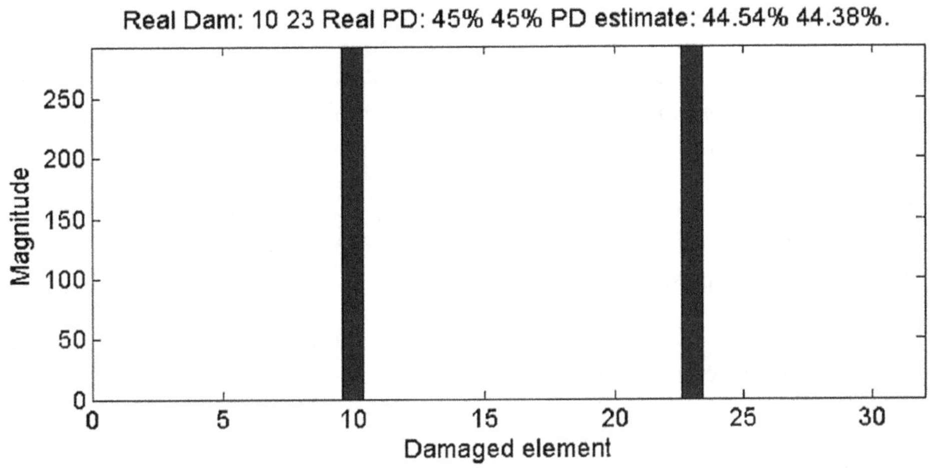

FUTURE RESEARCH DIRECTIONS

The numerical applications implemented give an alternative solution to real problems. However, it is necessary to continue the research to validate the numerical model with the experimental prototypes. In addition, other digital signal processing techniques and mathematical tools can be used in order to increase the accuracy of the Expert System.

CONCLUSION

In this section, the different steps used to build an Expert System based on Self Organizing Maps with capability to identify stiffness faults in pipelines and metallic frames has been explained. Dynamical response of the structures to different damage scenarios has been simulated and collected by solving the finite element model, which allow to create the base of knowledge used by the Expert System during its process of training and evaluation. Eventually, the Expert System has been implemented using Wavelet Transform, Principal Component Analysis and different descriptors from the original signal.

Finally, the accuracy of the Expert System has been tested and calibrated in evaluation mode with different damage scenarios. The best results of the ES were accomplished using TW and PCA during the processing section and the difference between the excitation and propagated signals during the pre-processing section. Additionally, the greater results were obtained using db6 and the detail coefficients to calculate Rew descriptor, using a THR of 0.5, which was used to train and evaluate the ES.

REFERENCES

Begambre, O. J. (2004). *Detecção de dano a partir da resposta dinâmica da estrutura: estudo analítico com aplicação a estruturas do tipo viga.* (Master dissertation). Retrieved from http://web.set.eesc.usp.br/producao/ \

BERRAIS, A. (1999). Artificial Neural Networks in Structural Engineering: Concept and Applications. *Engineering and Science*, *12*(1), 53–67.

Caicedo, J. M. (2011). Practical guidelines for the natural excitation technique (NexT) and the eigensystem realization algorithm (ERA) for modal identification using ambient vibration. *Experimental Techniques*, *35*(4), 52–58. doi:10.1111/j.1747-1567.2010.00643.x

Camacho, J. (2010). *Sistema experto para la monitorización de salud estructural mediante el reconocimiento de patrones: adaptación y validación numérica*. (Master dissertation). Retrieved from http://tangara.uis.edu.co/biblioweb/tesis/2010/136485.pdf

Camacho, J., & Villamizar, R. (2010). Adaptación del razonamiento basado en casos al problema de monitorización de salud estructural. Memorias Semana Técnica de Ingenierías eléctrica, electrónica y afines STIEEA.

Carden, E. P., & Fanning, P. (2004). Vibration based condition monitoring: A review. *Structural Health Monitoring*, *3*(4), 355–377. doi:10.1177/1475921704047500

Chen, B., Zhao, S. L., & Li, P. Y. (2014). Application of Hilbert-Huang Transform in Structural Health Monitoring: A State-of-the-Art Review. *Mathematical Problems in Engineering*.

Chou, J. H., & Ghaboussi, J. (2001). Genetic algorithm in structural damage detection. *Computers & Structures*, *79*(14), 1335–1353. doi:10.1016/S0045-7949(01)00027-X

Contreras, M., Vergara, C., Pereira, M., Colonia, J. D., & García, H. (2013). Pipeline integrity assessment and mitigation techniques in unstable soil conditions based on comparison of in line inspection results. *ASME 2013 International Pipeline Geotechnical Conference*. Pipeline Systems Division.

De Carvalho, J., Tita V.,& Lirani, J. (2003).Theoretical and experimental dynamic analysis of fiber reinforced composite beams. *Journal of the Brazilian Society of Mechanical Sciences and Engineering*, *25*(3), 306-310.

Dyke, S., & Caicedo, J. (2000). *Monitoring of a Benchmark Structure for Damage Identification. Department of Civil Engineering*. St Louis, MO: Washingtong University.

Dyke, S. J., Bernal, D., Beck, J., & Ventura, C. (2003). Experimental phase II of the structural health monitoring benchmark problem. In *Proceedings of the 16th ASCE engineering mechanics conference*.

Fan, W., & Qiao, P. (2011). Vibration-based damage identification methods: A review and comparative study. *Structural Health Monitoring*, *10*(1), 83–111. doi:10.1177/1475921710365419

Fayyadh, M. M., & Razak, H. A. (2011). Modified damage location indices in beam-like structure: Analytical study. *Scientific Research and Essays*, *6*(32), 6605–6621.

Guyan, R. J. (1965). Reduction of stiffness and mass matrices. AIAA Journal, 3(2), 380.

Hatch, M. R. (2001). *Vibration simulation using Matlab and Ansys*. Washington, DC: Chapman & Hall/CRC.

Hatch, M. R. (2001). *Vibration simulation using Matlab and Ansys*. Washington, DC: Chapman & Hall/CRC.

Ii'kaev, R. I., Zelesnev, V. E., Aleshin, V. V., & Klishin, G. S. (2005). *Numerical simulation of gas pipeline networks. Theory, computational implementation, an industrial application.* Moscow: Editorial URSS.

Inman, D. J. (2001). *Engineering vibration.* Prentice Hall Inc.

Johnson, E. A., Lam, H. F., Katafygiotis, L. S., & Beck, J. L. (2001). In F. Casciati & G. Magonette (Eds.), *A benchmark problem for structural health monitoring and damage detection. Structural Control for Civil and Infrastructure Engineering* (pp. 317–324). Singapore: World Scientific. doi:10.1142/9789812811707_0028

Jolliffe, I. T. (1972). Discarding Variables in a Principal Component Analysis I: Artificial Data. *Applied Statistics, 21*(2), 160–173. doi:10.2307/2346488

Jolliffe, I. T. (1973). Discarding Variables in a Principal Component Analysis II: Real Data. *Applied Statistics, 21*(2), 160–173. doi:10.2307/2346488

Jones D.E. & Holtz W.G. (1973). Expansive soils - The hidden disaster. *American Society of Civil Engineers, 43*(8), 49-51.

Khoo, L. M., Mantena, P. R., & Jadhav P (2004). Structural damage assessment using vibration modal analysis. *Structural Health Monitoring, 3*(2), 177–194.

Kidder, R. L. (1973). Reduction of structural frecuency equations. AIAA Journal, 11(6), 892.

Kohonen, T. (2001). Self-Organizing maps (3rd ed.). Berlin: Springer-Verlag. doi:10.1007/978-3-642-56927-2

Li, H. N., Yi, T. H., Ren, L., Li, D. S., & Huo, L. S. (2014). Reviews on innovations and applications in structural health monitoring for infrastructures. *Structural Monitoring and Maintenance, 1*(1), 1–45. doi:10.12989/smm.2014.1.1.001

Lin, N., & Qun, C. (2013, January). Structural Health Monitoring and Damage Detection Using Neural Networks. In *Proceedings of the 2013 Third International Conference on Intelligent System Design and Engineering Applications* (pp. 1302-1304). IEEE Computer Society. doi:10.1109/ISDEA.2012.307

Mujica, L., Vehí, J., & Rodellar, J. (2005). *Detección de impactos mediante razonamiento basado en conocimiento: Aplicación a una sección de ala de avión. 1Departament d'Electrònica, Informàtica i Automàtica.* Escola Politècnica Superior - Universitat de Girona.

Mujica, L. E. (2006). *A hybrid approach of knowledge-based reasoning for structural assessment.* (Doctoral dissertation). Retrieved from http://www.tdx.cat/bitstream/handle/10803/7737/tlemd.pdf?sequence=1

Mujica, L. E., Rodellar, J., Fernandez, A., & Guemes, A. (2010). Q-statistic and T2-statistic PCA-based measures for damage assessment in structures. *Structural Health Monitoring*, 1475921710388972.

Nestleroth, J. B. (2006) Pipeline in lines inspection. Challenges to NDT. *ECNDT 2006. 9th European Conference on NTD.* Berlin, Germany: ECNDT.

Sandoval, W. A. (2010). *Sistema experto para la evaluación del estado de deformación en tuberías de gas: diseño y validación numérica.* (Master dissertation). Retrieved from http://tangara.uis.edu.co/biblioweb/tesis/2010/134415.pdf

Sinou, J. J. (2009). A review of damage detection and health monitoring of mechanical systems from changes in the measurement of linear and non-linear vibrations. *Mechanical Vibrations: Measurement, Effects and Control*, 643-702.

Taha, M. R., Noureldin, A., Lucero, J. L., & Baca, T. J. (2006). Wavelet transform for structural health monitoring: A compendium of uses and features. *Structural Health Monitoring*, *5*(3), 267–295. doi:10.1177/1475921706067741

Tiratsoo, J. N. H. (1999). *Pipeline pigging technology* (2nd ed.). Houston, TX: Butterworth Heinemann.

Villamizar, R. (2005). *Robust control of systems subjected to uncertain disturbances and actuator dynamics*. (Doctoral dissertation). Retrieved from http://www.tdx.cat/bitstream/handle/10803/7729/Trvm.pdf?sequence=1

Wahab, A. M. M., & De Roeck, G. (1999). Damage Detection in Bridges using Modal Curvatures: Application to a Real Damage Scenario. *Journal of Sound and Vibration*, *226*(2), 217–235. doi:10.1006/jsvi.1999.2295

Wijewickreme, D. (2013). *Role of geotechnical engineering in assuring the integrity of buried pipeline systems*. Retrieved from http://dl.lib.mrt.ac.lk/bitstream/handle/123/8879/SBE-KN_1.pdf?sequence=1

Yan, Y. J., Cheng, L., Wu, Z. Y., & Yam, L. H. (2007). Development in vibration-based structural damage detection technique. *Mechanical Systems and Signal Processing*, *21*(5), 2198–2211. doi:10.1016/j.ymssp.2006.10.002

Yolken, H. T., & Matzkanin, G. A. (2008). Recent trends in structural health monitoring technologies. *The AMMTIAC Quarterly*, *3*(4), 3-6.

Yu, E., Kim, S. N., Park, T., & Lee, S. H. (2014). Detection of earthquake-induced damage in a framed structure using a finite element model updating procedure. *TheScientificWorldJournal*. PMID:24574888

Chapter 8
Statistical Approach to Structural Damage Diagnosis under Uncertainty

Shankar Sankararaman
SGT Inc., USA & NASA Ames Research Center, USA

Sankaran Mahadevan
Vanderbilt University, USA

ABSTRACT

This chapter presents a statistical methodology for structural damage diagnosis (detection, localization and estimation), in the context of continuous online monitoring. There are several sources of uncertainty such as physical variability, measurement uncertainty and model errors that affect structural damage diagnosis, and therefore, it may not be possible to precisely detect, localize, and estimate damage. Hence, a statistical approach can help to identify these sources of uncertainty, quantify their combined effect on diagnosis, and thereby, provide an estimate of the confidence in the results of diagnosis. Damage detection is based on residuals between nominal and damaged system-level responses, using statistical hypothesis testing whose uncertainty can be captured easily. Localization is based on the comparison of damage signatures derived from the system model. Both classical statistics-based methods and Bayesian statistics-based methods are investigated to quantify the uncertainty in all the three steps of diagnosis, i.e. detection, localization, and quantification. While classical statistics-based methods use the concept of least squares-based optimization, Bayesian methods make use of likelihood function and Bayes theorem. The uncertainties in damage detection, isolation and quantification are combined to quantify the overall uncertainty in diagnosis. The proposed methods are illustrated using the example of a structural frame.

INTRODUCTION

Structural Health Monitoring (SHM) refers to the process of developing and implementing a damage identification strategy for structural applications in civil, aerospace, and mechanical domains (Farrar & Worden, 2007). Structural health monitoring techniques play a vital role in different types of practical

DOI: 10.4018/978-1-4666-8490-4.ch008

engineering applications, because they can be used to monitor the performance of these applications and aid several types of operational decision-making activities. This is established through the use of a combination of strategically placed array of sensors and a variety of mathematical methods and techniques, all of which constitute the overall structural health monitoring system.

An ideal structural health monitoring system is expected to facilitate automated continuous monitoring and diagnosis, thereby enabling prognosis and online health management. Damage diagnosis consists of three major steps – damage detection, damage localization (or isolation), and damage quantification. The first step of damage detection attempts to detect abnormalities in the structure and finds out if there has been any structural damage or deterioration. The second step of localization focuses on identifying that component(s) of the overall structure that has been subjected to damage. The third step of damage quantification quantifies the extent of damage, for example, estimating the reduction of stiffness in a structural member.

Several structural systems are complex in nature and consist of several members and components. Usually, only a few response quantities are monitored by the SHM system, but the number of prospective damage candidates is very high. In such cases, it is almost impossible to achieve precise damage diagnosis. Further, the inputs to the system, the physical parameters, sensor measurements, etc., may all be uncertain. While some of these sources of uncertainty are aleatory (irreducible sources of uncertainty), some of them are epistemic in nature (reducible). Due to the presence of these uncertainties, the behavior of these structural systems is uncertain in nature; it is important to account for these uncertainties while modeling the corresponding structural system response. As a result, the model predictions are uncertain. In addition, the experimental data may not be certain due to the presence of sensor noise, bias, or gain. Therefore, when both uncertain experimental data and uncertain models are both used for damage diagnosis, it is natural that the results of diagnosis are also uncertain.

Sankararaman and Mahadevan (Sankararaman & Mahadevan, 2011, 2013) explain that it is important to quantify the effect of the different sources of uncertainty and estimate the confidence in damage diagnosis by quantifying the uncertainty in damage detection, localization, and estimation, and the overall uncertainty in diagnosis. Such quantification of uncertainty in damage diagnosis is an essential step to guide decision-making regarding the functioning of the overall structural system.

In several research articles, the uncertainty in diagnosis (Achenbach, 2000) has been previously addressed using non-destructive evaluation (NDE) techniques, by estimating the probability of detection (POD). In such POD calculations, nominally identical damage is introduced in a number of nominally identical specimens, and the number of successful detections is used to calculate the probability of detection. Such an approach is completely different from the goal in structural health monitoring where the focus is on one structural system that is being monitored. Further, it is not possible to perform repeated tests since the goal in SHM is to make use of monitoring data in order to diagnose damage that has been caused due to external sources.

System identification techniques have been pursued by several researchers for the purpose of damage isolation and quantification. Fundamentally, these methods could be viewed as statistical calibration problems where the model parameters are calibrated and the uncertainty in calibration can be expressed through confidence bounds in the parameter estimation. Several studies have investigated classical statistics-based methods (Dalai, Weyer, & Campi, 2005; Vuerinckx, Pintelon, Schoukens, & Rolain, 2001) and Bayesian methods (Ching, Beck, & Porter, 2006; Jiang & Mahadevan, 2008a; Peterka, 1981; Vanik, Beck, & Au, 2000) in system identification and structural health monitoring. These methods involve estimating all the system parameters simultaneously; this makes the computation time-consuming

Figure 1. Framework for structural health monitoring

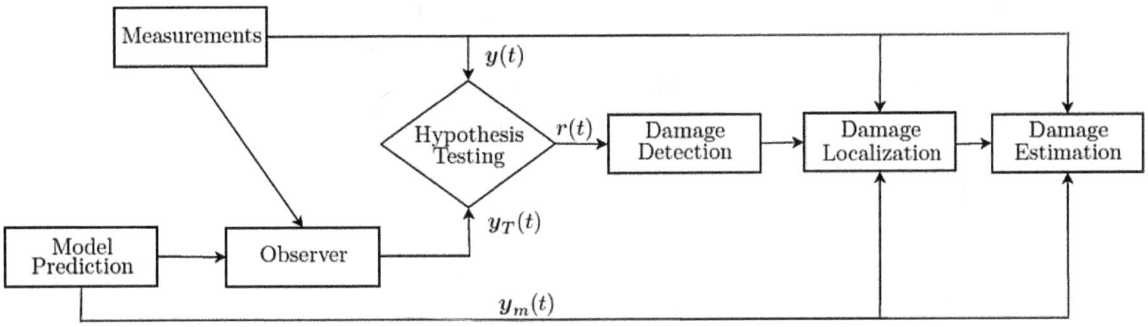

and thus not suitable for online diagnosis. System identification-based methods also combine damage isolation and damage quantification into one procedure; the individual contributions from different stages of diagnosis (detection, localization, and quantification) to the overall uncertainty in diagnosis are not separated in this procedure.

This chapter presents a systematic statistical methodology for structural damage diagnosis, where it is possible to account for the different sources of uncertainty and quantify the effect of these sources of uncertainty by estimating the uncertainty in damage detection, localization and estimation, and the overall uncertainty in diagnosis. To this end, both classical statistics-based methods (Sankararaman & Mahadevan, 2011) and Bayesian statistics-based methods (Sankararaman & Mahadevan, 2013) are discussed in this chapter. The differences between these two methods and the difference in their interpretations are also discussed.

A generic model-based framework, as shown in Figure 1, is used for structural health monitoring in this chapter. This model may be developed using first principles of physics or can be developed using data-driven techniques. Measurements available through sensors are used for damage diagnosis. The various steps of diagnosis, i.e. detection, localization and estimation (quantification) are clearly delineated as shown in Figure 1.

In this framework, a mathematical model (which may be developed using first principles of physics or using data-driven techniques) is used to predict the output ($\mathbf{y}_m(t)$) of the structural system, and this output is compared with the measurements ($\mathbf{y}(t)$) collected using sensors, to aid structural damage diagnosis. In the model, damage is represented by change in value of a particular model parameter. It is assumed that only one parameter is damaged at a given instant, and the focus of damage diagnosis is to identify that damage parameter and quantify the extent of damage by estimating how much the value of that damage parameter has changed.

Until damage is detected, a tracking method (such as Kalman filtering (Kalman & Bucy, 1961) or particle filtering (Arulampalam, Maskell, Gordon, & Clapp, 2002)) is used to estimate the state and the output of the system; the output estimated by the tracking method is denoted by $\mathbf{y}_T(t)$. Damage detection is performed by comparing $\mathbf{y}_T(t)$ against $\mathbf{y}(t)$, and the residuals ($\mathbf{r}(t)= \mathbf{y}_T(t)$ - $\mathbf{y}(t)$) are subjected to hypothesis testing. Then, the uncertainty in detection is computed using classical hypothesis testing and Bayesian hypothesis testing methods. In classical hypothesis testing (Sankararaman & Mahadevan, 2011), the choice of the significance level of testing is used to calculate the uncertainty in detection. In the Bayesian hypothesis testing-based approach, the likelihoods of two scenarios, i.e. "absence of true damage"' and "presence of true damage" are calculated and the uncertainty in the detection is quantified.

After damage is detected, tracking is completely stopped; from here on, only the model prediction ($y_m(t)$) is compared against experimental data ($y(t)$). Typically, both of these quantities are uncertain; the model prediction is uncertain due to the different sources of uncertainty associated with model parameters and model errors, and experimental data is uncertain due to the presence of sensor noise, bias and/or gain. It is important to account for these sources of uncertainty during damage detection, localization, and estimation.

Damage detection is first performed through hypothesis testing. Damage localization is performed in two steps. First, qualitative cause-effect relationships between model parameters and sensor measurements are used to isolate a prospective set of candidate damage parameters (Mosterman & Biswas, 1999). Second, statistical tools are used to estimate the uncertainty in localization, by considering this set of candidate damage parameters. In the classical statistics-based approach, a heuristic measure based on the least squares corresponding to each candidate damage parameter is used to quantify the uncertainty in damage isolation. In the Bayesian statistics-based approach, the probability of a particular parameter being faulty is calculated directly.

The third step of damage quantification is performed for each candidate damage parameter, using least squares minimization or likelihood maximization. It is possible to approach damage quantification using system identification techniques that focus on parameter estimation in dynamic systems. The uncertainty in damage quantification in calculated using statistical confidence intervals and confidence bounds (Beck, 1989; Dalai et al., 2005) in the classical statistics-based approach, and using Bayesian inference techniques (Ching et al., 2006; Peterka, 1981) in the Bayesian approach. Note that the latter approach directly calculates the probability distribution of the damage parameter while the former approach provides a simpler measure of uncertainty.

Finally, the overall uncertainty in diagnosis is calculated by combining the uncertainties in the three stages of diagnosis using the principle of total probability, and the overall probability distribution of the candidate damage parameters are calculated. These statistical methods for uncertainty quantification in damage detection, localization and quantification, and overall diagnosis uncertainty quantification are presented in detail in the following sections. Finally, the methods are illustrated using a numerical example.

DAMAGE DETECTION

Damage detection is based on comparison between healthy system response (predicted by the model, $\boldsymbol{y}_m(t)$) and the result of system tracking ($\boldsymbol{y}_T(t)$). 'Significant' deviation of $\boldsymbol{y}_T(t)$ from the expected healthy system performance indicates the presence of damage. The difference, i.e the residuals $\boldsymbol{r}(t)$ are averaged over a moving window and the mean (μ) of the residuals is subjected to a hypothesis test at a chosen significance level α, as follows:

$$H_0: \mu = 0 \text{ (No damage)} \tag{1}$$

$$H_1: \mu \neq 0 \text{ (Damage)} \tag{2}$$

The null hypothesis (H_0) in Eq. 1 states that the residuals are not significantly different from zero (indicating no presence of damage) while the alternate hypothesis (H_1) in Eq. 2 states that the residuals are significantly different from zero (indicating the presence of damage). Note that the tracking is stopped after damage detection, and only the raw measurements ($y_m(t)$) are used for damage localization and estimation.

Classical Statistics-Based Approach

The uncertainty in detection is calculated as the probability that the damage is true after having detected a fault. This is established using the choice of the significance level α in hypothesis test. The probabilities of different types of errors in hypothesis testing are well established in literature. The error of rejecting a correct null hypothesis is known as Type-I error while the error of not rejecting a false null hypothesis is known as Type-II error. The probability of Type-I error is equal to α and the probability of Type-II error is denoted by β. This information can be rewritten in terms of probability, as:

$$P\ (H_0 \text{ is decision} | H_0 \text{ is true}) = 1 - \alpha \tag{3}$$

$$P\ (H_1 \text{ is decision} | H_0 \text{ is true}) = \alpha \tag{4}$$

$$P\ (H_0 \text{ is decision} | H_1 \text{ is true}) = \beta \tag{5}$$

$$P\ (H_1 \text{ is decision} | H_1 \text{ is true}) = 1 - \beta \tag{6}$$

Suppose that the residuals calculated from the system have been subjected to statistical hypothesis testing at significance level α and damage has been detected. In other words, the decision is H_1, and the null hypothesis H_0 has been rejected. The confidence in damage detection can be calculated as the probability that the truth is H_1 subject to having rejected the hypothesis H_0, and this probability is written as P (H_1 is true| H_1 is decision). Let P_d denote this probability, and this can be calculated using Bayesian theorem as:

$$P_d = \frac{P\left(H_1 \textit{ is decision} \mid H_1 \textit{ is true}\right) * P\left(H_1 \textit{ is true}\right)}{P\left(H_1 \textit{ is decision} \mid H_1 \textit{ is true}\right) * P\left(H_1 \textit{ is true}\right) + P\left(H_1 \textit{ is decision} \mid H_0 \textit{ is true}\right) * P\left(H_0 \textit{ is true}\right)} \tag{7}$$

In the context of damage detection, equal probabilities may be assigned to the events corresponding to "H_0 is true" and "H_1 is true". Therefore, the probabilities for these two events are equal to 0.5, and hence:

$$P_d = \frac{1-\beta}{1-\beta+\alpha} \tag{8}$$

Though Bayes theorem has been used in this approach, note that the fundamental methodology for detection is still based on the classical hypothesis testing method.

Bayesian Statistics-Based Approach

In this method, residuals are assumed to be normally distributed, i.e. $r \sim N(\mu, \sigma^2)$, and the probability distribution of the mean μ is continuously updated with measurements. Let the prior distribution be N (ρ,

τ^2). If no additional information is available, Migon and Gamerman (Migon & Gamerman, 1999) suggest $\rho = 0$ and $\tau^2 = \sigma^2$. Then, using the observed residual values ($r_1, r_2, r_3 \ldots r_n$, calculated using the available data D), the posterior distribution of μ can be calculated to be normal with mean and variance as follows:

$$\frac{\frac{\rho}{\tau^2} + \frac{r_1 + r_2 \ldots + r_n}{\sigma^2}}{\frac{1}{\tau^2} + \frac{n}{\sigma^2}} \tag{9}$$

$$Posterior\ Variance\ of\ \mu = \left(\frac{1}{\tau^2} + \frac{n}{\sigma^2}\right)^{-1} \tag{10}$$

Damage detection can be achieved through the use of Bayes factor (B) which is defined as the ratio of likelihood of the two scenarios: "damage" and "no damage" as follows:

$$B = \frac{P(D \mid H_1)}{P(D \mid H_0)} \tag{11}$$

Jiang and Mahadevan (Jiang & Mahadevan, 2008b) derived the expression for the (natural) logarithm of Bayes factor, as:

$$\log B = -\frac{1}{2}\log(n+1) + \frac{(n \times \mu_R)^2}{2(n+1)\sigma^2} \tag{12}$$

In Eq. 6, μ_R stands for the mean of the observed residuals, i.e. $\mu_R = (r_1 + r_2 + r_3 \ldots + r_n) / n$. If the Bayes factor B is greater than 1, it implies that the data favors the hypothesis H_1, and hence suggests that there is damage. The confidence in damage detection, i.e. the probability that the fault is true can be calculated as $P(H_1|D)$. If the prior probabilities for $P(H_0)$ and $P(H_1)$ are chosen to be equal (implying that there is equal belief against the presence of damage and for the presence of damage), then,

$$P(H_1|D) = \frac{B}{B+1} \tag{13}$$

This metric gives a direct measure of the probability of damage and hence is easy to compute and interpret in comparison with the classical statistics-based metric. Further, the probabilities $P(H_0)$ and $P(H_1)$ can be updated continuously with measurements and thereby helps in tracking the uncertainty with acquisition of more measurements.

DAMAGE LOCALIZATION

Damage localization is based on the comparison between damage signatures calculated from the system model and the symbols calculated from the system measurements. First, the system model is used to derive cause-effect relationships between the model parameters (that correspond to damage parameters) and model outputs (that correspond to system measurements). These cause-effect relationships, collectively called as the damage signatures, describe the qualitative changes in the 0th and 1st derivatives of a system output as a result of a change in a system parameter. Sometimes, advanced modeling techniques such as bond graphs and temporal causal graphs have been used to generate such signatures (Moustafa, Mahadevan, Daigle, & Biswas, 2010; Sankararaman & Mahadevan, 2011). Alternatively, these signatures may also be generated through simpler finite differencing approaches using the model, as a means of approximating derivatives that are equivalent to fault signatures (Sankararaman & Mahadevan, 2013).

However, practical engineering systems usually have a large number of parameters that could become faulty, but only a small set of measurements. It may be difficult to exclusively isolate one particular damage parameter if several candidates have the same set of damage signatures. Instead, a set of prospective candidate damage parameters, Θ_i ($i = 1$ to m) may be suggested by the isolation procedure. This section associates a measure of uncertainty to each of the candidate damage parameters. Recall that only one of these parameters is the true damage parameter because of the assumption that multiple damages do not occur concurrently. Further, after the damage has occurred, the value of the damage parameter Θ_i has changed to a new unknown value that is denoted by θ_i.

In order to assess the uncertainty in damage localization, measurements of the system are necessary after damage has been detected. Let t_f denote the time at which damage was detected. Let N denote the number of measurements available post damage detection. Now it is necessary to consider all output quantities at once, as a vector. While $\boldsymbol{y}$ denotes the measurements, $\boldsymbol{y}_m^t\left(\theta_i\right)$ denotes the model prediction as a function of time t that ranges from time t_f until time t_f+n. Hereon, the model prediction needs to be considered as a function of the damage parameter because the value of the damage parameter has changed.

Classical Statistics-Based Approach

The uncertainty in damage localization is calculated using a heuristic least squares-based metric. Assuming that Θ_i is the correct damage parameter, all the other damage parameters are assumed be healthy (and their values are chosen to be nominal), and the least squares estimate of can be calculated by minimizing Eq. 14.

$$S\left(\theta\right) = \sum_{t=t_f}^{t_f+N} \left(\boldsymbol{y}_m^t\left(\theta_i\right) - \boldsymbol{y}^t\right)^T \left(\boldsymbol{y}_m^t\left(\theta_i\right) - \boldsymbol{y}^t\right) \tag{14}$$

Note that $y_m^t\left(\theta_i\right)$ and y^t are column vectors whose lengths are equal to the number of output quantities. Hence, the right hand side of Eq. 14 is a scalar.

Let the least square estimate be $\widehat{\theta}_i$ and $S\left(\widehat{\theta}_i\right)$ the corresponding least squares value. Then the probability that Θ_i is the true damage parameter can be heuristically said to be inversely proportional to $S\left(\widehat{\theta}_i\right)$ and the constant of proportionality by considering all prospective damage parameters and equating the total probability to unity. Thus, the uncertainty in localizing the damage to the damage parameter Θ_i can be expressed mathematically as:

$$P_{\Theta_i} = P\left(\Theta_i \textit{ is the true damage parameter}\right) \propto \frac{1}{S\left(\widehat{\theta}_i\right)} \tag{15}$$

Further, the constant of proportionality can be evaluated by imposing the constraint:

$$\sum_{i=1}^{i=m} P_{\Theta_i} = 1 \tag{16}$$

Bayesian Statistics-Based Approach

The Bayesian approach directly calculates the probability that a given damage parameter is faulty and provides an exact measure of uncertainty as against a heuristic measure provided by the classical statistics-based approach. Let A denote the event that Θ_i is the true damage parameter. The Baeysian approach directly calculates the probability of event A after observing the data from time t_f until time t_f+N. (D).

A prior probability P_i' is first assumed for each damage parameter to be the true damage parameter. A suitable assumption for this prior probability would be equal to *1/m* since any of the candidate damage parameters is equally likely to be the damage parameter. Then, this prior distribution is updated using Bayes' theorem to calculate a posterior probability P_i'' based on the following likelihood function:

$$\text{P}(D|A) = \text{P}(D \mid \Theta_i \text{ is the true damage parameter}) = L\left(\Theta_i\right) \tag{17}$$

This probability is the likelihood of event A, denoted by $L\left(\Theta_i\right)$. The calculation of this probability is done in two steps.

Let B denote the event that θ_i is the value of the damage parameter Θ_i. i.e., after damage has occurred. Then, the likelihood function corresponding to the event $A \cap B$ (the probability of observing the data conditioned on

1. Event A, i.e., Θ_i is the true damage parameter; and
2. Event B, i.e., the true damaged value is θ_i) can be calculated using the residuals.

In the context of statistical parameter estimation, it is common to assume a joint Gaussian probability distribution for the residuals (with zero mean and Σ covariance). Therefore, the likelihood can be written as (Pawitan, 2001):

$$P\left(D \mid A \cap B\right) = L\left(\Theta_i, \theta_i\right) \propto \prod_{t=t_f}^{t_f+N} \exp\left[-\left(\boldsymbol{y}_m^t\left(\theta_i\right) - \boldsymbol{y}^t\right)^T \Sigma^{-1}\left(\boldsymbol{y}_m^t\left(\theta_i\right) - y^t\right)\right] \tag{18}$$

The covariance matrix Σ is a square matrix whose size is equal to the number of output quantities. Hence, the right hand side of Eq. 18 is a scalar.

The joint likelihood function $L\left(\Theta_i, \theta_i\right)$ is then used to calculate $L\left(\Theta_i\right)$, which in turn, can be used to update P_i' using Bayes' theorem to calculate P_i''. Let $f'\left(\theta_i \mid \Theta_i\right)$ denote the probability density function (PDF) of the damage value θ_i conditioned on event that θ_i is the true damage parameter. The key is to calculate the posterior probabilities P_i'' and $f''\left(\theta_i \mid \Theta_i\right)$ using Bayes' theorem. While the former indicates the uncertainty in damage localization, the latter indicates the uncertainty in damage quantification. While the upcoming section focuses on the latter topic, i.e., the estimation of $f''\left(\theta_i \mid \Theta_i\right)$, the present section deals with the estimation of P_i''. It can be easily proved that (Sankararaman & Mahadevan, 2013):

$$P_i'' \propto P\left(D \mid A\right) \times P_i' \propto P_i' \times \int L\left(\Theta_i, \theta_i\right) f'\left(\theta_i \mid \Theta_i\right) d\theta_i \tag{19}$$

Note that there is a proportionality constant in Equation 19. This constant can be evaluated by imposing the constraint that the sum of the values of P_i'' corresponding to the m different damage parameters should be equal to unity. This condition is similar to that in Eq. 16, with the difference being that, now the posterior probabilities need to be normalized. Note that the notion of posterior probabilities was absent when Eq. 16 was presented in the classical statistics-based method. Therefore, the advantage of the Bayesian approach is that such posterior probabilities can be updated continuously with the availability of measurements, and therefore, the uncertainty in damage localization can be tracked continuously as a function of time.

DAMAGE ESTIMATION

This section discusses damage estimation and the calculation of the uncertainty in the damage quantification stage, by estimating the uncertainty in the value of damage parameter. The classical statistics-based approach calculates statistical confidence intervals on the value of damage parameter while the Bayesian statistics-based approach directly calculates the probability distribution of the value of the damage parameter.

Classical Statistics-Based Approach

Recall that the least squares estimate of the damaged value of the damage parameter Θ_i was obtained by minimizing the sum of squares in Eq. 14. The least squares estimate was denoted by $\hat{\theta}_i$ and the corresponding value of sum of squares was denoted by $S\left(\hat{\theta}_i\right)$. Using this information, it is possible to compute confidence bounds on θ_i, the value of the damage parameter Θ_i. The values of θ_i that satisfy the inequality in Eq. 20 belong to the α-level confidence bounds of θ_i. In several problems, there is an interval ($[\theta_{i,\alpha\text{-}min}, \theta_{i,\alpha\text{-}max}]$) within which all values of θ_i satisfy this inequality, and therefore, this interval is referred to as the α-level confidence interval.

$$S\left(\theta_i\right) \leq S_\alpha = S\left(\hat{\theta}_i\right)\left\{1 + \frac{p}{n-p} F^{\alpha}_{p,N-p}\right\} \tag{20}$$

In Eq. 20, N denotes the number of observations over which the least squares error was estimated and p denotes the number of parameters being estimated. In this case of single fault assumption, $p = 1$. $F^{\alpha}_{p,N-p}$ denotes the F-statistic, with estimation (numerator) degrees of freedom p and residual (denominator) degrees of freedom $N - p$ at significance level α (Seber & Wild, 1989).

The confidence bounds of θ_i can be graphically illustrated using Fig. 2. The minimum ($\theta_{i,\,\alpha\text{-}min}$) and maximum ($\theta_{i,\,\alpha\text{-}max}$) values that satisfy Eq. 20 are also indicated in this figure, and they can be solved using optimization techniques (Sankararaman & Mahadevan, 2011).

The minimum value $\left(\theta_{i,\alpha-min}\right)$ can be estimated by minimizing the expression $\left(S\left(\theta_i\right) - S_\alpha\right)^2$ by considering values of θ_i that are strictly smaller than $\widehat{\theta}_i$, i.e., $\theta_i \leq \hat{\theta}_i$. Similarly, the maximum value $\left(\theta_{i,\alpha-max}\right)$ can be estimated by minimizing the expression $\left(S\left(\theta_i\right) - S_\alpha\right)^2$ by considering values of θ_i that are strictly larger than $\hat{\theta}_i$, i.e., $\theta_i \geq \hat{\theta}_i$.

The α-level confidence interval on θ_i, i.e., $\left[\theta_{i,a-min}, \theta_{i,a-max}\right]$, needs to be interpreted carefully. It acknowledges that the true value of the damage parameter Θ_i is unknown but deterministic; there is a (1-α)% probability that this interval will contain the true value of the damage parameter. This confidence interval can be calculated at multiple levels of α, but this is not equivalent to estimating the probability distribution of θ_i. In fact, in the classical statistics approach, θ_i is an unknown but deterministic quantity,

Figure 2. Confidence bounds in damage estimation

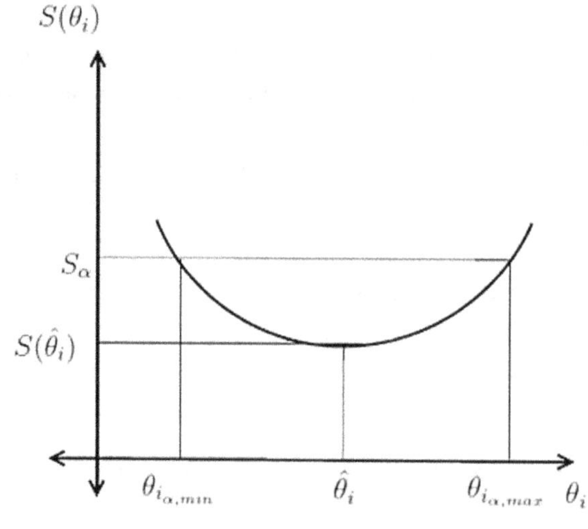

and hence cannot be assigned a probability distribution. However, with the availability of more data, the confidence interval can be reconstructed and its width may decrease with more acquisition of data, if the correct damage parameter is being estimated.

Bayesian Statistics-Based Approach

The Bayesian approach can assign probability distributions even to deterministic quantities (Calvetti & Somersalo, 2007) since it assigns probabilities based on the extent of belief. Therefore, it calculates the entire probability distribution of θ_i, the value of the damage parameter Θ_i. The likelihood function $L(\Theta_i, \theta_i)$ earlier defined in Eq. 18 can be used in Bayes' theorem and the posterior probability distribution of θ_i can be calculated as:

$$f''\left(\theta_i \mid \Theta_i\right) = \frac{L\left(\Theta_i, \theta_i\right) f'\left(\theta_i \mid \Theta_i\right)}{\int L\left(\Theta_i, \theta_i\right) f'\left(\theta_i \mid \Theta_i\right) d\theta_i} \tag{21}$$

Note that it would be meaningful to estimate the damaged value only if Θ_i is the true damage parameter. Therefore, the left hand side of the above equation is conditioned on the event that Θ_i is the true damage parameter. Further, it is also implied that damage has been detected; only then, isolation and estimation can be sequentially performed. Hence, the above posterior distribution is also conditioned on the event that damage has been detected.

In Eq. 21, $f'(\theta_i|\Theta_i)$ is the prior density function and represents the knowledge about θ_i before damage estimation. It is likely that there is no such "prior" knowledge, and a uniform distribution may be assumed, to begin with. Then, this distribution can be updated continuously with acquisition of more measurements. While Markov Chain Monte Carlo methods are commonly used to solve Eq. 21, advanced numerical integration techniques (McKeeman, 1962; Sankararaman, Ling, & Mahadevan, 2010) may be used to provide quicker results. Such numerical techniques are particularly useful since there is only one damage parameter that is being estimated.

The probability distribution of the value θ_i of the damage parameter Θ_i can be calculated, and updated continuously with acquisition of more measurements. For example, a first set of measurements can be used to update the uniform prior distribution to the corresponding posterior distribution. Then, this posterior distribution is used as a prior while updating using the second set of measurements. This procedure can be continued until a desired confidence level in damage quantification is attained. This convenience of continuously updating the uncertainty is a unique feature of the Bayesian method and is not possible when classical statistics-based techniques are used.

OVERALL UNCERTAINTY IN DIAGNOSIS

Having estimated the uncertainty in each of the individual steps of damage diagnosis, i.e., detection, localization, and quantification respectively, it is important to aggregate these three results to compute the overall uncertainty in damage diagnosis. In order to address this issue, consider the uncertainty estimated in the damage quantification step. The probability distribution of θ_i is, in fact, conditioned on two events: damage detection being correct and damage localization being correct. The goal in this section is to estimate the probability distribution of θ_i by accounting for the uncertainty in damage detection and damage localization, and the resultant distribution will be reflective of the overall uncertainty in diagnosis.

Recall that P_d denotes the confidence in damage detection, and P_i'' denotes the confidence in damage localization. While the former refers to the probability that the damage is true, the latter refers to the probability that Θ_i is the true damage parameter. The unconditional distribution of the value of the damage parameter, i.e., the unconditional distribution of θ_i can be estimated by considering three possible scenarios:

1. There is no damage, and hence, the structural health monitoring system triggered a false alarm. This event has a probability $(1 - P_d)$.
2. There is damage, but Θ_i is not the correct damage parameter. The probability for this event is equal to P_d times $\left(1 - P_i''\right)$.
3. There is damage and Θ_i is the correct damage parameter. The probability for this event is equal to P_d times P_i''.

In the first two cases, the probability distribution of θ_i is still nominal (i.e., no damage). Let $f(\theta_i|\text{healthy})$ denote this probability distribution. In the third case, the probability distribution of θ_i corresponds to the posterior distribution $f''\left(\theta_i \mid \Theta_i\right)$. Using the tools of total probability, the unconditional distribution of θ_i can be calculated as:

$$f(\theta_i) = (1 - P_d)\, f(\theta_i|\text{healthy}) + P_d * (1 - P_i) * f(\theta_i|\text{healthy}) + P_d * P_i * f''(\theta_i|\Theta_i) \quad (22)$$

Since the above unconditional distribution accounts for the uncertainty in damage detection, damage localization, and damage quantification, it is indicative of the overall uncertainty in diagnosis. This distribution also can be continuously estimated in time, with the acquisition of measurements. If the true damage parameter is being correctly estimated, then it is expected that the variance of this distribution will decrease with acquisition of measurements, thus indicating increasing confidence in diagnosis.

ILLUSTRATION USING A STRUCTURAL FRAME

Consider a simple two-storied structural frame, as shown in Figure 3. This section illustrates the classical statistics-based method and the Bayesian method for structural damage diagnosis, using this structural frame as a numerical example.

Description of the Structural Frame

The system has 6 parameters, m_1, m_2, k_1, k_2, D_1, D_2, that represent the masses, stiffness, and damping parameters of the two stories of the structural frame. The inputs to the system are forces at the two levels ($F_1(t)$ and $F_2(t)$) and the measured outputs are the accelerations at the two levels. The nominal values of these quantities are provided in Table 1.

Figure 3. (a) Structural frame (b) equivalent mechanical model

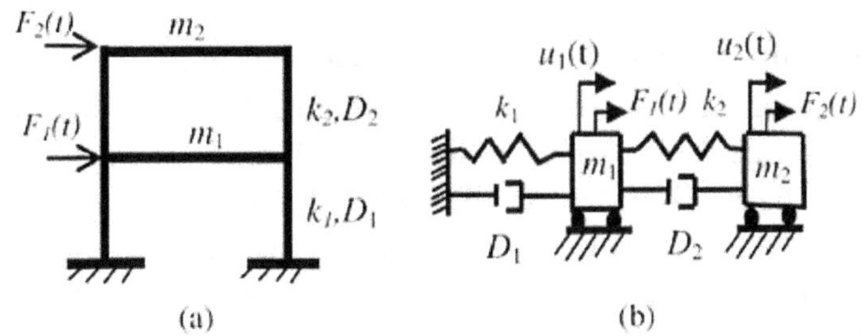

The masses of the first and second floors are assumed not to change. Hence, there are only 4 parameters that could be affected by damage. Using the damage signatures (cause-effect relationships between the model parameters and measurements), it is possible to qualitatively identify the story (floor) that is damaged, and hence the prospective damage parameters are the stiffness and damping of that particular story (floor).

Damage Simulation

In order to verify the effectiveness of the methodology, it is assumed that a known amount of damage is introduced during the normal operation of the structure. For the example of illustration, consider that the value of k_1 is reduced to 80% its nominal value at a particular time instant. The goal is to make use of the methods described in this chapter and diagnose this damage. For this purpose, sensors are used to measure the accelerations of both the stories. The sensors are not accurate and the measurement noise of the sensors is accounted for, through inclusion in the tracking method. Note that the parameters are assumed to be fully known in this numerical example, and the primary source of uncertainty arises from errors in the experimental data due to sensor noise.

Table 1. Two-story frame: parameter values

Parameter	Value
k_1	30700 N/m
k_2	44300 N/m
D_1	307 Ns/m
D_2	443 Ns/m
m_1	136 Ns²/m
m_2	66 Ns²/m
$F_1(t)$	75 sin(9t) N
$F_2(t)$	100 sin(9t) N

Damage Detection: Results

Damage detection is done through hypothesis testing, as explained in earlier in this chapter. Residuals are computed by differencing the observed sensor measurements and the output of tracking system. A moving window is used to calculate the mean of the residuals and this mean is subjected to hypothesis testing to detect damage. The classical statistics-based hypothesis testing detects damage with 95% confidence. Since the values of Type-I and Type-II errors in classical hypothesis testing do not change, the confidence in detection also do not change. In contrast, the Bayesian method facilitates quantification and updating of uncertainty from a 61% confidence at 5 seconds to 99.98% confidence at 5.05 seconds.

Measurements for Continuous Updating

Without loss of generality, the time at which damage is detected is denoted as $t = 0$ seconds. Once damage has been detected, tracking is not used. The measurements and model predictions are directly used for the purpose of quantifying the uncertainty in damage localization and damage quantification. For this purpose, three sets of data are considered, with each set consisting of ten seconds of acceleration data. The first set of data consists of measurements from $t = 0$ seconds until $t = 10$ seconds. The second set of data consists of measurements from $t = 10$ seconds until $t = 20$ seconds. Finally, the third set of data consists of measurements from $t = 20$ seconds until $t = 30$ seconds.

Note that continuous updating of uncertainty through these three sets of data is possible only using the Bayesian method. Therefore, continuous updating results are provided below, only for the Bayesian method. On the hand, the results using classical statistics-based method are provided by considering all the available data as a single set.

Damage Localization: Results

Damage localization, as explained earlier in the chapter, is performed in two steps. First, qualitative damage signatures are used to identify a list of candidate damage parameters. In this example, as stated, earlier, it is possible to uniquely identify the story (floor) that is faulty. Therefore, the candidate damage parameters are k_1 and D_1 respectively. The heuristic measures of uncertainty using classical statistics-based methods yield 98% and 2% probabilities for k_1 and D_1 to be the damage parameters respectively. On the other hand, the uncertainty in localization of k_1, using the Bayesian method, is calculated for the first data set as 99%, and updated using Bayes theorem for the following two sets to 100% and 100%.

Damage Estimation: Results

Damage estimation using the classical statistics procedure yields a 90% confidence interval of (24509 N/m, 24612 N/m). The Bayesian approach directly calculates the probability distribution of k_1 while computing the overall uncertainty in diagnosis; the resultant unconditional probability density function is shown in Figure 4. From Figure 4, it can be seen that (i) the maximum likelihood estimate (MLE) of k_1 matches closely with the true damage value, and (ii) the uncertainty in the estimate of k_1 decreases with acquisition of more measurements.

Figure 4. Continuous updating of damage diagnosis: PDF of k_1

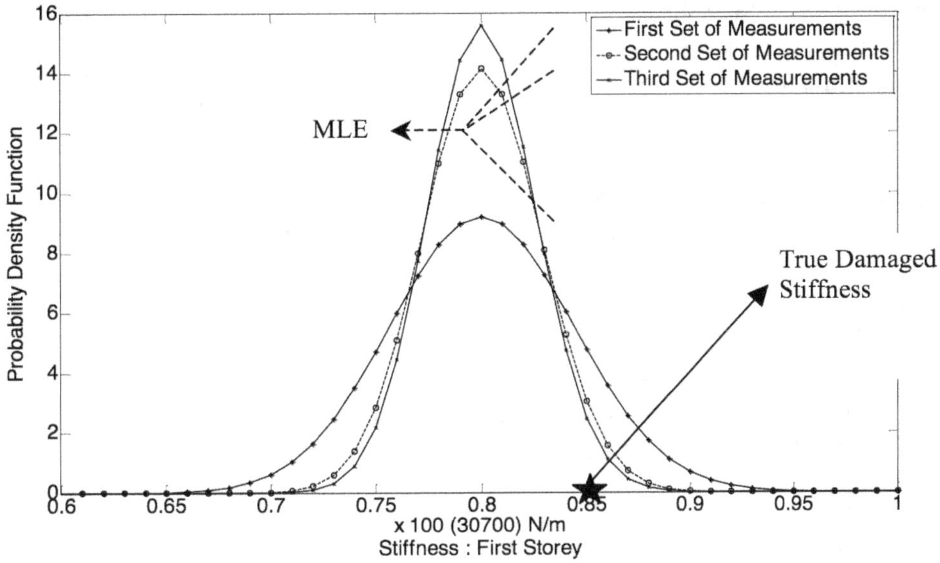

Extension to Multi-Story Frames and Other Systems

The above diagnosis procedure can be easily extended to multi-story frames. The advantage is that a qualitative localization procedure can easily isolate the damage to the accurate floor/story. In fact, different damage scenarios were considered for the five-story frame, and these were detected, localized and estimated satisfactorily by Sankararaman and Mahadevan (Sankararaman & Mahadevan, 2011). The easy extension to multi-story frames clearly indicates that this diagnosis methodology is applicable to practical structures. Further, the methodology has also been extended to health monitoring of different types of systems where field data has been used for online monitoring purposes (Daigle, Sankararaman, & Kulkarni, 2015).

SUMMARY

This chapter presented a statistical methodology for structural damage diagnosis. This methodology can account for the various sources of uncertainty in structural health monitoring and compute the uncertainty in damage detection, localization, and estimation. Further, the overall uncertainty in diagnosis can also be computed, thereby providing a measure of confidence in diagnosis.

Both classical statistics-based methods and Bayesian methods were presented in this chapter. Damage detection was based on statistical hypothesis testing. While the classical hypothesis testing procedure calculated the uncertainty in detection based on the significance level of testing, the Bayesian hypothesis testing procedure calculated the likelihood of "no damage" scenario to "damage" scenario, thereby providing a robust Bayesian metric to assess the confidence in detection. The uncertainty in localization was quantified using heuristic measures in the classical statistics-based approach while the Bayesian method directly quantified the probability that a given damage parameter is faulty. The uncertainty in damage quantification was expressed in statistical confidence intervals in the classical statistics-based approach while the Bayesian method gives the entire probability distribution of the damage parameter. Finally, the uncertainties in the three different steps were combined to determine the probability distribution of the damage parameter as a measure of the overall uncertainty in diagnosis. While the two methods are fundamentally different and hence, it may not be correct to compare the results from the two methods, it is evidently seen that the Bayesian method is easy to implement and interpret. Further, the Bayesian method also facilitates easy updating of uncertainty with acquisition of measurements.

REFERENCES

Achenbach, J. (2000). Quantitative nondestructive evaluation. *International Journal of Solids and Structures*, *37*(1-2), 13–27. doi:10.1016/S0020-7683(99)00074-8

Arulampalam, M. S., Maskell, S., Gordon, N., & Clapp, T. (2002). A tutorial on particle filters for online nonlinear/non-Gaussian Bayesian tracking. *Signal Processing. IEEE Transactions on*, *50*(2), 174–188. doi:10.1109/78.978374

Beck, J. (1989). Statistical system identification of structures. In *Fifth International Conference on Structural safety and reliability*, (pp. 1395–1402). New York: ASCE.

Calvetti, D., & Somersalo, E. (2007). *Introduction to Bayesian Scientific Computing: Ten Lectures on Subjective Computing*. Springer.

Ching, J., Beck, J., & Porter, K. (2006). Bayesian state and parameter estimation of uncertain dynamical systems. *Probabilistic Engineering Mechanics*, *21*(1), 81–96. doi:10.1016/j.probengmech.2005.08.003

Daigle, M., Sankararaman, S., & Kulkarni, C. (2015). Stochastic Prediction of Remaining Driving Time and Distance for a Planetary Rover. In *2015 IEEE Aerospace Conference*, Big Sky, MT. doi:10.1109/AERO.2015.7119144

Dalai, M., Weyer, E., & Campi, M. (2005). Parametric Identification of Nonlinear Systems : Guaranteed Confidence Regions. In *44th IEEE Conference on Decision and Control, and the European Control Conference*. Seville, Spain: IEEE. doi:10.1109/CDC.2005.1583191

Farrar, C. R., & Worden, K. (2007). An introduction to structural health monitoring. *Philosophical Transactions of the Royal Society A: Mathematical, Physical and Engineering Sciences, 365*(1851), 303–315.

Jiang, X., & Mahadevan, S. (2008a). Bayesian Probabilistic Inference for Nonparametric Damage Detection of Structures. *Journal of Engineering Mechanics*, *134*(10), 820–831. doi:10.1061/(ASCE)0733-9399(2008)134:10(820)

Jiang, X., & Mahadevan, S. (2008b). Bayesian Wavelet Methodology for Structural Damage Detection. *Structural Control and Health Monitoring*, *15*(7), 974–991. doi:10.1002/stc.230

Kalman, R. E., & Bucy, R. S. (1961). New results in linear filtering and prediction theory. *Journal of Fluids Engineering*, *83*(1), 95–108.

McKeeman, M. (1962). Algorithm 145: Adaptive numerical integration by Simpson's rule. *Communications of the ACM*, *5*(12), 604. doi:10.1145/355580.369102

Migon, H., & Gamerman, D. (1999). *Statistical Inference: An Integrated Approach*. London: Arnold.

Mosterman, P., & Biswas, G. (1999). Diagnosis of continuous valued systems in transient operating regions. *IEEE Transactions on Systems, Man, and Cybernetics. Part A, Systems and Humans*, *29*(6), 554–565. doi:10.1109/3468.798059

Moustafa, A., Mahadevan, S., Daigle, M., & Biswas, G. (2010). Structural and Sensor Damage Identification using the Bond Graph Approach. *Structural Control and Health Monitoring*, *17*(2), 178–197. doi:10.1002/stc.285

Pawitan, Y. (2001). *In All Likelihood: Statistical Modelling and Inference Using Likelihood* (1st ed.). New York: Oxford University Press.

Peterka, V. (1981). Bayesian Approach to System Identification. In *Trends and Progress in System Identification*. Oxford, UK: Pergamon Press. doi:10.1016/B978-0-08-025683-2.50013-2

Sankararaman, S., Ling, Y., & Mahadevan, S. (2010). Statistical inference of equivalent initial flaw size with complicated structural geometry and multi-axial variable amplitude loading. *International Journal of Fatigue*, *32*(10), 1689–1700. doi:10.1016/j.ijfatigue.2010.03.012

Sankararaman, S., & Mahadevan, S. (2011). Uncertainty quantification in structural damage diagnosis. *Structural Control and Health Monitoring*, *18*(8), 807–824. doi:10.1002/stc.400

Sankararaman, S., & Mahadevan, S. (2013). Bayesian methodology for diagnosis uncertainty quantification and health monitoring. *Structural Control and Health Monitoring*, *20*(1), 88–106. doi:10.1002/stc.476

Seber, G., & Wild, C. (1989). *Nonlinear Regression*. New York: John Wiley and Sons. doi:10.1002/0471725315

Vanik, M., Beck, J., & Au, S. (2000). Bayesian probabilistic approach to structural health monitoring. *Journal of Engineering Mechanics*, *126*(7), 738–745. doi:10.1061/(ASCE)0733-9399(2000)126:7(738)

Vuerinckx, R., Pintelon, R., Schoukens, J., & Rolain, Y. (2001). Obtaining Accurate Confidence Regions for the Estimated Zeros and Poles in System Identification Problems. *IEEE Transactions on Automatic Control*, *46*(4), 656–659. doi:10.1109/9.917672

KEY TERMS AND DEFINITIONS

ASCE Benchmark: Experimental data and simulations models obtained from the 4-story, 2-bay by 2-bay steel-frame scale-model structure located in the Earthquake Engineering Research Laboratory at the University of British Columbia (UBC). These experiments are intended to evaluate and compare several SHM methods. A finite element model is used to simulate damage scenarios.

Case Base Reasoning: Model of human reasoning in which the people rely on previous experiences when they need to solve problems, reusing solutions without thinking too much about the situation.

Discrete Wavelet Transform: Signal-processing technique used to capture both frequency and time location information. Discrete refers that wavelets basis are discretely sampled.

Pipe Structure: Test structure used to evaluate the effectiveness of Expert Systems to assess stiffness damage on pipelines. The pipeline dynamical model was implemented by using modal beam theory, mass-spring-damper system, finite element method and space-state analysis.

Principal Component Analysis: Statistical tool used for data dimensional reduction, pattern recognition, data compression and extraction where the main goal of PCA is to discern among a set of data which are significant for the system, which are redundant and which are just noise.

Self-Organizing Map: Type of unsupervised artificial neural network used to produce a discretized low-dimensional representation of the input space of the training samples. In this work, the SOM are used to cluster similar cases.

Structural Damage: Changes introduced into a system (intentionally or unintentionally), that adversely affect the current or future performance of that system. In this work, these changes are induced by reducing stiffness value in elements of the structure.

Chapter 9
Nonlinear Ultrasonics for Early Damage Detection

Rafael Munoz
Universidad de Granada, Spain

Guillermo Rus
Universidad de Granada, Spain

Nicolas Bochud
Universidad de Granada, Spain

Daniel J. Barnard
Iowa State University, USA

Juan Melchor
Universidad de Granada, Spain

Juan Chiachío Ruano
Universidad de Granada, Spain

Manuel Chiachío
Universidad de Granada, Spain

Sergio Cantero
Universidad de Granada, Spain

Antonio M. Callejas
Universidad de Granada, Spain

Laura M. Peralta
Universidad de Granada, Spain

Leonard J. Bond
Iowa State University, USA

ABSTRACT

Structural Health Monitoring (SHM) is an emerging discipline that aims at improving the management of the life cycle of industrial components. The scope of this chapter is to present the integration of nonlinear ultrasonics with the Bayesian inverse problem as an appropriate tool to estimate the updated health state of a component taking into account the associated uncertainties. This updated information can be further used by prognostics algorithms to estimate the future damage stages. Nonlinear ultrasonics allows an early detection of damage moving forward the achievement of reliable predictions, while the inverse problem emerges as a rigorous method to extract the slight signature of early damage inside the experimental signals using theoretical models. The Bayesian version of the inverse problem allows measuring the underlying uncertainties, improving the prediction process. This chapter presents the fundamentals of nonlinear ultrasonics, their practical application for SHM, and the Bayesian inverse problem as a method to unveil damage and manage uncertainty.

DOI: 10.4018/978-1-4666-8490-4.ch009

INTRODUCTION

Nonlinear ultrasonics can play a major role in industrial structural health monitoring, since one of its main features is the capability to detect early damage manifestations. The nonlinear constitutive properties of a material, in the sense of deviation from Hooke's law, are recently being shown to be orders of magnitude more sensitive to micro-damage than the linear ones. As a consequence, a number of effects arise in the propagation of the ultrasonic waves as they interact with the material nonlinearity (Ostrovsky & Johnson, 2001; Broda, Staszewski, Martowicz, & Silberschmidt, 2014), namely,

- **Higher Harmonics Generation:** When a sinusoidal wave is emitted, a signal with additional harmonics, multiples of the fundamental, is received.
- Hysteretic mechanical behavior, in the stress-strain relationship.
- Wave modulations:
    - In amplitude and phase when two sinusoidal frequencies are propagated.
    - **Cross-Modulation of Waves (Luxemburg-Gorky Effect):** Passing modulation from an amplitude modulated excitation to a simultaneous and initially pure sinusoidal excitation.
- Amplitude dependent resonant frequency shifts when the material is insonified.
- Attenuation that depends on the excitation amplitude.
- **Acoustic Conditioning:** An immediate offset on material property values when insonification is active. The initial transient period time is also referred as fast dynamics.
- **Relaxation Effects or Fast and Slow Dynamics:** When excitation vanishes, material properties has a partial fast recovery and a slow final recovery that last minutes to achieve the original values, before excitation.
- Subharmonic generation.

These nonlinear effects, which are originated in the materials' microscopic and mesoscopic structures, are separable from the linear propagation in the frequency domain by a variety of experimental configurations (Ostrovsky & Johnson, 2001; Zheng, Maev, & Solodov, 1999; Jhang, 2009). Mesoscopic size is the scale of early damage manifestations, and micro-damage has also been shown to be particularly invisible to linear techniques (Nicholson & Bouxsein, 2000). Some recent experimental observations suggest that hysteretic and/or *nonlinear mechanical properties* may be a key factor to quantify changes and could unveil details of the micro- and meso-structure; both the intrinsic nonlinearity in the material or that of an induced damage. In practice, micro-damage modes such as micro-cracks behave as ultrasonic sources at frequencies different from the excitation, making it a promising tool to clearly locate and quantify damage (Matikas T., 2010). Nonetheless, few studies have been performed on the relationship between crack density and level of nonlinearity (Haupert, et al., 2014; Renaud, Callé, Remenieras, & Defontaine, 2008).

Industry will always quest for an as early as possible detection of damage. The increasing complexity of materials and damage modes, along with early detection needs, is leading to a cutting-edge development and usage of nondestructive techniques. The challenge is the extraction of a weak damage signature within the received signals. This demanding evaluation of the material integrity is one of the drivers on the incorporation of theoretical models to discriminate these weak damage signatures, by means of solving a model-based inverse problem. The aim is to reproduce, by modeled simulations, the acquired signals of the test, using an iterative procedure, which progressively adjusts the values of the model parameters, material properties included. Damage is quantified in terms of the final set of material parameters and its numerical difference to the undamaged values. However, models, materials, experimentation and signal processing methods, are subject to uncertainty, in the sense of lack of knowledge or incomplete information, which may not be neglected, especially when this information is used to predict the remaining life of the material. A Bayesian inverse problem approach is necessary to deal with such uncertainty to allow prognosis tools to incorporate both evaluation result and its uncertainty. This is an important step because security standards in some industries are extremely high and all information must be included in the decision process.

The structure of the chapter begins presenting the fundamentals of nonlinear ultrasonics, focusing primarily on classical and nonclassical hysteretic nonlinearities. Next, nonlinear ultrasonic techniques of the main interest for industrial applications are described. An explanation of nonlinear evidences in metals and composites has been included, because of their relevance in industries with high security standards. Right after, the need for computational modeling to solve the inverse problem for evaluation is addressed; acquired data and theoretical models can be integrated strengthening the evaluation process to allow the solution of cases with weak damage signatures. Finally, the Bayesian inverse problem is explained, as the suitable link to prognosis tools.

FUNDAMENTALS

This section presents the most frequent mathematical descriptions of nonlinearity in the literature. We propose that nonlinearity can be understood from three different perspectives:

1. At the continuum mechanics level, where the constitutive laws that relate stress and strain deviate from Hooke's law,
2. At the theoretical level, or wave equation level, where the nonlinear parameters are directly related to the observable harmonics that appear in oscillatory movements such as ultrasonics, and
3. At the micro-mechanics level, in which atomistic, clapping micro-cracks, thermo-elastic effects and so on, are proposed hypothesis that may originate the observable nonlinearity at the other two levels.

As opposed to other reviews, here we start from the continuum mechanics level, since there is no consensus over the variety of mechanistic origins of nonlinearity.

The nonlinear behavior of the material is quantified in several *nonlinear material parameters*, being its characterization of major importance. The most frequent nonlinearity parameters from the different continuum mechanics nonlinear definitions are consistently derived, being all based on Taylor expansions of different magnitudes that appear in the constitutive equations. This expansion makes no assumption on the nature of the nonlinearity, which can be proposed *ad hoc*. This formulation serves as a basis to quantitatively understand and interpret experimental observations, since damage alters these mechanical properties.

The nonlinear constitutive equation is described here by making use of the Third Order Elastic Constants (TOEC), which are then linked to the nonlinear ultrasound propagation under the one dimensional P-wave restriction. Classical and nonclassical types of acoustic nonlinearities are developed below, showing their effect on common stress-strain, strain-time and strain amplitude-frequency spectrum relations. Finally, Westervelt nonlinear wave equation is referred to explain how nonlinearity couples with the dissipation phenomena in fluids.

Nonlinear Elasticity

In order to classify the nonlinear acoustic behavior of materials, the principles of nonlinear elasticity are introduced. The general constitutive equations are considered in tensile form, to derive the one-dimensional wave equation with nonlinear elastic coefficients.

The notation used here was introduced by (Norris, 1998). Lagrangian and Eulerian coordinates are defined as x_i and x_i^0, and displacements as $u_i = x_i - x_i^0$. The deformation gradient tensor is $F_{ij} = \partial x_i^0/\partial x_i$ and the associated Green-Lagrange strain tensor is $E_{ij} = 1/2\ (F_{ij}\, F_{ij}^{\mathrm{T}} - \delta_{ij})$ or,

$$E_{ij} = \frac{1}{2}\left(\frac{\partial u_i}{\partial x_j} + \frac{\partial u_j}{\partial x_i} + \frac{\partial u_k}{\partial x_i}\frac{\partial u_k}{\partial x_j}\right) \tag{1}$$

equation that gives the general kinematic relations. The last term is frequently and erroneously neglected in ultrasonic literature because small strains are considered. However, it should be taken into account to describe nonlinear effects, as harmonic generation, since geometric nonlinearity alone also contributes to it, even for an ideally Hookean elastic material. Equation (1) can be written with respect to either coordinate system. Working in Eulerian coordinates, the equation of motion is,

$$\rho\, u_{i,\,tt} = \sigma_{ij,j} \tag{2}$$

where ρ is the density, $u_i = (x_1, x_2, x_3, t)$ the displacement and σ_{ij} the Cauchy stress, being $i,j = 1,2,3$ the Cartesian components and noting that the compact notation of derivative is used ($v,_k = \partial v/\partial k$). The Cauchy stress is given by the constitutive equation, generally defined as,

$$\sigma_{ij} = \frac{1}{\det(F_{ij})} F_{ij} \frac{\partial W}{\partial E_{ij}} F_{ij}^T$$

Defining a relationship between strain energy W and strain E_{ij}, as a series expansion allows any degree of approximation to reality without loss of generality,

$$W = \underbrace{\frac{1}{2!} C_{ijkl} E_{ij} E_{kl}}_{linear} + \underbrace{\frac{1}{3!} C_{ijklmn} E_{ij} E_{kl} E_{mn}}_{nonlinear} + \ldots$$

where C_{ijkl} and C_{ijklmn} are the second order and third order stiffness matrices, accounting for the mechanical properties of the material, the first term describing the linear elasticity. As a consequence, the stress can be also decomposed into linear and nonlinear parts as,

$$\sigma_{ij} = \sigma_{ij}^L + \sigma_{ij}^{NL}$$

Table 1. Literature values of third-order elastic constants experimentally determined

TOEC	Muir (2009) AL7075 (GPa)	Stobbe (2005) AL7075 (GPa)	Dubuget et al. (1996) AL7075 (GPa)	Smith (1966) AL2S (GPa)
$\mathcal{A}$	-334.5	-351.2	~ -282	-408
$\mathcal{B}$	-125.35	-149.4	~ -179	-197
$\mathcal{C}$	60.5	-102.8	~ 53	-114

If all this is applied to an isotropic material, the nonlinear version of the Hooke's law can be written as,

$$\sigma_{ij} = \underbrace{\lambda E_{kk}\delta_{ij} + 2\mu E_{ij}}_{linear} + \underbrace{\left(\mathcal{A} + 4\mu\right)E_{ij}E_{jk} + \mathcal{B}E_{kp}E_{pk}\delta_{ij} + 2\left(\mathcal{B} + \lambda - \mu\right)E_{kk}E_{ij} + \left(\mathcal{C} - \lambda\right)(E_{kk}\delta_{ij})^2}_{nonlinear}$$

where λ and μ are the Lamé constants related to the Young modulus E, and Poisson ratio ν, as $\lambda = E\nu/[(1+\nu)(1-2\nu)]$ and $\mu = E/[2(1+\nu)]$, E_{ij} is the strain tensor, E_{kk} is the trace of the strain tensor, $E_{ij}E_{jk}$ is the square of the strain tensor, δ_{ij} is the Kronecker delta, and $\mathcal{A}$, $\mathcal{B}$ and $\mathcal{C}$ are often referred to as the nonlinear, third order elastic constants (Landau & Lifshitz, 1986). Note that many authors defined other third order elastic constants, by taking into account different combinations of invariants,

Murnaghan (1951):

$$l = \mathcal{B} + \mathcal{C},\ m = \frac{1}{2}\mathcal{A} + \mathcal{B},\ n = \mathcal{A}$$

Standard C_{ijk}:

$$c_{111} = 2\mathcal{A} + 6\mathcal{B} + 2\mathcal{C},\ c_{123} = 2\mathcal{C},\ c_{144} = \mathcal{B},\ c_{112} = 2\mathcal{B} + 2\mathcal{C},\ c_{456} = \frac{1}{4}\mathcal{A},\ c_{166} = \frac{1}{2}\mathcal{A} + \mathcal{B}$$

As an example, Table 1 illustrates some experimental values of third order elastic constants in terms of Landau relations, using experimental acousto-elasticity techniques for Aluminum (Muir, 2009) and by other authors for Aluminum 7075 and Aluminum 2S samples using finite amplitude methodology and various ultrasonic setups.

A solid is characterized by positive bulk and shear moduli but the signs of third order elastic constants are not univocal as shown in Table 1.

Nonlinear Propagation of Ultrasound

The nonlinear wave propagation is particularized to one dimensional P-wave and S-wave equation to extract nonlinearity parameter of first order β, which has been widely used in the literature. The second and hysteretic nonlinearity parameters are considered and characterized in the next sections. Finally Westervelt nonlinear equation is also described using the *B*/*A* parameter in fluids.

1. One dimensional longitudinal (P-wave) Equation.

The one-dimensional nonlinear wave equation for solids is derived from the continuum model above. The nonlinear wave equation is reduced here to the classical theory and its fundamental solution up to third- and fourth-order is provided below.

The kinematic relations applied to the one-dimensional case from Equation (1) can be written in the following equation (Zarembo & Krasil'Nikov, 1971):

$$\varepsilon_{ij} = \frac{1}{2}\left(u_{i,j} + u_{j,i} + u_{k,i}u_{k,j}\right) \tag{3}$$

Under 1D P-wave restriction we note that $u_2=u_3=0$, $u_1(x_1, x_2, x_3)=u_1(x_1)$, since particles only move in the 1-direction, and $\varepsilon_{ij}=0$, $\forall ij\neq 11$,

Kinematic relations:

$$\varepsilon_{11} = u_{1,1} + \frac{1}{2}u_{1,1}^2 \tag{4}$$

Nonlinear constitutive law:

$$\sigma_{11} = (\lambda + 2\mu)\varepsilon_{11} + [(\lambda + 2\mu) + (\mathcal{A} + 3\mathcal{B} + \mathcal{C})]\varepsilon_{11}^2 + O(\varepsilon_{11}^3) \tag{5}$$

(neglecting attenuation)

Dynamic equilibrium:

$$\rho u_{1,tt} = \sigma_{11,1} \tag{6}$$

2. Nonlinearity of first-order.

The case of the nonlinear wave equation up to the first-order nonlinearity (one nonlinear parameter β_H) is first considered, neglecting the term $O(\varepsilon_{11}^3)$ in the constitutive law. Therefore, Equations (4) to (6) are reduced to,

$$\rho\ddot{u}(x,t) = Mu''(x,t)\left(1 - 2\beta_H u'(x,t)\right) \tag{7}$$

where $u = u_1$, $\ddot{u}=u_{1,tt}$ and $u''=u_{1,11}$, $M = \lambda + 2\mu = \rho\, c_p^{\,2}$, and $c_p^{\,2}$ denotes the longitudinal wave velocity. Note that the nonlinear coefficient β_H have been introduced, labeled H since it corresponds to the definition by Hamilton and Blackstock (1998),

$$\beta_H = -\frac{3}{2} - \frac{\mathcal{A} + 3\mathcal{B} + \mathcal{C}}{\lambda + 2\mu}$$

Although Hamilton and Blackstock define this coefficient as negative, a positive version, β^+_H, is usually managed in literature, transforming the minus sign of the equation (7) into a positive one. The solution can be found using the perturbation theory (Ginsberg, 2008), which decomposes the wave displacement as,

$$u = u_0 + u_1 + \ldots \tag{8}$$

where u_0 and u_1 denote the zero-order and first-order perturbation solutions, respectively. The zero-order perturbation solution corresponds to the fundamental solution to the linear wave equation (that is, when β_H=0, linear case). When considering a monochromatic plane wave propagating in a semi-infinite non-linear elastic layer, the latter is given as,

$$u_0(x,t) = a(x)e^{i(kx-\omega t)}$$

where k is the wave number, $a(x)$ is the amplitude of fundamental harmonic and $\omega = 2\pi f$ is the angular frequency.

The first-order solution adds the deviation from the linear behavior and the nonlinear parameter of first order β_H, as a measure of how the material generates this deviation, can be determined for this special case of plane wave by iteration from Equations (7) and (8),

$$\ddot{u}_1 - c_p^2 u_1'' = 2c_p^2 \beta_H^+ u_0'' u_0'$$

The former expression has the form of a classical partial differential equation with an inhomogeneous part. Thus, a particular solution may be obtained by the method of variations of parameters,

$$u_1(x,t) = b(x)e^{2i(kx-\omega t)}$$

where $b(x)$ is the amplitude of the second harmonic. From that solution, assuming $a(x)=a$ and $b(x)=b$ as constants, the nonlinear parameter of first-order β^+_H is derived as,

$$\beta_H^+ = \frac{4b}{k^2 a^2 x} \tag{9}$$

A different definition of the nonlinear parameter of first order (β_G) was given by Guyer and Johnson (1999) assuming a different expansion of the constitutive law,

$$\sigma' = K\left(\varepsilon' + \beta_G \left(\varepsilon^2\right)' + O\left[\left(\varepsilon^3\right)'\right]\right) + A[\varepsilon, \dot{\varepsilon}]$$

where K is the compressional modulus, β_G is the nonlinear term and $A[\varepsilon, \dot{\varepsilon}]$ are the nonclassical hysteretic elastic elements, typically non-analytic. Note that it is only valid for fluids since it coincides with that of Hamilton if μ=0. This is used to explain both harmonics generation and frequency mixing. For instance, if two excitation frequencies, ω_a, ω_b are considered (with $\omega_a > \omega_b$), the solution of the wave equation yields to additional waves with frequencies $\omega_c = \omega_a \pm \omega_b$ with amplitude proportional to $\varepsilon_c \sim \varepsilon_a \varepsilon_b$, increasing with x. Their amplitudes ε_a, ε_b and ε_c, are related by,

$$\beta_G \approx \frac{\varepsilon_c}{x\varepsilon_a^2} \tag{10}$$

Other different nonlinear parameters of first order were deduced by Zarembo and Krasil'Nikov (1971), for the particular hypothesis of plane waves, obtaining the nonlinear longitudinal parameter of first order β^l_Z, defined as follows,

$$\beta_Z^l = 3\frac{\lambda + 2\mu}{\rho} + \frac{1}{\rho}\left(2\mathcal{A} + 6\mathcal{B} + 2\mathcal{C}\right) = -\frac{2(\lambda + 2\mu)}{\rho}\beta_H$$

and considering S-waves in the u_2-direction $u_2(x_1)$, and $u_1 = u_3 = 0$, the nonlinear transversal parameter β^τ_Z, defined as follows,

$$\beta_Z^\tau = \frac{\mu}{\rho} + \frac{1}{\rho}\left(\frac{\mathcal{A}}{2} + \mathcal{B}\right)$$

This definition establishes a relationship between nonlinear longitudinal and transversal parameters of first-order with the TOECs.

3. Nonlinearity of second-order.

The case of the nonlinear wave equation up to the second-order nonlinearity (two nonlinear parameter β_H and δ) is considered now. That means to neglect $O(\varepsilon_{11}^4)$ in the constitutive law. Therefore, as it was applied before, equations (4) to (6) are reduced to,

$$\ddot{u}(x,t) = c_p^2\left(1 + 2\beta_H^+ u'(x,t) + 3\delta\left(u'(x,t)^2\right)\right)u''(x,t)$$

The nonlinear coefficient of second-order δ can be derived in the same manner as the first-order one β_H, using the perturbation theory, resulting in,

$$u_2'' - \frac{1}{c_p^2}\ddot{u}_2 = -2\beta_H^+\left(u_0''u_1' + u_1''u_0'1\right) - 3\delta u_0''1(u_0')^2$$

being u_2 is the second-order perturbation solution. The solution for the δ parameter with monochromatic plane wave excitation is,

$$u_2(x,t) = c(x)e^{3i(kx-\omega t)}$$
$$\delta = \frac{6c}{k^3 a^3 x},$$
$$if\ \beta_H^2(k)x \ll |\delta|$$

4. Extension to nonclassical nonlinearity.

Recent studies show that a broad category of materials share nonclassical nonlinear elastic behavior. Manifestations of nonclassical nonlinearity include stress-strain hysteresis and discrete memory in quasi-static experiments, and specific dependencies of the harmonic amplitudes with respect to the driving amplitude in dynamic wave experiments, which are remarkably different from those predicted by the classical theory.

Nonclassical nonlinear effects are believed to be due to the presence of soft regions in hard materials (e.g., micro-cracks, micropores and soft bonding regions between material grains). They have successfully been reproduced by a model introduced by (Holcomb, 1981) and, later, by Guyer and McCall (1999), based on a Preisach-Mayergoyz (PM) space representation (Van Den Abeele K. E., Johnson, Guyer, & McCall, 1997) in analogy with the treatment of magnetic hysteresis. Differences between nonclassical and classical Landau-type (Landau & Lifshitz, 1986) nonlinear dynamic behavior include: a downshift of the resonance frequency, proportional to the resonance amplitude in the nonclassical case versus a quadratic amplitude dependence in the classical case; nonlinear attenuation versus amplitude independent attenuation; quadratic amplitude dependence of the third harmonic versus cubic in the classical case (Haupert, et al., 2011).

Given a harmonic excitation on a nonlinear constitutive material, the history of any variable (either the strain, stress or displacement) at any point will not necessarily be harmonic and can therefore be represented by,

$$\sigma = M\varepsilon\left(1 + \beta_G\varepsilon + \delta\varepsilon^2 + \ldots + \alpha\left[\varepsilon + \varepsilon\ \mathrm{sgn}(\dot{\varepsilon})\right]\right)$$

The term with α introduces the nonclassical behavior as shown in Figure 1, where the implications of the nonlinear classical and nonclassical contributions on the acoustic wave propagation are depicted, highlighting the effects in the stress-strain relations, the frequency spectrum and the deformation of the signal.

5. *B/A* parameter in fluids. Westervelt nonlinear propagation model.

In order to describe nonlinearity in fluids, it is possible to derive the classical nonlinear Westervelt equation, where the nonlinear parameter *B/A* is usually provided. This is originated in the Taylor series expansion of density in terms of sound pressure, instead of the stress,

Figure 1. Nonlinear contribution to constitutive equation
Adapted from Van Den Abeele, Johnson and Sutin (2000).

Linear	Nonlinear classic 1st order (β)	Nonlinear classic 2nd order (δ)	Nonlinear Hysteretic (α)

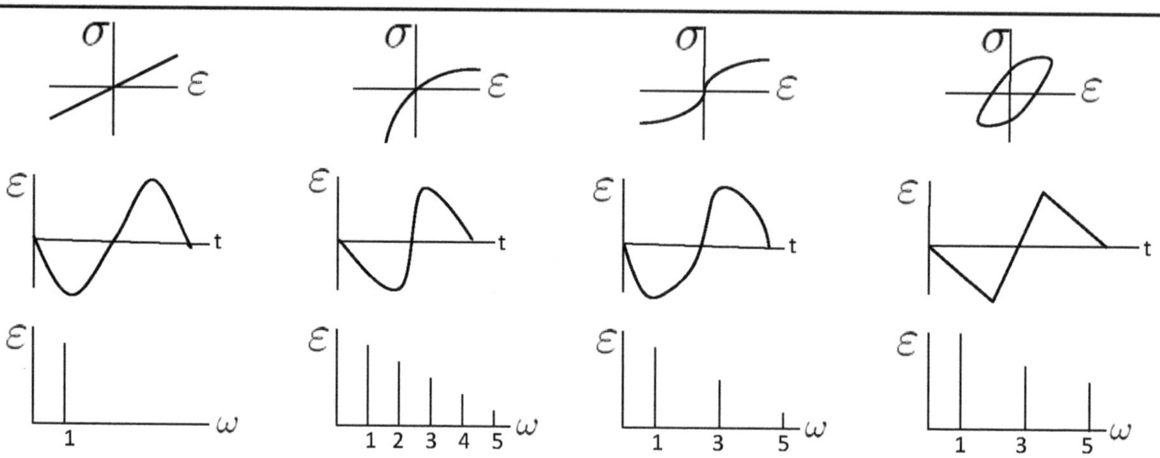

$$p = A\left(\frac{\rho'}{\rho_0}\right) + \frac{B}{2!}\left(\frac{\rho'}{\rho_0}\right)^2 + \frac{C}{3!}\left(\frac{\rho'}{\rho_0}\right)^3 + \ldots$$

By using the relation $c^2 = (\partial P/\partial \rho)$ where P is absolute pressure, $p=P-P_0$ sound pressure with P_0 equilibrium absolute pressure and ρ density respectively, the Westervelt equation can be obtained (Hamilton & Blackstock, 1998),

$$\nabla^2 p - \frac{1}{c_0^2}\frac{\partial^2 p}{\partial t^2} + \frac{\delta}{c_0^4}\frac{\partial^3 p}{\partial t^3} = \frac{\beta_W}{\rho_0 c_0^4}\frac{\partial^2 p}{\partial t^2}$$

where ρ_0, is the ambient density and $c_0 = (K/\rho)^{1/2}$, is the small signal sound speed and δ is the sound diffusivity. β_W is the nonlinear coefficient of first order as defined from Westervelt,

$$\beta_W = 1 + \frac{B}{2A}$$

where A and B are the coefficients of the Taylor expansion. Values of B/A has been calculated for selected fluids and liquefied gases. Their relation with TOECs are taken, accounting for the null value of the shear modulus in fluids,

$$\lambda = A\ \mathcal{A}\ \subsetneq = 0\ \mathcal{B} = -A\ \mathcal{C} = \frac{A-B}{2}$$

Table 2. Relationship between various nonlinear classic first order β from literature and TOEC, harmonics amplitudes, and B/A ratio

Nonlinear Classic 1st Order β	Relation with Third Order Elastic Constants	Harmonics Amplitudes and Strains	*B/A* Ratio (Fluids)
β_H	$-\frac{3}{2}-\frac{\mathcal{A}+3\mathcal{B}+\mathcal{C}}{\lambda+2\mu}$	$\frac{4b(x)}{k^2a(x)^2x}$	$-5-\frac{B}{2A}$
β_Z^l	$3\frac{\lambda+2\mu}{\rho}+\frac{1}{\rho}\left(2\mathcal{A}+6\mathcal{B}+2\mathcal{C}\right)$	-	$-\frac{2A+B}{\rho}$
β_Z^T	$\frac{\mu}{\rho}+\frac{1}{\rho}\left(\frac{\mathcal{A}}{2}+\mathcal{B}\right)$	-	-
β_G	$-\frac{3}{2}-\frac{\mathcal{A}+3\mathcal{B}+\mathcal{C}}{K}$	$\frac{\varepsilon_c}{x\varepsilon_a^2}$	-
β_W	$1+\frac{2\mathcal{C}+\mathcal{B}}{2\mathcal{B}}$	-	$1+\frac{B}{2A}$

Table 2 shows the equivalence between β_H, β_G, β_Z and β_W with the third order elastic constants, harmonic amplitudes, and *B*/*A* ratio for fluids.

NONLINEAR EXPERIMENTAL TECHNIQUES AND EVIDENCES

Experimental techniques are designed to detect experimental nonlinear behavior. The usual process is the discovery of some experimental evidence with a determined setup. The experimental setup and procedure are modified to magnify the evidence. Finally, theoretical models are modified to include the observed evidences in the simulated results. Meanwhile, experimental modification could lead to a new technique, and its posterior improvement.

Experimental Techniques

The following experimental techniques are briefly described,

- Finite Amplitude Technique (FA).
- Nonlinear Elastic Wave Spectroscopy (NEWS):
  - Nonlinear Resonant Ultrasound Spectroscopy (NRUS).
  - Nonlinear Wave Modulation Spectroscopy (NWMS) or Vibro-Acoustic Modulation (NUVM).
- Dynamic Acousto-Elasticity (DAE).

Finite Amplitude Technique

Measuring the amplitudes of generated harmonics is commonly referred to as the finite-amplitude (FA) method, initially developed by Breazeale and Thompson (1963). The nonlinear coefficients are usually determined by measuring the second-harmonic generation and sometimes higher harmonics, and can be used to characterize acoustic nonlinear properties of gases, liquids, and solids. For this technique, the through-transmission mode in immersion is usually preferred. Instead of using two transducers, it is opportune to replace the receiver by a needle hydrophone (with a nearly linear frequency response), in order to conveniently measure the second and higher-harmonics. A finite-duration burst of (nearly) pure tone - typically around 20 cycles long - is launched towards the specimen, and the progress of some stationary peaks near the end of the tone-burst is followed and selected to compute the Fast Fourier Transform (FFT), which then allows to obtain the second and higher-order harmonics amplitude.

Different wave types have been investigated with this technique, such as shear waves, surface acoustic waves (SAW), lamb waves (Matsuda & Shiro, 2014), but above all longitudinal waves. Initially, bulk exploration using nonlinear longitudinal waves has attracted the attention of many authors. In the case of shear waves in pure isotropic materials, the second harmonic is impossible (Zarembo & Krasil'Nikov, 1971), so early damage is detectable since it makes them arise. Bulk waves experiences need double-sided access to the specimen, while SAW and Lamb ones just need a single-side access. Nonewithstanding, the pulse-echo technique using longitudinal waves has slightly been assessed in fluids (Vander Meulen & Haumesser, 2008), and metals (Best, Croxford, & Neild, 2013; Fukuda, Nishihira, & Imano, 2007). In such a case, the rebound on a pressure release boundary in most NDT applications and the double interaction make the signal interpretation particularly difficult (Bender, Kim, Jacobs, & Qu, 2013).

The finite-amplitude technique has been shown to be useful for defect detection in ceramics (Sutin & Nazarov, 1995), concrete structures (Zarembo & Krasil'Nikov, 1971; Bruno, Gliozzi, Scalerandi, & Antonaci, 2009), composites (Matikas T., 2010), fatigue cracks in metals, such as steels, titanium, and aluminum alloys (Na, Cantrell, & Yost, 1996; Cantrell & Yost, 1994). Such defects are originated in internal stresses, micro-cracks, zero-volume disbonds, and usually precede the main cracking mechanisms and the failure of the material. Therefore, a considerable number of authors have performed laboratory experiments to show that cracks and imperfect interfaces can behave in a nonlinear fashion (Donskoy, Sutin, & Ekimov, 2001; Solodov, Krohn, & Busse, 2002), and have thus opened new opportunities to detect partially closed cracks that may not be identified by conventional linear methods.

The finite-amplitude method is a relatively straightforward technique to measure the harmonic peaks, and obtain the nonlinear coefficients of materials (Equation). The low complexity of the experimental installation could make of this method a low-cost and valuable technology for in-situ industrial applications. Nonetheless, a practical extraction of the harmonics requires numerous efforts in minimizing the nonlinear distortions from electronic devices and in optimizing the reproducibility of the experiment. Indeed, several factors such as the size of the gap between the specimen and the receiver or the geometrical dispersion of the transducers (inherently related to the focal distance) may have a drastic influence on the measured nonlinear elastic coefficients, and should be analyzed carefully.

NEWS: Nonlinear Elastic Wave Spectroscopy

Nonlinear Elastic Wave Spectroscopy (NEWS) denotes a group of different nonlinear ultrasonic techniques, which consist of measuring the variations of material parameters with amplitude changes on an ultrasonic excitation. Typical material parameters are the resonance frequencies, attenuation, fundamental frequency or the amplitude of new harmonics.

This group can be divided into two main families of techniques, Nonlinear Resonant Ultrasound Spectroscopy (NRUS), and Nonlinear Wave Modulation Spectroscopy (NWMS), also known as Nonlinear Ultrasound Vibro-Acoustic Modulation Technique (VMT or NUVM). It should be noted that some authors do not follow this classification, and they refer to resonant techniques for NRUS and frequency mixing techniques, parametric interaction or modulation techniques.

NRUS: Nonlinear Resonant Ultrasound Spectroscopy

NRUS is a resonance testing method used to evaluate nonlinear vibrational response of both intact and damaged materials. The technique was earlier applied to geomaterials (Johnson, Zinszner, & Rasolofosaon, 1996; Guyer & Johnson, 1999; Guyer, McCall, & Van Den Abeele, 1998), but has received significant attention in a wide range of applications such as fiber-reinforced polymer composites (Polimeno & Meo, 2009; Payan, Garnier, Moysan, & Johnson, 2007), concrete (Payan, Garnier, Moysan, & Johnson, 2007), and bone (Muller, et al., 2005; Muller, et al., 2006). Whereas in standard Resonance Ultrasound Spectroscopy (RUS) the resonance spectra of intact and damaged samples are compared, NRUS examines the excitation amplitude dependence of resonance frequencies, a result of the nonlinearity of the material under test, whether in an intact or damaged state.

The measurement concept is straightforward: couple a sample, such as a simple beam, to an excitation source and sweep the excitation frequency, for example, ±25-100Hz about a resonance mode peak frequency while measuring the response of the beam, repeating the measurement for various levels of excitation. It is common to sweep around the first bending mode resonance of simple beams, but higher order modes can be examined. Common excitation sources are piezoelectric or magnetostrictive actuators or audio speakers, while accelerometers, laser vibrometers or even phonograph needles can be used to monitor the samples vibrational response. A diagrammatic representation of a measurement setup is shown in Figure 2.

The typical response seen in NRUS experiments is a softening of the material with increasing excitation level, that is, the resonance frequency decreases with increased excitation, demonstrating a softening of the moduli with increasing strain. A typical response is shown in Figure 2 (a), (b) and (c); the response of an intact carbon fiber-reinforced polymer sample, 25mm wide x 150mm long, of 30 unidirectional plies with samples cut from the laminate plate along the 0°, 45° and 90° directions.

Apart from the usual response, fast dynamics (TenCate, 2011) has been additionally observed, with a variation in the response depending on the direction of the frequency sweep. Tracking a resonance peak with alternating periods of high strain and low strain input excitations, conditioning and relaxation effects (slow dynamics) have been reported. Johnson (2006) demonstrated the change in resonance response

Figure 2. Sample with the vibrational actuator and the accelerometer. Change in resonance peak for the first bending mode for increasing excitation amplitudes for sample long axis along 0° (case a), 45° (case b) and 90° directions (case c)
From Chakrapani, Barnard and Vinay (2014).

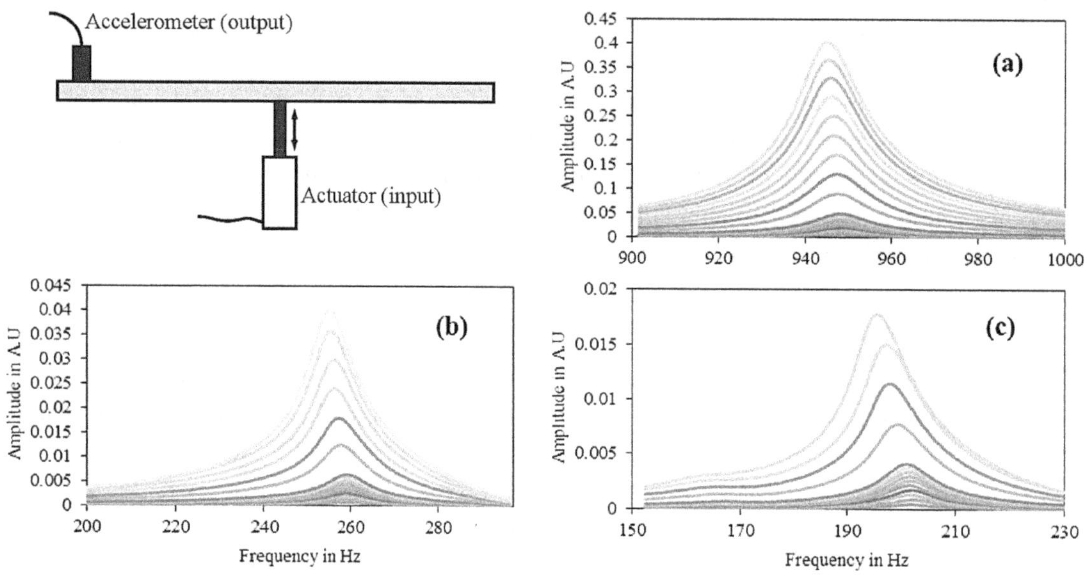

in concrete in the presence of damage, where the resonance frequency of a particular bending mode is decreased and the overall response broadened (damped) as a result of damage.

The resonance response as shown above is typically modified to show the change in resonance peak frequency normalized to the lowest excitation peak frequency ($\Delta f / f_0$) as a function of strain in order to allow comparison of the frequency response for different bending modes, boundary conditions and material conditions. This is accomplished converting velocity (or acceleration) to strain using,

$$\varepsilon \approx 0.219 \frac{T}{f\sqrt{12}} \left(\frac{1.875}{L/2} \right)^2 v$$

corresponding in this case to the response for a cantilever beam of half length, where v is the velocity determined by integrating the acceleration data over time, L and T are sample dimensions and f is the resonance frequency. Further, to determine attenuation or modal damping ratio (MDR) (Van Den Abeele, Sutin, Carmeliet, & Johnson, 2001), the MDR formula is used,

$$Q^{-1} = \frac{f_{hw}}{f}$$

where f_{hw} is the frequency width at half the peak amplitude of the resonant peak (f) at finite strain amplitude. MDR has been used by many researchers to show non-classical nonlinear responses. In the linear regime, MDR remains a constant. At higher strains, a linear increase in MDR can be observed for non-classical nonlinear materials (Chakrapani, Barnard, & Vinay, 2015).

Several of the examples shown demonstrate that, in undamaged materials, the intrinsic nonlinear response is significant, and it should be understood that this response is further modified by the experimental conditions, particularly temperature. Pasqualini et al. (2007) described the care required to achieve a sub-hertz frequency stability, where a thermal stability of 10mK was required. Haupert et al. (2011) noted observations where temperature changes over the course of a measurement lead to changes in the frequency and damping on the order of the intrinsic nonlinearity, and suggested an optimized NRUS method.

A simpler time-domain version of the method, called Nonlinear Reverberation Spectroscopy (NRS), has been developed (Van Den Abeele, Le Bas, Van Damme, & Katkowski, 2009). Once the steady state is reached next to a resonance frequency, the excitation is turned off, and the subsequent reverberation is recorded, that is, the decaying transient time signal. Linear materials show a constant frequency along the overall signal, whereas a gradual frequency increase is observed with nonlinear materials. The response is independent of the amplitude and frequency applied on the initial excitation, so, there is no need for frequency sweeps nor repetitions at different excitation amplitudes, since the decaying signal incorporates by itself diverse amplitude levels.

NRUS represents a laboratory method that has demonstrated significant sensitivity and utility in characterizing a wide variety of materials as well as damage in those materials. The equipment and sensors required are commonly found in many laboratory inventories and do not represent a significant cost to acquire and employ, particularly when the frequencies are in the audio range.

Nonlinear Wave Modulation Spectroscopy: Vibro-Acoustic Modulation Technique

NWMS or NUVM is one of the most time efficient nonlinear NDE method, which has been shown to be suitable for a straightforward application in field (Zagrai, Donskoy, Chudnovsky, & Golovin, 2008; Van Den Abeele, Johnson, & Sutin, 2000). It is based on monitoring the mixing interaction of two waves with different fundamental frequencies. Concretely, this technique consists of exciting the component with a low frequency (LF) continuous sine-wave (in the range of few kHz, sometimes even less), usually called pump wave or LF wave, generated by an ultrasonic probe or, alternatively, by a vibrational actuator or a speaker. In fact, in order to maximize its amplitude, a resonant frequency is often selected. Although this is not a requirement, it can be troublesome if a node matches a defect location, so that the experiment should be repeated with other pumping frequencies. At the same time, a high frequency (HF) continuous wave, often hundreds of kHz, is injected into the material, usually with an ultrasonic probe. This wave is amplitude modulated by the material nonlinearity, for example a defect, incorporating an amplitude variation with the period of the pump wave (see signal (R) in Figure 3). The resulting interaction is received by a piezoelectric probe, an accelerometer or a laser device. If no damage is present on a linear material, both monochromatic signals f_l and f_h are received, and the spectrum shows just the pumping and the high frequencies, where neither mixing nor modulation are present. In contrast, for a damaged material, the acoustic modulation is expressed in the spectrum in terms of sidebands and LF harmonics. Sidebands manifest as frequencies at both sides of the higher frequency ($f_h \pm nf_l$) (Equation (10)), which result from mixings of both fundamentals. On the other hand, LF harmonics are multiple harmonics of the pumping frequency (nf_l). All those interactions can also be slightly observed in a non-classical nonlinear material with no damage. The experiment can be repeated for several amplitudes of the pump wave, observing how sidebands and LF harmonics grow.

Figure 3. Emitted HF signal (E), LF signal and received amplitude modulated signal (R). Up-right: Frequency distribution of the received signal with sidebands ($f_h \pm nf_l$) at both sides of the HF frequency, f_h, and LF harmonics (nf_l)

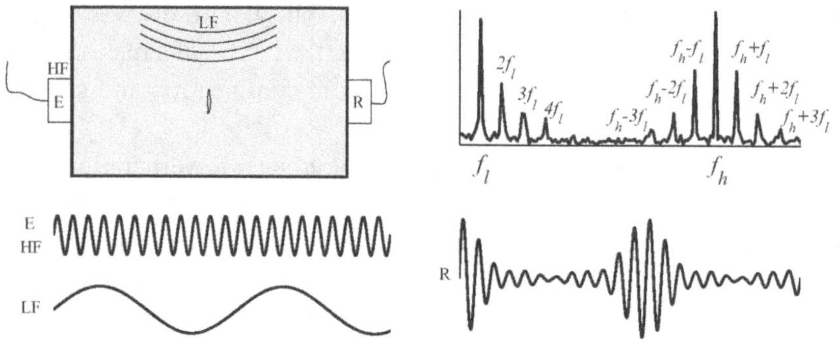

The nonlinear manifestations usually observed in classical or hysteretic materials are summarized in Table 3 (Van Den Abeele, Johnson, & Sutin, 2000; Zaitsev, Matveev, & Matveyev, 2011). It must be noted that the presence of a defect often multiplies such manifestations, at least for the nonlinear coefficient of first-order β.

In damaged materials, the modulation effect is explained in terms of changes in the rough contact area of the defects (Donskoy & Sutin, 1998; Broda, Staszewski, Martowicz, & Silberschmidt, 2014), affecting the elastic constants. The weakening of the defect contact, when tensile semicycle of the pump wave is acting, distorts the amplitude of several cycles of the high frequency wave, producing its amplitude modulation.

This technique allows detecting early presence of damage on the inspected component, just in the volume where wave interactions occur. An early detection means that changes can be observed at the mesoscale defect size, in contrast to linear techniques that are limited to the detection of defects that are larger than the wavelength. The larger the defect, the larger the nonlinear interactions, although a sound previous experience on the specific application must be accumulated in order to be able to quantify the level of damage. In that vein, attempts to establish a quantitative measure have been developed by proposing a Damage Index (DI) (Zagrai, Donskoy, Chudnovsky, & Golovin, 2008), which was defined as the difference between modulation indexes with and without damage. As a drawback, it should be mentioned that a precise location of damage is not available.

According to the literature, NUVM is typically applied using longitudinal waves, but it has also been extended to other wave types, such as SAW (Jhang, 2009; Ellwood, Stratoudaki, Sharples, Clark, & Somekh, 2014; Matikas & Aggelis, 2011), Lamb waves (Shkerdin & Glorieux, 2010; Jhang, 2009) or

Table 3. Amplitude of the LF harmonics and sidebands depending on the nonlinear nature. C(β,δ), a combination of the respective nonlinear parameters

	$2f_l$	$3f_l$	$f_l \pm f_h$	$f_l \pm 2f_h$
Classical	β	$\sim \varepsilon^3$	$\beta A_l A_h$	$C(\beta,\delta) A_l^2 A_h$
Hysteretic	β	$\sim \varepsilon^2$	$\beta A_l A_h$	$\alpha A_l A_h$

torsional waves (Ekimov, Didenkulov, & Kazakov, 1999). Given that nonlinearity manifests as the appearance of new harmonics that arise from the interaction, this technique does not demand a strict linear behavior of the instrumentation, and is even quite independent on environmental changes (in contrast to the FA technique). These additional features further support the in-field usage.

The so-called TR-NEWS and NEWS-TR techniques are further developments of NUVM based on time reversal methods (Fink, 1997) to solve its defect location incapacity, at the cost of more complex instrumentation. In TR-NEWS (Ulrich, Sutin, Guyer, & Johnson, 2008), each source (LF and HF) is separately emitted and the received signals are all time reversed and both simultaneously re-emitted from the original sources (reciprocal TR), producing a focusing effect on the detector location. This procedure is repeated moving the receiver at other locations, and thus enables to detect and image damage near the surface. In NEWS-TR (Gliozzi, Griffa, & Scalerandi, 2006; Goursolle, Callé, Dos Santos, & Bou Matar, 2007) modulated signals are received by an array of transducers, filtered to erase all parts due to linear interactions keeping only the nonlinear ones. Afterwards, the time-reversed versions of the signals are emitted in the material by each element of the array in a modeled simulation, producing a focusing effect in the location where the nonlinear interactions occur.

Another similar technique to NUVM is cross-modulation, based on the Luxemburg-Gorky effect (Zaitsev, Gusev, & Castagnede, 2002; Zaitsev & Matveev, 2006). Here the pump wave is initially amplitude modulated, so this wave has two frequencies, that of the pump itself (pump carrier f_p), and that of the amplitude change cycle (modulation frequency f_m). Being f_h the frequency of the ultrasonic probe, damage is expressed in terms of growing sidelobes around the central frequency of the probe wave ($f_h \pm f_m$). Higher sensitivity than NUVM is reported (Zaitsev, Matveev, & Matveyev, 2011) if the experiment is correctly tuned, along with higher flexibility in the frequency selection.

Parametric emission (Renaud, Callé, Remenieras, & Defontaine, 2008) and non-collinear mixing (Croxford, Wilcox, Drinkwater, & Nagy, 2009) are techniques based on the same physics, but at different ranges of operation. The first one consists of a frequency mixing of two pure harmonic waves with a slight frequency difference. After the nonlinear interactions, a LF wave of frequency (f_1-f_2) can be acquired. Non-collinear mixing is a mixing of two frequencies travelling in different directions that can produce mode conversions in the mixed signal, for instance, two incident S waves can generate a P wave. Finally, several contributions also analyze the phase of the modulated signals to detect damage, for instance, by means of the bi-spectral analysis (Courtney, Drinkwater, Neild, & Wilcox, 2008), and describe phase modulations in the received signals that are used to measure the nonlinear coefficient of first-order β (Vila, Vander Meulen, Dos Santos, Haumesser, & Bou Matar, 2004; Stratoudaki, et al., 2011). This phase modulation will be the key feature in DAE technique, discussed hereafter.

DAE: Dynamic Acousto-Elasticity

Dynamic Acousto Elastic Technique (DAE, also DAET) is another frequency mixing procedure. This technique is also based on the use of a pump and a high frequency waves, but relies on the phase modulation, usually measured in the time domain employing short pulses. DAE origins can be found in tests of ultrasonic propagation when the sample is statically compressed (SAE – Static acousto-elasticity) (Winkler & Liu, 1996). It was observed that the propagation velocity varied depending upon the different applied compression forces due to nonlinear changes in the elastic parameters, especially in the elastic modulus. Computing the changes of the ultrasonic times-of-flight (TOF) served as a measure to quantify the nonlinearity.

Figure 4. TOF measure oscillates over its reference value, that without LF, leading to oscillations of propagation velocity that depend on the actual applied strain by the LF pump. A larger hysteretic behavior can be detected when a crack is in the middle of the sound path.

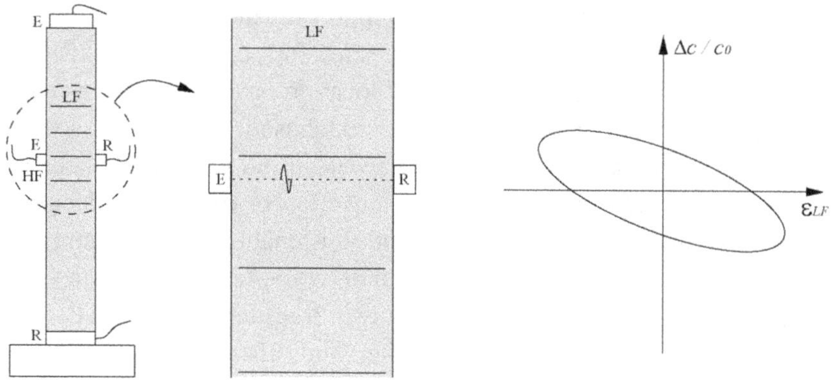

The underlying fundaments are similar to SAE but dynamically performed with smaller strain levels. The sample is subject to a compression-tension low-frequency cycle by means of a LF acoustic pump. Simultaneously, HF ultrasonic short pulses with constant time intervals are transmitted normally to the LF and received by another transducer on the opposite side (see Figure 4). The complete HF signal is acquired consisting of a repetition of pulses, so that their TOF (Rivière, Renaud, Guyer, & Johnson, 2013; Renaud, Talmant, Callé, Defontaine, & Laugier, 2011) and/or attenuation (Renaud, Riviere, Haupert, & Laugier, 2013; Renaud, Talmant, Callé, Defontaine, & Laugier, 2011) are measured. Given the large frequency difference between pump and HF wave, the HF pulse nearly finds the same compression state along its complete sound path and in the upper and lower part of the HF acoustic beam. A TOF measure without pump is first acquired and taken as the reference for the successive TOFs, to observe the variations of propagation velocity of the successive TOF's, once the pump is in place. Consequently, instantaneous nonlinear effects can be measured over the complete strain (LF) cycle. This *photographic* feature is one of the main peculiarities of DAE, since all other dynamic nonlinear techniques supply averaged information over the pump cycle. It should be noted that HF and LF signals must be sampled with a high acquisition rate in order to match each HF pulse with its corresponding compression state.

The hysteretic behavior of materials has clearly been described with DAE (Haupert, et al., 2014; Rivière, Renaud, Guyer, & Johnson, 2013), expressed in terms of a different curve followed by the velocity propagation along the tensile-compression pump cycle (Figure 4, right). The higher the pump amplitude, the larger the hysteresis cycle. The photographic feature allows to observe and measure some nonlinear effects that other techniques cannot distinguish; namely, acoustic conditioning and relaxation effects (Renaud, Riviere, Haupert, & Laugier, 2013). Haupert et al. (2014) presented an optimized procedure improving DAE sensitivity and reducing the specimen size, allowing the technique to be applied to weak nonlinear materials.

Bone studies have been the driver subject for the development of this lab technique (Renaud, Callé, Remenieras, & Defontaine, 2008; Moreschi, et al., 2011), followed by geomaterials (Renaud, Riviere, Haupert, & Laugier, 2013; Rivière, Renaud, Guyer, & Johnson, 2013) and incipient damage research on metals under fatigue (Haupert, et al., 2014) or stress corrosion cracking (Rivière, et al., 2014). Both dry contact and underwater experiences have been reported.

Evidences in Metals

Crack and micro-crack generation in metals can be originated by multiple causes. Fatigue is one of the most common sources of failure, along with thermal stress, ageing, or corrosion. Two stages in the process of micro-crack generation must be noted; the previous and posterior moments to micro-crack appearance. Nonlinear ultrasonics are useful to detect both phases. This section will focus on fatigue assessment before micro-crack appearance and contact nonlinearity for damage detection, independently on its generation mechanism. References for hardening, thermal stress, ageing, precipitation effects and corrosion previous to micro-crack appearance can be found in literature (Jhang, 2009).

Fatigue Evaluation

Fatigue in metals is a consequence of cyclic loading, and the subsequent growing number of dislocations, their spatial distribution within the material and their aggregation into crack. Dislocation dynamics includes elastic and microplastic effects. Specifically, when dislocations appear, they grow in density, move through the lattice and get accumulated at discontinuities like grain boundaries. This increasing accumulation generates micro-cracks, which gain size and finally provoke the structure collapse. This process is followed by a material hardening, which can only be detected at the end of the life by linear ultrasonics.

FA is particularly useful to detect/evaluate fatigue in metals, since a direct, even monotonic, relation between dislocation density and the measured nonlinear coefficient of first-order β has been reported. This effect can be detected at the early life of the material. Examples have been described, for instance, in a Ti-Al alloy (Matikas T., 2010), Al alloy and stainless steel (Na, Cantrell, & Yost, 1996; Cantrell J., 2009; Sagar, Das, Paridaa, & Bhattacharya, 2006). Experiences with other metals can be found, among others, quenched steels and stainless steel (Jhang, 2009), high temperature steels such as Inconel 718 (Barnard, Dace, & Buck, 1997) and other austenitic steels (Viswanath, et al., 2011). Hence, this rather simple technique enables to quantify the dislocation density at the actual life cycle of the material, by measuring the second harmonic amplitude and comparing the actual β coefficient to that initially obtained β_0. Ratios of around 180% have been described at the end of the life (Matikas T., 2010). A model to describe this relationship was proposed by Cantrell and Yost (1994), where the nonlinear parameter β was deduced from dislocation parameters. The values provided by this model were in good agreement with those experimentally obtained on their Al alloy.

Other wave types for fatigue assessment have also been proposed. At the early fatigue stages, shear wave second harmonic generation has been proven to provide a reliable signature of the dislocation content (Zarembo & Krasil'Nikov, 1971). A precise description of the studies conducted with Lamb waves and SAW is beyond the scope of this chapter, and the reader is referred to the reviews of Jhang (2009) and Broda (2014). Other nonlinear techniques have also been applied for evaluating fatigue, but they are mainly used to detect micro-cracks, further discussed in the next section, since they do not seem suitable for the first part of the life cycle.

Contact Nonlinearity Detection

Contact nonlinearity represents a planar discontinuity between two surfaces of a material, which can be associated to a simple contact between two samples, the contact of both sides of a micro-crack, two bonded or locally debonded layers, the material-transducer interface, etc.

Changes in stiffness between compressional and tensile part of the oscillation modify the ultrasonic signal after crossing the interface, leading to higher harmonics generation (Richardson, 1979), so that FA method can be suitable for crack and micro-crack detection.

NRUS technique in metals has been poorly investigated, mainly with the presence of large cracks and scarcely to detect micro-damage. Concerning the latter, Cheon et al. (2010) found experimental evidences for carbon steel CT showing a relationship between driving voltage slope and nonlinearity produced by fatigue micro-cracks. The higher the driving voltage slope, the larger the nonlinearity in the cracks. In a related proposal, Ohtani and Ishii (2012) started to detect differences in fatigued copper samples at around 70% of the life cycle.

Detection of fatigue cracks in steel pipes and pins have been studied with NUVM technique (Donskoy & Sutin, 1998), showing dependence between the nonlinear parameter β and the increase of fatigue cycles. Zagrai et al. (2008) proposed a damage index (DI) in stainless steel specimens providing information about crack appearance. Another nonlinear evidence with this technique is the size quantification of fatigue cracks in steel bars (Donskoy, Sutin, & Ekimov, 2001), which shows a linear dependence between modulation amplitude and vibration amplitude.

DAE technique has been recently applied to study cracks of certain size. In fatigued aluminum an elliptical hysteresis loop for the propagation velocity variation has been observed (Rivière, et al., 2014). The hysteresis was measurable along the whole crack, but it was much stronger for signals passing though the crack tip, rapidly decaying after crack tip towards a non hysteretic loop. The nonlinear coefficient of first-order β was also extracted and showed a similar variation. A parallel experience in steel was applied on a crack subject to corrosion (Haupert, et al., 2014), showing a higher nonlinearity in the crack tip, in terms of hysteresis in propagation velocity and the extracted nonlinear coefficients of first- and second-order, β and δ. These higher nonlinear effects on crack tips suggest that the application of the technique to micro-crack detection might be feasible.

Layer adhesion quality control has also been a topic of major interest, especially weak adhesions and zero volume disbonds, quite elusive to linear techniques. This problem is far to be solved and nonlinear acoustics opens an expectation to improve the evaluation of this defects. Resonance methods (Drewry, et al., 2009), FA technique (Barnard, Dace, Rehbein, & Buck, 1997), non-collinear mixing (Escobar-Ruiz, et al., 2014) or mixing of Lamb waves (Shkerdin & Glorieux, 2010) have also been used.

Evidences in Composites

The main causes of damage in carbon fiber-reinforced polymers (CFRP) are fatigue, impact loads, especially low velocity loads that imply damage that is invisible from surface, and high temperatures. The failure evolution when subject to tension fatigue loads usually consists of a) matrix cracking, with growing cracks tending to reach consecutive fiber layers, d) delamination between layers, especially of different orientation, from tip joining of matrix cracks, and c) fiber breakage. All these processes are followed by some characteristic stiffness degradation, strongly correlated with damage development and, so, with the whole life cycle. Compressive fatigue loading does not clearly show these three phases and initial stiffness loss is smaller. Consequently, early damage detection should address matrix cracking rather than delamination.

Most of the applications developed to detect damage in composites make uses of standard linear techniques, such as through-transmission under normal (i.e. for delaminations) and oblique incidence (i.e. for matrix cracks). Lamb waves have attracted the attention of many authors, since damage slow down their propagation. Recently, some nonlinear techniques have also been applied to this end, but they are still unusual.

FA technique in pulse-echo mode has been applied to generate C-scans of the second harmonic to detect delamination (Kawashima, et al., 2006) and of the fifth harmonic for impact damage detection (Krohn, Stoessel, & Busse, 2002). Hirsekorn et al. (2006) studied nine samples subjected to a gradually increasing impact damage using a nonlinear through-transmission technique, showing that fundamental frequency decreases with impact energy, and the distortion wave factor clearly increased in the samples with higher damage. Concerning NRUS technique, results show a higher resonant frequency change on each measurement with increasing number of fatigue loading cycles, especially during the last part of the life. Slow dynamics has also been detected for advanced stages of life (Van Den Abeele, Katkowski, & Mattei, 2006). In the same vein, experiments with samples subjected to high temperature induced damage have shown a quasi-linear dependence between the average crack density and the nonlinear parameter Γ, being Γ the slope of the resonant frequency change against the applied strain (Van Den Abeele, Le Bas, Van Damme, & Katkowski, 2009). NUVM technique has also been applied to study impact damage in a sandwich plate (Meo & Zumpano, 2005), showing that matrix cracking or fiber breakage can be related to the appearance of sidebands in the frequency spectrum, besides the harmonic generation. Aymerich (2010) obtained similar results for a CFRP sample subjected to a low velocity impact damage.

Toyama&Takatsubo (2004) measured the lamb wave propagation in CFRP under different static tensile strains, finding a nonlinear stress-strain behavior. The measured Young modulus with different static strains slightly but clearly decayed with increasing crack density, obtaining a good correlation between the change of both magnitudes.

Finally, Armitage&Wright (2013) commented that the scarce nonlinear experiences in literature are due to the difficulty to figure out which technique is the most adequate for which kind of damage, taking into account that damage from different origins may be present. To face this problem, they proposed a multi-technique approach to inspect CFRP, creating and testing a special device for this purpose.

NEED FOR THEORETICAL MODELS IN DAMAGE EVALUATION

Computational Modeling

The complexity of the nonlinear observations suggests comparing the experimental measurements with theoretical models, with the aim of extracting relevant quantitative information from damage. This strategy is being proven to be far more sensitive than applying heuristic signal processing techniques. For this purpose, numerous computational models have been proposed to solve the various wave equations derived from the fundamental models overviewed above. Since a deep description would be necessary to precisely introduce all those models, only a brief review of the most commonly applied methods is provided in Table 4.

Table 4. Brief overview of the main computational models

Classification	Approach	Characteristics
1D Finite Difference techniques	The second-order partial differential equations are approximated replacing derivatives with finite differences, which are sometimes combined with a recursive approximation based on the perturbation theory.	Solves isotropic homogeneous wave propagation. Accounts for various attenuation models. Solves classical nonlinearity (McCall K., 1994; Van Den Abeele K. E., 1996) and nonclassical, PM-space by a variety of approximations (McCall & Guyer, 1994; Van Den Abeele & Johnson, 1996; Gusev, 2000; Nazarov, Radostin, Ostrovsky, & Soustova, 2003; Meurer, Qu, & Jacobs, 2002).
1D layered Transfer Matrix Formalism	Piecewise semi-analytical solutions are derived for every layer and grouped matricially, which is sometimes combined with a recursive approximation based on the perturbation theory.	Yields solutions for transversal propagation across an arbitrary number of layers. Accounts for various attenuation models. Solves classical nonlinearity (Bochud, 2014; Yun, Miao, Zhang, Huang, & Wei, 2005)
2D/3D Finite Difference Time Domain (FDTD)	The wave equations are approximated by finite differences and discretized over a staggered grid, which is sometimes combined with Green functions.	Classical nonlinearity (KZK, Westervelt, Burgers) in fluids is solved by standard FDTD (Lee & Hamilton, 1995; Cleveland, Hamilton, & Blackstock, 1996; Hallaj & Cleveland, 1999). Alternative techniques (Local interaction simulation approach: LISA) (Delsanto, Hirsekorn, Agostini, Loparco, & Koka, 2002) solve heterogeneous media in conjunction with a spring model (Lee & Hamilton, 1995; Cleveland, Hamilton, & Blackstock, 1996; Hallaj & Cleveland, 1999; Van Den Abeele K. E., 1996; McCall K., 1994; McCall & Guyer, 1994; Barbieri, Meo, & Polimeno, 2009).
2D/3D Finite Difference and Finite Element for hysteretic nonlinearity	The variational form of the wave equations, discretized on the basis of element shape functions, is solved by minimizing an error function.	Implements crack clapping contact in transient FEM. Has a limited computational efficiency (Lee & Hamilton, 1995; Cleveland, Hamilton, & Blackstock, 1996; Hallaj & Cleveland, 1999; Köjl, Hurlebaus, & Gaul, 2004). Alternative formulations implement nonlinear boundary conditions to account for Hertzian contacts (Pecorari, 2003).

Model-Based Inverse Problem

Material health assessment through damage evaluation consists of obtaining information about abnormal behavior of the material using observations, which can further be used for condition-based maintenance, predictions about future damage, etc. In contrast to the forward problem (FP) where simulated responses are obtained using pre-defined values for a set of model parameters, an inverse problem (IP) is aimed at retrieving the model parameters, whose simulated response matches the experimental real response.

When particularizing to ultrasonic modeling for damage detection, a forward model is required to simulate the underlying physics of the ultrasound signal propagation across the material in presence of complex patterns of damage. On the other hand, the inverse procedure allows detecting, locating and evaluating the presence of damage in the material, after receiving the signals that have interacted with its internal structure.

Typical applications of FP's in industry consist of obtaining preliminary evaluations on the insonified region of the material before purchasing a specific transducer and scaling testing parameters for new inspection procedures, just to name a few. From the perspective of forward modeling problems, complex models may be preferred over simpler models because they can be argued to provide more realistic simulated responses. For inverse problems, however, this may lead to over-fitting of the data where the model is unnecessarily adjusted to fit the specific set of data used without necessarily obeying real physics. Then, the model does not generalize well when making predictions. On the other hand, solving the inverse problem by using simplistic physical models, or even, by ray tracing (as typically happens in industry), may lead to inaccurate predictions. Therefore, the challenge is *to infer a suit combination of parameters to be plugged into the model in order to reproduce the measurements* with an acceptably low error. Damage is revealed because this suitable set of model parameters (which includes material parameters and approaches simulation to real signals) is affected by variations from their undamaged values (Farim, Gallego, Bochud, & Rus, 2013; Rus, Wooh, & Gallego, 2007).

At this standpoint, a generic description of the IP way to proceed is presented:

- Perform the ultrasonic test to obtain experimental signals,
- Select a suitable forward model that depends on a number of model parameters (material properties included).
- Parametrization: select the model parameters to calculate and infer expected damage, fixing the rest to constant values.
- Make an initial hypothesis on the value of the model parameters, based on prior knowledge,
- Simulate the forward model to obtain a simulated signal at a specified location,
- Perform a comparison between the simulated and experimental signals at the same location, using some kind of metric to quantify the discrepancy,
- Propose a new set of parameters to minimize the discrepancy and control the convergence process by some kind of algorithm (genetic algorithm, gradient-based algorithms, Markov chain, etc.), until the pre-defined (typically user-defined) level of convergence is reached (see an example in Figure 5),
- Repeat the process for every location of interest,
- Finally, interpret the spatial distribution of damage according to the changes of parameter values along the different positions (see Figure 6 as example).

Figure 5. from Bochud (2014). Simulated signal (black) already fitted to the experimental one (red) with enough low error. Above the adjusted model parameters of the fitted simulated signal. The first six parameters are Young Moduli of the cross-plies and interfaces of a [0 90]$_{4S}$ impact-damaged CFRP laminate and the last three are their nonlinear coefficient of first order.

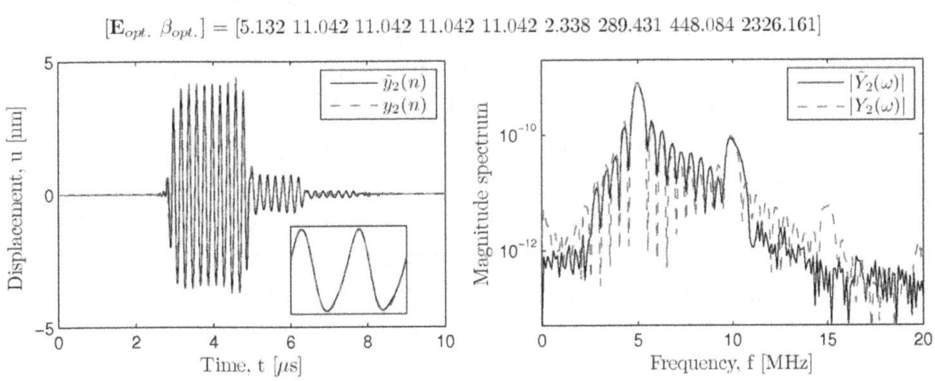

Figure 6. C-Scan showing the spatial distribution of the Young modulus of the second layer of a cross-ply CFRP laminate
From Bochud (2014).

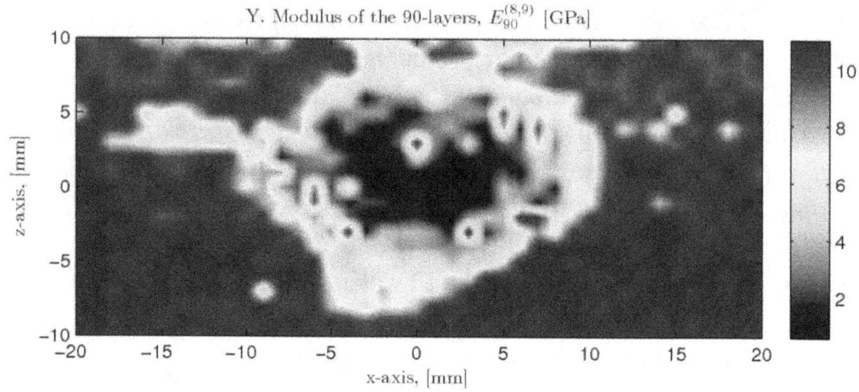

Bayesian Inverse Problem: Allowing Prognostics Tools to Manage Uncertainty

Reconstructing the values of the model parameters (Young modulus, attenuation or nonlinear parameters, etc.) has a limited meaning in presence of measurement errors, heterogeneity of properties within the specimen, and even the fact that the model used to idealize ultrasound signal transmission, is just an approximation of reality. A suitable option is provided by considering a probabilistic approach based on the Bayesian inverse problem, whereby the aforementioned uncertainty is accounted by considering the measured data and assuming a particular damage hypothesis within a specific model class. The result is not a specific single-valued parameter for each location, but some kind of "cloud" around a more likely value of each parameter, a probability distribution function (PDF). This means that data are not only considered, but the quality of the data in terms of information content. The quantified uncertainty can be further employed for making robust predictions since both, test errors and model errors, can be accounted within a Bayesian inverse problem framework. In the authors opinion, this is where resides the main benefit of Bayesian inverse problem to predictions of damage signature in the industry.

In this work, it is denoted by model class $\mathcal{M}$, the ultrasonic wave propagation model (UWP) which is hypothesized and idealized to simulate the experimental system. A certain model parameterization is enclosed within each damage hypothesis, in such a way that a subset of all parameters in the model is considered to have fixed values and only the rest of the parameters form the subset that is subject to calculation in the inverse problem. The model, its selected parametrization together with the prior information about model parameters, defines a specific model class. The prior information represents the initial relative plausibility of model parameters before any information from measurements is incorporated, as explained further below. Since different damage hypothesis can be formulated, different model classes will be available to solve the inverse problem, yielding different values of model parameters but also physically different sets of parameters.

For the purpose of parameter estimation, a probability-based description of the ultrasonic wave propagation (UWP) model described in Fundamentals section is needed. To this end, let consider a UWP model that is defined by a deterministic relationship $x\left(t\right) = h_z\left(t; \boldsymbol{u}, \boldsymbol{\theta}\right) : \mathbb{R}^{Ni} \times \mathbb{R}^{Nm} \rightarrow \mathbb{R}$, between the model input $\boldsymbol{u} \in \mathbb{R}^{Ni}$ and the model output $x(t) \in \mathbb{R}$, given a set of N_p uncertain model parameters $\boldsymbol{\theta} \in \Theta \subset \mathbb{R}^{Np}$. This model can be "embedded" stochastically (Beck, 2010) by adding a model-error term e, that represents the difference between the measured ultrasound signal $y(t)$ and the model-simulated signal as follows,

$$\underbrace{y(t)}_{measured\ signal} = \underbrace{h_z\left(t;\boldsymbol{u},\boldsymbol{\theta}\right)}_{\text{model}} + e(t) \tag{11}$$

The PDF chosen to model the error term e in Equation determines the probability model for the measured output y. For example, if the model-error term is assumed to be a zero-mean Gaussian distribution, i.e. $e(t) \sim \mathcal{N}(0,\sigma(t))$, then the observed output y will be also a Gaussian, as shown below,

$$e(t) = y(t) - h_z\left(t;\boldsymbol{u},\boldsymbol{\theta}\right) \sim \mathcal{N}(0,\sigma(t)) \Rightarrow y(t) \sim \mathcal{N}(h_z\left(t;\boldsymbol{u},\boldsymbol{\theta}\right),\sigma(t)) \tag{12}$$

where $\sigma(t)$ is the standard deviation of the white-noise-type error e, that here is interpreted as an independent time-varying sequence $\left\{e\left(t_1\right),e\left(t_2\right),\ldots,e\left(t_3\right)\right\}$. This assumption of white-noise-type is supported by the Principle of Maximum Information Entropy (PMIE) (Jaynes, 1989; Jaynes, 2003), which enables a rational way to establish a probability model for the model-error term such that it produces the largest uncertainty (largest Shannon entropy), the less favorable and informative case. Following the Gaussian PDF of $y(t)$ in, the stochastic UWP model can be defined as a function of model parameters $\boldsymbol{\theta} \in \Theta$,

$$p\left(y \mid \boldsymbol{u},\boldsymbol{\theta},\mathcal{M}\right) = k_1 \exp\left(-J\left(y,\boldsymbol{u},\boldsymbol{\theta}\right)\right) \tag{13}$$

In the last equation k_1 is a normalizing constant and $J\left(y\left(t\right),\boldsymbol{u},\boldsymbol{\theta}\right)$ is a misfit function between the simulated by the model and measured ultrasound signal, defined as follows,

$$J(y,\boldsymbol{u},\boldsymbol{\theta}) = \frac{1}{2}\int_0^T \left(\frac{y(t) - h_z(t;\boldsymbol{u},\boldsymbol{\theta})}{\sigma(t)}\right)^2 dt$$

To sum up, a discrete set of candidate Bayesian model classes,

$$\boldsymbol{M} = \{\mathcal{M}_1,\ldots,\mathcal{M}_j,\ldots,\mathcal{M}_{N_M}, N_M \in \mathbb{N}\}$$

can be obtained by considering different parameterizations for the stochastic UWP model and also different prior information on model parameters, $p\left(\boldsymbol{\theta} \mid \mathcal{M}_j\right)$.

Bayesian Inference of Model Parameters

The focus in Bayesian inversion (Tarantola, 2005) is to determine the posterior PDF of the model parameters $\boldsymbol{\theta}$ within the set $\Theta \subset \mathbb{R}^d$ of possible values according to the model class $\mathcal{M}_j$ which is accomplished using Bayes' Theorem as follows,

$$\underbrace{p(\boldsymbol{\theta} \mid \mathcal{D}, \mathcal{M}_j)}_{posterior} = k_2^1 p(\mathcal{D} \mid \boldsymbol{\theta}, \mathcal{M}_j) \underbrace{p(\boldsymbol{\theta} \mid \mathcal{M}_j)}_{prior} \quad (14)$$

being $\mathcal{D}$ the available experimental dataset. In the last equation, k_2 is a normalizing constant so that $p\left(\theta \mid \mathcal{D}, \mathcal{M}_j\right)$ represents a valid PDF,

$$\int_\Theta p(\boldsymbol{\theta} \mid \mathcal{D}, \mathcal{M}_j) d\boldsymbol{\theta} = k_2^1 \int_\Theta p(\mathcal{D} \mid \boldsymbol{\theta}, \mathcal{M}_j) p(\boldsymbol{\theta} \mid \mathcal{M}_j) d\boldsymbol{\theta} = 1$$

The resulting posterior PDF is interpreted as a measure of the relative plausibility of the values of model parameters $\boldsymbol{\theta}$ conditioned on the available information contained in the data $\mathcal{D}$. It is a measure of the uncertainty about the parameter values. The PDF $p(\mathcal{D} \mid \boldsymbol{\theta}, \mathcal{M}_j)$ is known as the *likelihood function*, and provides a measure of how well the UWP model specified by $\boldsymbol{\theta}$ within the model class $\mathcal{M}_j$, simulates the experimental dataset $\mathcal{D}$. In this work, the data set consists of an experimental sequence of N independent ultrasonic signals $\mathcal{D} = \left\{y_1\left(t\right), \ldots, y_N\left(t\right)\right\}$ measured over a spatial grid of N points covering a certain damaged area in the material. Hence, the likelihood function can be formulated as,

$$p(\mathcal{D} \mid \boldsymbol{\theta}, \mathcal{M}_j) = \prod_{i=1}^{N} p(y_i(t) \mid \boldsymbol{u}, \boldsymbol{\theta}, \mathcal{M}_j) \quad (15)$$

By substituting Equation into Equation the likelihood function can be finally expressed as,

$$p(\mathcal{D} \mid \boldsymbol{\theta}, \mathcal{M}_j) = k_3 \exp\left(-\sum_{i=1}^{N} J(y_i, \boldsymbol{u}, \boldsymbol{\theta})\right)$$

where k_3 is a normalizing constant. Note that, up to this point, the former probabilistic information (i.e., prior and posterior PDFs of parameters, likelihood function, etc.) is restricted to the assumption of a particular damage hypothesis of the system, according to the model class $\mathcal{M}_j$. However, when the set of model classes $\boldsymbol{M} = \left\{\mathcal{M}_1, \ldots, \mathcal{M}_j, \ldots, \mathcal{M}_{NM}\right\}$ is considered, the challenge is to calculate the plausibility of the j-th candidate model class for representing the system being measured through the ultrasounds (Chiachío, Chiachío, Saxena, Shankararaman, & Goebel, 2015; Yuen, 2010).

FUTURE TRENDS AND CONCLUSION

SHM is changing the mentality about maintenance policy in industry. The fusion of different technologies to build different strategies is coming to scene. The difference between acting after unexpected results of a nondestructive test and performing successive projections on the remaining life incorporating all the historical information is huge; any intervention can be planned beforehand, taking components to their last reasonable limits of performance. Sudden out-of-operation opportunity cost avoidance and more profitable investments are the main consequences, whereas security standard is even increased.

The continuous cycle represented by [nondestructive test – model-based evaluation with uncertainty management – prognosis – decision building] illustrates the convergence of different technologies; a trend with increasing attention aimed to obtain a sound degree of integration. This chapter has addressed the first two links of the chain as important steps to build something superior.

Early detection of damage is another key contribution to this end, and nonlinear ultrasonics a promising tool within this framework. The link between different kinds of damage and their nonlinear manifestations is subject to deeper insight, requiring a combination of fundamental and experimental research. Improved theoretical models are expected, for instance, to unveil the underlying mechanisms down to molecular dynamics, to better understand ultrasonic relaxation phenomena such as slow dynamics and hysteresis or third- and higher-order harmonics or to know how early damage shifts nonlinear behavior. Concerning techniques and their usage, it has been shown that only FA and frequency mixing techniques are suitable for an in-field application, since NRUS and DAE seem to be restricted to laboratory applications, at least under the current experimental configurations. This situation, along with the limited number of applications and their short development is going to burst applied research to widen nonlinear ultrasonic usage, looking for more useful results. Particular attention should be given to composite materials, since their structural complexity and damage diversity slow down advances. Therefore, new nonlinear inspection techniques and procedures are expected and, more important, a clarification of the suitable technique to detect any kind of damage. In fact, first attempts of combining several nonlinear techniques at once have been mentioned, shedding light on this current technique-damage confusion and the diversity of damage modes.

Concerning evaluation process, it has been shown that the imaging capacity of nonlinear techniques is subject to improvement. Most techniques are based on the processing of individual signals received in one or very few locations. New scanned versions of techniques could be developed to generate new nonlinearity imaging modalities, as could probably happen with wave mixing or DAE techniques, showing multidimensional plots of different harmonics in amplitude and phase (multiples of fundamental or frequency mixing), amplitude of material properties variation (propagation velocity, Young modulus, etc.), amplitude of hysteretic loop, amplitude of fast dynamics effect, etc.

Another line of development will be model-based inverse problem for evaluation, where finding the suitable theoretical and computational models to reduce computation times, depending on data quality, will be a key factor for industry to adopt it. The aim consists of achieving accurate long-term predictions of damage by fusing a model-based evaluation and SHM data. But, especially when techniques are forced around their limits, it is mandatory to manage uncertainty. Bayesian inverse problem has demonstrated to be an effective method to incorporate uncertainty to be further used to

1. Obtain robust predictions,
2. Scrutinize the behavior of a model to reproduce the SHM measurements, and finally
3. Rank the best-suited class of models.

Based on the inverse problem procedure, a real-time diagnostic tool using ultrasound-based SHM data would contribute to achieve condition-based maintenance technologies, which are fully impacting on critical engineering systems like nuclear engineering and aerospace, among others.

REFERENCES

Armitage, P., & Wright, C. (2013). Design, development and testing of multi-functional non-linear ultrasonic instrumentation for the detection of defects and damage in CFRP materials and structures. *Composites Science and Technology*, *87*, 149–156. doi:10.1016/j.compscitech.2013.07.014

Aymerich, F., & Staszewski, W. J. (2010). Impact damage detection in composite laminates using nonlinear acoustics. *Composites. Part A, Applied Science and Manufacturing*, *41*(9), 1084–1092. doi:10.1016/j.compositesa.2009.09.004

Barbieri, E., Meo, M., & Polimeno, U. (2009). Nonlinear wave propagation in damaged hysteretic materials using a frequency domain-based PM space formulation. *International Journal of Solids and Structures*, *46*(1), 165–180. doi:10.1016/j.ijsolstr.2008.08.025

Barnard, D. J., Dace, G. E., & Buck, O. (1997). Acoustic Harmonic Generation Due to Thermal Embrittlement of Inconel 718. *Journal of Nondestructive Evaluation*, *16*(2), 67–75. doi:10.1007/BF02683879

Barnard, D. J., Dace, G. E., Rehbein, D., & Buck, O. (1997). Acoustic harmonic generation at diffusion bonds. *Journal of Nondestructive Evaluation*, *16*(2), 77–89. doi:10.1007/BF02683880

Beck, J. (2010). Bayesian system identification based on probability logic. *Structural Control and Health Monitoring*, *17*(7), 825–847. doi:10.1002/stc.424

Bender, F., Kim, J.-Y., Jacobs, L., & Qu, J. (2013). The generation of second harmonic waves in an isotropic solid with quadratic nonlinearity under the presence of a stress-free boundary. *Wave Motion*, *50*(2), 146–161. doi:10.1016/j.wavemoti.2012.08.009

Best, S., Croxford, A., & Neild, S. (2013). Pulse-Echo Harmonic Generation Measurements for Non-destructive Evaluation. *Journal of Nondestructive Evaluation*, *32*(4), 1–11.

Bochud, N. (2014). *Signal processing-based identification of pathology using ultrasonics*. (Unpublished doctoral dissertation). University of Granada, Granada, Spain.

Breazeale, M., & Thompson, D. (1963). Finite-Amplitude ultrasonic waves in aluminum. *Applied Physics Letters*, *3*(5), 77–78. doi:10.1063/1.1753876

Broda, D., Staszewski, W. J., Martowicz, A. U., & Silberschmidt, V. V. (2014). Modelling of nonlinear crack-wave interactions for damage detection based on ultrasound-A review. *Journal of Sound and Vibration*, *333*(4), 1097–1118. doi:10.1016/j.jsv.2013.09.033

Bruno, C. L., Gliozzi, A. S., Scalerandi, M., & Antonaci, P. (2009). Analysis of elastic nonlinearity using the scaling subtraction method. *Physical Review B: Condensed Matter and Materials Physics*, *79*(6), 1–13. doi:10.1103/PhysRevB.79.064108

Cantrell, J., Thompson, D. O., & Chimenti, D. E. (2009). Nondestructive evaluation of metal fatigue using nonlinear acoustics. *AIP Conference Proceedings*, *1096*(1), 19–32. doi:10.1063/1.3114205

Cantrell, J. H., & Yost, W. T. (1994). Acoustic harmonic generation from fatigue-induced dislocation dipoles. *Philosophical Magazine A*, *69*(2), 315–326. doi:10.1080/01418619408244346

Chakrapani, S. K., Barnard, D. J., & Vinay, D. (2015). Influence of fiber orientation on the inherent acoustic nonliearity in carbon fiber reinforced composites. *The Journal of the Acoustical Society of America, 137*(2), 617–624. doi:10.1121/1.4906165 PMID:25697996

Cheon, Y.-M., Alam, M., & Kim, C. (2010). Nonlinear parameters for diagnosis of microscale cracks using a nonlinear resonant ultrasound spectroscopy (NRUS). *AIP Conference Proceedings, 1211*(1), 1439–1444. doi:10.1063/1.3362237

Chiachío, J., Chiachío, M., Saxena, A., Shankararaman, S., & Goebel, K. (2015). Bayesian model selection and parameter estimation for fatigue damage progression models in composites. *International Journal of Fatigue, 70*, 361–373. doi:10.1016/j.ijfatigue.2014.08.003

Cleveland, R., Hamilton, M., & Blackstock, D. (1996). Time-domain modeling of finite-amplitude waves in relaxing fluids. *The Journal of the Acoustical Society of America, 99*(6), 3312–3318. doi:10.1121/1.414983

Courtney, C. R., Drinkwater, B. W., Neild, S. A., & Wilcox, P. D. (2008). Factors affecting the ultrasonic intermodulation crack detection technique using bispectral analysis. *NDT & E International, 41*(3), 223–234. doi:10.1016/j.ndteint.2007.09.004

Croxford, A. J., Wilcox, P. D., Drinkwater, B. W., & Nagy, P. B. (2009). The use of non-collinear mixing for nonlinear ultrasonic detection of plasticity and fatigue. *The Journal of the Acoustical Society of America, 126*(5), 117–122. doi:10.1121/1.3231451 PMID:19894785

Delsanto, P., Hirsekorn, S., Agostini, V., Loparco, R., & Koka, A. (2002). Modeling the propagation of ultrasonic waves in the interface region between two bonded elements. *Ultrasonics, 40*(1-8), 605–610. doi:10.1016/S0041-624X(02)00183-X PMID:12160009

Donskoy, D., Sutin, A., & Ekimov, A. (2001). Nonlinear acoustic interaction on contact interfaces and its use for nondestructive testing. *NDT & E International, 34*(4), 231–238. doi:10.1016/S0963-8695(00)00063-3

Donskoy, D. M., & Sutin, A. M. (1998). Vibro-acoustic modulation nondestructive evaluation technique. *Journal of Intelligent Material Systems and Structures, 9*(9), 765–771. doi:10.1177/1045389X9800900909

Drewry, M., Smith, R., Phang, A., Yan, D., Wilcox, P., & Roach, D. (2009). Ultrasonic techniques for detection of weak adhesion. *Materials Evaluation, 67*(9), 1048–1058.

Dubuget, M., El Guerjouma, R., Dubois, S., Baboux, J., & Vincent, A. (1996). Characterization of the non-linear elastic properties of aluminium alloys using ultrasonic evaluation under load. *Materials Science Forum, 217-222*, pp. 951-956.

Ekimov, A., Didenkulov, I., & Kazakov, V. (1999). Modulation of torsional waves in a rod with a crack. *Journal of the Acoustical Society of America, 106*(3–1), 1289.

Ellwood, R., Stratoudaki, T., Sharples, S., Clark, M., & Somekh, M. (2014). Determination of the acoustoelastic coefficient for surface acoustic waves using dynamic acoustoelastography: An alternative to static strain. *The Journal of the Acoustical Society of America, 135*(3), 1064–1070. doi:10.1121/1.4864308 PMID:24606250

Escobar-Ruiz, E., Ruiz, A., Hassan, W., Wright, D. C., Collison, I. J., Cawley, P., & Nagy, P. B. (2014). Non-linear Ultrasonic NDE of Titanium Diffusion Bonds. *Journal of Nondestructive Evaluation*, *33*(2), 187–195. doi:10.1007/s10921-013-0217-5

Farim, A., Gallego, R., Bochud, N., & Rus, G. (2013). Model-based damage reconstructionin compositesfrom ultrasound transmission. *Composites. Part B, Engineering*, *45*(1), 50–62. doi:10.1016/j.compositesb.2012.09.003

Fink, M. (1997). Time reversed acoustics. *Physics Today*, *50*(3), 34–40. doi:10.1063/1.881692

Fukuda, M., Nishihira, M., & Imano, K. (2007). Real Time Detection of Second-Harmonic Components Generated from Plastic-Deformed Metal Rod Using Double-Layered Piezoelectric Transducer. *Japanese Journal of Applied Physics*, *46*(7B), 4529–4531. doi:10.1143/JJAP.46.4529

Ginsberg, J. H. (2008). Perturbation Methods. In M. H. Hamilton & D. T. Blackstock (Eds.), *Nonlinear Acoustics* (pp. 279–308). San Diego, CA: Academic Press.

Gliozzi, A., Griffa, M., & Scalerandi, M. (2006). Efficiency of time-reversed acoustics for nonlinear damage detection in solids. *The Journal of the Acoustical Society of America*, *120*(5), 2506–2517. doi:10.1121/1.2345955

Goursolle, T., Callé, S., Dos Santos, S., & Bou Matar, O. (2007). A two-dimensional pseudospectral model for time reversal and nonlinear elastic wave spectroscopy. *The Journal of the Acoustical Society of America*, *122*(6), 3220–3229. doi:10.1121/1.2799900 PMID:18247734

Gusev, V. (2000). Propagation of acoustic pulses in material with hysteretic nonlinearity. *The Journal of the Acoustical Society of America*, *107*(6), 3047–3058. doi:10.1121/1.429333 PMID:10875350

Guyer, R., & Johnson, P. (1999). Nonlinear mesoscopic elasticity: Evidence for a new class of materials. *Physics Today*, *52*(4), 30–36. doi:10.1063/1.882648

Guyer, R., McCall, K., & Van Den Abeele, K. (1998). Slow elastic dynamics in a resonant bar of rock. *Geophysical Research Letters*, *25*(10), 1585–1588. doi:10.1029/98GL51231

Hallaj, I. M., & Cleveland, R. O. (1999). FDTD simulation of finite-amplitude pressure and temperature fields for biomedical ultrasound. The Journal of the Acoustical Society of America. *The Journal of the Acoustical Society of America*, *105*(5), L7–L12. doi:10.1121/1.426776 PMID:10335650

Hamilton, M. F., & Blackstock, D. T. (1998). *Nonlinear acoustics* (Vol. 237). San Diego, CA: Academic press.

Haupert, S., Renaud, G., Riviere, J., Talmant, M., Johnson, P. A., & Laugier, P. (2011). High-accuracy acoustic detection of nonclassical component of material nonlinearity. *The Journal of the Acoustical Society of America*, *130*(5), 2654–2661. doi:10.1121/1.3641405 PMID:22087892

Haupert, S., Rivière, J., Anderson, B., Ohara, Y., Ulrich, T., & Johnson, P. (2014). Optimized dynamic acousto-elasticity applied to fatigue damage and stress corrosion cracking. *Journal of Nondestructive Evaluation*, *33*(2), 226–238. doi:10.1007/s10921-014-0231-2

Hirsekorn, S., Rabe, U., & Arnold, W. (2006). Characterization and evaluation of composite laminates by nonlinear ultrasonic transmission measurements. *Proceedings of the 9th European Conference on NDT (ECNDT 2006), CD-ROM BB-103*. Berlin, Germany.

Holcomb, D. J. (1981). Memory, relaxation, and microfracturing in dilatant rock. *Journal of Geophysical Research: Solid Earth (1978-2012), 86*(B7), 6235-6248.

Jaynes, E. T. (1989). Papers on Probability, Statistics and Statistical Physics. (R. Rosenkrantz, Ed.). Dordrecht, The Nederlands: Kluwer Academic Publishers - Springer.

Jaynes, E. T. (2003). *Probability Theory: The Logic of Science* (G. Bretthorst, Ed.). New York, NY: Cambridge University Press. doi:10.1017/CBO9780511790423

Jhang, K. Y. (2009). Nonlinear ultrasonic techniques for nondestructive assessment of microdamage in material: A review. *International Journal of Precision Engineering and Manufacturing*, *10*(1), 123–135. doi:10.1007/s12541-009-0019-y

Johnson, P. A. (2006). Nonequilibrium Nonlinear Dynamics in Solids: State of the Art. In P. P. Delsanto (Ed.), *Universality of Nonclassical Nonlinearity* (pp. 49–69). New York, NY: Springer. doi:10.1007/978-0-387-35851-2_4

Johnson, P. A., Zinszner, B., & Rasolofosaon, P. N. (1996). Resonance and elastic nonlinear phenomena in rock. *Journal of Geophysical Research: Solid Earth (1978-2012), 101*(B5), 11553-11564.

Kawashima, K., Murase, M., Yamada, R., Matsushima, M., Uematsu, M., & Fujita, F. (2006). Nonlinear ultrasonic imaging of imperfectly bonded interfaces. *Ultrasonics*, *44*(Supplement), e1329–e1333. doi:10.1016/j.ultras.2006.05.011 PMID:16782154

Köjl, M., Hurlebaus, S., & Gaul, L. (2004). Finite element simulation of non-destructive damage detectionwith higher harmonics. *NDT & E International*, *37*(3), 195–205. doi:10.1016/j.ndteint.2003.09.003

Krohn, N., Stoessel, R., & Busse, G. (2002). Acoustic non-linearity for defect selective imaging. *Ultrasonics, 40*(1–8), 633-637.

Landau, L. D., & Lifshitz, E. M. (1986). *Theory of Elasticity*. Oxford, UK: Elsevier Butterworth-Heinemann.

Lee, Y., & Hamilton, M. (1995). Time-domain modeling of pulsed finite-amplitude sound beams. *The Journal of the Acoustical Society of America*, *97*(2), 906–917. doi:10.1121/1.412135

Matikas, T. (2010). Damage Characterization and Real-Time Health Monitoring os Aerospace Materials Using Innovative NDE Tools. *Journal of Materials Engineering and Performance*, *19*(5), 751–760. doi:10.1007/s11665-009-9532-5

Matikas, T. E., & Aggelis, D. G. (2011). New Trends in Materials Nondestructive Characterization Using Surface Acoustic Wave Methodologies. In *Ultrasonic Waves* (pp. 251-282). In Tech. Retrieved from http://www.intechopen.com/books/ultrasonic-waves/new-trends-in-materials-nondestructive-characterization-using-surface-acoustic-wave-methodologies

Matsuda, N., & Shiro, B. (2014). Frequency Dependence of Second-Harmonic Generation in Lamb Waves. *Journal of Nondestructive Evauation*, *33*(2), 169–177. doi:10.1007/s10921-014-0227-y

McCall, K. (1994). Theoretical study of nonlinear elastic wave propagation. *Journal of Geophysical Research, 99*(B2), 2591–2600. doi:10.1029/93JB02974

McCall, K., & Guyer, R. (1994). Equation of state and wave propagation In hysteretic nonlinear elastic materials. *Journal of Geophysical Research, 99*(B12), 23887–23897. doi:10.1029/94JB01941

McCall, K., & Guyer, R. (1999). A new theoretical paradigm to describe hysteresis, discrete memory and nonlinear elastic wave propagation in rock. *Nonlinear Processes in Geophysics, 3*(2), 89–101. doi:10.5194/npg-3-89-1996

Meo, M., & Zumpano, G. (2005). Nonlinear elastic wave spectroscopy identification of impact damage on a sandwich plate. *Composite Structures, 71*(3), 469–474. doi:10.1016/j.compstruct.2005.09.027

Meurer, T., Qu, J., & Jacobs, L. J. (2002). Wave propagation in nonlinear and hysteretic media - a numerical study. *International Journal of Solids and Structures, 39*(21-22), 5585–5614. doi:10.1016/S0020-7683(02)00366-9

Moreschi, H., Callé, S., Guerard, S., Mitton, D., Renaud, G., & Defontaine, M. (2011). Monitoring trabecular bone microdamage using a dynamic acousto-elastic testing method. *Proceedings of the Institution of Mechanical Engineers. Part H, Journal of Engineering in Medicine, 225*(3), 282–295. PMID:21485329

Muir, D. D. (2009, May). *One-sided ultrasonic determination of third order elastic constants using angle-beam acoustoelasticity measurements.* (Doctoral Dissertation). Georgia Institute of Technology. Retrieved from http://hdl.handle.net/1853/29766

Muller, M., Sutin, A., Guyer, R., Talmant, M., Laugier, P., & Johnson, P. (2005). Nonlinear resonant ultrasound spectroscopy (NRUS) applied to damage assessment in bone. *The Journal of the Acoustical Society of America, 118*(6), 3946–3952. doi:10.1121/1.2126917 PMID:16419838

Muller, M., Tencate, J., Darling, T., Sutin, A., Guyer, R., Talmant, M., & Johnson, P. (2006). Bone micro-damage assessment using non-linear resonant ultrasound spectroscopy (NRUS) techniques: A feasibility study. *Ultrasonics, 44*, e245–e249. doi:10.1016/j.ultras.2006.06.043 PMID:16876843

Murnaghan, F. (1951). *Finite deformation of an elastic solid.* London, UK: Chapman & Hall.

Na, J. K., Cantrell, J. H., & Yost, W. T. (1996). Linear and Nonlinear Ultrasonic Properties of Fatigued 410Cb Stainless Steel. In D. O. Thompson & D. E. Chimenti (Eds.), *Review of Progress in Quantitative Nondestructive Evaluation* (pp. 1347–1352). New York, NY: Plenum Press. doi:10.1007/978-1-4613-0383-1_176

Nazarov, V. E., Radostin, A., Ostrovsky, L., & Soustova, I. (2003). Wave Processes in Media with Hysteretic Nonlinearity. Part I. *Acoustical Physics, 49*(3), 344–353. doi:10.1134/1.1574363

Nicholson, P., & Bouxsein, M. (2000). Quantitative ultrasound does not reflect mechanically induced damage in human cancellous bone. *Journal of Bone and Mineral Research, 15*(12), 2467–2472. doi:10.1359/jbmr.2000.15.12.2467 PMID:11127211

Norris, A. (1998). Finite Amplitude Waves in Solids. In M. F. Hamilton & D. T. Blackstock (Eds.), *Nonlinear Acoustics* (pp. 263–277). San Diego, CA: Academic Press.

Ohtani, T., & Ishii, Y. (2012). Nonlinear Resonant Ultrasound Spectroscopy (NRUS) applied to fatigue damage evaluation in a pure copper. *AIP Conference Proceedings*, *1474*, 203–206. doi:10.1063/1.4749331

Ostrovsky, L. A., & Johnson, P. A. (2001). Dynamic nonlinear elasticity in geomaterials. *Rivista del Nuovo Cimento*, *24*(7), 1–46.

Pasqualini, D., Heitmann, K., TenCate, J., Habib, S., Higdon, D., & Johnson, P. (2007). Nonequilibrium and nonlinear dynamics in Berea and Fontainebleau sandstones: Low-strain regime. *Journal of Geophysical Research. Solid Earth*, *112*(B01204), 1–16.

Payan, C., Garnier, V., Moysan, J., & Johnson, P. (2007). Applying nonlinear resonant ultrasound spectroscopy to improving thermal damage assessment in concrete. *The Journal of the Acoustical Society of America*, *121*(4), EL125–EL130. doi:10.1121/1.2710745 PMID:17471756

Pecorari, C. (2003). An extension of the spring model to nonlinear interfaces. *AIP Conference Proceedings*, *657*, 49–56. doi:10.1063/1.1570118

Polimeno, U., & Meo, M. (2009). Detecting barely visible impact damage detection on aircraft composites structures. *Composite Structures*, *91*(4), 398–402. doi:10.1016/j.compstruct.2009.04.014

Renaud, G., Callé, S., Remenieras, J. P., & Defontaine, M. (2008). Exploration of trabecular bone nonlinear elasticity using time-of-flight modulation. *IEEE Transactions on Ultrasonics, Ferroelectrics, and Frequency Control*, *55*(7), 1497–1507. doi:10.1109/TUFFC.2008.825 PMID:18986939

Renaud, G., Riviere, J., Haupert, S., & Laugier, P. (2013). Anisotropy of dynamic acoustoelasticity in limestone, influence of conditioning, and comparison with nonlinear resonance spectroscopy. *The Journal of the Acoustical Society of America*, *133*(6), 3706–3718. doi:10.1121/1.4802909 PMID:23742326

Renaud, G., Talmant, M., Callé, S., Defontaine, M., & Laugier, P. (2011). Nonlinear elastodynamics in micro-inhomogeneous solids observed by head wave based dynamic acoustoelastic testing. *The Journal of the Acoustical Society of America*, *130*(6), 3583–3589. doi:10.1121/1.3652871 PMID:22225015

Richardson, J. M. (1979). Harmonic generation at an unbonded interface-I. Planar interface betwwen semi-infinite elastic media. *International Journal of Engineering Science*, *17*(1), 73–85. doi:10.1016/0020-7225(79)90008-9

Rivière, J., Remillieux, M., Ohara, Y., Anderson, B., Haupert, S., Ulrich, T., & Johnson, P. (2014). Dynamic Acousto-Elasticity in a Fatigue-Cracked Sample. *Journal of Nondestructive Evaluation*, *33*(2), 216–225. doi:10.1007/s10921-014-0225-0

Rivière, J., Renaud, G., Guyer, R., & Johnson, P. (2013). Pump and probe waves in dynamic acousto-elasticity: Comprehensive description and comparison with nonlinear elastic theories. *Journal of Applied Physics*, *114*(5), 054905. doi:10.1063/1.4816395

Rus, G., Wooh, S., & Gallego, R. (2007). Processing of ultrasonic array signals for characterizing defects. Part I: signal synthesis. Part II: Experimental work. *IEEE Transactions on Ultrasonics, Ferroelectrics, and Frequency Control*, *54*(10), 2129–2145. doi:10.1109/TUFFC.2007.508 PMID:18019251

Sagar, S. P., Das, S., Paridaa, N., & Bhattacharya, D. (2006). Non-linear ultrasonic technique to assess fatigue damage in structural steel. *Scripta Materialia*, *55*(2), 199–202. doi:10.1016/j.scriptamat.2006.03.037

Shkerdin, G., & Glorieux, C. (2010). Nonlinear clapping Modulantion of lamb modes by normally closed delamination. *IEEE Transactions on Ultrasonics, Ferroelectrics, and Frequency Control*, *57*(6), 1426–1433. doi:10.1109/TUFFC.2010.1561 PMID:20529717

Smith, S. (1966, May). Random-loading fatigue crack growth behavior of some aluminum and titanium alloys. *Fifth Pacific Area National Meeting of the American Society for Testing and Materials: Structural Fatigue in Aircraft. STP 404*, (pp. 74-100). American Society for Testing and Materials (ASTM). doi:10.1520/STP48341S

Solodov, I., Krohn, N., & Busse, G. (2002). CAN: An example of nonclassical acoustic nonlinearity in solids. *Ultrasonics*, *40*(1), 621–625. doi:10.1016/S0041-624X(02)00186-5 PMID:12160013

Stobbe, D. M. (2005, August). *Acoustoelasticity in 7075-T651 Aluminum and Dependence of Third Order Elastic Constants on Fatigue Damage.* (Master's thesis). Georgia Institute of Technology. Retrieved from http://hdl.handle.net/1853/7184

Stratoudaki, T., Ellwood, R., Sharples, S., Clark, M., Somekh, M. G., & Collison, I. J. (2011). Measurement of material nonlinearity using surface acoustic wave parametric interaction and laser ultrasonics. *The Journal of the Acoustical Society of America*, *129*(4), 1721–1728. doi:10.1121/1.3560945 PMID:21476629

Sutin, A., & Nazarov, V. (1995). Nonlinear acoustic methods of crack diagnostics. *Radiophysics and Quantum Electronics, 38*(3-4), 109-120.

Tarantola, A. (2005). *Inverse Problem Theory and Methods for Model Parameters Estimation*. Philadelphia, PA: SIAM. doi:10.1137/1.9780898717921

TenCate, J. A. (2011). Slow Dynamics of Earth Materials: An Experimental Overview. *Pure and Applied Geophysics*, *168*(12), 2211–2219. doi:10.1007/s00024-011-0268-4

Toyama, N., & Takatsubo, J. (2004). An investigation of non-linear elastic behavior of CFRP laminates and strain measurement using Lamb waves. *Composites Science and Technology*, *64*(16), 2509–2516. doi:10.1016/j.compscitech.2004.05.007

Ulrich, T. J., Sutin, A. M., Guyer, R. A., & Johnson, P. A. (2008). Time reversal and non-linear elastic wave spectroscopy (TR NEWS) techniques. *International Journal of Non-linear Mechanics*, *43*(3), 209–216. doi:10.1016/j.ijnonlinmec.2007.12.017

Van Den Abeele, K., Johnson, P., & Sutin, A. (2000). NEWS techniques to discern material damage, Part I: NWMS (Nonlinear Wave Modulation Spectroscopy). *Research in Nondestructive Evaluation*, *12*, 17–30. doi:10.1080/09349840009409646

Van Den Abeele, K., Katkowski, T., & Mattei, C. (2006). Nonlinear Acoustic and Ultrasonic NDT of Aeronautical Components. *AIP Conference Proceedings*, *838*(1), 75–78. doi:10.1063/1.2210321

Van Den Abeele, K., Le Bas, P. Y., Van Damme, B., & Katkowski, T. (2009). Quantification of material nonlinearity in relation to microdamage density using nonlinear reverberation spectroscopy: Experimental and theoretical study. *The Journal of the Acoustical Society of America*, *126*(3), 963–972. doi:10.1121/1.3184583 PMID:19739709

Van Den Abeele, K., Sutin, A., Carmeliet, J., & Johnson, P. (2001). Microdamage diagnostics using nonlinear elastic wave spectroscopy (NEWS). *NDT & E International*, *34*(4), 239–248. doi:10.1016/S0963-8695(00)00064-5

Van Den Abeele, K. E. (1996). Elastic pulsed wave propagation in media with second or higher order nonlinearity. Part I. Theoretical framework. *The Journal of the Acoustical Society of America*, *99*(6), 3334–3345. doi:10.1121/1.414890

Van Den Abeele, K. E., & Johnson, P. A. (1996). Elastic pulsed wave propagation in media with second or higher order nonlinearity. Part II. Simulation of experimental measurements on Berea sandstone. *The Journal of the Acoustical Society of America*, *99*(6), 3346–3352. doi:10.1121/1.414891

Van Den Abeele, K. E., Johnson, P. A., Guyer, R. A., & McCall, K. R. (1997). On the quasi-analytic treatment of hysteretic nonlinear response in elastic wave propagation. *The Journal of the Acoustical Society of America*, *101*(4), 1885–1898. doi:10.1121/1.418198

Vander Meulen, F., & Haumesser, L. (2008). Evaluation of B/A nonlinear parameter using an acoustic self-calibrated pulse-echo. *Applied Physics Letters*, *92*(21), 214106. doi:10.1063/1.2937838

Vila, M., Vander Meulen, F., Dos Santos, S., Haumesser, L., & Bou Matar, O. (2004). Contact phase modulation method for acoustic nonlinear parameter measurement in solid. *Ultrasonics*, *42*(1-9), 1061–1065. doi:10.1016/j.ultras.2003.12.024 PMID:15047429

Viswanath, A., Purna Chandra Rao, B., Mahadevana, S., Parameswaran, P., Jayakumar, T., & Raj, B. (2011). Nondestructive assessment of tensile properties of cold worked AISI type 304 stainless steel using nonlinear ultrasonic technique. *Journal of Materials Processing Technology*, *211*(3), 538–544. doi:10.1016/j.jmatprotec.2010.11.011

Winkler, K., & Liu, X. (1996). Measurements of third-order elastic constants in rocks. *The Journal of the Acoustical Society of America*, *100*(3), 1392. doi:10.1121/1.415986

Yuen, K. (2010). Recent developments of Bayesian model class selection and applications in civil engineering. *Structural Safety*, *32*(5), 338–346. doi:10.1016/j.strusafe.2010.03.011

Yun, Y., Miao, G. Q., Zhang, P., Huang, K., & Wei, R. J. (2005). Nonlinear acoustic wave propagation in one-dimensional layered system. *Physics Letters. [Part A]*, *343*(5), 351–358. doi:10.1016/j.physleta.2005.05.092

Zagrai, A., Donskoy, D., Chudnovsky, A., & Golovin, E. (2008). Micro- and macroscale damage detection using the nonlinear acoustic vibro-modulation technique. *Research in Nondestructive Evaluation*, *19*(2), 104–128. doi:10.1080/09349840801931817

Zaitsev, V., Gusev, V., & Castagnede, B. (2002). Luxemburg-Gorky effect retooled for elastic waves: A mechanism and experimental evidence. *Physical Review Letters*, *89*(10), 105502. doi:10.1103/PhysRevLett.89.105502 PMID:12225205

Zaitsev, V. Y., & Matveev, L. (2006). Microinhomogeneous medium, elastic nonlinearity, linear dissipation, amplitude dependent dissipation. *Russian Geology and Geophysics*, *47*(5), 694–709.

Zaitsev, V. Y., Matveev, L., & Matveyev, A. (2011). Elastic-wave modulation approach to crack detection: Comparison of conventional modulation and higher-order interactions. *NDT & E International*, *44*(1), 21–31. doi:10.1016/j.ndteint.2010.09.002

Zarembo, L., & Krasil'Nikov, V. (1971). Nonlinear phenomena in the propagation of elastic waves in solids. *Physics-Uspekhi*, *13*(6), 778–797. doi:10.1070/PU1971v013n06ABEH004281

Zheng, Y., Maev, R., & Solodov, I. (1999). Nonlinear acoustic applications for material characterization: A review. *Canadian Journal of Physics*, *77*(12), 927–967. doi:10.1139/cjp-77-12-927

KEY TERMS AND DEFINITIONS

Computational Model Uncertainty: Computational models introduce new errors and assumptions on the digitalized calculation method, so that simulated signals accumulate theoretical and computational uncertainties.

Data Uncertainty: The test is subject to generate signals with noise and all kind of uncontrolled oscillations originated in test instrumentation, materials and performance. They could lead to misinterpretations, especially when looking for early and weak defects.

Material Nonlinearity: The fact that the internal structure of the material behaves as a nonlinear mechanical system, being stress the input and strain the output. In a linear material, the strain is directly proportional to the applied stress.

Mesoscale Level: The size scale level between 10^{-8}-10^{-9}m and 10^{-6}m, above atomic and molecule size, where matter discontinuity and aggregation generates local fluctuations on the parameters around an average value.

Nonlinear System: Given two possible inputs A and B on a system which lead to the respective outputs a and b, the system is linear if an input, that is a linear combination of A and B, $mA+nB$, yields to an output that is the same linear combination of the individual outputs $ma+nb$. Otherwise, nonlinear.

Theoretical Model Uncertainty: Theoretical models are not complete. They are based on assumptions and simplifications that can lead to failures in their predicting behavior capacity.

Theoretical vs. Computational Model: The same theoretical model can be implemented by different computational models, introducing different calculation strategy, assumptions, discretization modes, errors, etc.

Chapter 10

Fatigue Crack Growth Analysis and Damage Prognosis in Structures

Shankar Sankararaman
SGT Inc., USA & NASA Ames Research Center, USA

You Ling
General Electric Global Research Center, USA

Sankaran Mahadevan
Vanderbilt University, USA

ABSTRACT

This chapter describes a computational methodology for fatigue crack growth analysis and damage prognosis in structures. This methodology is applicable to a variety structural components and systems with complicated geometry and subjected to variable amplitude multi-axial loading. Finite element analysis is used to address complicated geometry and calculate the stress intensity factors. Multi-modal stress intensity factors due to multi-axial loading conditions are combined to calculate an equivalent stress intensity factor using a characteristic plane approach. Crack growth under variable amplitude loading is modeled using a modified Paris law that includes retardation effects. During cycle-by-cycle integration of the crack growth law, a Gaussian process surrogate model is used to replace the expensive finite element analysis, thereby significantly improving computational effort. The effect of different types of uncertainty – physical variability, data uncertainty and modeling errors – on crack growth prediction is investigated. The various sources of uncertainty include, but not limited to, variability in loading conditions, material parameters, experimental data, model uncertainty, etc. Three different types of modeling errors – crack growth model error, discretization error and surrogate model error – are included in analysis. The different types of uncertainty are incorporated into the framework for calibration and crack growth prediction, and their combined effect on crack growth prediction is computed. Finally, damage prognosis is achieved by predicting the probability distribution of crack size as a function of number of load cycles, and this methodology is illustrated using a numerical example of surface cracking in a cylindrical component.

DOI: 10.4018/978-1-4666-8490-4.ch010

IINTRODUCTION

Structural systems and their components are often subjected to cyclic loads, leading to fatigue, crack initiation and crack growth. System models are coupled with crack growth models to study crack growth propagation and thereby predict the crack growth as a function of number of loading cycles. This, in turn, can be used in life prediction and reliability evaluation of the system. The process of fatigue crack growth is affected by many sources of variability, such as loading, material properties, geometry and boundary conditions. Therefore it is appropriate to describe the crack size after a certain number of load cycles through a probability distribution that accounts for these different sources of uncertainty.

Fracture mechanics and probabilistic fracture mechanics are extensive areas of research and numerous studies have focused on developing computational techniques probabilistic crack growth and life prediction. Probabilistic crack growth analysis has been applied to both metals (e.g. Johnson and Cook (Johnson & Cook, 1985), Maymon (Maymon, 2005)) and composites (Hwang & Han, 1986; Peng et al., 2013)). Practical applications of these methods include nuclear structural components (Yagawa, Kanto, Yoshimura, Machida, & Shibata, 2001), helicopter gears (Patrick et al., 2007), gas turbine engines (Tryon, Cruse, & Mahadevan, 1996), and aircraft components (S Sankararaman, Ling, & Mahadevan, 2011). These developments have led to software for probabilistic fracture mechanics (McClung, Enright, Millwater, Leverant, & Hudak Jr, 2004), and several commercial software tools such as DARWIN (Wu, Enright, & Millwater, 2002), and other software tools that build probabilistic analysis around well-established codes such as AFGROW (Grell & Laz, 2010) and FASTRAN (Newman Jr, Brot, & Matias, 2004).

This chapter a presents a computational methodology for modeling fatigue crack growth analysis and predicting crack growth in order to aid damage prognosis and life prediction in structural metallic systems. This methodology is directly useful for system design since it is ideal to choose design parameters that maximize the life expectancy of the system. Further, the methodology also facilitates risk assessment and management, inspection and maintenance scheduling and operational decision-making.

Researchers have pursued two different kinds of methodologies for fatigue life prediction. The first method is based on material testing (to generate S-N, $\varepsilon - N$ curves) and use of an assumed damage accumulation rule. In this method, specimens are subjected to repeated cyclic loads under laboratory conditions. Hence the results are specific to the geometry of the structure as well as the nature of loading. Further, the performance of these components under field conditions is significantly different from laboratory observation, due to various sources of uncertainty accumulating in the field that render experimental studies less useful. Hence, this methodology cannot be used directly to predict the fatigue life of practical applications wherein complicated structures subjected to multi-axial loading.

The second method for fatigue life prediction is based on rigorous principles of fracture mechanics and crack growth analysis. A crack growth law is assumed and the progressive growth of the crack is modeled. However, this is not straightforward. To begin with, small crack growth is not completely understood and cannot be modeled easily. Further, fatigue crack growth is a stochastic process and there are different types of uncertainty – physical variability, data uncertainty and modeling errors, associated with it. Uncertainty appears at different stages of analysis and the interaction between these sources of uncertainty cannot be modeled easily. Further, practical applications consist of structures with complicated geometry and may be subjected to multi-axial variable amplitude loading; this requires the use of expensive finite element analysis to calculate stress intensity factors that are needed for crack growth prediction. The computational framework developed Sankararaman et al. (S Sankararaman et al., 2011) addresses several of the above challenges; the primary goal of this chapter explains this framework in detail, and explains how the above challenges are overcome and illustrates the overall methodology using a numerical example.

Two common problems are encountered when the initial crack size is small. First it is difficult to measure the initial crack size if the limit of the inspection technique is reached. The second problem is that the commonly used crack growth models (such Paris' law) are applicable only when the initial crack size is beyond micro-scale. In practice, empirically assumed initial crack in the long size region is used (Merati & Eastaugh, 2007). However, this leads to a very conservative fatigue life prediction if the actual initial crack size is small. The concept of equivalent initial flaw size (EIFS) was developed about 30 years ago to address the two problems above (Liu & Mahadevan, 2009). It is defined such that the prediction of a long crack growth model which starts from EIFS is in agreement with the actual fatigue crack growth (Fawaz, 2003). Note that EIFS is not a physical quantity pertaining to the component, but a hypothetical flaw size that can be used as an initial crack size in long crack growth models. The equivalent initial flaw size depends on the material and geometrical properties of the specimen, and Sankararaman et al. (Shankar Sankararaman, Ling, & Mahadevan, 2010; Shankar Sankararaman, Ling, Shantz, & Mahadevan, 2011) developed methods for estimating EIFS based on statistical methods; the prior distribution of EIFS can be estimated based on Liu and Mahadevan (Liu & Mahadevan, 2009) and updated using Bayes' theorem. Having estimated the EIFS, it can be used along with a long crack growth model to assist crack propagation.

As mentioned earlier, in each cycle of crack growth, it is necessary to calculate the stress intensity factors using finite element analysis. The presence of multi-axial loading leads to multi-modal stress intensity factors that are combined using the characteristic plane approach developed by Liu and Mahadevan (Liu & Mahadevan, 2005, 2007), and an equivalent stress intensity factor is computed. The presence of variable amplitude loading is addressed using Wheeler's retardation model that is used in conjunction with the long crack growth model. It is practically impossible to use finite element analysis in every cycle of crack growth analysis, and therefore, an inexpensive Gaussian process surrogate model is trained using few runs of the finite element analysis and used in lieu of the original finite element model, thereby significantly reducing the computational effort.

An important feature of this chapter is the systematic treatment of the various sources of uncertainty that affect crack growth analysis and damage prognosis. The most commonly considered sources of uncertainty is natural variability in loading conditions, material and geometric properties, and crack growth model parameters. However, there are other sources of data uncertainty and model errors that significantly affect crack growth prediction. The different types of error and uncertainty appear at different stages of analysis, and may combine in nonlinear, nested, or iterative manner. Further, some of the errors are deterministic (e.g., finite element discretization error) while some others are stochastic (e.g., material properties). It is essential to systematically account for all these sources of error and uncertainty in order to ensure robustness in model parameter calibration and uncertainty quantification. This chapter accounts for the different sources of variability, data uncertainty, model uncertainty and errors that affect fatigue crack growth analysis, and develops a methodology to systematically quantify the overall uncertainty in the overall crack growth model prediction (damage prognosis).

This is accomplished through the use of a Bayesian methodology that can help in connecting the various models, the associated sources of uncertainty and available experimental data. Based on a principles of conditional probability and total probability, a dynamic Bayesian network is developed for this purpose and used for two purposes (Shankar Sankararaman, McLemore, Mahadevan, Bradford, & Peterson, 2013):

1. In an inverse problem (Shankar Sankararaman et al., 2011) where the uncertainty in the model parameters is quantified using experimental evidence; and
2. In a forward problem (Shankar Sankararaman, Ling, Shantz, & Mahadevan, 2009) that aids in probabilistic prediction.

In the inverse problem of calibration, all the system parameters and model errors are inferred based on experimental data. In the context fatigue, three types of data may be available:

1. Crack detected and size measured after a particular number of loading cycles;
2. Crack detected after a particular number of loading cycles, but size not measured; and
3. Crack not detected after a particular number of loading cycles.

Using such experimental data, Bayesian inference is used to calibrate the parameters. Since the number of parameters may be high, an initial screening is performed using global sensitivity analysis and only sensitive parameters are chosen for updating.

Having inferred the various parameters and model error terms, their posterior distributions are propagated through the crack growth model to compute the probability distribution as a function of number of loading cycles in the forward problem. Probabilistic damage prediction aids in effective prognosis under uncertainty, and helps to forecast the remaining useful life of the structural/mechanical system under study.

The steps of the proposed methodology can be summarized as follows:

1. Connect different models – finite element model, surrogate model, crack growth law, etc. efficiently through a Bayes network;
2. Quantify the errors in each model, treating deterministic and stochastic errors differently;
3. Include the different sources of uncertainty – physical variability, data uncertainty, and model uncertainty in the Bayesian network;
4. Use global sensitivity analysis to select calibration parameters;
5. Collect inspection data and update the parameters in multiple models;
6. Include different cases of inspection such as crack not detected, crack detected but size not measured, and crack detected with size measurement; and
7. Predict the probability distribution of crack size as a function of number of loading cycles.

The following sections describe the above steps in detail. First, the crack growth modeling framework is presented, and then, the various sources of uncertainty are explained in detail. This is followed by the description of the model calibration methodology; the use of global sensitivity analysis for parameter screening and the use of dynamic Bayesian networks for calibration are explained. Finally, the methodology is illustrated using a numerical example.

CRACK GROWTH MODELING FRAMEWORK

Different models such as the modified Paris' law (for long crack growth analysis), Wheeler's retardation model (for variable amplitude loading), characteristic plane approach (to calculate an equivalent stress intensity factor in the presence of multi-axial loading) are connected together to predict the crack growth as a function of number of load cycles. Note that these models are used only for the purpose of illustration and any other combination of appropriate models may be used.

Long Crack Growth Law

Consider any long crack growth law used to describe the relationship between da/dN and ΔK, where N represents the number of cycles, a represents the crack size and ΔK represents the stress intensity factor range. For the sake of illustration, consider modified Paris' law (Paris, Gomez, & Anderson, 1961) that includes Wheeler's retardation model (Wheeler, 1972) as:

$$\frac{da}{dN} = \varphi^r C\left(\Delta K\right)^n \left(1 - \frac{\Delta K_{th}}{\Delta K}\right)^m \tag{1}$$

In Eq. 1, ΔK_{th} refers to the threshold stress intensity factor range and φ^r refers to the retardation parameter which can be defined as:

$$\varphi^r = \begin{cases} \left[\dfrac{r_{p,i}}{a_{OL} + r_{p,OL} - a_i}\right]^\lambda & if\ a_i + r_{p,i} < a_{OL} + r_{p,OL} \\ 1 & if\ a_i + r_{p,i} \geq a_{OL} + r_{p,OL} \end{cases} \tag{2}$$

In Eq. 2, a_{OL} is the crack length at which the overload is applied, a_i is the current crack length, $r_{p,OL}$ is the size of the plastic zone produced by the overload at a_{OL}, $r_{p,i}$ is the size of the plastic zone produced at the current crack length a_i, and λ is the curve fitting parameter for the original Wheeler model termed the shape exponent. Song. et al. (Song, Sheu, & Chang, 2001) observed that crack growth retardation actually takes place within an effective plastic zone. Hence the size of the plastic zone can be calculated in terms of the applied stress intensity factor (K) and yield strength (σ) as:

$$r_p = \alpha \left(\frac{K}{\sigma}\right)^2 \tag{3}$$

where α is known as the effective plastic zone size constant which is obtained experimentally(Song et al., 2001). Eq. 2 and Eq. 3 can be used in combination with Eq. 1 under the assumption of small-scale plasticity, where the plastic zone size is estimated using linear-elastic fracture mechanics (LEFM).

The expressions in Eq. 2 and Eq. 3 can be combined with Eq. 1 and used to calculate the crack growth as a function of number of cycles. In each cycle, the stress intensity factor can be expressed as a function of the crack size (a), loading (L) and angle of orientation (χ). Hence, the crack growth law in Eq. 1 can be rewritten as

$$\frac{da}{dN} = g\left(a, L, \chi\right) \tag{4}$$

Equivalent Initial Flaw Size

Note that the long crack growth model is not applicable to the short crack growth regime and the concept of an equivalent initial flaw size (EIFS) was proposed to bypass short crack growth analysis and make direct use of a long crack growth law for fatigue life prediction. The equivalent initial flaw size θ, i.e., the initial condition of the differential equation in Eq. 1, can be calculated from material properties (ΔK_{th}, the threshold stress intensity factor and σ_f, the fatigue limit) and geometric shape factor (Y) as derived by Liu and Mahadevan (Liu & Mahadevan, 2009):

$$\theta = \frac{1}{\pi}\left(\frac{\Delta K_{th}}{Y\sigma_f}\right)^2 \tag{5}$$

By integrating the expression in Eq. 1 starting from θ, the number of cycles (N) to reach a particular crack size a_N can be calculated as:

$$N = \int dN = \int_{\theta}^{aN} \frac{1}{\varphi^r C\left(\Delta K\right)^n \left(1 - \frac{\Delta K_{th}}{\Delta K}\right)^m} da \tag{6}$$

In the uncertainty quantification procedure, the expression for EIFS in Eq. 5 is used as a prior distribution and updated using Bayes' theorem.

Equivalent Stress Intensity Factor

The stress intensity factor range ΔK in Eq. 6 can be expressed as a closed form function of the crack size for specimens with simple geometry subjected to constant amplitude loading. However, this is not the case in many mechanical components, where ΔK depends on the loading conditions, geometry and the crack size. Further, if the loading is multi-axial (for example, simultaneous tension, torsion and bending), then the stress intensity factors corresponding to three modes need to be taken into account. This can be accomplished using an equivalent stress intensity factor. If K_I, K_{II}, K_{III} represent the mode-I, mode-II and mode-III stress intensity factors respectively, then the equivalent stress intensity factor K_{eqv} can be calculated using a characteristic plane approach proposed by Liu and Mahadevan (Liu & Mahadevan, 2005) as:

$$K_{eqv} = \frac{1}{B}\sqrt{\left(k_1\right)^2 + \left(\frac{k_2}{s}\right)^2 + \left(\frac{k_3}{s}\right)^2 + A\left(\frac{k_H}{s}\right)^2} \tag{7}$$

In Eq. 7, k_1, k_2, k_3 are the parameters associated with mode I, II, and III loading, respectively. k_H is related to hydrostatic stress. s is the ratio of Modes II and I stress intensity factor at a specific crack growth rate (da/dN). A and B are material parameters. The characteristic plane approach is applicable only when the crack surface can be approximated to be planar. The use of the characteristic plane approach for crack growth prediction under multi-axial variable amplitude loading has been applied to cracks in railroad wheels (Liu & Mahadevan, 2005) and validated earlier with several data sets (Liu, Liu, & Mahadevan, 2007; Liu & Mahadevan, 2007).

Each cycle in the integration of Eq. 6 involves the computation of ΔK_{eqv} using a finite element analysis owing to

1. Complicated geometry, and
2. Variable amplitude, multi-axial loading.

Repeated evaluation of this finite element analysis renders this integration extremely expensive. Hence, it is computationally more affordable to substitute the finite element model with an inexpensive surrogate model (also known as response surface model). Different kinds of surrogate models (polynomial chaos, support vector regression, relevance vector regression, and Gaussian Process interpolation) were explored and the Gaussian process (GP) modeling technique has been found to have the least error for this type of problem (Shankar Sankararaman et al., 2010). A few runs of the finite element analysis are used to train the GP surrogate model and then, this GP model is used to predict the stress intensity factor for other crack sizes and loading cases (for which finite element analysis has not been carried out).

Gaussian Process Surrogate Modeling

Suppose that there are m training points, $\boldsymbol{x}_1, \boldsymbol{x}_2, \boldsymbol{x}_3 \ldots \boldsymbol{x}_m$ of a d-dimensional input variable vector (the input variables being the crack size and loading conditions in this case), yielding the resultant observed random vector $Y(\boldsymbol{x}_1), Y(\boldsymbol{x}_2), Y(\boldsymbol{x}_3) \ldots Y(\boldsymbol{x}_m)$ (the equivalent stress intensity factor, in this case). R is the m x m matrix of correlations among input variables at the training points. Under the assumption that the parameters governing both the trend function ($f^T(\boldsymbol{x}_i)$, at each training point) and the covariance (λ) are known, the expected value and the variance of the Gaussian process at an untested location $\boldsymbol{x}^*$ is calculated as in Eq. 8 and Eq. 9 respectively.

$$Y^* = E\left[Y\left(\boldsymbol{x}^*\right) \mid Y\right] = f^T\left(\boldsymbol{x}^*\right)\beta + r^T\left(\boldsymbol{x}^*\right)R^{-1}\left(Y - F\beta\right) \quad (8)$$

$$\sigma^2_{Y^*} = Var\left[Y\left(\boldsymbol{x}^*\right) \mid Y\right] = \lambda\left(1 - r^T R^{-1} r\right) \quad (9)$$

In Eq. 8 and Eq. 9, F is a matrix containing the rows of trend functions $f^T(\boldsymbol{x}_i)$ (i = 1 to m), r is the vector of correlations between $\boldsymbol{x}^*$ and each of the training points, β represents the coefficients of the regression trend. A constant trend function ($f^T(\boldsymbol{x}_i)$) and an exponential covariance function (λ) that decays as a function of the squared distance between two points are used. See Rasmussen (Rasmussen, 2004) for details of the Gaussian process modeling technique. Sankararaman et al. (Shankar Sankararaman et al., 2010, 2011) describe in detail how the Gaussian process surrogate model can be constructed and used to predict the equivalent stress intensity factor in cycle-by-cycle integration of the crack growth law, thereby calculating the crack size (A) as a function of number of load cycles (N).

Figure 1. Framework for crack growth analysis

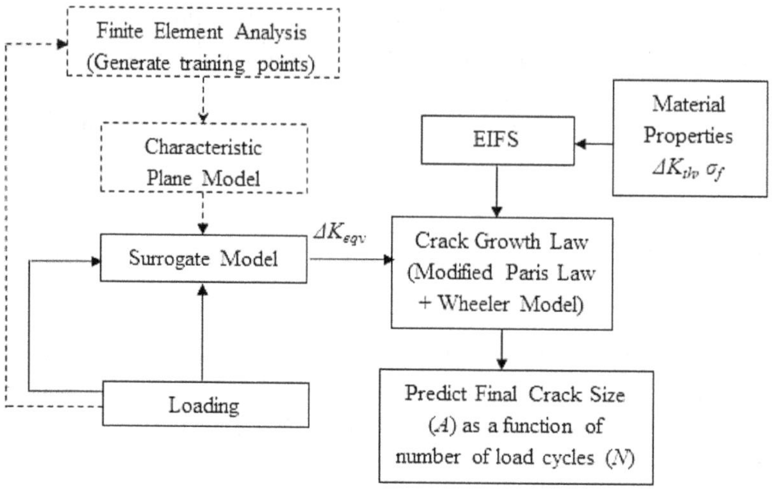

Algorithm for Crack Growth Modeling

The entire procedure for the adopted crack growth analysis is summarized in Figure 1. Note that the inite element analysis and the construction of surrogate model are performed offline, i.e. before the start of crack growth analysis. Crack propagation analysis is done only with the surrogate model.

The algorithm shown in Figure 1 for crack propagation analysis is deterministic and does not account for errors and sources of uncertainty. The next section discusses different sources of uncertainty associated with each of the blocks in Figure 1.

UNCERTAINTY IN CRACK GROWTH ANALYSIS

This section proposes methods to include the different sources of uncertainty in the crack growth analysis algorithm shown in Figure 1. These sources of uncertainty can be classified into three different types – physical variability, data uncertainty and model uncertainty - as shown in Figure 2.

Figure 2 shows the different sources of error and uncertainty considered in this chapter for the sake of illustration of the proposed methodology. There are several other sources of uncertainty that are not considered here. The physical variability in geometry, boundary conditions and some material properties (modulus of elasticity, friction coefficient, Poisson ratio, yield stress, ultimate stress, etc.) is assumed to be less significant and hence these quantities are treated to be deterministic. However, these can be included in the proposed framework by constructing different FEA models (for different geometry, boundary conditions, and material properties) and use multiple runs of these FEA models to train the Gaussian process surrogate model. Then, these parameters can also be treated as inputs to the surrogate model and sampled randomly in the uncertainty quantification procedure explained later in this chapter. Further, there may be several other types of model errors that arise due to

1. The assumption of a planar crack;
2. The assumption of linear elastic behavior and use of LEFM;

Figure 2. Sources of uncertainty in crack growth analysis

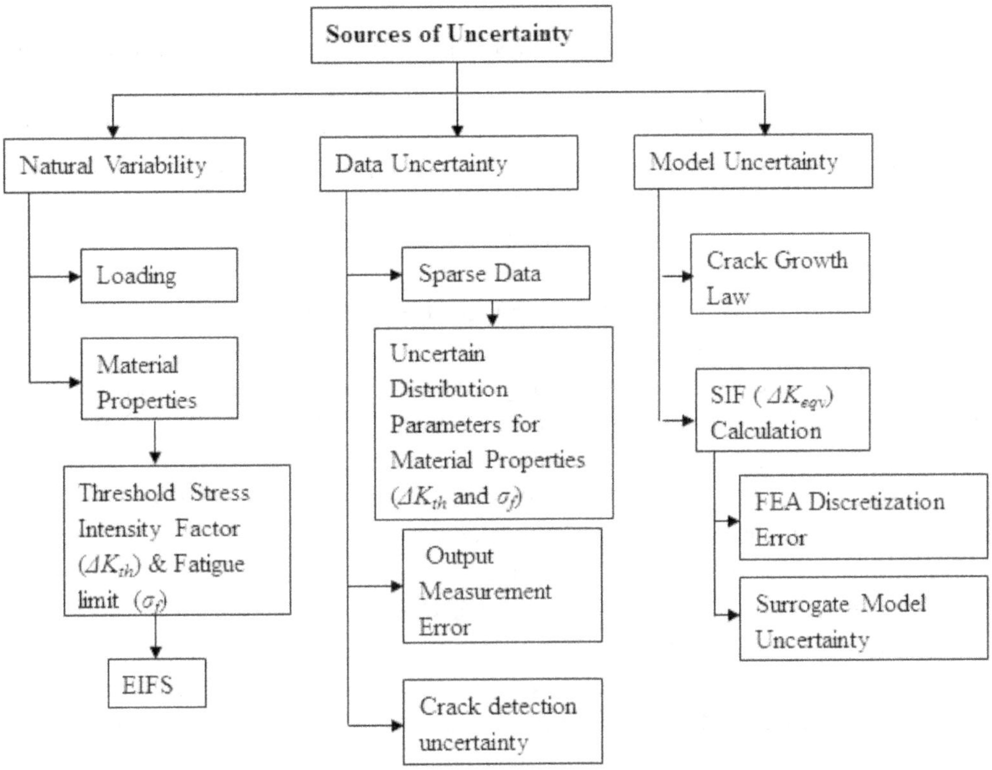

3. The use of a characteristic plane approach (to calculate an equivalent stress intensity factor) and parameters of the characteristic plane model in Eq. 7;
4. Errors in the estimation of retardation coefficient φ^r in Wheeler's model, etc.

These errors are not considered in this chapter because computational methods for the quantification of these errors are yet to be developed. The focus of this chapter is develop a methodology that can systematically account for the various sources of uncertainty and errors, and predict the crack size as a function of number of loading cycles. The error and uncertainty terms considered in this chapter adequately illustrate the overall methodology.

The following subsections discuss each source of uncertainty considered in this chapter and propose methods to handle them.

Physical or Natural Variability

Physical variability refers to the natural variability or fluctuations in the environment, test procedure, instruments, observer, etc. Hence, repeated observations of the same physical quantity do not yield identical results. The physical variability in loading conditions and material properties are considered during analysis. As mentioned earlier, the variability in other material properties such as modulus of elasticity, Poisson ratio, etc. is not considered since their effect on the final crack growth prediction is insignificantly small.

Data Uncertainty

As mentioned earlier, field inspections are used to collect crack growth data and this information is used to calibrate the distribution of EIFS. The different types of uncertainty in this procedure – crack detection uncertainty and measurement errors – are discussed. Also, the probability distributions of some material properties are inferred using data from laboratory experiments. These data may be sparse and cause uncertainty regarding the probability distribution type and parameters. This type of uncertainty is also discussed.

Crack Detection Uncertainty and Output Measurement Errors

For an in-service component, non-destructive inspection (NDI) is commonly used for damage detection. Several metrics could be used to evaluate the performance of NDI, such as probability of detection (POD), flaw size measurement accuracy, and false call probability (FCP). These criteria are developed from different methods, and they are used to evaluate different aspects of NDI performance. However, Zhang and Mahadevan (Zhang & Mahadevan, 2001) showed that these quantities can be mathematically related. POD and FCP can be derived from size measurement accuracy, which measures the difference between actual values and observed values of the crack size.

In the context of calibration, inspection results may be of three different types: (a) crack is not detected; (b) crack is detected but size not measured; and (c) crack is detected and size is measured. The concept of POD can be employed for the first two cases and for the case of detecting a crack and also measuring its size, size measurement accuracy can be employed. Based on the above consideration, size measurement accuracy can be used to quantify the uncertainty in experimental crack growth data, with the following expression determined by regression analysis (Zhang & Mahadevan, 2001):

$$a_m = \beta_0 + A.\beta_1 + \varepsilon_m \tag{10}$$

In Eq. 10, a_m is the measured flaw size; A is the actual flaw size; β_0 and β_1 are the regression coefficients; ε_m represents the unbiased measurement error, assumed as a normal random variable with zero mean and standard deviation σ_m. The value of σ_m is different for each inspection technique used.

Sparse Data for Characterizing Input Random Variables (ΔK_{th} and $\Delta\sigma_f$)

This section proposes a general methodology to characterize uncertainty in the input data, from which statistical distributions need to be inferred. In the numerical example later in this chapter, this method is implemented using experimental data available in the literature to characterize the distribution of threshold stress intensity factor (ΔK_{th}) and fatigue limit ($\Delta\sigma_f$).

Consider a random variable X whose statistics are to be determined from experimental data, given by $\boldsymbol{x} = \{x_1, x_2 .. x_n\}$. The likelihood of the parameters P can be calculated from first principles as (Shankar Sankararaman & Mahadevan, 2011):

$$L(\boldsymbol{P}) \propto \prod_{i=1}^{n} f_X\left(x_i \mid (\boldsymbol{P})\right) \tag{11}$$

where $f_X(x|\boldsymbol{P})$ is the probability density function of the random variable X conditioned on the parameters $\boldsymbol{P}$. For example, suppose that the random variable X follows a normal distribution, then the parameters ($\boldsymbol{P}$) of this distribution, i.e. mean and variance of X can be estimated from the entire data set $\boldsymbol{x}$.

Typically, the maximum likelihood estimates (by maximizing of Eq. 11) of the parameters ($\boldsymbol{P}$) are alone considered. In the aforementioned example of a normal distribution, the maximum likelihood estimates of the mean and the standard deviation are calculated. However, due to sparseness of data, these estimates of these parameters ($\boldsymbol{P}$) may not be accurate. As mentioned by Sankararaman and Mahadevan (Shankar Sankararaman & Mahadevan, 2011), the likelihood function is used to construct the entire PDF of the distribution parameters $\boldsymbol{P}$ as in Eq. 12. Let $f_P(\boldsymbol{p})$ denote the joint probability density of the parameters $\boldsymbol{P}$; by choosing a uniform prior density $f'_P(\boldsymbol{p}) = h$, Bayes' theorem reduces to:

$$f_P(\boldsymbol{p}) = \frac{L(\boldsymbol{P})h}{\int hL(\boldsymbol{P})d\boldsymbol{P}} = \frac{L(\boldsymbol{P})}{\int L(\boldsymbol{P})d\boldsymbol{P}} \tag{12}$$

Note that the uniform prior density function can be defined over the entire admissible range of the parameters $\boldsymbol{P}$. For example, the mean of a normal distribution can vary in $(-\infty, \infty)$ while the standard deviation can vary in $(0, \infty)$. The probability distribution of the parameters ($\boldsymbol{P}$) can be calculated by solving Eq. 12, either by using Markov Chain Monte Carlo-based methods or by using advanced numerical integration methods.

For each instance of a set of parameters ($\boldsymbol{P}$), X is defined by a realization of the chosen distribution type whose density function is given by $f_X(x|\boldsymbol{P})$. However, because the parameters ($\boldsymbol{P}$) themselves are stochastic and have a joint probability density function $f_P(\boldsymbol{P})$, X is defined by a family of distributions. In the aforementioned example of the normal distribution, for every given value of the mean and the standard deviation, there exists a unique normal probability distribution. Hence, by varying the mean and the standard deviation, the random variable X can be represented using a family of normal distributions.

The unconditional probability density function of X ($f_X(x)$) can be calculated by multiplying the conditional probability density function $f_X(x|\boldsymbol{P})$ of the variable X, and the joint probability density function $f_P(\boldsymbol{P})$, of the parameters $\boldsymbol{P}$ as (Shankar Sankararaman & Mahadevan, 2011):

$$f_X(x) = \int f_X(x \mid \boldsymbol{P})\, f_P(\boldsymbol{P})d\boldsymbol{P} \tag{13}$$

The integral in Eq. 13 can be evaluated through quadrature techniques or through Monte Carlo sampling. This can be achieved by selecting a sample of $\boldsymbol{P}$, and then selecting one sample of X corresponding to the previously sampled $\boldsymbol{P}$, and repeating the entire procedure to obtain multiple samples of X from which the unconditional probability density function of X ($f_X(x)$) can be calculated. This unconditional density function of X accounts for the sparse data as it includes the uncertainty in the estimate of the distribution parameters. In this chapter, this method has been used to characterize the uncertainty in the threshold stress intensity factor (ΔK_{th}) and fatigue limit ($\Delta\sigma_f$).

Model Uncertainty and Errors

It can be easily seen that different types of models (finite element model, characteristic plane model, surrogate model, crack growth model, retardation model, etc.) have been used to simulate crack growth behavior, and each of these models has its own error/uncertainty. In the discussion below, all the errors except the error due to the use of the characteristic plane approach and the uncertainty in the calculation of the retardation coefficient are considered. Some of them are deterministic while others are stochastic; the two types of errors need to be treated in different ways.

Uncertainty in Crack Growth Model

More than 20 different crack growth laws (e.g., Paris law, Foreman's equation, Weertman's equation) have been proposed in the literature. The mere presence of many such different models explains that none of these models can be applied universally to all fatigue crack growth problems. Each of these models has its own limitations and uncertainty..

The uncertainty in crack growth model can be subdivided into two different types: crack growth model error and uncertainty in model coefficients. If ε_{cg} is used to denote the crack growth model fitting error, then the modified Paris law (including the Wheeler retardation term) can be rewritten as in Eq. 14. Note that this error term is assumed to represent the difference between the model prediction and the experimental observation. Conventionally, in statistical model fitting, a normal error term is added to the model. Instead, the crack model error is represented using a lognormal and hence multiplicative term as shown in Eq. 14.

$$\frac{da}{dN} = \varphi^r C\left(\Delta K\right)^n \left(1 - \frac{\Delta K_{th}}{\Delta K}\right)^m \varepsilon_{cg} \tag{14}$$

An estimate of the variance of ε_{cg} can be obtained while calibrating the model parameters based on the available inspection data. The model coefficients in Paris law are *C*, *m* and *n*, and the uncertainty in these parameters can be represented through probability distributions.

Discretization Error in Finite Element Analysis

Theoretically, an infinitesimally small mesh size will lead to the exact solution but this is difficult to implement in practice. Hence, finite element analyses are carried out at a particular mesh size and the error in the solution, caused due to discretization needs to be quantified. Several methods are available in literature (Ainsworth & Oden, 1997; I. Babuska & WC. Rheinboldt, 1978) but many of them quantify some surrogate measure of error to facilitate adaptive mesh refinement. The Richardson extrapolation (RE) method has been found to come closest to quantifying the actual discretization error (SA. Richards, 1997).

Let h denote the mesh size used in the finite element analysis and Ψ the corresponding finite element prediction. Let Ψ^T denote the "true" solution of the finite element analysis which is obtained as h

tends to zero. According to the generalized Richardson extrapolation (SA. Richards, 1997), the relation between h and Ψ can be expressed as:

$$\Psi^T = \Psi + Ah^p \tag{15}$$

In Eq. 15, p is the order of convergence, and A is the polynomial coefficient. In order to estimate the true solution Ψ^T, three different mesh sizes ($h_1 < h_2 < h_3$) are considered and the corresponding finite element solutions (Ψ_1, Ψ_2, Ψ_3) are calculated. Eq. 15 has three unknowns p, A, and Ψ^T, which can be estimated based on the three mesh solutions. Mesh doubling/halving is commonly done to simplify the equations. If $r = h_3/h_2 = h_2/h_1$, then the discretization error (ε_h) and the true solution can be calculated as:

$$\Psi^T = \Psi_1 - \varepsilon_h; \varepsilon_h = \frac{\Psi_2 - \Psi_1}{r^p - 1}; p = \frac{\log\left(\dfrac{\Psi_3 - \Psi_1}{\Psi_2 - \Psi_1}\right)}{\log\left(r\right)} \tag{16}$$

The solutions Ψ_1, Ψ_2, Ψ_3 are dependent on the inputs (loading, current crack size, aspect ratio and angle of orientation) to the finite element analysis and hence the error estimates are also functions of these input variables. For each set of inputs, a corresponding discretization error (ε_h) is calculated and this error is subtracted from the finite element solution (using the finest mesh h_1) to estimate the "true" solution Ψ^T. The corrected solutions and the corresponding sets of inputs are used as training points for the surrogate model.

Uncertainty in the Surrogate Model Output

Several finite element runs for some combination of input-output variable values are used to train the Gaussian process surrogate model. Then, this surrogate model is used in cycle-by-cycle crack growth analysis to evaluate the stress intensity factor for other untrained combinations of input variable values. The residuals of the GP model form a Gaussian random field (Rasmussen, 2004). In each cycle, the error in the predicted ΔK_{eqv} is sampled using this random normal field and added to the GP model prediction.

(Note: The GP model is used as a surrogate for a deterministic finite element model, and the variance of the residual field accounts only for the uncertainty in approximating the original finite element model output with a surrogate model).

The stress intensity factor calculated by the surrogate model (mean and variance of the prediction in Eq. 8 and Eq. 9 respectively) is used in the crack growth equation to predict the crack size as a function of number of cycles. The following section incorporates all these sources of uncertainty into the calibration methodology, to estimate the crack growth model parameters.

MODEL PARAMETER CALIBRATION

This section explains the Bayesian calibration technique used to infer the probability distributions of the model parameters and modeling errors using experimental data. The calibration parameters are assigned prior distributions, and these distributions are updated after collecting experimental evidence corresponding to the model output (crack inspection after a particular number of loading cycles).

Dynamic Bayesian Network

There are several possible quantities that can be updated using the given procedure. These include:

1. Equivalent initial flaw size;
2. Parameters of modified Paris's law (*C, n, m*);
3. Error (ε_{cg}) of the modified Paris law;
4. Material properties such as threshold stress intensity factor (ΔK_{th}), and fatigue limit (σ_f);
5. Variance of the output (crack size after *N* cycles) measurement error (ε_m), etc.

All of these parameters can be connected in a Bayes network as shown in Fig. 3.

Note that this is a dynamic Bayes network, which connects the variables in one load cycle (*i*) to the next load cycle (*i+1*).

Though it is theoretically possible to calibrate all the parameters simultaneously, a multi-dimensional calibration procedure is computationally intensive. Also, in several situations, only a few parameters contribute effectively to the overall uncertainty in the final model prediction. It is advantageous to identify only such "important" parameters and calibrate them. To this end, global sensitivity analysis is used to quantify the contribution of the various sources of uncertainty to the overall uncertainty in the crack growth prediction, and the results of global sensitivity analysis are used to identify the parameters that need to be calibrated.

Figure 3. Dynamic Bayesian network for fatigue damage prognosis

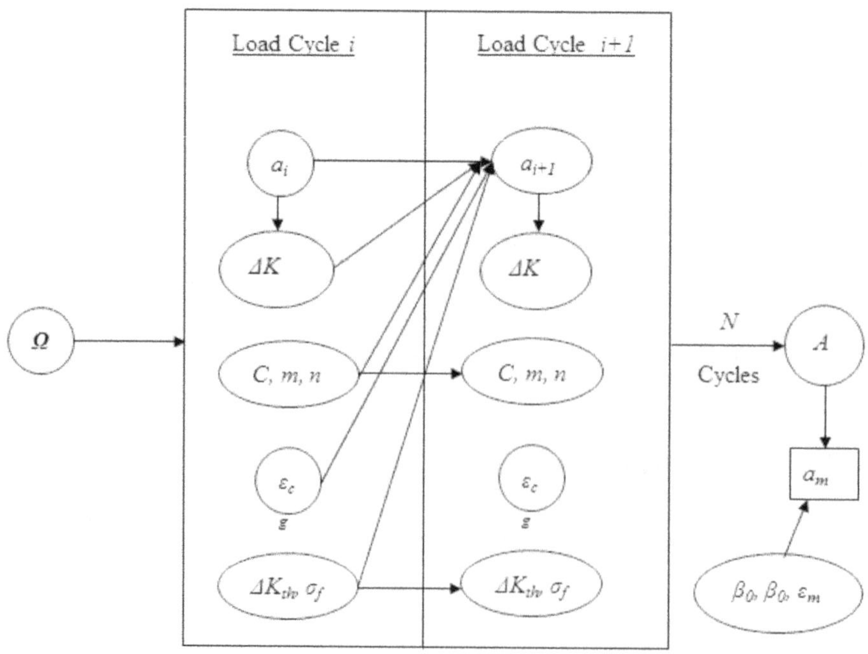

Global Sensitivity Analysis

Sankararaman et al. (Shankar Sankararaman et al., 2009) used a reduction of variance-based approach to identify the important contributors of uncertainty in crack growth analysis. However, this involves the "freezing" of one source of uncertainty and this approach is local with respect to where the uncertain variable is fixed. Saltelli et al. (Saltelli et al., 2008) argue that local sensitivities are not sufficient to study the contributions of multiple sources of uncertainty to the overall prediction uncertainty and it is necessary to purse a global sensitivity analysis approach for this purpose. Note that this approach is based only on second moment calculations, and calculates the effect of the variance of an input quantity on the variance of the output quantity. Consider a model given by:

$$Y = G\left(X_1, X_2, X_3 \ldots X_n\right) \tag{17}$$

Note that the *X's* and the *Y* are input-output pairs for a generic model. In the context of the crack growth problem, the output is the crack size at the end of a particular number of cycles, and the inputs include all quantities that affect that crack size prediction (equivalent initial flaw size, loading, crack growth parameters, model errors, etc.).

The first-order effect index of a variance X_i is given by

$$S_i^1 = \frac{V_{X_i}\left(E_{X_{-i}}\left(Y \mid X_i\right)\right)}{V\left(Y\right)} \tag{18}$$

In Eq. 18, $E_{X_{-i}}\left(Y \mid X_i\right)$ denotes the expectation of the output Y by fixing the variable X_i at a particular value and considering random variations of all other variables (denoted by $X_{\sim i}$). The output of this calculation is conditioned on where the variable X_i is fixed. Hence, Eq. 18 calculates the variance of this expectation by considering the random variation in X_i and fixing it at random values. The calculation of Eq. 18 requires an outer Monte carlo loop to calculate the variance and an inner Monte carlo loop to calculate the expectation. However, Saltelli et al. (Saltelli et al., 2008) provide single loop Monte carlo methods to evaluate Eq. 18.

The first-order effect calculated in Eq. 18 gives an estimate of the contribution of the variable X_i to the uncertainty in *Y*, without considering the effects of the other variables $X_{\sim i}$, as their contribution is averaged. The sum of first order indices of all variables is always less than or equal to unity; equality holds when the model *G* is additive with respect to all the variables. The contribution of the variable X_i in combination with all other variables is known as the total effects index and can be calculated as:

$$S_i^T = 1 - \frac{V_{X_{-i}}\left(E_{X_i}\left(Y \mid X_{-i}\right)\right)}{V\left(Y\right)} \tag{19}$$

The terms in Eq. 19 are similar to that in Eq. 18; there are two notable differences though. First, the variance is calculated in the inner loop and the expectation of the variance is calculated in the outer loop.

Second, in the inner loop, all quantities other than X_i (denoted by $X_{\sim i}$) are fixed at a particular quantity to calculate the variance by considering random variations in X_i., and then the outer loop considers random variations in all other quantities ($X_{\sim i}$).

The sum of the total effects indices of all variables is always greater than or equal to unity; equality holds when the model G is additive with respect to all variables. (In this case, the first-order effects indices are equal to the total effects indices).

Let X_i's denote the various sources of uncertainty and stochastic error terms; deterministic errors are corrected before sensitivity analysis and not included. The quantities that have a high influence on the variance of the final crack growth prediction are alone selected for calibration. While this screening procedure is based on second-moment calculation, the model calibration and validation techniques explained in the following sections consider the entire probability distribution for all quantities.

Bayesian Inference

Let $\boldsymbol{\Omega}$ denote the vector of quantities that are selected for calibration. Assume that there is a set of m experimental data points, i.e. crack inspection after N loading cycles. Each inspection may produce three possible results:

1. No crack detected;
2. Crack detected but size not measured; and
3. Crack detected and size measured to be a_m.

This is the first set of data, referred to as the calibration data. Bayesian updating is a three step procedure:

Prior probability distributions are assumed for each of the parameters. The prior distribution of EIFS is calculated using Eq. 5. The crack growth model error (ε_{cg}) has zero mean and the standard deviation is assumed to have a non-informative prior. The prior distributions of crack growth law parameters are obtained from Liu and Mahadevan (Liu & Mahadevan, 2009).

The likelihood of $\boldsymbol{\Omega}$ is defined as being proportional to the probability of observing the given data conditioned on $\boldsymbol{\Omega}$. (Note that the likelihood is meaningful only up to a proportionality constant. The mass of a likelihood function, in general, is not equal to unity and hence, "the likelihood $\boldsymbol{\Omega}$" of must not be confused with "the probability of $\boldsymbol{\Omega}$" or the "probability density of $\boldsymbol{\Omega}$".)

The prior and likelihood of $\boldsymbol{\Omega}$ are multiplied and normalized to calculate the posterior probability distribution. Finally, the joint distribution of $\boldsymbol{\Omega}$ is used to calculate the marginal distributions of the individual parameters.

In step 2, the likelihood of $\boldsymbol{\Omega}$ is calculated as the probability of observing the data conditioned on $\boldsymbol{\Omega}$ and hence is a function of $\boldsymbol{\Omega}$. For every given $\boldsymbol{\Omega}$, a Monte Carlo analysis is required for the calculation of likelihood along the following steps:

1. Construct the Gaussian process surrogate model as explained in this chapter. This is done through finite element analysis and then by calculating the discretization error in each of the runs. The discretization errors are added to the solutions of finite element analysis and used to train the Gaussian process surrogate model. Hereon, the GP model can be used to calculate the stress intensity factor as a function of crack size, loading, aspect ratio and angle of orientation.

For a given $\boldsymbol{\Omega}$

2. Generate a loading history (N cycles) based on the statistics of loading.
3. Use the deterministic prognosis methodology in Figure 1 to calculate the final crack size at the end of N_i cycles.
4. Repeat steps 2 and 3 and calculate the probability distribution of crack size at the end of N_i (for $i = 1$ to m) cycles. Let this distribution be denoted by $f(a)$. Use Eq. 10 to calculate $f(a_m|a)$. This probability density function can be used to calculate the likelihood of $\boldsymbol{\Omega}$.

If no crack is detected, then the likelihood function can be calculated as:

$$L(\boldsymbol{\Omega}) \propto 1 - \iint f(a_m \mid a) f(a \mid N_i, \boldsymbol{\Omega}) da_m \tag{20}$$

If a crack is detected but size not measured, the likelihood function can be calculated as:

$$L(\boldsymbol{\Omega}) \propto \iint f(a_m \mid a) f(a \mid N_i, \boldsymbol{\Omega}) da \; da_m \tag{21}$$

If a crack is detected and the size is measured to be A_i, then the likelihood function can be calculated as:

$$L(\boldsymbol{\Omega}) \propto \int f(a_m = A_i \mid a) f(a \mid N_i, \boldsymbol{\Omega}) da \tag{22}$$

Note that a_m and a vary in the interval $(0, \infty)$.

If different inspections give any of the three types of crack information, the corresponding likelihood function for each data point is calculated using Eq. 17, 18 or 19, and then all the likelihood functions are multiplied to get the overall likelihood function for the entire inspection data set. Finally, the likelihood is multiplied with the prior and normalized to calculate the joint posterior distribution of $\boldsymbol{\Omega}$, and this joint distribution can be marginalized to calculate the individual distributions.

FATIGUE DAMAGE PROGNOSIS AND RELIABILITY ANALYSIS

This section presents a sampling based strategy to combine all the different sources of uncertainty and thereby quantify the overall uncertainty in crack growth prediction as a function of number of loading cycles (N). The various steps in this procedure are outlined here.

1. Use the Gaussian process surrogate model created in the previous section for damage prognosis also, in order to compute the equivalent stress intensity factor.
2. Generate a loading history. First, randomly select a block length and then randomly select a maximum amplitude value and a minimum amplitude value for that particular block. Repeat the process till the number of cycles (N) is reached.

3. Sample a value of $\boldsymbol{\Omega}$ from the posterior probability distribution calculated after Bayesian inference in the previous section.
4. Use the deterministic procedure for crack growth analysis to calculate the final crack size at the end of N cycles. However, in each loading cycle, the stress intensity factor calculated from the GP model is a random normal variable and hence generate a random sample of stress intensity factor in each cycle. Also, the crack growth model error (ε_{cg}) is sampled in every cycle.

In this algorithm, Step 1 is a deterministic step while Steps 2 - 4 are probabilistic. Using this algorithm, the crack size after N cycles can be calculated for a particular load history that was generated in Step 2. Using Monte Carlo Sampling, Steps 2 - 4 can be repeated again and again, each leading to a final crack size at the end of N cycles. This can be used to characterize the distribution of final crack size at the end of N cycles. By varying N, the distribution of final crack size can be obtained as a function of the number of cycles (N). This information can be used to calculate the reliability of the structural component as a function of number of load cycles. Suppose that the component is supposed to have failed if the crack size is greater than a critical crack size (A_c), then the probability of failure can be calculated as a function of load cycles.

Note that the above methodology for uncertainty quantification is applicable to any model-based technique for fatigue crack growth analysis and damage prognosis. Thus any appropriate combination of fatigue crack growth models can be used instead of the models used.

NUMERICAL EXAMPLE

A two-radius hollow cylinder with an elliptical surface crack in fillet radius region is chosen for illustrating the proposed methodology, as shown in Fig. 4.

The commercial finite element software ANSYS is used to build and analyze the finite element model. A sub-modeling technique is used near the region of the crack for accurate calculation of the

Figure 4. Surface crack in a cylindrical structural component

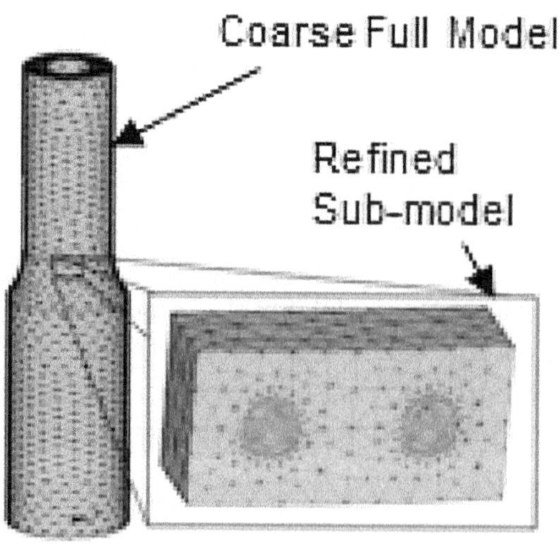

Table 1. Material properties

Aluminium 7075- T6	
Modulus of Elasticity	72 GPa
Poisson Ratio	0.32
Yield Stress	450 MPa
Ultimate Stress	510 MPa

Table 2. Geometric properties

Cylinder Properties	
Length	152.4 mm
Inside Radius	8.76 mm
Outside Radius	14.43 mm
Outside Radius	17.78 mm

stress intensity factor. As mentioned earlier, the material and geometric properties of the specimen are assumed to be deterministic and are listed in Table 1 and Table 2 respectively.

A variable amplitude multi-axial (bending and torsion) loading is considered here for the sake of illustration. A block loading history is considered (sample shown in Fig. 5). The block length is assumed to be a random variable and the maximum and minimum amplitudes in each block are also treated as random variables. To generate each block of loading, first the block length is randomly sampled and then the maximum and minimum amplitudes for that block are randomly sampled. The entire loading history is generated by repeating this process and creating several successive blocks to occupy the duration of interest. All the different types of uncertainty in Fig. 2 are included in this numerical example, as shown in Table 3.

This finite element model is run for 10 different crack sizes and 6 different loading cases and these results are used to train the Gaussian process surrogate model for the calculation of the stress intensity factor. In each run of the finite element analysis, the discretization error is calculated and the corrected solution in turn is used to train the Gaussian process surrogate model. The maximum discretization error was found to be about 2% of the finest mesh solution. If the discretization errors were found to be

Figure 5. Sample loading history

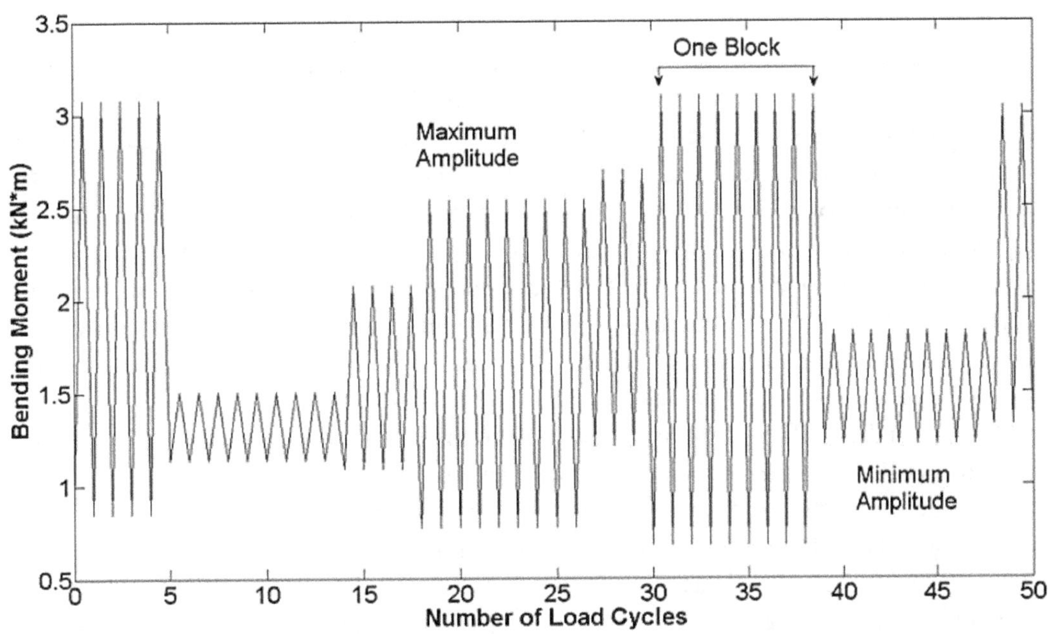

Table 3. Types of uncertainty

Type of Uncertainty	Quantities	Details
Natural Variability	Loading	Random block loading Block length ~ Uniform (0, 500) Max. Amplitude ~ $N(\mu_{max},\sigma_{max})$ Max. Amplitude ~ $N(\mu_{min},\sigma_{min})$ μ_{max} ~ Uniform (1.355, 0.263) kNm σ_{max} ~ Uniform (0.455, 0.130) kNm μ_{min} ~ Uniform (2.710, 0.260) kNm σ_{min} ~ Uniform (0.455, 0.130) kNm
	Material properties (used to calculate prior distribution of EIFS)	(ΔK_{th} *and* $\Delta\sigma_f$) – Distribution type, mean, and standard deviation based on (Liu & Mahadevan, 2009)
Data Uncertainty	Sparse data	Distribution parameters of (ΔK_{th} and $\Delta\sigma_f$) are uncertain → Assumed coefficient of variation for mean = 0.1 and standard deviation = 0.1 → Use Eq. 13 to calculate PDFs of ΔK_{th} and $\Delta\sigma_f$ → Use Eq. 5 to calculate PDF of EIFS (θ)
	Inspection data (crack not detected, crack detected but size not measured, and crack size measured)	Considered all three cases both for calibration data set and validation data set
Model Uncertainty	Crack growth model	Statistics of parameters based on (Liu & Mahadevan, 2009) and lognormal error term (ε_{cg}) with unit mean and standard deviation equal to 0.05; in the numerical example, EIFS (θ), parameter C and error ε_{cg} are calibrated with above distributions as priors.
	Discretization error	Calculated using Richardson extrapolation
	Surrogate model error	Calculated based on Eq. 9 and included in the uncertainty propagation

very large, then it indicates a coarsely resolved calculation, which may not be suitable in the context of model calibration.

Global sensitivity analysis was used to calculate the first order (individual) and total (combined) effects of the various crack growth parameters such as EIFS (θ), model parameters (C, m), crack growth model error (ε_{cg}), threshold stress intensity factor (ΔK_{th}), loading, etc. It was observed that the sensitivity of the parameters changed with time and hence, it was necessary to calculate the sensitivity indices as a function of number of cycles. After a significant number of loading cycles, the parameters EIFS θ, model parameter C, and crack growth model error (ε_{cg}) were found to be the most significant terms. The first-order and total effects sensitivities of these quantities are shown in Figures 6-8. Recall that the first-order effects describe the contribution of a particular variable to the uncertainty in the crack size, by itself, whereas the total effects describe the overall contribution of a particular variable to the uncertainty in the crack size, in combination with all other variables.

The sensitivity indices of the other parameters were insignificant (less than 0.001) and hence only three parameters (EIFS θ, model parameter C, and crack growth model error ε_{cg}) are considered to have a dominant influence on the uncertainty in the crack growth prediction in comparison with the other parameters. Hence, these three are chosen to be the calibration parameters.

Experimental evidence of 100 data points (inspection) was simulated (based on assumed true distributions for the inference quantities) and used for calibration. 100 components were inspected at different

Figure 6. Sensitivity indices of equivalent initial flaw size θ

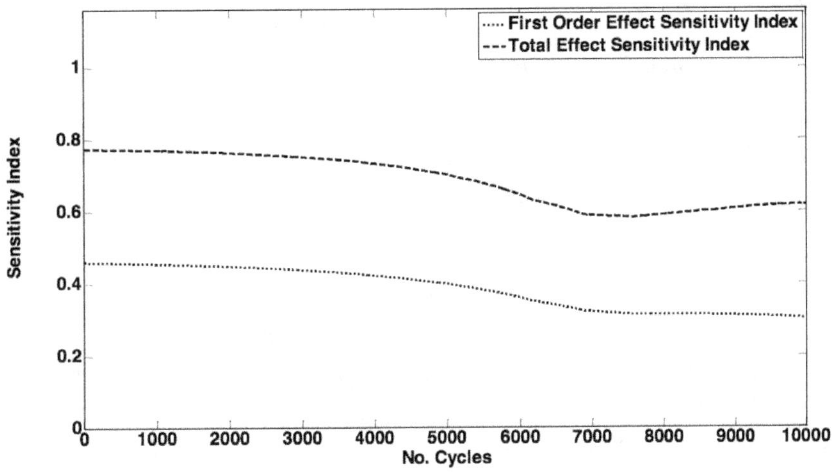

Figure 7. Sensitivity indices of model parameter C

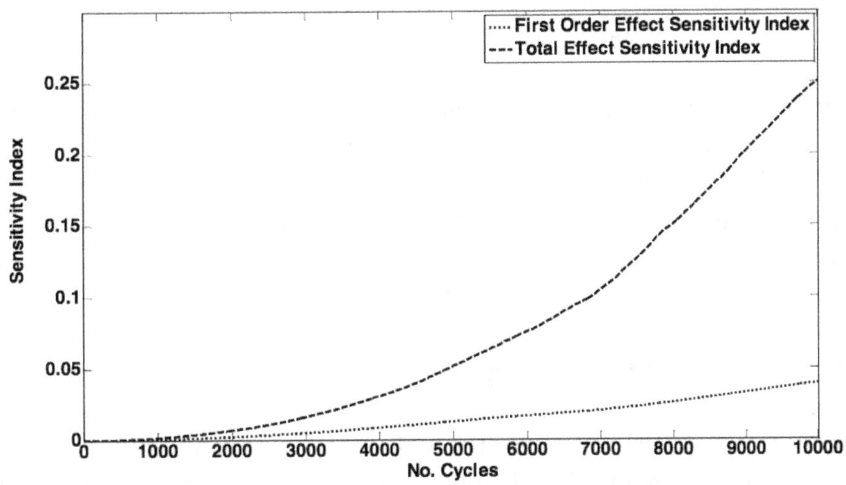

Figure 8. Sensitivity indices of crack growth model error εcg

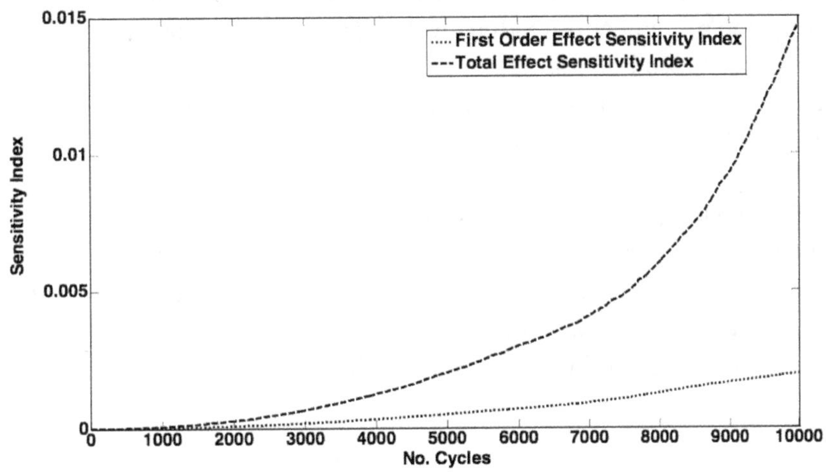

Table 4. Results of calibration

Quantity	True Statistics		Prior Statistics		Posterior Statistics	
	Mean	**Std. Dev.**	**Mean**	**Std. Dev.**	**Mean**	**Std. Dev.**
EIFS (θ) (mm)	0.3810	0.0254	0.5126	0.1146	0.3788	0.0222
C (m/cycle)	4.0 E -13	0.4 E -13	6.5 E -13	4.0 E -13	5.7 E-13	3.1 E -13
ε_{cg} (mm)	1	0.05	1	0.1	1.0	0.05

time instants; no crack was detected in one component, the crack size was measured in 49 components, and the crack size was not measured in 50 components. This experimental data is used to calculate the statistics of the parameters using the likelihood function (Eq. 20 – 22) and Bayes theorem.

Note that Eq. 20 – 22 calculate the joint likelihood of the inference quantities (EIFS θ, model parameter C, and crack growth model error ε_{cg}) and hence, Bayes theorem gives the joint probability density function of these quantities. Hence, in addition to the marginal posterior distributions, the correlations between these quantities can also be estimated using the joint density function.

The Bayesian calibration methodology is applicable to any combination of prior PDF and likelihood function; there are no restrictions on the distribution type. However, if the calculation of the likelihood function is non-trivial, as seen in this numerical example, the posterior distributions are non-parametric and may not conform to any particular distribution type. Hence, it is necessary to calculate the posterior distributions using numerical techniques; the mean and the standard deviation of the prior distribution and the posterior distribution of the inferred quantities are given in Table 4. Since this numerical example was simulated, it is possible to verify the results of the methodology by comparing against the "true" distributions of the parameters that were used for simulation.

The Bayesian approach will be applicable even if the mean of the prior distribution is one or two orders of magnitude away from the truth as long as the uncertainty in the prior is large enough to contain the truth. In other words, the Bayesian approach will work if there is any overlap between the prior distribution and the truth.

(If there is no overlap, then the posterior PDF will be zero everywhere. Note that the posterior distribution is proportional to the prior times the likelihood. In the "region of the truth", the prior will be zero as there is no overlap. Outside the "region of the truth", the likelihood will be zero.)

Having calibrated the parameters, the posterior distributions can be used to predict the distribution of crack size as a function of number of loading cycles. The mean, median and 90% prediction bounds of the final crack size are shown in Figure 7.

This information can then be used for reliability analysis to predict the failure probability as a function of number of loading cycles, as shown in Figure 10. A critical crack size of 2.54 mm (approximately 0.1 inch) is assumed for the purpose of illustration and the probability of failure is estimated as a function of number of load cycles and plotted in Figure 10. From Figure 10, it is seen that the probability of failure is negligible for about 3500 load cycles and it gradually increases after 4000 cycles.

There are two reasons for the observed increase in increase of failure probability. Firstly, the crack is growing in size and secondly, the uncertainty in the estimated crack size also increases with each loading cycle. After 10000 cycles of loading, the probability of failure is approximately equal to 0.01.

Figure 9. Mean, median, and 90% bounds

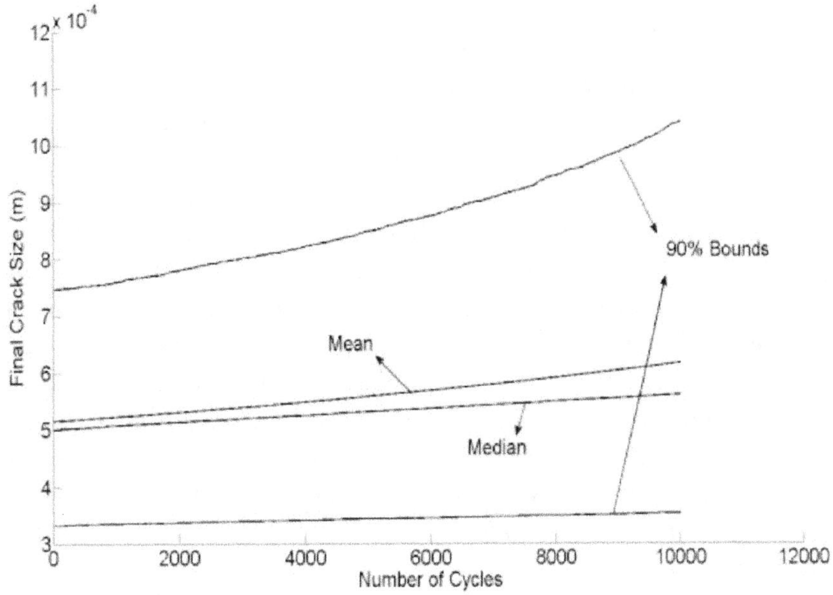

Figure 10. Probability of failure vs. no. load cycles

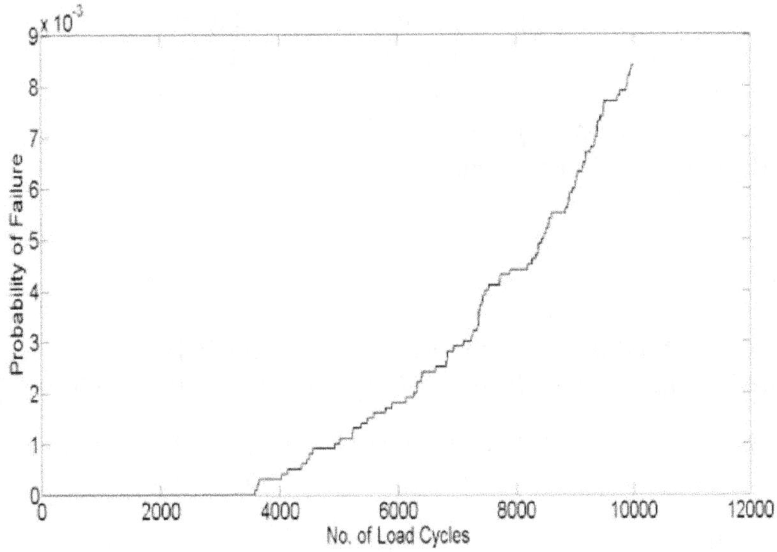

CONCLUSION

This chapter presented a computational methodology for modeling crack growth in structures with complicated geometry and multi-axial variable amplitude loading conditions. In order to bypass small crack calculations and use a long crack growth model for crack growth prediction, the concept of equivalent initial flaw size was introduced. Finite element analysis was used for the calculation of the multi-modal stress intensity factors and these were then integrated to compute an equivalent stress intensity that can be used in the aforementioned crack growth law. However, it may not be possible to use finite element analysis in every cycle of crack growth prediction, and therefore, few runs of the finite element model were used to train a Gaussian process surrogate model, thereby reducing the computational effort. The effect of variable amplitude loading conditions was addressed through the use of a retardation factor.

The different sources of uncertainty – physical variability, data uncertainty, and model error/uncertainty – were explained in detail, and included in the crack growth analysis. Different types of model errors – discretization error, crack growth model error, surrogate model error – were considered explicitly. Deterministic errors were corrected where they arose and stochastic errors were included by using random samples during uncertainty propagation.

The concept of dynamic Bayesian network was then used to integrate all of the above models (long crack growth model, finite element model, Gaussian process surrogate model, retardation model, etc.) and the various sources of uncertainty. Then, this Bayesian network can aid in both calibrating the model parameters and errors, and predicting the crack size as a function of number of loading cycles. Bayesian inference was used to calibrate the parameters of different models using inspection results (crack size after number of cycles, including cases where no crack was detected and crack was detected but size not measured). After calibration, a sampling-based scheme was used to predict the crack size as a function of the number of cycles. This framework has also been extended to analyze three dimensional cracks in metals (Hombal, Ling, Wolfe, & Mahadevan, 2012; Hombal & Mahadevan, 2013). This methodology is useful for design of structural components and systems where the focus is to maximize life expectancy and minimize cost through efficient and accurate reliability analysis.

REFERENCES

Ainsworth, M., & Oden, J. T. (1997). A posteriori error estimation in finite element analysis. *Computer Methods in Applied Mechanics and Engineering*, *142*(1), 1–88. doi:10.1016/S0045-7825(96)01107-3

Babuska, I., & Rheinboldt, W. C. (1978). A posterori error estimates for the finite element method. *International Journal for Numerical Methods in Engineering*, *12*(10), 1597–1615. doi:10.1002/nme.1620121010

Fawaz, S. A. (2003). Equivalent initial flaw size testing and analysis of transport aircraft skin splices. *Fatigue & Fracture of Engineering Materials & Structures*, *26*(3), 279–290. doi:10.1046/j.1460-2695.2003.00637.x

Grell, W. A., & Laz, P. J. (2010). Probabilistic fatigue life prediction using AFGROW and accounting for material variability. *International Journal of Fatigue*, *32*(7), 1042–1049. doi:10.1016/j.ijfatigue.2009.12.001

Hombal, V. K., Ling, Y., Wolfe, K. A., & Mahadevan, S. (2012). Two-stage planar approximation of non-planar crack growth. *Engineering Fracture Mechanics*, *96*, 147–164. doi:10.1016/j.engfracmech.2012.07.017

Hombal, V. K., & Mahadevan, S. (2013). Surrogate modeling of 3D crack growth. *International Journal of Fatigue*, *47*, 90–99. doi:10.1016/j.ijfatigue.2012.07.012

Hwang, W., & Han, K. S. (1986). Fatigue of composites—fatigue modulus concept and life prediction. *Journal of Composite Materials*, *20*(2), 154–165. doi:10.1177/002199838602000203

Johnson, G. R., & Cook, W. H. (1985). Fracture characteristics of three metals subjected to various strains, strain rates, temperatures and pressures. *Engineering Fracture Mechanics*, *21*(1), 31–48. doi:10.1016/0013-7944(85)90052-9

Liu, Y., Liu, L., & Mahadevan, S. (2007). Analysis of subsurface crack propagation under rolling contact loading in railroad wheels using FEM. *Engineering Fracture Mechanics*, *74*(17), 2659–2674. doi:10.1016/j.engfracmech.2007.02.012

Liu, Y., & Mahadevan, S. (2005). Multiaxial high-cycle fatigue criterion and life prediction for metals. *International Journal of Fatigue*, *27*(7), 790–800. doi:10.1016/j.ijfatigue.2005.01.003

Liu, Y., & Mahadevan, S. (2007). Threshold stress intensity factor and crack growth rate prediction under mixed-mode loading. *Engineering Fracture Mechanics*, *74*(3), 332–345. doi:10.1016/j.engfracmech.2006.06.003

Liu, Y., & Mahadevan, S. (2009). Probabilistic fatigue life prediction using an equivalent initial flaw size distribution. *International Journal of Fatigue*, *31*(3), 476–487. doi:10.1016/j.ijfatigue.2008.06.005

Maymon, G. (2005). Probabilistic crack growth behavior of aluminum 2024-T351 alloy using the "unified"approach. *International Journal of Fatigue*, *27*(7), 828–834. doi:10.1016/j.ijfatigue.2004.12.003

McClung, R. C., Enright, M. P., Millwater, H. R., Leverant, G. R., & Hudak, S. J. Jr. (2004). A software framework for probabilistic fatigue life assessment of gas turbine engine rotors. *Journal of ASTM International*, *1*(8), 11563. http://www.astm.org/cgi-bin/scholar.cgi?JAI11563 doi:10.1520/JAI11563

Merati, A., & Eastaugh, G. (2007). Determination of fatigue related discontinuity state of 7000 series of aerospace aluminum alloys. *Engineering Failure Analysis*, *14*(4), 673–685. doi:10.1016/j.engfailanal.2006.02.016

Newman, J. C. Jr, Brot, A., & Matias, C. (2004). Crack-growth calculations in 7075-T7351 aluminum alloy under various load spectra using an improved crack-closure model. *Engineering Fracture Mechanics*, *71*(16), 2347–2363. doi:10.1016/j.engfracmech.2004.01.004

Paris, P. C., Gomez, M. P., & Anderson, W. E. (1961). A rational analytic theory of fatigue. *The Trend in Engineering*, *13*(1), 9–14.

Patrick, R., Orchard, M. E., Zhang, B., Koelemay, M. D., Kacprzynski, G. J., Ferri, A. A., & Vachtsevanos, G. J. (2007). An integrated approach to helicopter planetary gear fault diagnosis and failure prognosis. In *Autotestcon, 2007 IEEE* (pp. 547–552). IEEE. Retrieved from http://ieeexplore.ieee.org/xpls/abs_all.jsp?arnumber=4374266

Peng, T., Saxena, A., Goebel, K., Xiang, Y., Sankararaman, S., & Liu, Y. (2013). A novel Bayesian imaging method for probabilistic delamination detection of composite materials. *Smart Materials and Structures*, *22*(12), 125019. doi:10.1088/0964-1726/22/12/125019

Rasmussen, C. E. (2004). Gaussian processes in machine learning. In Advanced Lectures on Machine Learning (pp. 63–71). Springer. Retrieved from http://link.springer.com/chapter/10.1007/978-3-540-28650-9_4 doi:10.1007/978-3-540-28650-9_4

Richards, S. A. (1997). Completed Richardson extrapolation in space and time. *Communications in Numerical Methods in Engineering*, *13*(7), 573–582. doi:10.1002/(SICI)1099-0887(199707)13:7<573::AID-CNM84>3.0.CO;2-6

Saltelli, A., Ratto, M., Andres, T., Campolongo, F., Cariboni, J., Gatelli, D. … Tarantola, S. (2008). Global Sensitivity Analysis : The Primer. John Wiley and Sons.

Sankararaman, S., Ling, Y., & Mahadevan, S. (2010). Statistical inference of equivalent initial flaw size with complicated structural geometry and multi-axial variable amplitude loading. *International Journal of Fatigue*, *32*(10), 1689–1700. doi:10.1016/j.ijfatigue.2010.03.012

Sankararaman, S., Ling, Y., & Mahadevan, S. (2011). Uncertainty quantification and model validation of crack growth prediction. *Engineering Fracture Mechanics*, *78*(7), 1487–1504. doi:10.1016/j.engfracmech.2011.02.017

Sankararaman, S., Ling, Y., Shantz, C., & Mahadevan, S. (2009). Uncertainty quantification in fatigue damage prognosis. In *Annual conference of the prognostics and health management society* (pp. 1–13). Retrieved https://www.phmsociety.org/sites/phmsociety.org/files/phm_submission/2009/phmc_09_51.pdf

Sankararaman, S., Ling, Y., Shantz, C., & Mahadevan, S. (2011). Inference of equivalent initial flaw size under multiple sources of uncertainty. *International Journal of Fatigue*, *33*(2), 75–89. doi:10.1016/j.ijfatigue.2010.06.008

Sankararaman, S., & Mahadevan, S. (2011). Likelihood-based representation of epistemic uncertainty due to sparse point data and/or interval data. *Reliability Engineering & System Safety*, *96*(7), 814–824. doi:10.1016/j.ress.2011.02.003

Sankararaman, S., McLemore, K., Mahadevan, S., Bradford, S. C., & Peterson, L. D. (2013). Test resource allocation in hierarchical systems using Bayesian networks. *AIAA Journal*, *51*(3), 537–550. doi:10.2514/1.J051542

Song, P.-S., Sheu, B.-C., & Chang, L. (2001). A Modified Wheeler Model to Improve Predictions of Crack Growth Following a Single Overload. *JSME International Journal Series A*, *44*(1), 117–122. doi:10.1299/jsmea.44.117

Tryon, R. G., Cruse, T. A., & Mahadevan, S. (1996). Development of a reliability-based fatigue life model for gas turbine engine structures. *Engineering Fracture Mechanics*, *53*(5), 807–828. doi:10.1016/0013-7944(95)00138-7

Wheeler, O. E. (1972). Spectrum loading and crack growth. *Journal of Fluids Engineering*, *94*(1), 181–186.

Wu, Y.-T., Enright, M. P., & Millwater, H. R. (2002). Probabilistic methods for design assessment of reliability with inspection. *AIAA Journal*, *40*(5), 937–946. doi:10.2514/3.15143

Yagawa, G., Kanto, Y., Yoshimura, S., Machida, H., & Shibata, K. (2001). Probabilistic fracture mechanics analysis of nuclear structural components: A review of recent Japanese activities. *Nuclear Engineering and Design*, *207*(3), 269–286. doi:10.1016/S0029-5493(01)00337-5

Zhang, R., & Mahadevan, S. (2001). Fatigue reliability analysis using nondestructive inspection. *Journal of Structural Engineering*, *127*(8), 957–965. doi:10.1061/(ASCE)0733-9445(2001)127:8(957)

Chapter 11
Prognostics Design for Structural Health Management

J. Chiachío
University of Granada, Spain

S. Sankararaman
SGT Inc., USA & NASA Ames Research Center, USA

M. Chiachío
University of Granada, Spain

A. Saxena
SGT Inc., USA & NASA Ames Research Center, USA

K. Goebel
NASA Ames Research Center, USA

ABSTRACT

The chapter describes the application of prognostic techniques to the domain of structural health and demonstrates the efficacy of the methods using fatigue data from a graphite-epoxy composite coupon. Prognostics denotes the in-situ assessment of the health of a component and the repeated estimation of remaining life, conditional on anticipated future usage. The methods shown here use a physics-based modeling approach whereby the behavior of the damaged components is encapsulated via mathematical equations that describe the characteristics of the components as it experiences increasing degrees of degradation. Mathematical rigorous techniques are used to extrapolate the remaining life to a failure threshold. Additionally, mathematical tools are used to calculate the uncertainty associated with making predictions. The information stemming from the predictions can be used in an operational context for go/no go decisions, quantify risk of ability to complete a (set of) mission or operation, and when to schedule maintenance.

INTRODUCTION

Prognostics is a core element of Prognostics and Health Management (PHM), which sets out to actively monitor and manage assets based on their state of health as opposed to scheduling periodic inspection and maintenance based on statistics of mean-time-to-failure or similar information. Prognostics is the science of determining the remaining useful life (RUL) of a component or subsystem given the current degree of wear or damage, the component's load history, and anticipated load and environmental conditions. A quantification of the degree of a component's wear or damage and the estimate of end-of-life (EOL) gives decision makers important information about the structural health of a system. This information

DOI: 10.4018/978-1-4666-8490-4.ch011

can be used for risk reduction in go/no-go decision, cost reduction through the scheduling of maintenance as-needed, and improved asset availability. Prognostics employs techniques that are often based on an analysis of historical data or an analysis of the fault modes and the modeling of the physics of both the component itself as well as the attributes that characterize the fault. Damage progression models include the effects of damage accelerators or stressors (such as load or environmental conditions). Algorithms that estimate the remaining life use estimation techniques that propagate the anticipated degradation into the future and provide as output the time where the component does no longer meet its desired functionality.

FUNDAMENTALS

Predictive information about a component fault/damage can be a valuable resource in determining an appropriate course of action to avoid failures. Potential of prognostics in positively contributing to safety and improving life-cycle costs is equally relevant to existing legacy systems and new system designs. Legacy systems adopt additional sensing and processing with a potentially high price of retrofitting and additional validation and/or certification costs to gain extended system life and safety factor. New system designs can significantly reduce these costs if prognostics and health management are adopted early in the design to facilitate a more optimal sensor placement for observability and coverage. This, however, requires integration of health management design into the systems engineering process. The following section briefly discusses various design considerations involved in design and development of PHM.

Design Considerations for Prognostics Health Management

With the term "design" one means in general a method of conceiving and planning to realize a solution to a particular engineering problem. This entails formulating a plan and drawing up a scheme how to realize the plan. Generally, the design process is partitioned into the stages

1. Analysis,
2. Concept, and
3. Synthesis.

One needs to start out with having a good understanding of the underlying goals. That is, an understanding of what the prognostic system is meant to accomplish. This could be a reduction of life-cycle cost, or a safety improvement, or an optimization of uptime to better guarantee mission availability or commercial dispatch readiness. The design process is shown in Figure 1 and explained below.

During the *Analysis* phase, requirements for the overall system are created based on the stated goals. Requirements for prognostic system design are guided by the end use of prognostics (Saxena, Celaya, Saha, Saha, Roychoudhury, *et al.*, 2010). There are different motivations for prognostics, such as improving safety, reducing costs, or increasing availability. However, to realize an effective health management system, requirements for prognostics and structural health monitoring (SHM) should be developed with those of the system. This requires forward thinking about what components or subsystems need monitoring and accordingly determine an appropriate sensor network design, which then can be integrated into the structure design in a more cost-effective manner. The application scenario dictates what mitigation action will be most beneficial given a prognostic estimate of remaining-life-of a structure. At the same

Figure 1. Design process of PHM solution
(Goebel, 2014).

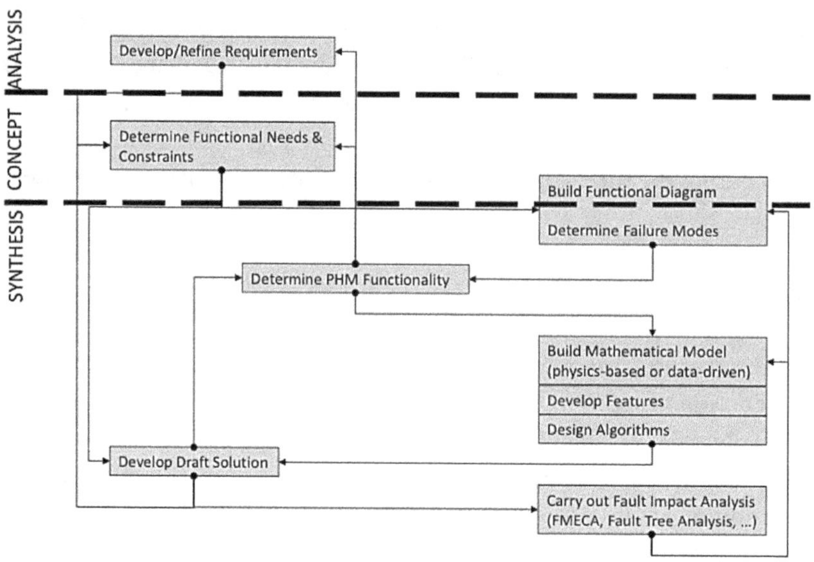

time the benefit achieved by using prognostics also depends on prediction performance. Decisions based on poor predictions can be detrimental in many ways. Late predictions (analogous to false negative) may result in system failures, whereas early predictions (analogous to false positive) trigger unnecessary maintenance actions.

From an engineering design perspective some key factors driving the requirements for prognostics include:

1. Maximum allowable probability of failure of the prognostic system to bound the risk of loss due to failure,
2. Maximum tolerable probability of proactive maintenance to bound unnecessary maintenance,
3. Lead time to specify the amount of advanced warning needed for appropriate actions, and
4. Required confidence to specify when prognosis is sufficiently good to be incorporated into the decision making process.

Keeping these factors in mind the top level system requirements can be translated into prognostic requirements. For instance, the risk posture assumed in a system determines how accurate predictions need to be without increasing the risk of loss beyond acceptable limits. A detailed discussion into deriving prognostics requirements from top level system goals is described in (Saxena *et al.*, 2012). These requirements are generally specified in terms of prediction performance that prognostics must satisfy for desired level of safety or cost benefit.

Next, the functional needs and constraints are evaluated during the *Concept* phase. These will drive a potential solution for which a fault impact analysis will be conducted. This analysis will be used in conjunction with an understanding of the functional operation of the solution approach to define PHM functionality. To that end, functional diagrams representing the desired functions of the subsystem can be used to capture these properties.

During the *Synthesis* phase, the functionality is encapsulated in appropriate models and algorithms that characterize the core of the PHM solution. Checking the degree of fulfillment of requirements may result in changes to the solution before a final choice is made. A clear understanding and representation of the functions that have to be carried out is needed to provide the framework for capturing both how a system operates and how it can fail. It further provides the manifestations of the failure, its consequences, and its impact on the structure as a whole. As the design moves through the synthesis stage and the design takes on shape, the functional diagram is augmented with the failure modes of the physical system. The design is continually analyzed to assess the degree to which it satisfies its requirements. This analysis may evaluate (depending on the goals) observability (i.e., to what degree the detrimental effect can be "seen"); diagnosability (i.e., the ability of the designed PHM system to isolate an anomalous condition to a specific single failure mode or to a group of potential failures called an ambiguity group); coverage (ability of the PHM system to provide indication that the occurrence of a given failure is imminent through the use of prognostics or diagnostics of upstream failures); choice of a method for realizing the observations (e.g., through appropriate sensor placement and information processing); choice of a form of suitable output (status indicators, alarms, reports, etc.); and determination of intended mitigation actions (e.g., report only, decision support, reflexive reconfiguration, repair, etc.). An account of design considerations in the context of integrated vehicle health management is delineated in (SAE, ARP6407).

When PHM methods are applied to structures, which are often subject to a variety of fault modes (cracks, voids, delamination, bondline failures, corrosion, etc.), dedicated sensors and sensor networks are required for observability as no single sensor type can cover all. Composite structures present more challenges due to multiple simultaneous fault modes and their effects in the anisotropic material. Therefore, it can be prohibitively expensive to cover all fault modes in large structures. Well-designed health management solutions carefully identify potential targets for diagnostics and prognostics to optimize costs and benefits. To that end it is a common practice to identify "hot spot" locations and critical fault modes to prioritize targets for prognostics. For instance, from a design perspective it may be decided that all identified hot spots are monitored for fault detection but only some critical ones are considered targets for prognostics. A common analysis tool often employed to identify risks and their consequences is *Failure Modes, Effects, and Criticality Analysis* (FMECA). Using FMECA system designers identify system components that are most likely to cause failures and therefore prioritize monitoring and prognostic targets.

Realizing PHM functions typically includes the need to represent the components of interest with a model, which might be either based on the physical operation of the system or derived from operational data that adequately characterizes the system. This is done after a concept has been chosen. Signal processing methods are employed to derive features from the sensor information that are correlated with the expected response from the model to arrive at an assessment of the health state. Some of the data sources that are utilized include sensor information from sensors that are placed on the components. As mentioned earlier, the sensor information needs to be processed to filter out noise and to derive features that are optimally correlated with the health information of interest. SHM literature covers a wide variety of modeling and sensing solutions for metallic as well as composite structures. A majority of these methods, however, focus on detection techniques and largely stop at damage localization and damage severity estimation. While most sensing and instrumentation commonly used for damage detection and identification in structures is also sufficient for prognostics, there are some unique elements in measurement that must be covered as well. Some of these aspects are discussed below in the context of prognostics.

Sensing for Prognostics

Prognostics is a natural extension of structural health monitoring where the user is not only concerned with detecting, locating, and identifying a fault mode but also determining the remaining time at hand before a failure occurs. It is a continuous process and predictions must be regularly updated as the system health state changes due to further usage. Therefore, operationally prognostics requires frequent measurement updates to continuously estimate structural health and revise predictions of remaining life. Consequently SHM sensors, that can interrogate the system periodically and assess any change in fault severity, are needed for prognostics. While non-destructive inspections can provide valuable trending information, this chapter focuses on largely automated SHM-based prognostics that makes use of permanently installed sensors built into the structure design. Prognostics broadly requires two categories of inputs from monitoring sensors:

1. Damage response measurement;
2. Operational and environmental loading measurement.

The choice of sensing methods or the sensors is primarily guided by what needs to be measured, but is significantly influenced by practical constraints given any application. For example weight, volume and power demand, and cost constraints for sensors onboard airspace systems results in a different sensor choice than for industrial building or bridge structures. Sensor locations are chosen such that the expected type of damage produces known, observable, and statistically significant effects in features derived from the measurements at these locations, which is often determined through numerical simulations or physical tests. Low-level local response caused by damage (e.g., cracks opening and closing) must be separated from large-amplitude global response such as that caused by aerodynamic loads on aircraft by determining required sensitivity and dynamic range through analysis or experimentation.

The following section describes these sensing requirements and common methodologies currently used. Additionally, the need for ground truth measurement for prognostic model development is discussed.

Damage Response Measurement

Prognostic models need information from sensors about degradation or damage in the structures through kinematic (strains, accelerations, etc.) response of the structure. These sensors are also called health monitoring sensors that can detect damage and damage growth, localize damage, assess damage severity, and identify damage types. A variety of sensing technologies has recently surfaced in addition to traditionally used strain and acceleration measurement methods. Among relatively new methods, SHM sensors based on piezoelectric or PZT discs are particularly useful for health monitoring in the context of prognostics. In pitch-catch mode they are used in active interrogation to assess type, location, and degree of damage (Chang *et al.*, 2007; Y. Liu *et al.*, 2010). Integrated into the structure on the surface or embedded into it, PZT sensors allow regular assessment of structural integrity through dynamic strain measurements. A time history of these measurements can inform damage growth model to make predictions of when the structure is expected to reach its failure threshold. Other similar embedded technologies include optical fiber based Fiber Bragg Gratings (FBG), vacuum tube networks, and micro-electro-mechanical systems (MEMS), to name a few.

System Loading Measurement

The other class of sensors characterizes loadings on the system. This includes both operational loading and environmental loading, both of which have an impact on how fast any given damage type grows. These include measurements of static and dynamic loads that the system has experienced and also the environment it has been exposed to. As mentioned in the previous section, this could be achieved by stress monitoring at the key load bearing points identified as prognostic targets. Stresses can be indirectly measured using sensors like strain gauges or be inferred from load calculations using static/dynamic models of the structure and knowledge of the load profile. To characterize environmental loadings such as moisture, temperature, acidity (or pH), etc., the duration of exposure must be monitored. The effect of these factors is then incorporated in the damage growth models used for prognostics.

Damage Ground Truth Measurement

During design and development stages, fault modeling requires access to run-to-failure time series data to understand damage growth characteristics and for model validation. Such data are often difficult to come by as they are rarely collected in practice due to lack of adequate sensing or due to the fact that expensive and critical systems are not allowed to run-to-failure due to safety and cost considerations. Therefore, carefully designed experiments such as accelerated aging are often used to collect such data in representative operational conditions. Another aspect of data collection required for prognostics is measurement of ground truth through non-destructive evaluation (NDE) methods. Commonly used NDE techniques such as X-rays, thermography, ultrasonic A, B, C-Scans etc. are some of the common methods for ground truth measurement.

Performance Evaluations

Within the health management context there are a variety of prediction methods that are generally used to make predictions about system behavior (Saxena *et al.*, 2014). Prognostics is one of those methods that predicts remaining useful life of a specific component or system based on its usage history inferred from monitoring data and expected future load profile. A variety of prognostic performance evaluation metrics have been defined (Saxena, Celaya, *et al.*, 2010b), but as described in (Saxena *et al.*, 2014), prognostic performance must account for three attributes of performance under the presence of uncertainty, namely; Correctness, Timeliness, and Confidence. Correctness deals with accuracy and spread measures of predictions when compared to observed outcomes. Timeliness accounts for how quickly a prediction algorithm produces its outputs, in comparison to the fault growth rate and the time required for enabling a mitigation action. Confidence, on the other hand, is a measure of uncertainty in a prognostic algorithm's output, which is key in making decisions about possible mitigating actions. These attributes are measured through the prediction horizon, the so-called α-λ accuracy (see Equation 1), and relative accuracy measures as described in (Saxena, Celaya, *et al.*, 2010b). These measures not only assess accuracy of predictions but also track performance change with time as predictions are periodically updated.

Prognostics metrics are expressed on a time versus RUL graph, as shown in Figure 2 and Figure 3. Predicted RULs are plotted against time of prediction. The straight line with negative slope represents the true RUL that decreases linearly as time progresses. Ideally predictions should stay on this line as

Figure 2. Illustration of determining Prognostic Horizon metric for two prognostics estimations (Source: Saxena et.al. 2010a).

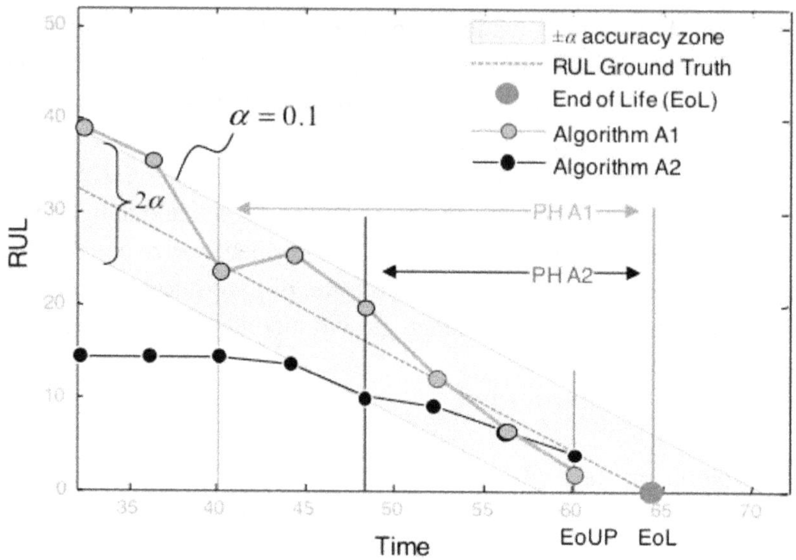

they are updated with time or stay within the accuracy zones specified by parameter α. An advanced version of these metrics considers RUL predictions in the form of PDFs rather than point estimates. These metrics are further parameterized by β, that specifies the minimum acceptable probability of overlap between the predicted RUL and the α accuracy bounds to consider a prediction accurate. Both α and β are design parameters to be determined based on application scenario. Based on these principles, three prognostic measures are briefly described below and further used in the example developed further below.

- **Prognostics Horizon (PH):** Determines the maximum advance warning a prediction algorithm can provide with desired confidence. It is defined as the maximum predicted RUL (or the first prediction) that falls within desired accuracy (α) bounds with an overlap probability larger than parameter β. It can be expressed as the largest RUL value that is accurate to the desired α level. Mathematically, it is defined as $PH = EOL - t_p^*$, such that

$$t_p^* = \min\left\{t \geq 0 : RUL\left(t_p^*\right) \in desired\ accuracy\ \alpha\right\}.$$

- ***α-λ* Performance:** Characterizes how well an algorithm's performance (accuracy and precision) improves as EOL approaches. Improvement is measured against desired accuracy level (α) expressed as a percentage of RUL at any specified time (λ).

$$\alpha - \lambda\ accuracy = \begin{cases} 1 & if\ prediction\ falls\ inside\ the\ \alpha\ accuracy\ cone \\ 0 & otherwise \end{cases} \quad (1)$$

- **Relative Accuracy:** Quantifies how accurate an algorithm is at any given time relative to the EOL. It is computed as RUL error normalized by actual RUL at any given time. It captures the notion that accuracy of prediction becomes more critical as EOL approaches.

Figure 3. Illustration of computation of α-λ accuracy for two prognostics estimations
(Source: Saxena et.al. 2010a).

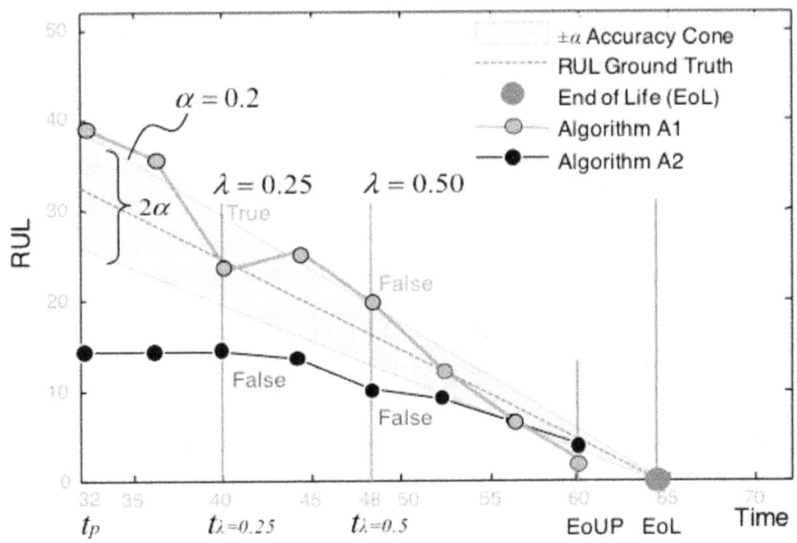

Prognostics Framework

The framework for prognostics hinges on the availability of several elements that are needed for a complete solution. These include a model (typically for both the nominal behavior as well as for the damage behavior), a quantifiable criterion for what constitutes failure, an algorithm to propagate the model to the damage threshold, and a mechanism to deal with uncertainty.

Modeling

As mentioned above, underlying any prediction requires a model that describes how the component of interest behaves under nominal conditions and how it will evolve as it experiences wear or a fault condition. To that end, one needs to represent that process in the form of mathematical equations. This "model" may be derived from laws of physics, encapsulated by empirical relations, or learned from data. Models can also be formed using a combination of these approaches. For example, parts of a component that are well-understood may be constructed using physics-based models, with unknown physical parameters learned from data using appropriate system identification techniques.

Modeling of physics can be accomplished at different levels, for example micro and macro levels. At the micro level, physical models are embodied by a set of dynamic equations that define relationships, at a given time or load cycle, between damage (or degradation) of a component and environmental and operational conditions under which the component is operated. Since measurements of critical damage properties (such as delamination in a composite structure) are rarely available, sensed system parameters have to be used to infer the damage properties. Micro-level models need to account for the assumptions and simplifications in the uncertainty management.

In contrast, macro-level models are characterized by a somewhat abridged representation that makes simplifying assumptions, thus reducing the complexity of the model (typically at the expense of accuracy). An example is a lumped parameter model which assumes that the attributes of the component have idealized behavior and the non-ideal characteristics are characterized with equivalent elements that suffice for a first-order approximation. When such a system is designed well, it will often (but certainly not always) result in satisfactory results, depending on performance requirements. It should be noted that for complex systems (e.g., a structure with complex geometry and with many different materials), even a macro-level model may be rather time-consuming and labor intensive. The resulting simplifications may need to be accounted for via explicit uncertainty management, which may pose more performance limitations than for a micro-level model.

Nominal System Modeling

The component model describes the characteristics of the component under nominal conditions. Ideally, such a model should be able to factor in the effects of operational and environmental conditions as well as any other conditions that cause different system response under nominal conditions. The model should also be able to adapt to changes of the system that are not considered abnormal. To that end, the system model could learn system behavior from examples, for instance using machine-learning techniques, or it could integrate domain expertise and be implemented using rules.

Damage Propagation Modeling

A damage propagation model describes how the damage is expected to grow in the future. It should, similar to the nominal model, account for operational and environmental conditions as well as any other conditions that have an impact on the damage. Depending on the fault mode, damage propagation may exhibit different symptoms and it may then be necessary to consider dedicated damage propagation models for different fault modes.

Algorithms

The role of the prognostic algorithm is applying the damage propagation model into the future. It needs to ensure that it properly considers the effects of environmental and operational conditions, as well as how to account for the different sources of uncertainty. Depending on the implementation, the damage propagation model and the prognostic algorithm may not be separable. However, for the general case, the damage propagation model and the prognostic algorithm will be treated as separate.

Uncertainty Management

Since prognostics deals with predicting the future behavior of engineering components and systems, it is important to understand that there are several sources of uncertainty that influence future behavior. Therefore, an efficient design for prognostics needs to rigorously account for all these sources of uncertainty to predict the remaining useful life. In the recent past, several researchers have developed different types of methods for uncertainty quantification in prognostics. For example, the Damage Prognosis project (Farrar & Lieven, 2007) at Los Alamos National Laboratory discussed uncertainty in fatigue crack growth by considering different sources of variability, in the context of structural health monitoring. Sankararaman *et al.* (Shankar Sankararaman *et al.*, 2009) developed a computational approach to account for variability, data uncertainty, and model uncertainty in crack growth prognosis. However, many of these methods are based on offline testing, i.e., testing before and after system operation, and not for condition-based prognostics and monitoring, i.e., prediction during system operation.

Developing methods for uncertainty quantification in the context of condition-based monitoring is challenging because uncertainty methods are generally computationally expensive, whereas condition-based monitoring and prognostics require real time computational power and results for decision-making. An important aspect of condition-based monitoring is the prediction of remaining useful life, and several publications (Engel *et al.*, 2000; Saha & Goebel, 2008) have discussed the importance of quantifying the uncertainty in remaining useful life prediction. Sankararaman *et al.* (S. Sankararaman *et al.*, 2014) discussed sampling methods and analytical methods for estimating the uncertainty in the remaining useful life prediction for prognostics.

Interpreting Uncertainty in Prognostics

Though probabilistic methods have been widely used for uncertainty quantification and management in various engineering applications, the interpretation of probability is, at times, not straightforward. There are two major interpretations of probability: frequentist versus subjective. Frequentist probability assignments are based on the existence of inherent variability and are suitable only in the context of truly random experiments. Traditional statistical principles are primarily based on the concepts of frequentist probability. On the other hand, subjective probabilities express the degree of the analyst's belief regarding a particular statement and can be assigned even in the absence of inherent variability. Principles of Bayesian statistics are based on the concept of subjective probability and are widely used in a variety of engineering applications today.

In the context of health management, the frequentist interpretation of uncertainty is applicable only for testing-based methods. In other words, when several nominally identical specimens are tested, there is inherent variability across them and probabilistic methods are used to quantify this variability. On the other hand, in condition-based prognostics the focus is typically on one particular unit; at any time-instant, this particular unit is in a given state without any associated variability. The uncertainty regarding this state is simply reflective of the degree of belief. Bayesian tracking methods (Kalman filtering, particle filtering, etc.) used for state estimation are called "Bayesian" not only because they use Bayes' theorem for state estimation but also because they are based on subjective probabilities. Similarly, the future loading conditions are also subjective estimates; hence, the resultant remaining useful life also needs to be interpreted subjectively. Thus, it can be clearly seen that the subjective interpretation of uncertainty is consistent across the entire domain of condition-based prognostics (S Sankararaman, 2014).

Sources of Uncertainty

To begin with, it is important to understand the various sources of uncertainty that affect calculations in prognostics. It has been conventional to classify the different sources of uncertainty into aleatory (physical variability) and epistemic (lack of knowledge), where epistemic uncertainty consists of data uncertainty and model uncertainty. However, in the case of condition-based monitoring, there is only one particular system being monitored, and not multiple realizations of a population, and therefore, it is not meaningful to discuss variability. For example, the system is at a particular state at any time instant and there is nothing variable about it. Variability would need to be accounted for only in the case of reliability-testing methods where multiple components/systems are tested. Therefore, the sources of uncertainty in prognostics are classified into:

- **Modeling Uncertainty:** Predicting the future is the most important aspect of prognostics, and typically, a physics-based model or a data-driven model is used for predicting future behavior. This model is usually represented using state-space equations. Modeling uncertainty represents the difference between the predicted response and the true response (which can neither be known nor measured accurately), and comprises of several parts such as model parameters, model class, process noise, etc.
- **Present Uncertainty:** The first step of prognostics is to estimate the condition/state of the component/system at any time instant. Output data (collected through sensors) is used to estimate the current state and many filtering approaches are able to estimate the state, and calculate the uncertainty associated with the state estimate.
- **Future Uncertainty:** The most important source of uncertainty in the context of prognostics is due to the fact that the future is unknown, i.e. both the loading and operating conditions are not known precisely. The future behavior (i.e., the response of the system to the loading and operating conditions) needs to be estimated using a model; the usage of a model imparts additional uncertainty as explained earlier.

Uncertainty-Related Activities in PHM

Having explained the various sources of uncertainty, it is important to define a series of activities that can be collectively used to address these sources of uncertainty in a systematic manner. In the context of prognostics and health management, uncertainty has been discussed from the perspectives of quantification, representation, and management, in various publications (Orchard *et al.*, 2008; Tang *et al.*, 2009). While all the three are different processes, they are often confused with each other and interchangeably used (Celaya *et al.*, 2012). For the purpose of clarity, four different uncertainty-related activities are detailed below:

1. **Uncertainty Representation and Interpretation:** The first step is the representation of uncertainty, which may be guided by the choice of the modeling and simulation framework. There are several methods for uncertainty representation that vary in the level of granularity and detail. A probabilistic framework is used for uncertainty representation in this chapter, and the interpretation of this uncertainty is discussed shortly.
2. **Uncertainty Quantification:** The second step of uncertainty quantification deals with identifying and characterizing the various sources of uncertainty that can prospectively affect the prognostic calculations.
3. **Uncertainty Propagation:** The third step of uncertainty propagation is most relevant to prognostics, since it accounts for all the uncertainties previously quantified and uses this information in order to predict the future behavior, remaining useful life, and the associated uncertainty.
4. **Uncertainty Management:** The fourth and final step is uncertainty management, and it is unfortunate that, in several articles, the term "Uncertainty Management" has been used instead of uncertainty quantification and/or propagation. As a result, there are few publications that directly

address the issue of uncertainty management. In general, uncertainty management is a term used to refer to different activities that aid in managing uncertainty in condition-based maintenance during real-time operation. There are several aspects of uncertainty management. One aspect of uncertainty management attempts to answer the question: "Is it possible to improve the uncertainty estimates?" The answer to this question lies in identifying which sources of uncertainty are significant contributors to the uncertainty in the RUL prediction. Another aspect of uncertainty management deals with how uncertainty-related information can be used in the decision-making process.

A PROGNOSTICS DESIGN SOLUTION

A general computational framework for prognostics and health management is presented in this section. Let us consider an engineering component or system such that, at any generic time-instant t_p, it is desired to perform prognostics and predict the remaining useful life. Figure 4 provides a conceptual scheme of the proposed prognostics framework, which consists of the following three sub-problems:

1. Present state estimation;
2. Future state prediction;
3. RUL computation.

In order to address the above problems, a generic time-dependent, state-space modeling framework is firstly presented. This framework needs to account for the different sources of uncertainty as explained in the previous section, and therefore, the state-space models also need to account for the various sources of uncertainty.

Stochastic System Modeling for Prognostics

Consider a general engineering component/system whose damage response can be modeled as

$$\frac{dx(t)}{dt} = g\left(t, x(t), u(t), \theta\right) + v(t) \tag{2}$$

where $x(t) \in \mathbb{R}^{n_x}$ is the state vector (that depicts the state of the system at any time instant), $\theta \in \mathbb{R}^{n_\theta}$ is the parameter vector (parameters of the model), $u(t) \in \mathbb{R}^{n_u}$ is the input vector (loadings, environmental conditions, operating conditions, etc.), $v(t) \in \mathbb{R}^{n_x}$ is the model error vector (used to represent unmod-

Figure 4. Prognostics architecture; a description of the mathematical symbols is provided further below.

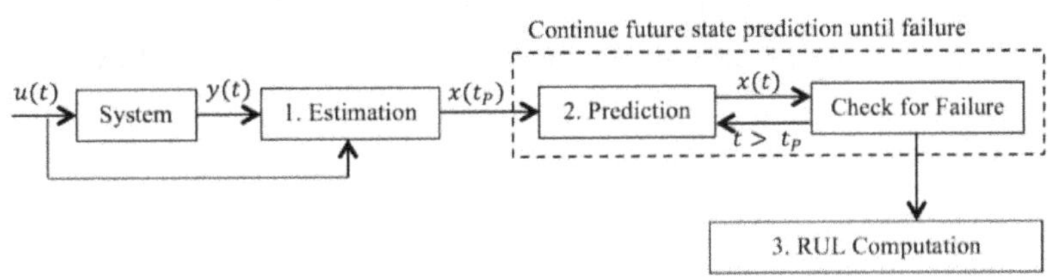

eled random variations in the engineering component/system), and $g \in \mathbb{R}^{n_x}$ is the state transition equation.

The above differential equation simply predicts how the state of a system continuously progresses as a function of time, loading conditions, model parameters, etc. Typically, in the context of prognostics and health management, a state variable or a collection of state variables are used to represent the health state of the system. Usually, the health of a system deteriorates as a function of time, and therefore, it can be estimated by quantifying the states of the system; hence, tracking the states is essential to tracking the health of the system.

As discussed in the last section, there are several sources of uncertainty that affect the performance of the system, and all of these sources of uncertainty can be included in this modeling framework. For example, certain characteristics of the system are represented using parameters and it may not be able to know these quantities precisely; therefore the parameters ($\boldsymbol{\theta}$) are uncertain. In addition, external input factors (u) such as loading conditions, operating conditions, etc. may not be known precisely or cannot be controlled. In fact, all of the variables in the above state-space modeling framework are uncertain and need to be modeled using probability distributions. As mentioned earlier, in many cases, such uncertainty must not be confounded with variability, particularly in the context of condition-based prognostics. Due to these different sources of uncertainty, the state estimates available through the estimation procedure will also become uncertain. It is important that the state estimation procedure rigorously accounts for these sources of uncertainty and accurately computes their effect on the state estimate. As a result, the state estimation procedure represents the system state using probability distributions.

State estimation is accomplished by continuously collecting measurements of the output variables through sensors, where these measurements are available until the time of prediction denoted by t_P. Then, along with the above state prediction model, an output-measurement model can also be defined as:

$$y(t) = h\left(t, x(t), u(t), \theta\right) + w(t) \tag{3}$$

where $y(t) \in \mathbb{R}^{n_y}$ is the output vector (measurements available through sensors until $t = t_P$), $w(t) \in \mathbb{R}^{n_y}$ is the measurement noise vector, and $h \in \mathbb{R}^{n_y}$ is the output equation.

Using recursively Equations (2) and (3), the state estimate at time of prediction $x(t_P)$ can be calculated using an accurate state estimation procedure, as shown further below.

Bayesian State Estimation

As stated before, many problems in a variety of areas of science and engineering can be expressed using a state space model, as defined by Equations (2) and (3). In general, for the purpose of prognostics, this state-space model is usually converted into a discrete-time system by using numerical integration over a sufficiently small time step Δt, leading to a hidden Markov model, as follows:

$$x_n = g_n\left(x_{n-1}, u_n, \theta\right) + v_n \tag{4}$$

$$y_n = h_n\left(x_n, u_n, \theta\right) + w_n \tag{5}$$

where $x_n = x(n \cdot \Delta t)$ and $y_n = y(n \cdot \Delta t)$ are the latent (or hidden) state and the measured response of the system at time $n \cdot \Delta t$; and $u_n = u(n \cdot \Delta t)$, $v_n = v(n \cdot \Delta t)$ and $w_n = w(n \cdot \Delta t)$ are the input vector, model error and measurement error at time $t = n \cdot \Delta t$, respectively. The sequence of states $\left\{x_n; n \in \mathbb{N}\right\}$ is assumed to follow a hidden Markov process, while the observations $\left\{y_n; n \in \mathbb{N}\right\}$ are conditionally independent given the process $\left\{x_n; n \in \mathbb{N}\right\}$. From now on, time will be denoted by the index n since it is sufficient to indicate which time is being considered.

The goal of Bayesian state estimation is to sequentially estimate the PDF of the sequence of latent states $x_{0:n} \triangleq \left\{x_0, \ldots, x_{n-1}, x_n\right\}$ given a sequence of measurements up to time n, denoted by $y_{0:n} \triangleq \left\{y_0, \ldots, y_{n-1}, y_n\right\}$. This PDF is typically known as the *posterior*, expressed as $p\left(x_{0:n} \mid y_{0:n}, \theta\right)$, since it is obtained from the last available state estimation, known as the prior PDF, $p\left(x_{0:n-1} \mid y_{0:n-1}, \theta\right)$. Hereinafter, and without loss of generality, input parameters u_n will be dropped from the formulation for the sake of simpler notation.

In the state space modeling literature, the posterior PDF is commonly studied through one of its marginals, the so-called filtering distribution, $p\left(x_n \mid y_{0:n}, \theta\right)$. To obtain this distribution, a two-stage recursive approach is typically formulated:

1. The prediction stage, in which the one-step ahead predictive PDF $p\left(x_n \mid y_{0:n-1}\right)$ is estimated based on available data up to time $n-1$ using Total Probability Theorem, and
2. The updating stage, in which the former PDF is updated to get the posterior PDF $p\left(x_n \mid y_{0:n}\right)$ based on a new measurement at time n, using Bayes' Theorem. That is:

$$\begin{aligned} p\left(x_n \mid y_{0:n-1}, \theta\right) &= \int p\left(x_n \mid x_{n-1}, \theta\right) p\left(x_{n-1} \mid y_{0:n-1}, \theta\right) dx_{n-1} \\ p\left(x_n \mid y_{0:n}, \theta\right) &= \frac{p\left(y_n \mid x_n, \theta\right) p\left(x_n \mid y_{0:n-1}, \theta\right)}{\int p\left(y_n \mid x_n, \theta\right) p\left(x_n \mid y_{0:n-1}, \theta\right) dx_n} \end{aligned} \tag{6}$$

Continually iterating between the equations in (6) constitutes the basis of sequential state estimation. However, this recursive scheme is only a theoretical solution given that, in general, the integrals involved in these equations cannot be calculated analytically, except for some especial cases. The first special case is when Equations (4) and (5) involve linear functions and errors v_n and w_n are Gaussians. In this case, an analytical solution can be obtained for the recursions in (6) using the well-known Kalman filter (Kalman, 1960). The other case is when the state variable x_n can be represented by a discrete set of values, where grid-based methods (Arulampalam *et al.*, 2002) can be applied to efficiently solve the system in (6). For the general case, Sequential Monte Carlo (SMC) methods (Doucet *et al.*, 2001) can be used to efficiently solve these iterations avoiding the calculation of the aforementioned multidimensional integrals, as shown further below.

When the interest is instead in obtaining the posterior PDF $p\left(x_{0:n} \mid y_{0:n}\right)$, then Bayes' Theorem can be applied to update the last posterior a time $n-1$, as follows:

$$\begin{aligned} p\left(x_{0:n} \mid y_{0:n}, \theta\right) = & \; p\left(x_{0:n} \mid y_n, y_{0:n-1}, \theta\right) \\ \propto & \; p\left(y_n \mid x_{0:n}, y_{0:n-1}, \theta\right) p\left(x_{0:n} \mid y_{0:n-1}, \theta\right) \\ \propto & \; p\left(y_n \mid x_n, \theta\right) p\left(x_n \mid x_{n-1}, \theta\right) \underbrace{p\left(x_{0:n-1} \mid y_{0:n-1}, \theta\right)}_{last\ updating} \end{aligned} \tag{7}$$

In equations (6) and (7) it is assumed that $p\left(x_n \mid x_{0:n-1}\theta\right) = p\left(x_n \mid x_{n-1}, \theta\right)$ since Equation (4) describes a Markov model of order one, i.e., the process is conditionally dependent on the past sequence only through the last state. In addition, it is assumed that the probability models for error terms v_n and w_n are given, and as a consequence, the PDFs for the state transition equation and observation equation are prescribed. For example, when v_n and w_n are modeled as zero-mean Gaussian distributions, which is supported by the Principle of Maximum Information Entropy (PMIE) (Jaynes, 1983), then the transition equation and the observation equation in (4) and (5) can be expressed as Gaussian distributions, as:

$$p\left(x_n \mid x_{n-1}, \theta\right) = \left((2\pi)^{n_x} \mid \Sigma_{v_n} \mid\right)^{-\frac{1}{2}} \exp\left(-\frac{1}{2}\left(x_n - \overline{x}_n\right)^T \Sigma_{v_n}^{-1}\left(x_n - \overline{x}_n\right)\right) \tag{8}$$

$$p\left(y_n \mid x_n, \theta\right) = \left((2\pi)^{n_y} \mid \Sigma_{w_n} \mid\right)^{-\frac{1}{2}} \exp\left(-\frac{1}{2}\left(y_n - \overline{y}_n\right)^T \Sigma_{w_n}^{-1}\left(y_n - \overline{y}_n\right)\right) \tag{9}$$

where $\overline{x}_n = g_n\left(x_{n-1}, u_n, \theta\right)$ and $\overline{y}_n = h_n\left(x_n, u_n, \theta\right)$, being $\Sigma_{v_n} \in \mathbb{R}^{n_x x n_x}$ and $\Sigma_{w_n} \in \mathbb{R}^{n_y x n_y}$ the covariance matrices of the model error and the measurement error, respectively. As shown in Equations (6) and (7), the two last probability models $p\left(x_n \mid x_{n-1}, \theta\right)$ and $p\left(y_n \mid x_n, \theta\right)$ are fundamental elements in the Bayesian state estimation problem explained in this section.

Joint State-Parameter Estimation

To better estimate the current damage state of the system, and more importantly, to reduce the uncertainty of the future predictions about the system, a followed solution is to sequentially update both states and model parameters using the sequential state estimation framework presented above. In doing so, the system equations g_n and h_n are expected to provide more accurate predictions as long as model parameters are allowed to learn from data. To deal with the Bayesian state estimation problem for both states and parameters, an augmented state $z_n = \left(x_n, \theta_n\right) \in \mathbb{R}^{n_x + n_\theta}$ is usually defined so that the same fundamental equations defined in (7) can be redefined to obtain the posterior PDF for the augmented state, i.e., $p\left(z_{0:n} \mid y_{0:n}\right)$, as follows:

$$\begin{aligned} p\left(z_{0:n} \mid y_{0:n}\right) = \; & p\left(z_{0:n} \mid y_n, y_{0:n-1}\right) \\ \propto \; & p\left(y_n \mid z_{0:n}, y_{0:n-1}\right) p\left(z_{0:n} \mid y_{0:n-1}\right) \\ \propto \; & p\left(y_n \mid z_n\right) p\left(z_n \mid z_{n-1}\right) p\left(z_{0:n-1} \mid y_{0:n-1}\right) \end{aligned} \tag{10}$$

where

$$p\left(z_n \mid z_{n-1}\right) = p\left(x_n \mid x_{n-1}, \theta_n\right) p\left(\theta_n \mid \theta_{n-1}\right) \tag{11}$$

$$p\left(y_n \mid z_n\right) = p\left(y_n \mid x_n, \theta_n\right) \tag{12}$$

A key problem that typically arises when sequentially updating the joint state-parameter vector $z_{0:n} = \left\{x_{0:n}, \theta_{0:n}\right\}$ as an augmented state is the non-dynamics nature of θ, which makes it difficult to obtain the required PDF $p(\theta_n \mid \theta_{n-1})$ and therefore, to explore the space of parameters. A common solution consists in adding an independent random perturbation ξ_n to the set of updated parameters at time $n-1$, before evolving to the next state at time n, i.e., $\theta_n = \theta_{n-1} + \xi_n$. This induces a Markovian-type artificial dynamics (Gordon *et al.*, 1993; J. Liu & West, 2001) to model parameters, from which the PDF $p\left(\theta_n \mid \theta_{n-1}\right)$ is prescribed. For example, in the common case of ξ_n being modeled as a white noise (zero-mean Gaussian), the PDF for artificial evolution of model parameters can be obtained as:

$$p\left(\theta_n \mid \theta_{n-1}\right) = \mathcal{N}\left(\theta_{n-1}, \Sigma_{\xi_n}\right) \tag{13}$$

where $\Sigma_{\xi_n} \in \mathbb{R}^{n_\theta \times n_\theta}$ is the covariance matrix of the *random walk* defined in the last equation. Observe that by using this method the model parameters are virtually time evolving although they are essentially not dependent on time. This time varying imposes a loss of information in θ over time (i.e. larger spread) since additional uncertainties are artificially added to model parameters, which ultimately influences the precision of the state estimation. There exist several methods in the literature to overcome this drawback, being the most popular those that impose some kind of shrinkage over Σ_{ξ_n} as long as new data is collected (J. Liu & West, 2001). An efficient method in this direction has been recently proposed by Daigle and Goebel (M. J. Daigle & Goebel, 2013) consisting in modifying the variance of the random walk by adding a negative scalar proportional to the relative difference between the actual and the target spread of the marginal posterior $p\left(\theta_{n,j} \mid y_{0:n}\right)$, with $\theta_{n,j}$ the jth component of the parameter vector θ_n, as follows:

$$\sigma^2_{\xi_{n,j}} = \sigma^2_{\xi_{n-1,j}} \left(1 - P_j^* \frac{RMAD\left(\theta_{n,j}\right) - RMAD_j^*}{RMAD\left(\theta_{n,j}\right)}\right) \tag{14}$$

where $\mathrm{RMAD}\left(\theta_{n,j}\right)$ is the *Relative Median Absolute Deviation* of $p\left(\theta_{n,j} \mid y_{0:n}\right)$, $RMAD_j^*$ is the required or target RMAD for $p\left(\theta_{n,j} \mid y_{0:n}\right)$, and $P_j^* \in \left[0,1\right]$ is a constant that tunes the speed of convergence to $RMAD_j^*$. The reader is referred to (M. J. Daigle & Goebel, 2013) for details about the optimal choice for P_j^* and $RMAD_j^*$. The term $\mathrm{RMAD}(\theta_{n,j})$ can be readily calculated using samples from the marginal posterior $\tilde{\theta}_{n,j} = \left\{\theta_{n,j}^{(1)},\ldots,\theta_{n,j}^{(k)},\ldots,\theta_{n,j}^{(K)}\right\} \sim p\left(\theta_{n,j} \mid y_{0:n}\right)$, as follows:

$$RMAD\left(\theta_{n,j}\right) = \frac{median\left(\left\{\left|\theta_{n,j}^{(k)} - median\left(\tilde{\theta}_{n,j}\right)\right|\right\}_{k=1}^{K}\right)}{median\left(\tilde{\theta}_{n,j}\right)} \tag{15}$$

The proposed method for joint state-parameter estimation is presented as Algorithm 1 at the end of this section.

Future State Prediction

Having calculated the probability distribution of the state estimate at the time of prediction n, the next step for prognostics is to predict the distribution of future states of the system ℓ steps forward in time (measured using time-indices) in absence of new measurements.

In this context, the resultant distribution is known as the $\ell-$step-ahead predictive PDF, defined mathematically as $p\left(z_{n+\ell} \mid y_{0:n}\right)$ where $\ell \in \mathbb{N} > 1$. Using the up-to-date information of the system based on the filtering distribution $p\left(z_n \mid y_{0:n}\right)$ and the dynamics of the system, which is encapsulated in the state transition equation defined in Equation (11), the $\ell-$steps-ahead prediction of the system health state $p\left(z_{n+\ell} \mid y_{0:n}\right)$ can be formulated using Total Probability Theorem as follows:

$$\begin{aligned} p\left(z_{n+\ell} \mid y_{0:n}\right) &= \int_{\mathcal{Z}} p\left(z_{n+\ell} \mid z_{n:n+\ell-1}\right) p\left(z_{n:n+\ell-1} \mid y_{0:n}\right) dz_{n:n+\ell-1} \\ &= \int_{\mathcal{Z}} \prod_{l=n+1}^{n+\ell} p\left(z_l \mid z_{l-1}\right) p\left(z_n \mid y_{0:n}\right) dz_{n:n+\ell-1} \end{aligned} \tag{16}$$

It should be noted that the last multidimensional integral is analytically intractable, however it can be readily approximated by the use of sampling-based methods, as shown further below in this section.

Defining EOL and RUL

At this juncture, once the problem of future state prediction has been assessed, the next natural step for prognostics is to estimate the EOL and RUL of the engineering component/system. To this end, a mathematical description of the aforementioned terms EOL and RUL is previously introduced. For the sake of generality, these concepts are explained here using continuous time notation instead of the discrete-index notation, useful from the implementation perspective. However, recall that there exist a one-to-one mapping between continuous time t and discrete time-index n (e.g., $t = n \cdot \Delta t$, being Δt a sufficiently small time-step) and thus the EOL and RUL could also be expressed in terms of discrete time.

Let define the region $\mathcal{A}$ of acceptable performance as a subspace of the joint state-parameter-inputs space $\mathcal{Z} \subset \mathbb{R}^{n_x} \times \mathbb{R}^{n_\theta} \times \mathbb{R}^{n_u}$. This region can be defined by means of a set of n_c constrains or thresholds, $C_{EOL} = \{c_i\}_{i=1}^{n_c}$, where each threshold $c_i \in C_{EOL}$ is a function that maps a given point $(x(t), \theta(t), u(t)) \in \mathcal{Z}$ into the Boolean domain $\mathcal{B} \triangleq \{0,1\}$, i.e. $c_i : \mathcal{Z} \to \mathcal{B}$, being $c_i(x(t), \theta(t), u(t)) = 1$ when the constrain is satisfied and 0 otherwise (Shankar Sankararaman *et al.*, 2013). These individual constrains may be combined into a single threshold function $T_{EOL} : \mathcal{Z} \to \mathcal{B}$ that can be mathematically defined as:

$$T_{EOL}(x(t), \theta(t), u(t)) = \begin{cases} 1 & if\ 0 \in \{c_i(x(t), \theta(t), u(t))\}_{i=1}^{n_c} \\ 0 & otherwise \end{cases} \tag{17}$$

In other words, $T_{EOL} = 1$ when any of the constraints is violated, and 0 otherwise. Therefore, the End-of-Life at the time of prediction t_p , denoted by $EOL(t_p)$, can be calculated as the earliest time $t \geq t_p$ for which $T_{EOL} = 1$. In mathematical terms:

$$EOL(t_p) = \inf\{t \in \mathbb{R}^+ : t \geq t_p \wedge T_{EOL}(x(t), \theta(t), u(t)) = 1\} \tag{18}$$

In Figure 5, a schematic illustration of the trajectory of a general state $x(t)$ is provided along with the region $\mathcal{A}$ of acceptable performance (in gray color). Note that the state is propagated forward in time from the time of prediction t_p until the boundary of the acceptable performance region is reached. The first "hitting" time t at which that happens defines the EOL. Having computed the EOL, the remaining useful life at the time of prediction can be readily calculated as:

$$RUL(t_p) = EOL(t_p) - t_p \tag{19}$$

Note from Equation (19) and preceding equations that an estimation of RUL at an arbitrary time of prediction depends on the following quantities:

- State estimate at time of prediction, $x(t_p)$.
- Future loading conditions $u(t)$, where $t \geq t_p$. Both environmental and operating conditions may be described along with loading conditions by the vector $u(t)$.
- Future values for model parameters, $\theta(t)$.
- Future values for model error, $v(t)$.

For the purpose of RUL prediction, all of the above quantities are functionally independent of each other; hence, RUL becomes a functionally dependent quantity, as shown in Figure 6. Let $X = (X_1, X_2, \ldots, X_m)$

Figure 5. Conceptual two-dimensional example of system trajectory to EOL

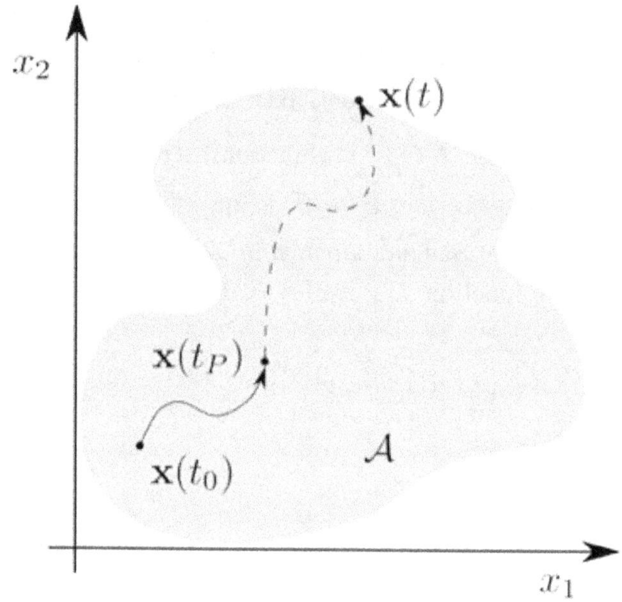

denote the vector of all of the above quantities, where m is the length of the vector X, i.e., the number of quantities influencing the RUL prediction. Then, the RUL can be expressed analytically by:

$$RUL = f(X) \tag{20}$$

In view of Equation (20), for every realization of X it is possible to compute the corresponding realization of *RUL*. However, as explained earlier, all of these quantities are uncertain and therefore, *RUL* is unavoidably uncertain. In order to account for the uncertainty in the remaining useful life estimation, it is necessary to propagate the uncertainty in X through f. It has been recently demonstrated that such uncertainty propagation is non-trivial even when the state-space models are linear and the uncertain variables follow Gaussian distributions. This is because the combination of the state-space model and the threshold function will always render f as a non-linear function (S. Sankararaman *et al.*, 2014). Hence, it is necessary to resort to statistical approaches in order to estimate the probability distribution of RUL. These methods can be roughly classified into three types: sampling-based methods, analytical methods, and hybrid methods. While some of them may calculate the cumulative distribution function (CDF) of RUL, denoted by $F_{RUL}(t) = P(RUL \leq t)$, other methods directly generate samples from it and use these samples to construct an estimation of the probability density function of RUL, denoted by $p(RUL)$. These methods are summarized below.

Sampling-Based Methods

The most intuitive method for uncertainty propagation is to make use of Monte Carlo simulation (MCS). The fundamental concept of MCS (Caflisch, 1998) is to generate pseudo-random number that is uniformly distributed on the interval [0, 1]; therefore the CDF of X is inverted to generate several random

Figure 6. Scheme for RUL computation

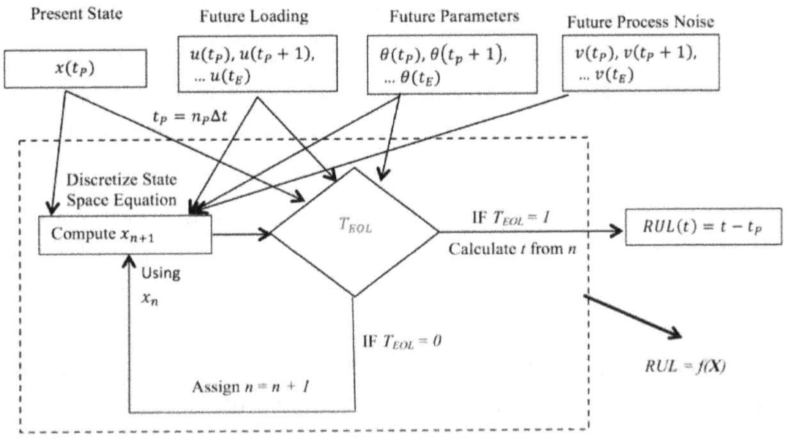

realizations of X whereby the corresponding random realizations of *RUL* are computed. Then the CDF $F_{RUL}\left(t\right)$ is calculated as the proportion of realizations such that $RUL \leq t$. The generation of each realization requires one evaluation/simulation of Equation (20). Several thousands of realizations may often be needed to calculate the entire CDF, especially for very high/low values of t. Error estimates for the CDF in terms of the number of simulations are well-known in the literature, see for example (Haldar & Mahadevan, 2000). Alternatively, the PDF *p(RUL)* can be computed based on the available samples from Equation (20) using kernel density estimation (Rosenblatt & others, 1956).

More sophisticated methods such as particle filtering (PF) (Gordon *et al.*, 1993) can be used to simply propagate the resultant "particles" into the future. Such approach is only slightly different from MCS where the difference resides in that the particles have their own weights. In order to accurately capture the entire probability distribution of RUL, it may be necessary to employ large amounts of "particles" which necessarily increases the computational cost. Henceforth, such an approach may not be suitable for online health monitoring and prognostics in complex systems. This is why it is necessary to further investigate new efficient sampling-based methods that can significantly improve the computational effort. Some of these sampling-based methods are:

- **Importance Sampling:** This algorithm (Glynn & Iglehart, 1989) does not generate random realizations of X from the original distribution. Instead, random realizations are generated from a proposal density function. Statistics of *RUL* are estimated and then corrected based on the values from the original density and the proposal density.
- **Adaptive Sampling:** This method (Bucher, 1988) is an advanced sampling technique where the efficiency of importance sampling is continuously improved by updating the proposal density function based on the information obtained after evaluating f for a few samples. Two classes of adaptive sampling methods are multi-modal sampling (Karamchandani *et al.*, 1989) and curvature-based sampling (Wu, 1994). It has been demonstrated (Haldar & Mahadevan, 2000) that adaptive sampling techniques can accurately estimate the failure probability using hundreds of samples whereas traditional Monte Carlo techniques typically require several hundreds of thousands of samples.

- **Stratified Sampling:** In this sampling approach, the overall domain of *X* is split into multiple sub-domains and samples are independently drawn from each sub-domain. The process of dividing the overall domain into multiple sub-domains is referred to as stratification. This method is applicable when subpopulations within the overall population are significantly different.
- **Latin Hypercube Sampling:** This is a sampling method commonly used in design of computer experiments (Loh, 1996). When sampling a function of m_1 variables, the range of each variable is divided into m_2 equally probable intervals, thereby forming a rectangular grid. Then, sample positions are chosen such that there is exactly one sample in each row and column of this grid. Each resultant sample is then used to compute a corresponding realization of *RUL*, and thereby the PDF *p(RUL)* can be calculated.
- **Unscented Transform Sampling:** Unscented transform sampling (M. Daigle *et al.*, 2012; Van Zandt, 2001) is a sampling approach that focuses on estimating the mean and variance of *RUL*, instead of the entire probability distribution. Certain pre-determined sigma points are selected in the *X*-space and these sigma points are used to generate corresponding realizations of *RUL*. Using weighted averaging principles, the mean and variance of *RUL* are calculated.
- **Subset Simulation:** Subset Simulation is an adaptive stochastic simulation approach originally conceived to compute small failure probabilities of engineering systems (Au & Beck, 2001). The conceptual idea behind Subset Simulation is to represents a small failure probability as a product of larger probabilities. Recently, (M. Chiachio *et al.*, 2014) have proposed a new algorithm called "PFP-SubSim" that exploits the benefits of Subset Simulation into the prognostics procedure, and has demonstrated high efficiency for the prognostics of rare-events.

Analytical Methods

A class of analytical methods was developed in the field of structural engineering to perform uncertainty propagation without the need to employ repetitive evaluations. These methods facilitate a quick (in terms of number of evaluations of function *f*) and efficient (reasonably accurate) calculation of the probability distribution of *RUL*.

- **First Order Second Moment Method:** This method (Dolinski, 1983) uses only the mean and variance of all the uncertain quantities and the first-order Taylor's series expansion of *f*, in order to calculate the mean and variance of the response quantity *RUL*.
- **First Order Reliability Method:** This method calculates the CDF function $F_{RUL}(t)$ by linearizing *f* around the so-called most probable point (Haldar & Mahadevan, 2000; M Hohenbichler & Rackwitz, 1983; Shankar Sankararaman *et al.*, 2013). By repeating this calculation for multiple values of *t*, the entire CDF can be obtained. While this approach is an approximation, it has been demonstrated that it can estimate the CDF with reasonable accuracy in many practical applications (Haldar & Mahadevan, 2000; Shankar Sankararaman *et al.*, 2013).
- **Inverse First Order Reliability Method:** This method is the inverse of the first-order reliability method, i.e., it calculates the value of *t* that corresponds to a given value of *β* such that $F_{RUL}(t) = \beta$. By repeating this approach for several values of *β*, the entire cumulative distribution function of *RUL* can be easily calculated, thereby estimating the uncertainty in *RUL*.

- **Second Order Reliability Method:** The second order reliability method (Der Kiureghian *et al.*, 1987) improves the estimate of the first-order reliability method through a quadratic approximation of *f*, instead of using a linear approximation. There are different types of quadratic approximations and corresponding second-order reliability estimates, thereby leading to a variety of computational methods as those proposed by Kiureghian et al. (Der Kiureghian *et al.*, 1987), Tvedt (Tvedt, 1989), Hohenbichler and Rackwitz (Michael Hohenbichler & Rackwitz, 1988), etc.

Hybrid Methods

In addition to the sampling-based methods and analytical methods, there are also a class of methods called hybrid methods that combine both sampling and the use of analytical tools. Typically these methods are of the form of surrogate models where few samples of X and the corresponding values of *RUL* are computed and further used by basis functions (or interpolating functions) in order to facilitate the evaluation of f at untrained locations of X. Such surrogate modeling approaches include regression techniques (Haldar & Mahadevan, 2000) polynomial chaos expansion (Najm, 2009), kriging (Dwight & Han, 2009), among others.

Summary

In summary, while the aforementioned uncertainty propagation methods have been used in different types of engineering applications, it is still necessary to investigate their applicability to prognostics. Further, note that uncertainty propagation is still a challenging problem in several practical applications due to the limitations of the aforementioned methods. It is rarely possible to accurately calculate the actual probability distribution of RUL by analytical methods. Accurate calculation is possible when using infinite samples by Monte Carlo sampling. Any other method (even the use of a limited, finite number of samples) will lead to uncertainty in the estimated probability distribution, and this additional uncertainty is referred to as prediction-method uncertainty. It is possible to decrease (and maybe, eventually eliminate) this type of uncertainty either by using advanced probability techniques or powerful computing power.

Algorithms for Prognostics

Having discussed a variety of methods for prognostics, this section presents two algorithms for sequential state estimation and RUL prediction, respectively. Both of them are widely used by the prognostics community for their versatility and ease of implementation.

As explained before, the Bayesian state estimation methodology presented earlier in this section is analytically intractable except some especial cases mentioned above (Arulampalam *et al.*, 2002). An alternative for the general case of both non-linear and non-Gaussian state-space models, and non-discretizable state variable, is by the use of Sequential Monte Carlo (SMC) methods (Arulampalam *et al.*, 2002; Cappé *et al.*, 2007; Doucet *et al.*, 2001; Gordon *et al.*, 1993) for drawing samples from the posterior PDF $p\left(z_{0:n} \mid y_{0:n}\right)$ avoiding the calculation of the multidimensional integrals in Equations (6) and (7). A comprehensive treatment of SMC methods can be found in (Doucet *et al.*, 2001). As a special case of SMC methods, particle filters (PF) (Gordon *et al.*, 1993) emerge as the state-of-art of SMC algorithms for sequential state estimation. These methods share a common principle of approximating the required posterior PDF through a set of N discrete samples or *particles* $\left\{z_{0:n}^{(i)}\right\}_{i=1}^{N}$ with associated set of *weights* $\left\{\omega_{n}^{(i)}\right\}_{i=1}^{N}$, such that

$$p\left(z_{0:n} \mid y_{0:n}\right) \approx \sum_{i=1}^{N} \omega_{n}^{(i)} \delta\left(z_{0:n} - z_{0:n}^{(i)}\right) \tag{21}$$

where δ is the Dirac delta.

In practice, it is not possible to efficiently sample from the true posterior $p\left(z_{0:n} \mid y_{0:n}\right)$, given that it is seldom known exactly. To overcome this problem, a Sequential Importance Sampling (SIS) approach can be adopted. To this end, an importance density PDF $q\left(z_{0:n} \mid y_{0:n}\right)$ defined over a support region including that of the true posterior PDF can be conveniently defined so that samples can be efficiently generated from it. Thus, to compensate for the difference between the importance density and the true posterior density, the unnormalized particle weights are computed as:

$$\hat{\omega}_{n}^{(i)} = \frac{p\left(z_{0:n}^{(i)} \mid y_{0:n}\right)}{q\left(z_{0:n}^{(i)} \mid y_{0:n}\right)} \tag{22}$$

where $\omega_{n}^{(i)} = \dfrac{\hat{\omega}_{n}^{(i)}}{\sum_{i=1}^{N} \hat{\omega}_{n}^{(i)}}$, $i = 1,\ldots,N$.

It follows that the choice of the importance density is crucial for the particle filter algorithm, so that a bad election may lead to a poor algorithm performance (i.e. larger variances of the particle-based estimation). There is a vast literature dealing with optimal choice of importance density (Arulampalam *et al.*, 2002; Doucet *et al.*, 2001) however in most of the practical applications the importance density is conveniently chosen as $q\left(z_{0:n} \mid y_{0:n}\right) = q\left(z_{0:n} \mid y_{0:n-1}\right)$, so that it can be factorized as $q\left(z_{0:n} \mid y_{0:n-1}\right) = q\left(z_{n} \mid z_{n-1}\right) q\left(z_{0:n-1} \mid y_{0:n-1}\right)$ (Arulampalam *et al.*, 2002). Thus, by substituting Equation (10) into Equation (22), the unnormalized weight for the ith particle at time n can be rewritten as:

$$\hat{\omega}_{n}^{(i)} \propto \underbrace{\frac{p\left(z_{0:n-1}^{(i)} \mid y_{0:n-1}\right)}{q\left(z_{0:n-1}^{(i)} \mid y_{0:n-1}\right)}}_{\omega_{n-1}^{(i)}} \frac{p\left(z_{n}^{(i)} \mid z_{n-1}^{(i)}\right) p\left(y_{n} \mid z_{n}^{(i)}\right)}{q\left(z_{n}^{(i)} \mid z_{n-1}^{(i)}\right)} \tag{23}$$

Typically, the PDF $q\left(z_{n} \mid z_{n-1}\right)$ in Equation (23) is chosen to coincide with the state transition equation $p\left(z_{n} \mid z_{n-1}\right)$ (Gordon *et al.*, 1993; Tanizaki & Mariano, 1998) since it is easy to evaluate. In this case the last equation simplifies to

$$\hat{\omega}_n^{(i)} \propto \omega_{n-1}^{(i)} p\left(y_n \mid z_n^{(i)}\right) \tag{24}$$

and the resulting algorithm is commonly known as "bootstrap filter", after (Gordon *et al.*, 1993).

A well-known and unavoidable problem when using particle filters based on SIS is the weight degeneracy, by which the normalized probability mass tends to concentrate over a decreasing number of particles as long as the number of iterations increases (Arulampalam *et al.*, 2002; Cappé *et al.*, 2007). A common solution is to introduce a resampling step whenever a significant degeneracy is observed (e.g., by quantifying the effective sample size, ESS, (Kong *et al.*, 1994)). By the resampling, the particles with higher weights are replicated while those with low weights are removed, using the normalized weights as probabilities of selection. After the resampling step, the normalized importance weights are reset to $1/N$. A pseudocode implementation for a SIS particle filter with resampling and parameter regeneration is provided in Algorithm 1.

Finally, given the current state estimation based on particle filters, a filtered estimation of the predictive PDF $p(z_{n+\ell} \mid y_{0:n})$ can be obtained by replacing $p\left(z_n \mid y_{0:n}\right)$ in Equation (16) by its particle-filter approximation (as a marginal of the updating PDF in Equation 21) as follows:

$$\begin{aligned} p\left(z_{n+\ell} \mid y_{0:n}\right) &\approx \int_{\mathcal{Z}} \prod_{l=n+1}^{n+\ell} p\left(z_l \mid z_{l-1}\right) \sum_{i=1}^{N} \omega_n^{(i)} \delta\left(z_{0:n} - z_{0:n}^{(i)}\right) dz_{n:n+\ell-1} \\ &= \sum_{i=1}^{N} \omega_n^{(i)} \int_{\mathcal{Z}} p\left(z_{n+1} \mid z_n^{(i)}\right) \prod_{l=n+2}^{n+\ell} p\left(z_l \mid z_{l-1}\right) dz_{n+1:n+\ell-1} \end{aligned} \tag{25}$$

Note that the last equation cannot be solved analytically, however it can be sampled by drawing one conditional sample trajectory $z_{n+1:n+\ell}^{(i)} = \left\{z_{n+1}^{(i)}, z_{n+2}^{(i)}, \ldots, z_{n+\ell}^{(i)}\right\}$ from each of the N multidimensional integrals appearing in Equation (25). This can be readily achieved by conditional sampling, using the one-step transition equation defined in Equation (11), i.e.: first sample $z_{n+1}^{(i)}$ using the aforementioned state transition equation conditional on a sample from the filtered state $z_n^{(i)}$, i.e., $z_{n+1}^{(i)} \sim p\left(z_{n+1} \mid z_n^{(i)}\right)$; then sample the succeeding state conditional on the previous sample, i.e. $z_{n+2}^{(i)} \sim p\left(z_{n+2} \mid z_{n+1}^{(i)}\right)$; finally, repeat the same process until the target time $n+\ell$ is reached. Finally, an estimate of the $\ell-$step predictive ahead PDF can be expressed as

$$p\left(z_{n+\ell} \mid y_{0:n}\right) \approx \sum_{i=1}^{N} \omega_n^{(i)} \delta\left(z_{n+\ell} - z_{n+\ell}^{(i)}\right) \tag{26}$$

where $\omega_n^{(i)}$ is the weight of ith particle at time of prediction n.

Algorithm 1. SIS particle filter with parameter adaptation

At $n = 0$

Generate $\left\{x_0^{(i)}, \theta_0^{(i)}\right\}_{i=1}^{N}$, sampling from prior PDFs $p(x_0)$ and $p(\theta)$, respectively.

Assign the initial weights: $\left\{\omega_0^{(i)} = 1/N\right\}_{i=1}^{N}$

At $n \geq 1$

Set initial covariance for artificial dynamics: $\Sigma_{\xi_0} = diag\left(\sigma_{\xi_{0,1}}^2, \ldots, \sigma_{\xi_{0,n_\theta}}^2\right)$

for $i = 1$ to N

Sample from Eq.: $\theta_n^{(i)} \sim p\left(\theta \mid \theta_{n-1}^{(i)}\right)$

Sample from Eq.: $x_n^{(i)} \sim p\left(x_n \mid x_{n-1}^{(i)}, \theta_n^{(i)}\right)$

Set $z_n^{(i)} = \left(x_n^{(i)}, \theta_n^{(i)}\right)$ and $z_{0:n}^{(i)} = \left\{z_{0:n-1}^{(i)}, z_n^{(i)}\right\}$

Update weights according to Eq.: $\hat{\omega}_n^{(i)} \propto \omega_{n-1}^{(i)} p\left(y_n \mid z_n^{(i)}\right)$

end for

for $i = 1$ to N

Normalize $\omega_n^{(i)} \leftarrow \hat{\omega}_n^{(i)} \Big/ \sum_{j=1}^{N} \hat{\omega}_n^{(j)}$

end for

for $j = 1$ to n_θ

Sample $\left\{\tilde{\theta}_{n,j}^{(k)}\right\}_{k=1}^{K} \sim p\left(\theta_{n,j} \mid y_{0:n}\right) \approx \sum_{i=1}^{N} \omega_n^{(i)} \delta\left(\theta_{n,j} - \theta_{n,j}^{(i)}\right)$

continued on following page

Algorithm 1. Continued

Compute $\mathrm{RMAD}\left(\theta_{n,j}\right)$ according to Eq.

Update variance of random walk: $\sigma^2_{\xi_{n,j}} = \sigma^2_{\xi_{n-1,j}}\left(1 - P_j^* \dfrac{\mathrm{RMAD}\left(\theta_{n,j}\right) - RMAD_j^*}{\mathrm{RMAD}\left(\theta_{n,j}\right)}\right)$

end for

set $\Sigma_{\xi_n} = diag\left(\sigma^2_{\xi_{n,1}}, \cdots, \sigma^2_{\xi_{n,n_\theta}}\right)$

if EES $< N_T$

$\left\{z_n^{(i)}, 1/N\right\}_{i=1}^{N} \leftarrow \mathrm{Resample}\left\{z_n^{(i)}, \omega_n^{(i)}\right\}_{i=1}^{N}$

end if

Having estimated the current state of the system and the future state $\ell-$steps ahead, Algorithm 2 can be used for predicting the RUL and EOL of the system. Using the updated weights of the particles at time of prediction $n = n_p$, a probabilistic estimation of the EOL can be obtained as:

$$p\left(EOL_{n_p} \mid y_{0:n_p}\right) \approx \sum_{i=1}^{N} \omega_{n_p}^{(i)} \delta\left(EOL_{n_p} - EOL_{n_p}^{(i)}\right) \tag{27}$$

where $\omega_{n_p}^{(i)}$ is the normalized weight of the ith particle at time of prediction n_p. From the EOL estimation of the ith particle, denoted as $EOL_{n_p}^{(i)}$, the remaining useful life for that particle can be readily obtained as $RUL_{n_p}^{(i)} = EOL_{n_p}^{(i)} - n_p$. By adopting the same normalized weights than those used for EOL estimation, a probability-based estimation of RUL can be obtained as:

$$p\left(RUL_{n_p} \mid y_{0:n_p}\right) \approx \sum_{i=1}^{N} \omega_{n_p}^{(i)} \delta\left(RUL_{n_p} - RUL_{n_p}^{(i)}\right) \tag{28}$$

The details of the method for EOL and RUL prediction are summarized in Algorithm 2.

PROGNOSTICS EXAMPLE ON COMPOSITE MATERIALS

The proposed framework is exemplified using SHM data obtained from a set of carefully designed run-to-failure fatigue experiments in cross-ply graphite-epoxy laminates. This example is motivated by the

Algorithm 2. EOL and RUL prediction

Inputs: $\left\{\left(x_{n_p}^{(i)}, \theta_{n_p}^{(i)}\right), \omega_{n_p}^{(i)}\right\}_{i=1}^{N}$

Outputs: $\left\{EOL_{n_p}^{(i)}, RUL_{n_p}^{(i)}, \omega_{n_p}^{(i)}\right\}_{i=1}^{N}$

$n \leftarrow n_p$

for $i = 1$ to N

while $T_{EOL}\left(x_n^{(i)}, \theta_n^{(i)}\right) = 0$

Sample: $\theta_{n+1}^{(i)} \sim p\left(\theta_{n+1} \mid \theta_n^{(i)}\right)$;

Sample: $x_{n+1}^{(i)} \sim p\left(x_{n+1} \mid x_n^{(i)}, \theta_{n+1}^{(i)}\right)$;

$\left(x_n^{(i)}, \theta_n^{(i)}\right) \leftarrow \left(x_{n+1}^{(i)}, \theta_{n+1}^{(i)}\right)$

$n \leftarrow n + 1$

end while

$EOL_{n_p}^{(i)} \leftarrow n$

$RUL_{n_p}^{(i)} = EOL_{n_p}^{(i)} - n$

end for

high dispersion that can be observed when degradation data are obtained from fatigue tests in composite materials, even for tests of nominally identical specimens. This highlights the need of integrating the SHM system within a Bayesian prognostic framework to deal with RUL prediction along with its quantified uncertainty. In this exercise, Torayca T700G unidirectional carbon pre-preg material was used for $15.24\ [\mathrm{cm}] \times 25.4\ [\mathrm{cm}]$ coupons with dog-bone geometry and $\left[0_2 / 90_4\right]_s$ stacking sequence, whose mechanical properties are listed in Table 1 (at this point, see only the nominal values in the 3rd column). A notch ($5.1\ [\mathrm{mm}] \times 19.3\ [\mathrm{mm}]$) was created in these coupons to induce damage modes others than matrix-cracks, such as delamination, thereby introducing additional sources of uncertainty and then demonstrating the proposed framework under more realistic conditions.

Description of Experiment Set-Up and SHM Data

SHM data for this numerical example consist of periodically measurements for laminate stiffness reduction, as a measure of macro-scale manifestation of damage, along with micro-crack density and delamination area, as micro-scale manifestation of damage. Fatigue tests were conducted under load-controlled tension-tension cyclic loading with a maximum applied load of $31.13\ [\mathrm{KN}]$, a frequency $f = 5\ [\mathrm{Hz}]$, and a stress ratio $R = 0.14$ (relation between the minimum and maximum stress for each fatigue cycle). Monitoring data were collected from a network of 12 piezoelectric (PZT) sensors SMART Layer® from Acellent Technologies Inc. using Lamb wave signals and three triaxial strain-gages. Additionally, periodic X-rays were taken to visualize and characterize subsurface damage features, in particular, the micro-cracks density. This information was then used to develop a mapping between PZT raw signals and micro-cracks density, as reported in Larrosa and Chang (Larrosa & Chang, 2012). More details about these tests are reported in the Composite dataset, NASA Ames Prognostics Data Repository (Saxena *et al.*, 2013). Damage data used in this example correspond to laminate L1S19 in (Saxena *et al.*, 2013).

Table 1. Nominal values and prior uncertainty for main parameters used in calculations; parameters selected for sequential updating are identified in parentheses. The nominal values for fitting and error parameters have been defined through initial fitting tests. See Table 2 for nomenclature description.

Type	Parameter	Nominal Value	Units	Prior PDF
Mechanical	$E_1\ (\theta_1)$	127.55	GPa	LogN(ln(127.55), 0.1)
	$E_2\ (\theta_2)$	8.41	GPa	LogN(ln(8.41), 0.1)
	G_{12}	6.20	GPa	LogN(ln(6.20), 0.1)
	ν_{12}	0.31	--	LogN(ln(0.31), 0.1)
	G_{23}	2.82	GPa	LogN(ln(2.82), 0.1)
	$d\ (\theta_3)$	$1.5\cdot10^{-4}$	m	LogN(ln($1.5\cdot10^{-4}$), 0.1)
Fitting	$\alpha\ (\theta_4)$	1.8	--	LogN(ln(1.8), 0.2)
	A	10^{-4}	--	LogN(ln(10^{-4}), 0.2)
Error	$\sigma_{v1}\ (\theta_5)$	--	cracks/(m × cycle)	U(0.5,1.5)
	$\sigma_{v2}\ (\theta_6)$	--	--	U(0.001,0.003)

Damage Model

The rate of change of internal damage per cycle can be evaluated by the modified Paris' law (Nairn & Hu, 1992) as follows:

$$\frac{d\rho}{dn} = A(\Delta G)^{\alpha} \tag{29}$$

where ρ is the matrix micro-crack density, and α and A are fitting parameters. The term ΔG is the increment in energy release rate G evaluated for the maximum and minimum stress in the cyclic load series: $\Delta G = G\big|_{\sigma_{max}} - G\big|_{\sigma_{min}}$. For a given axial tension σ_x, the energy released by the formation of a new crack between two existing cracks can be calculated using the following expression:

$$G\big|_{\sigma_x} = \frac{\sigma_x^2 h}{2\rho t_{90}}\left(\frac{1}{E_x^*(2\rho)} - \frac{1}{E_x^*(\rho)}\right) \tag{30}$$

where h and t_{90} are the laminate and 90°-sublaminate half-thickness, respectively. The term $E_x^*(\rho)$, as a function of crack density ρ, is the effective laminate Young's modulus due to the current damage state. Several micro-damage mechanic models from the literature (Talreja & Singh, 2012) can be adopted to evaluate $E_x^*(\rho)$, e.g. *shear-lag models* (Garrett & Bailey, 1977; Highsmith & Reifsnider, 1982) *variational models* (Hashin, 1985), and *crack opening displacement* based models (Gudmundson & Weilin, 1993; Lundmark & Varna, 2005), among others. The shear-lag approach is adopted in this work for being simpler and well suited for symmetric cross-ply laminates (J. Chiachio *et al.*, 2015), which is the laminate type used for this study. Following this approach, $E_x^*(\rho)$ can be obtained as (Joffe & Varna, 1999):

$$E_x^* = \frac{E_{x,0}}{1 + a\frac{1}{2\bar{l}}R(\bar{l})} \tag{31}$$

In the last equation, $E_{x,0}$ is the initial (undamaged) longitudinal Young's modulus of the laminate; $\bar{l}$ is the half crack-spacing normalized by the 90° sub-laminate thickness, i.e., $\bar{l} = l/\rho t_{90}$ where $l = 1/2\rho$; and a is a known function of the laminate mechanical and geometrical properties, defined as follows:

$$a = \frac{E_2 t_{90}}{E_1 t_0}\left(h\frac{\nu_{12}\cdot(1-\nu_{12})}{t_{90} + t_0\frac{E_2}{E_1}}\right)\frac{1-\nu_{12}^2}{1-\nu_{12}^2\frac{E_2}{E_1}} \tag{32}$$

Regarding the initial Young's modulus of the laminate $E_{x,0}$, it can be approximated for this example using a simple rule of mixtures as

$$E_{x,0} = \frac{t_0 E_1 + t_{90} E_2}{t_0 + t_{90}} \tag{33}$$

The function $R(\bar{l})$ in Equation (31) is known as the *average stress perturbation function* (Joffe & Varna, 1999), and it is defined as:

$$R(\bar{l}) = \frac{2}{\xi} \tanh(\xi \bar{l}) \tag{34}$$

where

$$\xi^2 = G_{23}\left(\frac{1}{E_2} + \frac{1}{\lambda E_1}\right) \tag{35}$$

and $\lambda = h_0 / h_{90}$. For the sake of clarity, a description of the terms involved in Equations (32) and (35) is provided in Table 2.

At this standpoint, it is easy to understand that a closed-form solution for the resulting Paris' Law equation is difficult to obtain. This is due to the complexity of the expression for ΔG, which involves the underlying micro-damage mechanics model for the computation of $E_x^*(\rho)$. To overcome this draw-

Table 2. Nomenclature description for main geometric and mechanical parameters

Level	Parameter	Description
Laminate	E_x	Longitudinal Yong's modulus
	E_x*	Effective long. Young's modulus
	h	Laminate half-thickness
	t_{90}	90º-sublaminate half-thickness
	t_0	0º-sublaminate thickness
Ply	E_1	Longitudinal Young's modulus
	E_2	Transverse Young's modulus
	ν_{12}	In-plane Poisson ratio
	G_{12}	In-plane shear modulus
	G_{23}	Out-of-plane shear modulus
	d	Ply thickness

back, the resulting differential equation in (29) can be solved by approximating the derivative using "unit-cycle' finite differences by assuming that damage evolves cycle-to-cycle as:

$$\rho_n = \rho_{n-1} + A\left(\Delta G\left(\rho_{n-1}\right)\right)^{\alpha} \tag{36}$$

where ρ_n denotes the matrix micro-cracks density at cycle n.

Stochastic Embedding of Damage Models

In this section, the PF-based prognostic framework presented above is specialized to the fatigue damage modeling approach explained in the last section. To this end, let consider that the progression of damage is studied by focusing on the matrix-cracks density ρ_n, and the normalized effective stiffness, defined as $D_n = E_x^* / E_{x,0}$, describing a joint state transition equation of two components $g = \left(g_1, g_2\right)$ as follows:

$$\begin{aligned} x_{1,n} &= \rho_n = \underbrace{g_1\left(\rho_{n-1}, u_n, \theta\right)}_{\text{Eq. 36}} + v_{1,n} \\ x_{2,n} &= D_n = \underbrace{g_2\left(\rho_n, u_n, \theta\right)}_{\text{Eq. 31}} + v_{2,n} \end{aligned} \tag{37}$$

where $x_n = \left(x_{1,n}, x_{2,n}\right) \in \mathbb{R}^2$ is the system response at cycle *n*, and $v_n = \left(v_{1,n}, v_{2,n}\right) \in \mathbb{R}^2$ is the model error vector for the overall system. Subscripts 1 and 2 correspond to each of the damage subsystems, namely, matrix-crack density and normalized effective stiffness, respectively. A key concept here is the consideration of model errors $v_{1,n}$ and $v_{2,n}$ as stochastically independent, even though the models corresponding to the damage subsystems g_1 and g_2 are mathematically related, as shown in Equation (37). It follows that the covariance operator Σ_{v_n} in Equation (8) is a diagonal matrix, i.e., $\Sigma_{v_n} = diag\left(\sigma_{v_{1,n}}, \sigma_{v_{2,n}}\right)$, where $\sigma_{v_{1,n}}$ and $\sigma_{v_{2,n}}$ are the corresponding standard deviations of errors $v_{1,n}$ and $v_{2,n}$ respectively. Therefore, the state transition equation of the overall system can be readily expressed as a product of univariate Gaussians, as:

$$p\left(x_n \mid x_{n-1}, \theta\right) = p\left(D_n \mid \rho_n, \theta\right) p\left(\rho_n \mid \rho_{n-1}, \theta\right) \tag{38}$$

where

$$\begin{aligned} p\left(\rho_n \mid \rho_{n-1}, \theta\right) &= \frac{1}{\sqrt{2\pi}\sigma_{v_{n,1}}} \exp\left(-\frac{\left(\rho_n - g_1\left(\rho_{n-1}, \theta\right)\right)^2}{2\sigma_{v_{n,1}}^2}\right) \\ p\left(D_n \mid \rho_n, \theta\right) &= \frac{1}{\sqrt{2\pi}\sigma_{v_{n,2}}} \exp\left(-\frac{\left(D_n - g_2\left(\rho_n, \theta\right)\right)^2}{2\sigma_{v_{n,2}}^2}\right) \end{aligned} \tag{39}$$

In the last equations the model parameters θ are selected among the complete set of mechanical and geometrical parameters describing Equations (29) to (35) (see Table 1) through a Global Sensitivity Analysis following the methodology proposed by (Saltelli *et al.*, 2008). The ply properties $\{E_1, E_2, t\}$ together with the fitting constant α emerged as sensitive parameters to model output uncertainty (J. Chiachio *et al.*, 2015). To the last cited selection, the standard deviations of the model errors $v_{1,n}$ and $v_{2,n}$ are added since they are uncertain a priori, resulting in $\theta = \left(\alpha, E_1, E_2, t, \sigma_{v_{1,n}}, \sigma_{v_{2,n}}\right) \in \mathbb{R}^6$. The rest of parameters are fixed at any point within their range of variation, (e.g. the mean value) without significantly influencing the output uncertainty.

To allow the incorporation of SHM measurement to the stochastic system, the following measurement equation is considered:

$$y_n = x_n + w_n \tag{40}$$

where $y_n = \left(y_{1,n}, y_{2,n}\right) \equiv \left(\hat{\rho}_n, \hat{D}_n\right) \in \mathbb{R}^2$ are measurements for both matrix-cracks density and normalized effective stiffness respectively, and $w_n = \left(w_{1,n}, w_{2,n}\right) \in \mathbb{R}^2$ is the vector of measurement errors. As stated before, the PMIE is used to choose w_n as Gaussian PDF with covariance matrix Σ_{w_n}. Thus, the *measurement equation* defined in Equation (40) can be expressed in probabilistic terms as:

$$p(y_n \mid x_n) = \left((2\pi)^{n_y} \mid \Sigma_{w_n} \mid\right)^{-\frac{1}{2}} \exp\left(-\frac{1}{2}\left(y_n - x_n\right)^T \Sigma_{w_n}^{-1}\left(y_n - x_n\right)\right) \tag{41}$$

Since the measurements of each subsystem (micro-cracks and stiffness loss) are considered as stochastically independent, then $\Sigma_{w_n} = diag\left(\sigma_{w_{1,n}}, \sigma_{w_{2,n}}\right)$. Here $\sigma_{w_{1,n}}$ and $\sigma_{w_{2,n}}$ are the standard deviation of the corresponding measurement errors $w_{1,n}$ and $w_{2,n}$ respectively. Thus, the measurement equation defined in (41) can be readily expressed as:

$$p\left(y_n \mid x_n\right) = p\left(\hat{\rho}_n \mid \rho_n\right) p\left(\hat{D}_n \mid D_n\right) \tag{42}$$

where

$$\begin{aligned} p\left(\hat{\rho}_n \mid \rho_n\right) &= \frac{1}{\sqrt{2\pi}\sigma_{w_{1,n}}} \exp\left(-\frac{\left(\hat{\rho}_n - \rho_n\right)^2}{2\sigma_{w_{1,n}}^2}\right) \\ p\left(\hat{D}_n \mid D_n\right) &= \frac{1}{\sqrt{2\pi}\sigma_{w_{2,n}}} \exp\left(-\frac{(\hat{D}_n - D_n)^2}{2\sigma_{w_{2,n}}^2}\right) \end{aligned} \tag{43}$$

Finally, by substituting the Equation (42) into (24), the formula for updating the particle weights leads to the next expression:

$$\omega_n^{(i)} \propto \omega_{n-1}^{(i)} p\left(\hat{\rho}_n \mid \rho_n^{(i)}\right) p\left(\hat{D}_n \mid D_n^{(i)}\right) \tag{44}$$

Results for RUL Prediction

This section presents RUL prediction results on the composite fatigue example and also illustrates the use of prognostics metrics for prediction performance evaluation. As fatigue data become available, the prognostic algorithm makes predictions of when micro-cracks density and the normalized effective stiffness reach their respective thresholds, defining the EOL and RUL for this experiment.

Every time new data arrive, the damage variables $x_n = \left(\rho_n, D_n\right)$ together with model parameters θ_n are updated using Algorithm 1 with $N = 5000$ particles. The scaling variables $RMAD_j^*$ and P_j^* (see Algorithm 1) are fixed to $0.3RMAD_{0,j}$ and 0.001, respectively, where $RMAD_{0,j}$ is calculated according to Equation (15) using samples from the marginal prior $p(\theta_j)$, $j = 1,\ldots,6$. The prior PDFs of model parameters are specified in Table 1. The diagonal elements of the covariance matrix $\Sigma_{\xi_{0,j}}$ (recall Equation (13)) are appropriately selected through some initial test runs and set at 0.5% of the 5th-95th inter-percentile range of the prior PDFs for each of the jth component of θ. For this example, the systematic importance resampling (SIR) version of the SIS algorithm is adopted whereby the resampling step is run every time new data are collected, hence $ESS = N$. Damage states are initialized at $x_0 = \left(\rho_0, D_0\right)$, being $\rho_0 = 0.1$ $\left[cracks/mm\right]$ and $D_0 = 1$ (dimensionless). The standard deviation of the measurement error parameters are set to $\sigma_{w_{1,n}} = 0.05$ $\left[cracks/mm\right]$ and $\sigma_{w_{2,n}} = 0.01$, taking them as known. At each time n, the updated damage states are further propagated into the future using the models to compute the remaining useful life of the laminate, calculated as $RUL_n = EOL_n - n$, following the methodology described above. The region of acceptable performance is defined in this example as $\mathcal{A} \triangleq \left\{(\rho, D) \in [0, 0.42] \times [0.88, 1]\right\} \subset \mathbb{R}^2$, where ρ is expressed in $\left[cracks/mm\right]$ and D is dimensionless.

Figure 7 shows the results for damage state estimation and multi-step ahead prediction at fatigue cycles $n = \{1, 3, 8\} \times 10^4$ for both, micro-cracks density and relative stiffness loss. While predictions are generally updated at regular intervals, for the sake of conciseness and clarity only three predictions are presented here. The solid black markers indicate the measurement data available until time of prediction n. For comparison purposes, the upcoming data are also represented using hollow markers. The multi-step ahead predictions for both damage features are shown in solid lines. Additionally, two credibility interval ranges, 25%-75% (dark dashed lines) and 5%-95% (light dashed lines), are included to show the uncertainty in the predicted estimates. Since predictions are updated at regular intervals, the uncertainty reduction is clearly revealed by comparing the credibility intervals at increasing prediction times.

The results for remaining useful life are plotted in Figure 8, where RUL predictions made at different times (about every 10,000 cycles) are plotted with corresponding uncertainties represented as error bars.

Figure 7. Example results for sequential state estimation and multi-step ahead prediction made at prediction times $n = \{1,3,8\} \times 10^4$

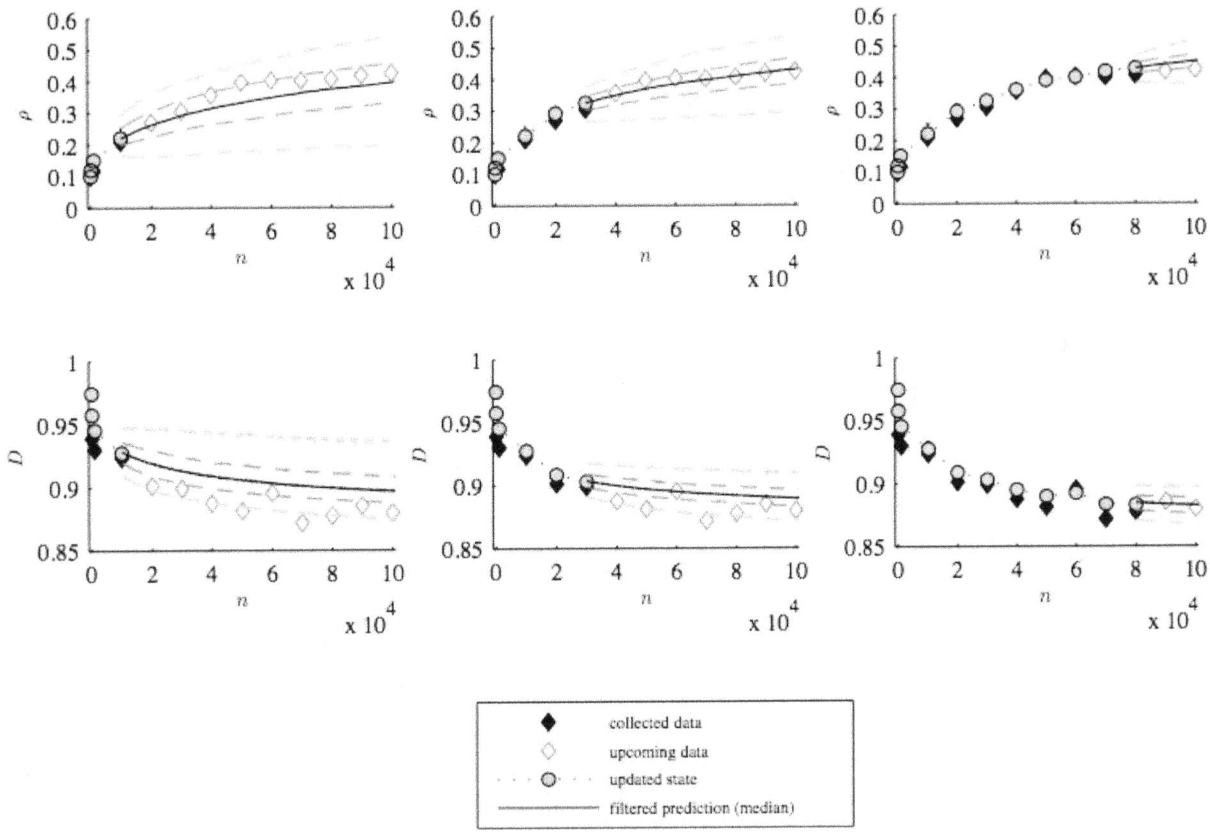

Two cones of accuracy at 10% and 20% of true RUL, denoted as RUL*, are included to interpret performance with regards to prediction accuracy and precision.

Observe that the RUL prediction is significantly inaccurate for the first few predictions during the early stages of the fatigue process, which is attributed to the time lapse required for SHM data to train model parameters. From this period, not only the prediction precision clearly improves with time (values closer to RUL* line), but also the prediction uncertainty gradually tends to diminish. This visualization allows assessment of how prediction performance changes over time in terms of *correctness* (accuracy and precision). Also, by assessing how quickly performance converges within desired accuracy levels one can estimate the prediction horizon, as defined in Fundamentals. In particular for this example, the PH for 0.2 as α-accuracy is $PH=8\cdot10^4-1\cdot10^3=7.9\cdot10^4$ cycles.

Furthermore, the error bars and their pattern of shrinkage contribute to the confidence in predictions vis-à-vis and their usage in decision making. It is also noteworthy in Figure 8 that accuracy seems to depart from true RUL at the final stages of the fatigue process, which indicates that the model and its variance structure do not fully capture the damage dynamics towards the end of the damage process. Such behavior has been previously reported in (Saxena, Celaya, *et al.*, 2010a) and may be related to the difficulty of obtaining adequate amount of measurements to account for behaviors when the damage process has reached asymptotic growth behavior.

Figure 8. Using RUL vs. time plot to assess Remaining Useful Life prediction performance for composites under fatigue aging

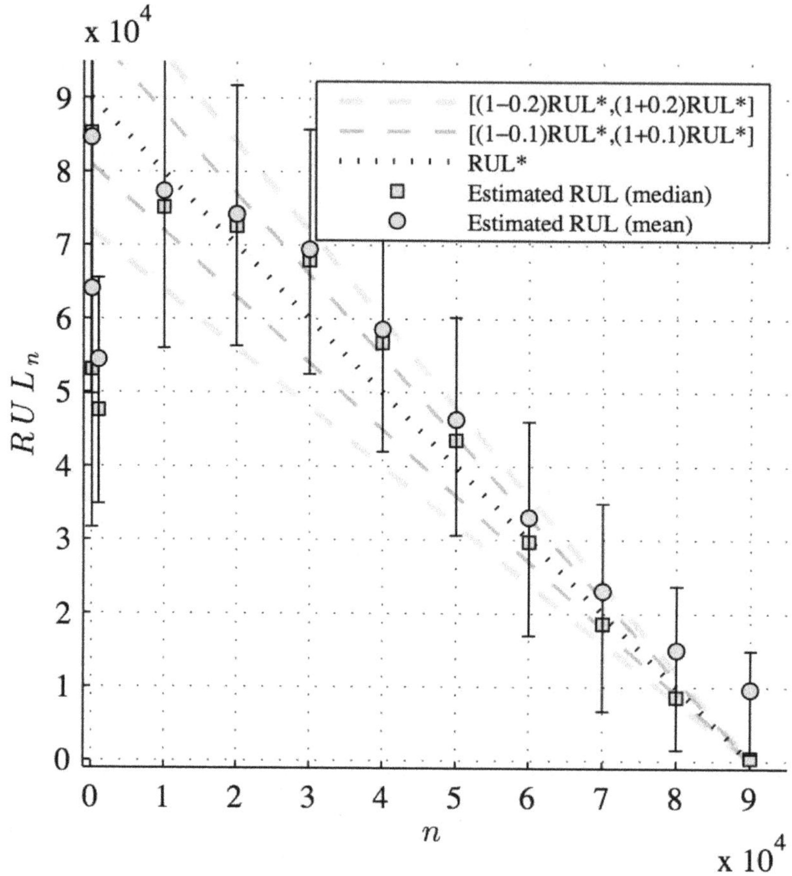

CONCLUSION

This chapter presented a generic framework for prognostics and remaining useful life prediction. First, an architecture for prognostics was developed; in this architecture, prognostics was broken down into the three important problems of state estimation, future state prediction, and remaining useful life prediction. Stochastic state-space system models were used to depict the degradation of system health that was in turn represented using the states of the system. These states were estimated using output measurements and the estimated states were used to forecast state values and hence, the health of the system. The end of life was defined using a multi-dimensional threshold function and the time-instant at which the system reaches end of life was calculated.

A key feature of the above methodology was a systematic treatment of the various sources of uncertainty. Since prognostics deals with the prediction of future events, it is important to understand that it is almost impossible to make precise predictions due to the various uncertainties that cloud the future. To being with, the interpretation of uncertainty was discussed in detail. The traditional frequentist approach is not suitable for understanding uncertainty in condition-based prognostics and it is necessary to adopt a subjective approach for interpreting uncertainty in this context. Therefore, a Bayesian approach that is based on the concept of subjective probability is extremely suitable for investigating uncertainty in prognostics.

The various sources of uncertainty in prognostics include system inputs (loading, environmental, operational conditions), system parameters, state-space models, etc. As a result of these sources of uncertainty, the state estimates, future state predictions, and the remaining useful life are all uncertain. It is necessary to understand that these quantities are simply dependent on the different sources of uncertainty and hence, it is important to systematically quantify the effect of the various sources of uncertainty on prognostics and remaining useful life prediction. This was accomplished through the use of Bayesian state estimation that can compute the probability distribution of the system states by accounting the different sources of uncertainty. Then, it was illustrated that, in order to accurately calculate the probability distribution of remaining useful life, the prediction of remaining useful life needs to be posed as an uncertainty propagation problem that can be solved using a variety of statistical techniques. Some of these methods were briefly discussed, and algorithms were provided for both state estimation and RUL prediction.

REFERENCES

Arulampalam, M. S., Maskell, S., Gordon, N., & Clapp, T. (2002). A tutorial on particle filters for online nonlinear/non-Gaussian Bayesian tracking. *Signal Processing. IEEE Transactions on, 50*(2), 174–188. doi:10.1109/78.978374

Au, S. K., & Beck, J. L. (2001). Estimation of small failure probabilities in high dimensions by Subset Simulation. *Probabilistic Engineering Mechanics, 16*(4), 263–277. doi:10.1016/S0266-8920(01)00019-4

Bucher, C. G. (1988). Adaptive sampling—an iterative fast Monte Carlo procedure. *Structural Safety, 5*(2), 119–126. doi:10.1016/0167-4730(88)90020-3

Caflisch, R. E. (1998). Monte carlo and quasi-monte carlo methods. *Acta Numerica, 1998*, 1–49. doi:10.1017/S0962492900002804

Cappé, O., Godsill, S. J., & Moulines, E. (2007). An overview of existing methods and recent advances in sequential Monte Carlo. *Proceedings of the IEEE, 95*(5), 899–924. doi:10.1109/JPROC.2007.893250

Celaya, J., Saxena, A., & Goebel, K. (2012). *Uncertainty Representation and Interpretation in Model-based Prognostics Algorithms based on Kalman Filter Estimation*. Paper presented at the Proceedings of the Annual Conference of the PHM Society.

Chang, F.-K., Markmiller, J. F. C., Ihn, J.-B., & Cheng, K. Y. (2007). A Potential Link from Damage Diagnostics to Health Prognostics of Composites Through Built-in Sensors. *Transaction of the ASME, 129*(December). doi:10.1115/1.2730530

Chiachio, J., Chiachio, M., Saxena, A., Sankararaman, S., Rus, G., & Goebel, K. (2015). Bayesian Model Selection and Parameter Estimation for Fatigue Damage Progression Models in Composites. *International Journal of Fatigue, 70*, 361–373. doi:10.1016/j.ijfatigue.2014.08.003

Chiachio, M., Chiachio, J., Saxena, A., & Goebel, K. (2014). An efficient simulation framework for prognostics of asymptotic processes- a case study in composite materials. *2nd European Conference of the Prognostics and Health Management Society, Nantes, France2014. EPHM 20141.*

Daigle, M., Saxena, A., & Goebel, K. (2012). *An Efficient Deterministic Approach to Model-based Prediction Uncertainty Estimation.* Paper presented at the Annual Conference of the Prognostics and Health Management Society.

Daigle, M. J., & Goebel, K. (2013). Model-based prognostics with concurrent damage progression processes. *Systems, Man, and Cybernetics: Systems. IEEE Transactions on, 43*(3), 535–546.

Der Kiureghian, A., Lin, H.-Z., & Hwang, S.-J. (1987). Second-order reliability approximations. *Journal of Engineering Mechanics, 113*(8), 1208–1225. doi:10.1061/(ASCE)0733-9399(1987)113:8(1208)

Dolinski, K. (1983). First-order second-moment approximation in reliability of structural systems: Critical review and alternative approach. *Structural Safety, 1*(3), 211–231. doi:10.1016/0167-4730(82)90027-3

Doucet, A., De Freitas, N., & Gordon, N. (2001). *Sequential Monte Carlo methods in practice.* Springer Verlag. doi:10.1007/978-1-4757-3437-9

Dwight, R. P., & Han, Z.-H. (2009). *Efficient uncertainty quantification using gradient-enhanced kriging.* Paper presented at the 11th AIAA Non-Deterministic Approaches Conference, Reston, VA.

Engel, S. J., Gilmartin, B. J., Bongort, K., & Hess, A. (2000). *Prognostics, the real issues involved with predicting life remaining.* Paper presented at the Aerospace Conference Proceedings. IEEE. doi:10.1109/AERO.2000.877920

Farrar, C. R., & Lieven, N. A. (2007). Damage prognosis: the future of structural health monitoring. *Philosophical Transactions of the Royal Society A: Mathematical, Physical and Engineering Sciences, 365*(1851), 623-632.

Garrett, K., & Bailey, J. (1977). Multiple transverse fracture in Latex translation cross-ply laminates of a glass fibre-reinforced polyester. *Journal of Materials Science, 12*(1), 157–168. doi:10.1007/BF00738481

Glynn, P. W., & Iglehart, D. L. (1989). Importance sampling for stochastic simulations. *Management Science, 35*(11), 1367–1392. doi:10.1287/mnsc.35.11.1367

Goebel, K. (2014). PHM Design Process. NASA Ames Research Center.

Gordon, N., Salmond, D., & Smith, A. (1993). Novel approach to nonlinear/non-Gaussian Bayesian state estimation. *IEEE-Proceedings-F, 140*, 107–113.

Gudmundson, P., & Weilin, Z. (1993). An analytic model for thermoelastic properties of composite laminates containing transverse matrix cracks. *International Journal of Solids and Structures, 30*(23), 3211–3231. doi:10.1016/0020-7683(93)90110-S

Haldar, A., & Mahadevan, S. (2000). *Probability, reliability, and statistical methods in engineering design.* John Wiley & Sons, Incorporated.

Hashin, Z. (1985). Analysis of cracked laminates: A variational approach. *Mechanics of Materials, 4*(2), 121–136. doi:10.1016/0167-6636(85)90011-0

Highsmith, A., & Reifsnider, K. (1982). Stiffness-reduction mechanisms in composite laminates. *Damage in composite materials. ASTM Special Technical Publication, 775*, 103–117.

Hohenbichler, M., & Rackwitz, R. (1983). First-order concepts in system reliability. *Structural Safety, 1*(3), 177–188. doi:10.1016/0167-4730(82)90024-8

Hohenbichler, M., & Rackwitz, R. (1988). Improvement of second-order reliability estimates by importance sampling. *Journal of Engineering Mechanics, 114*(12), 2195–2199. doi:10.1061/(ASCE)0733-9399(1988)114:12(2195)

Jaynes, E. T. (1983). *Papers on Probability, Statistics and Statistical physics* (R. D. Rosenkrantz, Ed.). Kluwer Academic Publishers.

Joffe, R., & Varna, J. (1999). Analytical modeling of stiffness reduction in symmetric and balanced laminates due to cracks in 90 layers. *Composites Science and Technology, 59*(11), 1641–1652. doi:10.1016/S0266-3538(99)00025-1

Kalman, R. E. (1960). A new approach to linear filtering and prediction problems. *Journal of Fluids Engineering, 82*(1), 35–45.

Karamchandani, A., Bjerager, P., & Cornell, C. (1989). *Adaptive importance sampling*. Paper presented at the International Conference on Structural Safety and Reliability.

Kong, A., Liu, J. S., & Wong, W. H. (1994). Sequential imputations and Bayesian missing data problems. *Journal of the American Statistical Association, 89*(425), 278–288. doi:10.1080/01621459.1994.10476469

Larrosa, C., & Chang, F. (2012). Real time in-situ damage classification, quantification and diagnosis for composite structures *Proceedings of the 19th International Congress on Sound and Vibration* (Vol. 15).

Liu, J., & West, M. (2001). *Combined Parameter and State Estimation in Simulation-Based Filtering* (pp. 197–223). Sequential Monte Carlo Methods in Practice. doi:10.1007/978-1-4757-3437-9_10

Liu, Y., Mohanty, S., & Chattopadhyay, A. (2010). Condition Based Structural Health Monitoring and Prognosis of Composite Structures under Uniaxial and Biaxial Loading. *Journal of Nondestructive Evaluation, 29*(3), 181–188. doi:10.1007/s10921-010-0076-2

Loh, W.-L. (1996). On Latin hypercube sampling. *Annals of Statistics, 24*(5), 2058–2080. doi:10.1214/aos/1069362310

Lundmark, P., & Varna, J. (2005). Constitutive relationships for laminates with ply cracks in in-plane loading. *International Journal of Damage Mechanics, 14*(3), 235–259. doi:10.1177/1056789505050355

Nairn, J., & Hu, S. (1992). The initiation and growth of delaminations induced by matrix microcracks in laminated composites. *International Journal of Fracture, 57*(1), 1–24. doi:10.1007/BF00013005

Najm, H. N. (2009). Uncertainty quantification and polynomial chaos techniques in computational fluid dynamics. *Annual Review of Fluid Mechanics, 41*(1), 35–52. doi:10.1146/annurev.fluid.010908.165248

Orchard, M., Kacprzynski, G., Goebel, K., Saha, B., & Vachtsevanos, G. (2008). *Advances in uncertainty representation and management for particle filtering applied to prognostics*. Paper presented at the Prognostics and Health Management, 2008. PHM 2008. International Conference on. doi:10.1109/PHM.2008.4711433

Rosenblatt, M. et al. (1956). Remarks on some nonparametric estimates of a density function. *Annals of Mathematical Statistics*, *27*(3), 832–837. doi:10.1214/aoms/1177728190

Saha, B., & Goebel, K. (2008). *Uncertainty management for diagnostics and prognostics of batteries using Bayesian techniques*. Paper presented at the Aerospace Conference. doi:10.1109/AERO.2008.4526631

Saltelli, A., Ratto, M., Andres, T., Campolongo, F., Cariboni, J., Gatelli, D., & Tarantola, S. (2008). *Global Sensitivity Analysis: The Primer*. Wiley-Interscience.

Sankararaman, S. (2014). (Already published). Significance, interpretation, and quantification of uncertainty in prognostics and remaining useful life prediction. *Mechanical Systems and Signal Processing*. (52)228-247

Sankararaman, S., Daigle, M., Saxena, A., & Goebel, K. (2013). *Analytical algorithms to quantify the uncertainty in remaining useful life prediction*. Paper presented at the Aerospace Conference. doi:10.1109/AERO.2013.6496971

Sankararaman, S., Daigle, M. J., & Goebel, K. (2014). Uncertainty Quantification in Remaining Useful Life Prediction Using First-Order Reliability Methods. *Reliability. IEEE Transactions on*, *63*(2), 603–619.

Sankararaman, S., Ling, Y., Shantz, C., & Mahadevan, S. (2009). *Uncertainty Quantification in Fatigue Damage Prognosis*. Paper presented at the Annual Conference of the PHM Society, PHM09, San Diego, CA.

Saxena, A., Celaya, J., Saha, B., Saha, S., & Goebel, K. (2010a). Metrics for Offline Evaluation of Prognostic Performance. *International Journal of the PHM Society*, *1*(1), 20.

Saxena, A., Celaya, J., Saha, B., Saha, S., & Goebel, K. (2010b). Metrics for Offline Evaluation of Prognostic Performance. *International Journal of Prognostics and Health Management*, *1*(1), 21.

Saxena, A., Celaya, J., Saha, S., Saha, B., Roychoudhury, I., & Goebel, K. (2010). *Requirements Specification for Prognostics Performance – An Overview*. Paper presented at the AIAA Infotech @ Aerospace, Atlanta, GA.

Saxena, A., Goebel, K., Larrosa, C. C., & Chang, F.-K. (2013). *CFRP Composites Dataset*. Retrieved from http://ti.arc.nasa.gov/project/prognostic-data-repository

Saxena, A., Roychoudhury, I., Celaya, J., Saha, B., Saha, S., & Goebel, K. (2012). *Requirement Flowdown for Prognostics Health Management*. Paper presented at the AIAA Infotech@Aerospace, Garden Grove, CA. doi:10.2514/6.2012-2554

Saxena, A., Shankararaman, S., & Goebel, K. (2014, July, 2014). *Performance Evaluation for Fleet-based and Unit-based Prognostic Methods.* Paper presented at the 2nd European Conference of the PHM Society, Nantes, France.

Talreja, R., & Singh, C. V. (2012). *Damage and failure of composite materials.* Cambridge University Press. doi:10.1017/CBO9781139016063

Tang, L., Kacprzynski, G. J., Goebel, K., & Vachtsevanos, G. (2009). *Methodologies for uncertainty management in prognostics.* Paper presented at the Aerospace conference.

Tanizaki, H., & Mariano, R. S. (1998). Nonlinear and non-Gaussian state-space modeling with Monte Carlo simulations. *Journal of Econometrics, 83*(1), 263–290. doi:10.1016/S0304-4076(97)80226-6

Tvedt, L. (1989). *Second order reliability by an exact integral.* Springer. doi:10.1007/978-3-642-83828-6_26

Van Zandt, J. R. (2001). *A more robust unscented transform.* Paper presented at the International Symposium on Optical Science and Technology.

Wu, Y.-T. (1994). Computational methods for efficient structural reliability and reliability sensitivity analysis. *AIAA Journal, 32*(8), 1717–1723. doi:10.2514/3.12164

KEY TERMS AND DEFINITIONS

Coverage: A quantitative measure expressed as the ratio of failures detected to the entire theoretically-detectable failure population.

Diagnosability: A measure for how well system faults can be detected and isolated without ambiguity given a system's definition and it sensors.

Diagnostics: Detection of anomaly or a fault and determining the possible locations and/or causes.

Failure Threshold: A design parameter specified to indicate maximum threshold beyond which a component is not expected to perform its intended function within specifications and/or is outside safe limits.

Fault Isolation: The process of determining the location of a fault to the extent necessary to effect repair.

Observability: A measure for how well internal states of a system can be inferred by knowledge of its external outputs measured through sensors.

Prognostics: Prediction of a system's future health states, degradation, and remaining useful life (RUL) based on its usage history and anticipated future load profile.

Chapter 12
An Implementation of a Complete Methodology for Wind Energy Structures Health Monitoring

E. Zugasti
UPC-CoDaLab, Spain & IK4-IKERLAN, Spain

J. Anduaga
IK4-IKERLAN, Spain

L. E. Mujica
UPC-CoDaLab, Spain

F. Martinez
IK4-IKERLAN, Spain

ABSTRACT

In this chapter a complete methodology for a SHM damage detection solution is explained, and how this is validated in a laboratory tower model. Several methodologies are proposed for the typical process of SHM. Starting with sensor placement (the best possible sensor locations are found), selecting the more representative data, classifying the different environmental and operational conditions, applying a damage detection methodology, including sensor fault detection. The paradigm of damage detection can be tackled as a pattern recognition problem (comparison between the data collected from the structure without damages and the current structure in order to determine if there are any changes). There are lots of techniques that can handle the problem. In this work, accelerometer data is used to develop statistical data driven approaches for the detection of damages in structures. As the methodology is designed for wind turbines, only the output data is used to detect damage; the excitation of the wind turbine is provided by the wind itself or by the sea waves, being those unknown and unpredictable.

DOI: 10.4018/978-1-4666-8490-4.ch012

INTRODUCTION

Motivation

The world energy crisis has made the humanity find new energy sources. The development of renewable energy sources is a key research driver as the greenhouse effect is affecting the global warming problem. Among the renewable energy technologies, the wind energy is thought to be a cheap and clean energy source. Wind turbines are doing the job of creating energy from wind, and the maturity, and cost competitiveness makes the wind energy one of the best clean energy sources nowadays.

In order to harvest more energy through higher efficiencies and due to cost-effective considerations, the size of the wind turbine has increased over the years. The size is not the only thing that has changed, nowadays the wind farms are moving towards the sea. Taking into account the height of the turbine and the rough sea conditions, the wind industry is now facing new problems. Among these new problems, we can find:

1. It is difficult to perform inspection and maintenance work considering the height and location of the turbine. This can be difficult for the maintenance worker, and quite risky.
2. The marine sea conditions and the durability of the structure in the water, and specially the substructure.
3. How to transport the energy to the shore.

To improve safety considerations, to minimize down time, to reduce the frequency of sudden breakdowns and associated huge maintenance and logistic costs and to provide reliable power generation, the wind turbines must be continuously monitored to ensure that they are in good condition. Among all the monitoring systems, a Structural Health Monitoring (SHM) system is of primary importance because structural damage may induce catastrophic damage to the integrity of the system. A reliable SHM system, low cost and integrated into the wind turbine system may reduce wind turbine costs and make wind energy more affordable. The SHM information gathered could be used in a condition - based maintenance program to minimize the time needed for inspection of components, prevent unnecessary replacement of components, prevent failures and it allows utility companies to be confident of power availability.

Wind Energy Structures Health Monitoring Solutions

SHM and damage detection has been widely studied in the last two decades. The more complex the structures are, the more complex damage detection systems are needed. Vibrational methods are highly used in wind energy SHM solutions. Some global methodologies have been developed to determine the integrity of wind turbines. For example, Rolfes et al. proposed an integrated approach by using wireless sensors (eight biaxial accelerometers and two strain gauges) to detect all the possible vibrations on the structure. For the undamaged condition: dynamic stress, measured by strain gauges, is proportional to dynamic velocity, which can be derived from measured acceleration signals. This relation is valid when sensor locations are in line with the eigenform occurring. The sensor locations have to be chosen according to the maximum eigenform amplitudes and the maximum stress locations. If damage occurs, the relation between the dynamic velocities of these sensors is not the same, and an alert about the existence of damage should be created.

Adams et al explain how using modal properties and a statistical model, a damage can be detected by using vibrational data. The process starts with the best location for sensors (accelerometers) continuing with the feature extraction form those signals and finally calculating the frequency shift and the Modal Assurance Criterion discordance of the signal to be analyzed.

Kraemer and Fritzen use almost the same sensor distribution as Rolfes but the damage detection method is different. They use some statistical methods: Null Space damage detection and Auto Regresive (AR) modeling. Besides, they highlight the importance of the sensor fault identification, so as to know if the readings obtained from the sensors are well or not. They also try to locate damage using a finite element model.

Online damage detection is an important area where neural networks have been used. Oberholster and Heyns proposed a methodology for monitoring the online condition of axial flow fan blades with the use of neural networks. Results from a stationary experimental modal analysis of the structure were used for identifying global blade mode shapes and their corresponding frequencies. It was demonstrated that it was possible to classify damage for several fan blades by using neural networks with online vibration measurements from sensors not necessarily installed on the damaged blades themselves.

Friedmann et al introduce different types of damage detection methods for wind turbine, some of them are vibration based and others are not. Concerning about the section dealing with vibrational analysis, it is interesting how they use the random decrement method to eliminate the random input variable from the data, and how this is done in the smart sensor. That way, the smart sensor sends less information to the central processing unit. This article also includes a literature review of methods used for extracting health parameters from modal properties as well as the topic Operational Modal Analysis (OMA). The measurement system developed makes use of one of those OMA methods and combines it with the Random Decrement method to design a data acquisition system that can be implemented on a smart sensor network. As examples, the application of this system to the model of a wind turbine as well as to a full scale pedestrian bridge is described.

Smarsly et al present a SHM system for wind-turbines. This paper describes the design and prototype implementation of a multi-agent software paradigm for a self-managing SHM system. An integrated monitoring concept is proposed for continuous assessment of structural performance and safety as well as self-diagnostics and management of the SHM system. It gives a very good overview and detailed execution of how damage detection SHM system should be autonomous, and which steps should take into account. The proposed concept was validated with the continuous real-time monitoring of a wind turbinc for almost a one-year period. This long term validation experiment was able to demonstrated that the system he introduced was capable of reliably detecting anomalies and system malfunctions such as sensor breakdowns or temporarily unavailable resources.

Kraemer recently reported about vibrational methods for SHM, and how they can be used to achieve an almost complete schema of Structural Health Monitoring. He uses data from an offshore wind turbine in order to: detect damage (proved with more than one method), detect whether a sensor is faulty or not, localize the damage using a model-updating method, and finally the load characterization.

Main Contributions of the Chapter

The review of the state of the art revealed numerous and diverse damage detection methods. It is an active field, many types of sensors are available, and several approaches are being developed based on different principles. In general, the literature demonstrates that there is no universal optimum method for damage detection. Additionally, no algorithm has been proposed yet, which can be applied universally to identify any type of damage in any type of structure. There is no unique solution to the damage detection problem. The idea is to find the solution by many ways, also by combining different algorithms. For each problem, there will be a specific solution for it.

In this way, the authors of this chapter are especially interested in algorithms that can work properly using output only data. This is the case of big structures, such as, bridges, wind turbines, or buildings. These structures are very difficult to excite, so, usually, environmental sources are taken as excitation sources (wind, cars, or other vibrations) and only the response of the structure is stored.

The complete methodology presented in this chapter is based on Stochastic Subspace Identification (SSI) as strategy for damage detection. This identification applies Null-Space analysis of the Hankel matrix, which is able to detect changes in the dynamic response of the structure. This method is based on the work done by Basseville et al and Fritzen et al, It is a vibrational time-domain method, so no data conversion is needed. Considering the needs of a WES, this method is, from the point of view of authors, the most suitable because it only uses output data; the environmental excitation is difficult to measure. The method is not new; several works have been presented, all of them with good and promising results.

The main contribution of the chapter is the analysis, development and application of different techniques in each step proposed by Baseville, Fritzen and Kramer. For instance: Original signals are preprocessed using Principal Component Analysis (PCA) or/and some feature selection algorith as (Unsupervised minimum Redundancy Maximum Relevance (UmRMR)). For damage detection, some damage indicators were developed and compared. A combination of algorithms based on Modal Assurance Criterion (MAC) are used for optimal sensor placement. Fuzzy C-means clustering is used for classification of different EOC. Finally, the whole complete methodology is tested in a tower model, similar to those of a wind turbine.

SMART SYSTEM

The Smart System presented in Figure 1, represents the methodology proposed by Kramer and Fritzen. The first step is to find the most interesting places for sensing. Once the structure is instrumented, structural responses to different excitation and conditions are stored in a database. From that database, the newest data are selected and classified according to the Environmental and Operational Conditions (EOC). Afterwards, these data are compared with an undamaged pattern and a damage indicator is calculated. If this value is in the range defined as "healthy structure", the next measurement is checked. If the indicator is out of range, the integrity of sensors is analyzed. It has to be taken into account that the life of a sensor will probably be shorter than the life of the structure. If one of those does not work correctly, the wrong data is eliminated, and the indicator is estimated again. Otherwise, if the sensors work

Figure 1. Smart system for damage identification in WES

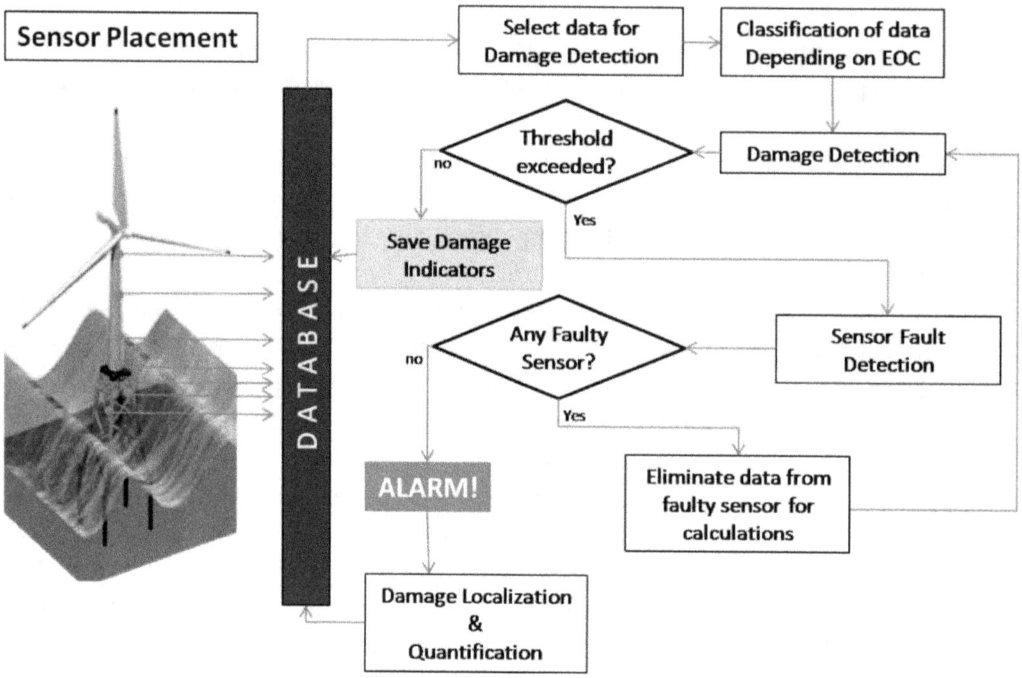

fine, the alarm is activated. In most of the cases, it is enough to detect damage and there is not need of locating it, or quantifying it. After ringing the alarm, next steps can be executed.

To find the best locations for sensing, a sensor placement algorithm named SEAMAC (Sensor Elimination Algorithm using Modal Assurance Criterion) is used. To select the most interesting data, Feature Selection and Extraction methods are used: UmRMR (Unsupervised minimum Redundancy Maximum Relevance) and PCA (Principal Component Analysis). The compensation of the environmental and operational conditions is tackled as a clustering problem by using Fuzzy C-Means algorithm. The damage detection itself is made using a pattern recognition algorithm based on statistical changes in the sensors (NullSpace). Finally, the integrity of the sensors is checked using Mutual Information.

SENSOR PLACEMENT

The optimal sensor location is found using the Sensor Elimination by Modal Assurance Criterion (SEAMAC) method implemented in FemTools. The Modal Assurance Criterion (MAC) [16] is commonly used to compare mode shapes, and it measures the likehood of the vibration modes, on another words, how similar the mode shapes are. Each term in the MAC-matrix measures the squared cosine of the angle between two mode shapes: $MAC_{i,j} = \cos^2\Theta_{i,j}$ and is calculated as:

$$MAC_{i,j} = \frac{\left|\{\Psi_i\}^t \{\Psi_j\}\right|^2}{\left(\{\Psi_i\}^t \{\Psi_i\}\right)\left(\{\Psi_j\}^t \{\Psi_j\}\right)} \tag{1}$$

where Ψ is the modal shape.

When $MAC_{i,j} = 0\%$, the i-th and j-th mode shapes are orthogonal to each other. Hence, the mode shapes are most linearly independent when all off-diagonal terms of MAC are zero. The other way around, a $MAC_{i,j} = 100\%$, the modes are totally the same.

The SEAMAC algorithm is a sensor elimination algorithm. This method tries to find a measurement point configuration that minimizes MAC values in off-diagonal terms. This means that it places the sensors where vibration modes can be well distinguished. The method SEAMAC is based on eliminating iteratively one by one those degrees of freedom (or Finite Element Nodes) that show lower impact on the MAC values. This iterative process stops when a default MAC matrix is obtained, high values in the diagonal terms and low values in off-diagonal terms.

The SEAMAC is able to find the location of the best sensors, but it is not able to define the amount of sensors. For that purpose, a criterion based on AutoMAC is applied. A sensor is placed in each node of the Finite Element Model of the structure. This is the best possible solution concerning to information, but it is impossible to apply. The MAC-matrix for this configuration is calculated (AutoMAC). SEAMAC algorithm is used to find a set of possible "best configurations" of sensors, and their MAC-matrices are calculated. That configuration which the sum of the off-diagonal terms of the resulting MAC is the closest to the double of the value of the best possible solution (AutoMAC) is selected as the best sensor placement.

MATHEMATICAL BACKGROUND FOR DAMAGE DETECTION ALGORITHM

The damage detection method used for this solution is the NullSpace Damage Detection method. It is based on Stochastic Subspace Identification (SSI) methods, and uses vibration responses from the structure as inputs. The system identification matrix used for this damage detection method is the Hankel Matrix.

Hankel Matrix

A way of estimating a SSI model is based on the Hankel matrix. This section is focused on details about this estimation and the application of the estimation matrix to the damage detection paradigm. For this work, we will only use the readings of the sensors (output only system). Based on the work done, the stochastic response from a system as a function of time will be considered as:

$$Y = \left[y_1, y_2, \ldots, \ldots, y_m\right] \tag{2}$$

The concept of subspace identification for linear systems, which was applied to modal analysis of structures is based on the definition of the Hankel matrix. This matrix may be calculated in two ways corresponding to either the covariance-driven or data-driven subspace identification algorithms.

$$H_{p,q} = \begin{bmatrix} \Lambda_1 & \Lambda_2 & \cdots & \Lambda_q \\ \Lambda_2 & \ddots & \ddots & \vdots \\ \vdots & \ddots & \ddots & \vdots \\ \Lambda_{p+1} & \cdots & \cdots & \Lambda_{p+q} \end{bmatrix}, \quad (3)$$

where p, q are user-defined parameters and Λ_i represents an unbiased estimate of the correlation matrix at time lag obtained from the definition of the correlation estimate [21].

The Block Hankel matrix defined in SSI is simply a gathering of a family of matrices that are created by shifting the original data matrix. Different Λ_i may be estimated from a set of N output data samples y_k (as taken from (3)) as:

$$\Lambda_i \simeq \left(\frac{1}{N-i-1}\right)\sum_{k=1}^{N-i} y_{k+i} y_k^T \quad (4)$$

This Hankel matrix may be factorized into two subspaces if Singular Value Decomposition is used. From these subspaces, especially from the left kernel subspace, the modal information may be extracted.

From these matrices, it is possible to obtain the modal parameters and system matrices. This is explained in detail. However, for damage detection case, the modal parameters are not interesting, and system matrices are not needed. The main goal is to detect whether the structure has changed or not. For that purpose, the Hankel matrix is the basic tool that contains all the dynamic properties of the structure.

NullSpace Extraction

The base mathematical foundations about modal identification on structures were presented in the previous section. This work is not focused on identifying the modal parameters of the structure. Instead, only relative changes of characteristic features are necessary for structural damage assessment. For this purpose, a method based on the null subspace (NullSpace) concept of Hankel matrices can be used. To find the NullSpace, singular-value decomposition (SVD) on the Hankel matrix is performed. In linear algebra, the SVD is a factorization of a matrix, with many useful applications in signal processing and statistics, and it is defined as:

$$H_{p,q} = U_H S_H V_H^T, \quad (5)$$

where U_H is an unitary matrix, S_H is a rectangular diagonal matrix with nonnegative real numbers on the diagonal, and V^T_H (the conjugate transpose of V) is an unitary matrix. The diagonal entries $S_{i,j}$ of S are known as the singular values of $H_{p,q}$. The columns of U_H and the columns of V_H are called the left-singular vectors and right-singular vectors of $H_{p,q}$, respectively. The NullSpace of the Hankel matrix ($U_{H,}$) is a matrix that makes the next property to be true:

$$U_{H0}^{T} H_{p,q} = 0\,. \quad (6)$$

This null hypothesis is the base of the damage detection method. Basically, the healthy structure data should achieve the null hypothesis. So, using a U_{H0} from a learning state of the structure and applying it to a healthy response, the p-value of the hypothesis testing should be small, while applying it to a damaged response, this p-value should be higher. U_{H0} contains the maximum number of independent column vectors that span the column null space of H. In order to find it, we have to find which singular values (S_H) are equal to zero, and take the left-hand singular vectors corresponding to those null singular values.

The basic model is found using this NullSpace, but in order to give variability to the model, more data are needed. So, a learning phase is needed to use several healthy data and give variability to the basic model.

Residuals

Using the idea that has been shown in the previous section; if the structure is undamaged, the product between NullSpace (U_{H0}) and a Hankel matrix calculated using data coming from a healthy structure (shown in Equation (6)) should be equal or really close to zero because both should have the same (or at least similar) NullSpace. If damage occurs, this product should be different from zero. The result of that multiplication is called residual matrix,

$$R_{i,j} = U_{H0}^{T} H_{i,j}\,, \quad (7)$$

where $R_{i,j}$ is the residual matrix, U_{H0} is the NullSpace of the first Healthy dataset (see next section) and $H_{i,j}$ is the Hankel matrix of a different input. Finally, this residual matrix is vectorized rearranging the columns of the R matrix in one vector:

$$\zeta = vec\left(R_{i,j}\right). \quad (8)$$

Damage Indicator

The information given by the residual matrix needs to be quantified. A Damage Indicator (DI) is a value that indicates the existence of damage; if it is low, there is no damage, on the contrary, if it is high, damage is present in the structure. DI calculation has two steps. First, in the learning phase the variance of the structure is calculated. In the second step, that variance is used to know if the structure is still inside the limits of being healthy. The transformation from the residual to DIs gives the result as scalar values, that way it is easier to determine if there is damage or not, just by looking to a number.

In the first step, using the different residuals $\left(\zeta\right)$ of the undamaged structure, the covariance estimate is constructed. This covariance estimate is a big matrix, and it is calculated by

$$\Sigma = \frac{1}{n-1} \sum_{n} \zeta_n \zeta_n^T\,, \quad (9)$$

where n is the number of inputs in the learning phase.

For the second step, a residual vector that is going to be evaluated is needed. For each evaluation residual vector, the will be applied to have a scalar value that is able to tell whether damage exists. The Damage Indicator is given by

$$DI = \zeta_{n+t}\Sigma^{-1}\zeta_{n+t}, \tag{10}$$

where ζ_{n+t} is the residual of the dataset to be evaluated.

DAMAGE DETECTION METHODOLOGY

In this section the algorithm itself and the different steps of the process are explained. The algorithm has two phases, the learning one and the detection one. In the first phase, the data used is data from a healthy structure and the healthy model is constructed. The bases for constructing the model are explained in the next sections. In this part the algorithm learns how the structure behaves in a healthy state. In the second one, the input data has to be evaluated. The model constructed in the learning phase is applied to the incoming data so as to determine if the structure is still in a healthy state or not.

Learning Phase

The Learning process has n inputs (datasets) and 2 outputs. The inputs are the healthy signals coming from the accelerometers, and the outputs are the NullSpace (U_{H0}) and the variability factor (Σ), which changes depending on the Damage Indicator used. Both outputs are explained in the next sections. The number of healthy datasets needed (n) for the learning process is not a fixed number; the more datasets used for learning, the more robust it is, and less false positives appear. Figure 2 shows the scheme of the learning process where:

- Y_k is the k-th input or dataset. This input is a matrix containing all the measurements of each sensor. The dimension of Y is shown in Equation 1.11 where i is the number of sensors and j the number of samples.

$$Y = \begin{bmatrix} y_{1,1} & y_{1,2} & \cdots & y_{1,j-1} & y_{1,j} \\ y_{2,1} & y_{2,2} & \cdots & y_{2,j-1} & y_{2,j} \\ \vdots & \vdots & \ddots & \vdots & \vdots \\ \vdots & \vdots & \ddots & \vdots & \vdots \\ y_{i,1} & y_{i,2} & \cdots & y_{i,j-1} & y_{i,j} \end{bmatrix}$$

- H_k is the Hankel matrix corresponding to the Y_k input; it corresponds to the k-th input.
- U_{H0} is the NullSpace of the H_l Hankel Matrix.
- ζ_k is the k-th residual; it corresponds to the k-th input.

Figure 2. Learning phase schematic

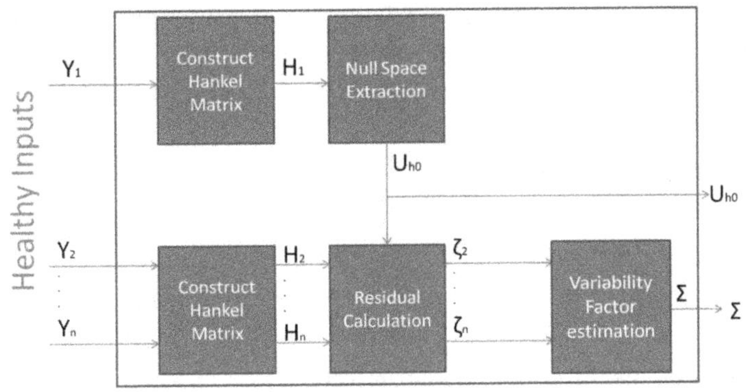

- Σ is the variability factor

There are n healthy inputs but the first one is used to find the NullSpace, and the rest are used to find the variability factor. On one hand, using the first dataset, its Hankel matrix is determined, and later the NullSpace matrix for that first input is determined using the Singular Value Decomposition.

On the other hand, the other inputs are used to determine the variance of the healthy structure. After finding the Hankel matrix for each of the inputs, defined in the next section, the residuals for each input are calculated. Finally, depending on the chosen Damage Indicator the variability factor will be found in one way or another. From this Learning process, the NullSpace matrix (U_{H0}), and the Variability Factor (Σ) are stored for the detection process. These two variables are the ones that construct the model.

Evaluation Phase

In the detection phase, *m*-(*n*+1) are the inputs to be evaluated. In real online monitoring system, this value is proportional to time the structure is operating, so it can be very large. The other 2 inputs are coming from the learning process, and are the variables that construct the model (U_{H0}, Σ). As output m-(n+1) Damage Indicator values are generated (one by each input). The value of the Damage Indicator indicates us whether the structure is damaged or not. The variables that take part in this process are the same as the ones used in the learning phase.

The detection is done almost the same way as the second part of the learning process. Instead of calculating the variability factor, it is applied in order to determine if the dataset is still inside the limits that the variability explains. First of all, the Hankel matrices are constructed, one for each incoming

Figure 3. NullSpace evaluation phase

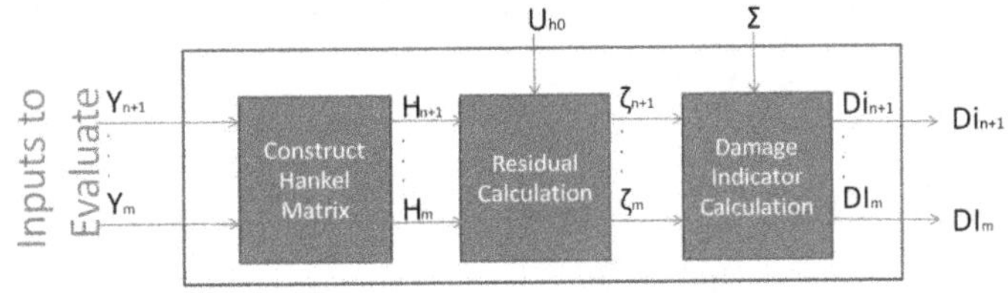

data. These Hankel matrices are multiplied with the NullSpace matrix U_{H0} coming from the learning process in order to determine the residual ζ . Once again, there is one residual for each input dataset. Finally each residual is evaluated using the variability factor, and the Damage Indicator of the structure is determined. These DIs are later plotted in order to have a visual diagnostic tool.

Feature Selection and Extraction

Toivola and Hollmén define Feature Extraction as "In pattern recognition and in image processing, feature extraction is a special form of dimensionality reduction", and Feature Selection as "In machine learning and statistics, feature selection, also known as variable selection, is the process of selecting a subset of relevant features for use in model construction".

Sometimes a huge amount of information can reduce the effectiveness of classification algorithms. Some of the columns of data attributes assembled for building and testing a model may not contribute meaningful information to the model. Some of these data may actually detract from the quality and accuracy of the model.

Datasets with many attributes may contain groups of correlated attributes. These attributes may actually be measuring the same underlying feature. Their presence can skew the logic of the algorithm and affect the accuracy of the model.

Wide data (many attributes) generally presents processing challenges. Model attributes are the dimensions of the processing space used by the algorithm. The higher the dimensionality of the processing space, the higher the computation cost involved in algorithmic processing. To minimize the effects of noise, correlation and, high dimensionality, dimensional reduction is sometimes a desirable preprocessing step. In this work, this goal is achieve by means of

1. **Feature Extraction:** Combining attributes into a new reduced set of features by using Principal Component Analysis (PCA), and
2. **Feature Selection:** Selecting the most relevant attributes by using Unsupervised minimum Redundancy Maximum Relevance (UmRMR).

The preprocessing method proposed is a mixture of both, feature selection and extraction methods. In first step, the most informative sensors are found and selected applying all the healthy datasets to the UmRMR. Afterwards, the healthy dataset with the assigned sensors is used to calculate the PCA model. In this way, the channels (or sensors) are selected and the features are extracted. Finally, the dataset (T_i) used for the damage detection method is obtained after selecting the appropriate sensors (given by UmRMR outputs (Y_i)) and projecting them into the PCA model (see Figure 4). In both phases (learning and evaluation), the algorithm for damage detection based on NullSpace explained before is applied to the transformed data.

It should be highlighted that using a preprocessing that extracts features using PCA firstly and later selects features using UmRMR has no sense, since PCA extracts and selects the most relevant components.

Figure 4. Data preprocessing step

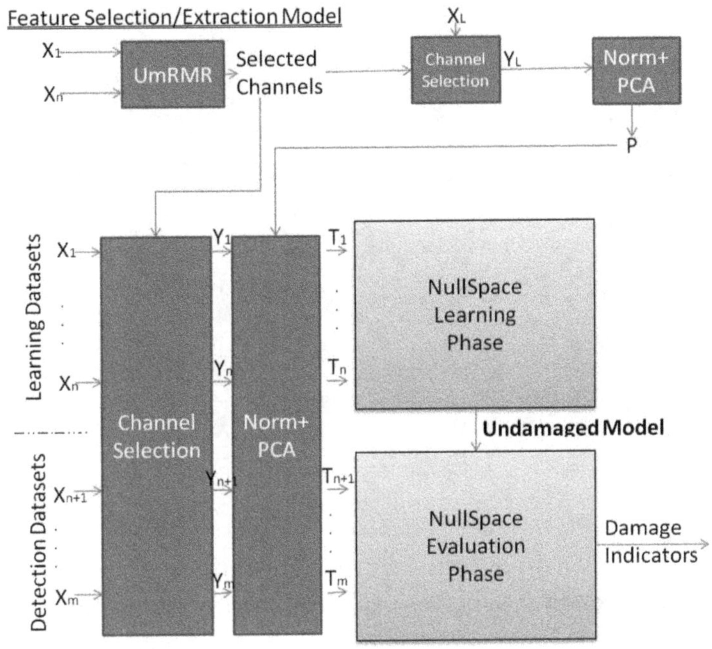

ENVIRONMENTAL AND OPERATIONAL CHANGES COMPENSATION

Environmental and Operational Conditions (EOC) play an important role when dealing with long term monitoring, because they complicate the damage detection procedure and sometimes it is impossible to detect damage if these are not taken into account. In fact, the variations in the input signals due to the EOC are bigger than the variations due to damage. Therefore, its influence should be compensated. Depending on the chosen strategy, the EOC compensation methods can be classified in those using the measurements of the EOC itself or the ones that do not use these EOC data.

In this System, the clustering solutions have been used in order to compensate the Environmental and Operational Changes. Among clustering techniques, it can be found: Hard clustering solutions, the ones that each dataset only belongs to one cluster, and the Soft (or fuzzy) clusters, where each dataset belongs in a certain degree to a cluster. In this work, the fuzzy solution has been used, specifically fuzzy C-means clustering.

This Fuzzy clusterization method is able to find cluster centers by optimization of an objective function. For a given EOC, each dataset belongs to every cluster in a certain degree. These degrees are used in order to create a healthy baseline for that EOC. So, for each dataset there is a healthy baseline, which is calculated using the Hankel matrices in the center of the clusters and the distance between the center EOC and the current EOC.

The application of the clustering method to the damage detection method is performed keeping the fuzzy nature of the clustering method. This means that for each different EOC, a compensated value of the different baselines is used to find an estimated value of the baseline for that exact EOC value. In this case, the learning process modelizes how the structure behaves for different environmental conditions.

Figure 5. EOC compensation general

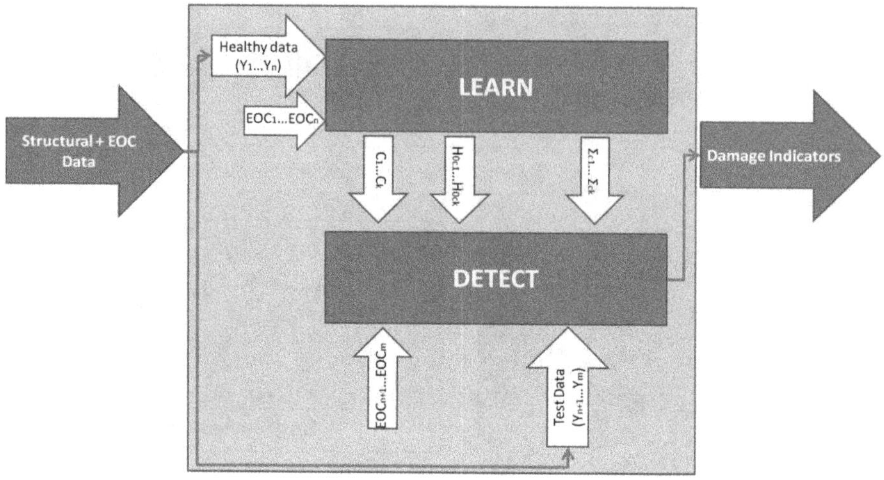

Since this method is supervised, environmental information is also needed. Once the EOC is determined, the process for damage detection is executed as explained in previous sections.

Learning Phase

In this phase, data corresponding to the healthy structure is used. The first step is to perform the primitive NullSpace. Using the damage indicators from the primitive NullSpace results and the Environmental and Operational data for each dataset, the fuzzy classification is performed, and the centers are calculated.

In Cluster Application block, centers (C_k) calculated in the previous block (fuzzy c-means clustering) are used. Along the centers, the Hankel matrices used in the first phase (H_x) and the environmental conditions (EOC_x) are used as inputs to the next block (Cluster Application). Its goal is to estimate a covariance matrix $\hat{\Sigma}$ for each cluster.

Figure 6. EOC compensation learning

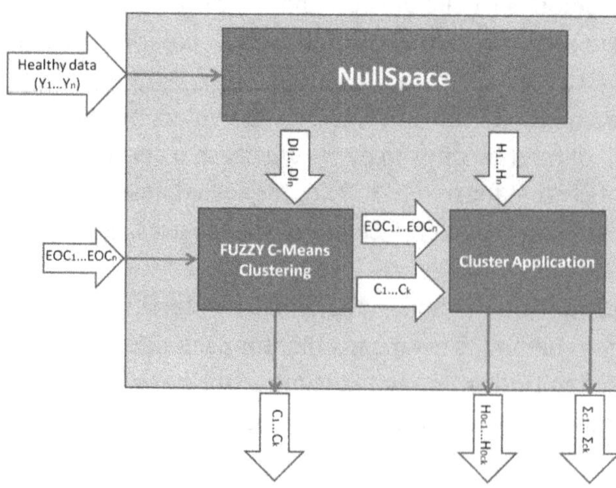

For each environmental condition, its healthy condition is estimated just by applying the distance to the centers. A Hankel matrix is estimated by weighting it with the distance to the different centers. That way, a baseline Hankel matrix for each different environmental condition is calculated. The reference Hankel matrix is estimated as follows:

$$H_{sX} = \sum_{k=1}^{k} \mu_{i,j} H_{0Ck} , \tag{12}$$

being ($\mu_{i,j}$) the distance between the center k and the EOC_x of the current data; k the number of centers of the cluster; and H_{0Ck} the Hankel matrix of the center k. This means, that the healthy condition is estimated using the distances to the center of the clusters, and not just applying the nearest information of the closest center. In this way the fuzzy nature of the classification is preserved and used in order to have a more precise solution.

A covariance matrix (see Equation (9)) per cluster center is calculated. Each residual will construct the covariance matrix of the closest center, being n the number of data sets that correspond to each center. In this case, there is no covariance estimate for each EOC. So the closest Covariance matrix is applied to the result of the estimated NullSpace and the Data Hankel Matrix. The outputs from the learning phase are: cluster centers (C_k), healthy Hankel matrices (H_{0Ck}) for the centers and, Covariance estimates $\hat{\Sigma}_{ck}$. All these variables are inputs in detection phase.

Evaluation Phase

In this phase, it is decided whether the data set belongs to healthy structure or not. Cluster centers (C_k), are used for classification of EOC. Healthy Hankel matrices (H_{0Ck}) and actual Hankel Matrices are used

Figure 7. EOC compensation evaluation

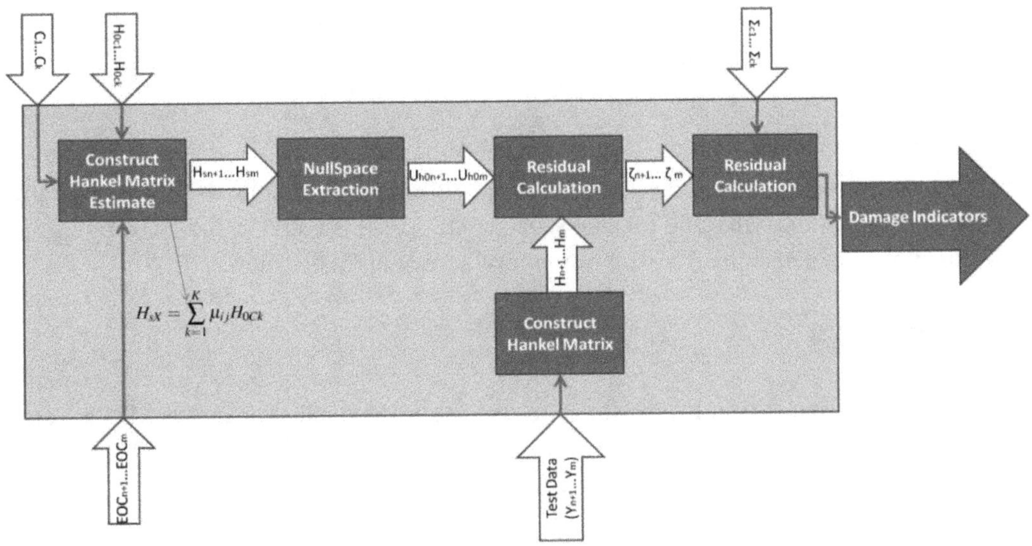

to calculate the NullSpace (U_{H0}) as done in learning phase. From the residual and covariance matrices are also calculated, one for each center. Finally, the Damage Indicator is found as follows:

$$DI = \zeta_{n+1} \hat{\Sigma}_{CX}^{-1} \zeta_{n+1} \tag{13}$$

SENSOR FAULT DETECTION

The integrity checking of the sensors is an important functionality part on an SHM solution. The alarm for the damage detection can only be activated when the integrity of the sensors has been validated. Otherwise, false alarms can be generated. In that way, it can be said that sensor fault detection is an important part of damage detection itself. There is not proper damage detection without being certain that the information gathered is reliable. When the sensors are placed in a structure, they work correctly, but usually the lifespan of a structure is larger than the sensors lifetime. In this chapter, a Sensor Fault Detection (SFD) method based on Mutual Information (MI) is described.

The basic idea is that the MI of any sensor comparing to another one, should remain the same if the sensors are working properly. Mutual Information measures the mutual dependence of the two variables. So basically, in a learning phase, the algorithm learns how the MI of each sensor versus all the rest is; for later to check whether the Mutual Information of the sensors has changed.

Mutual Information Matrix

Let us define the Mutual Information Matrix (MI_{mat}) as a matrix that has as elements, all the MI values between all the sensors in the structure. These values are able to give information about how dependent is each sensor of the others. For the k-th experiment (or dataset) this matrix is defined

$$MI_{mat} = \begin{bmatrix} MI_{1,2} & MI_{1,2} & \cdots & MI_{1,n} \\ MI_{2,1} & \ddots & \cdots & \vdots \\ \vdots & \cdots & \ddots & \vdots \\ MI_{n,1} & MI_{n,2} & \cdots & MI_{n,n} \end{bmatrix}, \tag{14}$$

where $MI_{i,j}$ denotes the MI between the i-th sensor and the j-th sensor. It is noted that this matrix is symmetric because the MI value of the sensor i with the sensor j, is the same as the value of the sensor j with the sensor i. Besides the MI value for one sensor with itself, is equal to the Entropy of that sensor. The size of this matrix is determined by the number of sensors in the structure (n).

Sensor Fault Indicator

One possibility to visualize the faulty sensor is to use the relative change in the MI as a sensor fault indicator (SFI):

$$SFI^k = \frac{\left| MI^k_{mat} - MI^{ref}_{mat} \right|}{MI^{ref}_{mat}}, \tag{15}$$

where k is the MI matrix of the actual data set and the upper index *Ref* represents the baseline MI matrix. If any sensor is damaged, the resulting matrix should show a change in the SFI referring to the faulty sensor.

Learning Phase

In the learning phase of the sensor fault detection methodology, all the MI matrices obtained from the healthy structure are used to build the baseline.

In Figure 8, the learning phase of the algorithm in a block diagram is shown. The objective of this part is to know how the MI between sensors changes when no damage is present in the sensors. The different blocks are divided as: the first block, some signal characteristics are extracted, the maximum and minimum values of each sensor, and these maximum and minimum values are stored for the detection phase. These values are used in order to be sensitive to the gain error. Once the signal is preprocessed, the reference MI matrix is estimated. The Mutual Information value for each sensor pair is estimated, and saved in different matrices, one per experiment. Once all the MI values for each pair of sensor in each different state are known, the reference Mutual Information matrix is calculated using the mean value of all the learning datasets. On the other hand, the matrix of sensor fault indicator of each dataset is calculated using the equation (15). Again the mean value of the SFIs is calculated to have the reference SFI.

Evaluation Phase

The block diagram of the Detection Phase is shown in Figure 9. In the detecting phase all the matrices calculated from the current dataset are compared to the baseline ones to see differences between them and infer about the state of the sensors.

The incoming dataset is limited using the limiters from the learned signal. Once the signal is preprocessed, the MI matrix is calculated and subsequently its SFI (by using the reference MI matrix from the learning phase). Finally, the SFI matrix is transformed into a vector and compared with the reference

Figure 8. SFD learning

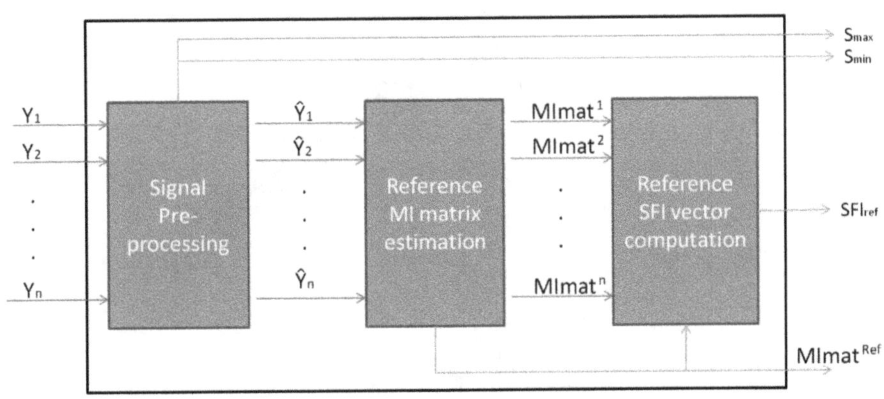

Figure 9. SFD evaluation

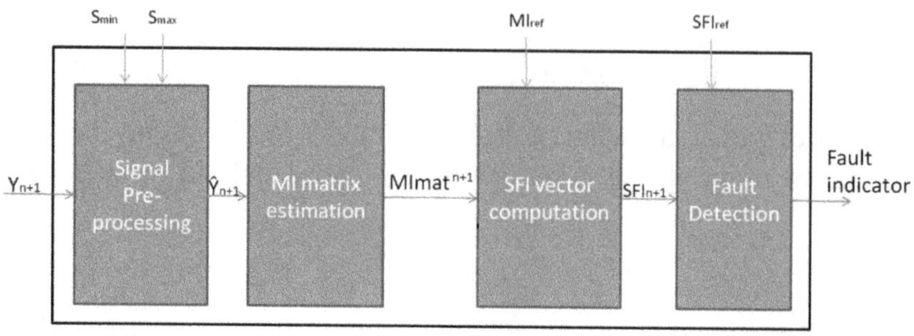

SFI to calculate the fault indicator. This fault indicator is a vector that contains a fault indicator value for each sensor. If the value of the Fault indicator is much bigger than the other values, means that the sensor is faulty.

CASE STUDY

The system was tested in a tower model, similar to those of a wind turbine (see Figure 10). The structure is 2.7m high; the top piece is 1m long and 0.6m width. The tower is composed of three sections joined with bolts, while the jacket is composed with several sections, all of them joined with bolts. The damage is introduced as a crack in one of these sections. As it can be seen in the Figure 10, on the top of the structure a modal shaker is located. This shaker simulates the nacelle mass and the environmental effects of the wind over the whole structure. Applying an electrical signal to the shaker, (white noise) the vibration needed to excite the structure is created. The simulation of different wind speeds is also simulated with this shaker, by changing the amplitude of the input electrical signal. In order to simulate

Figure 10. Tower model and damage

the nacelle orientation changes, the top part of the structure spins 0°, 30°, and 90°; 0° is the basic position shown in Figure 10.

The vibration modes of the structure are important in order to know the behavior of the structure, and how data should be acquired. In this tower, an Operational Modal Analysis was performed in order to have a good quality Finite Element Model. In this case, the first 10 modes are analyzed, which are located under 100 Hz. To simulate different wind speeds, the amplitude of the input wave that goes to the shaker was changed, increasing its amplitude. The shaker has an amplifier, that way the input signal can be amplified before it arrives to the shaker itself. This amplifier has as input the white noise signal, and as output the input signal amplified depending on the gain the user has adopted. For this test, 3 different amplitudes were used: one, two and three. Where one is the signal itself, and three the input signal amplified by means of three.

RESULTS

Sensor Placement

To find the best locations with SEAMAC, a Finite Element Model of the tower is built. Firstly, the best possible solution of the SEAMAC algorithm is found by using the AutoMAC, that puts a sensor in each node and calculates the MAC. That best possible solution is shown in Figure 11, and the sum of the off-diagonal terms is 303%. So, we have to find a value that is more or less the double of this value, reducing the number of sensors in the structure. After running SEAMAC for different amount of sensors, the best result is found using 8 sensors, where the off- diagonal terms sum 607% and the result is shown in Figure 12. Finally, the placement for the selected sensors is shown in Figure 13.

Figure 11. Best possible solution of SEAMAC (AutoMAC)

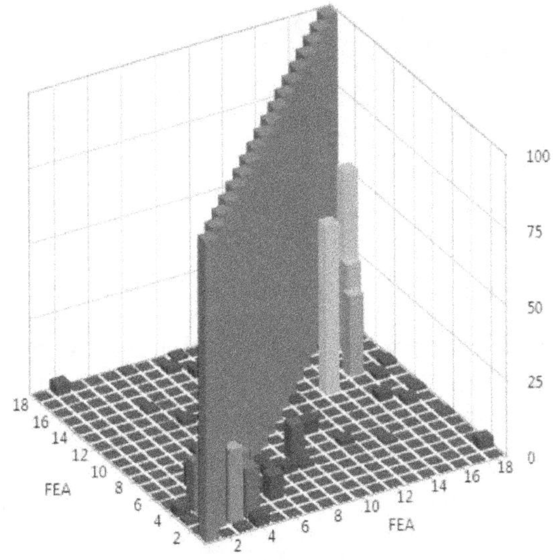

Figure 12. MAC for 8 sensors

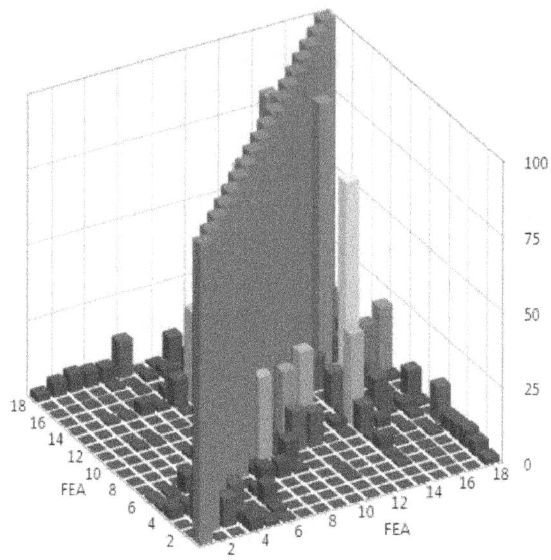

Figure 13. Sensor placement for tower

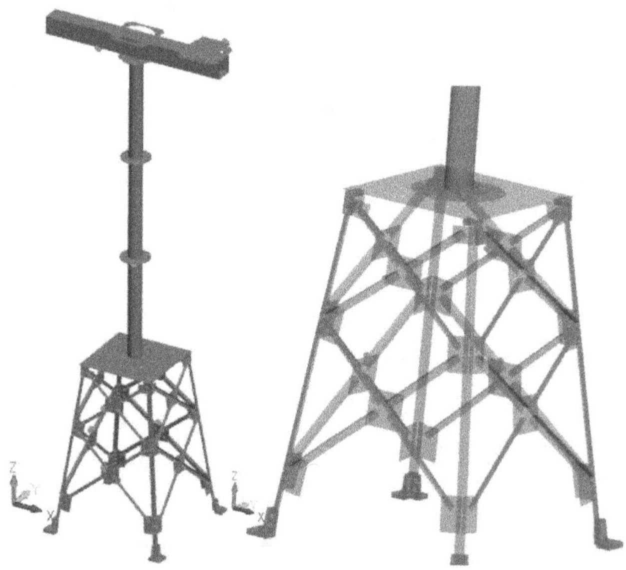

Damage Detection

The damages introduced are real 5mm cracks in a jacket section at different locations (see Figure 10). The dimensions used for the creation of the Hankel matrix are p=q=60. In the learning phase a non-original healthy bar was used as well as original bars. This means that two undamaged versions of a healthy bar were used in the learning phase, one of them the original and a replica, so that the bar change was not detected in the process of damage detection, that way we were focused on the damage, and not to the bar change. Seven different damages were seeded; each damage was named "Dx.y" where "D" means "damage", "x" and "y" represents the position of the damaged bar.

The experiment is performed as follows: first data is acquired while the structure is in healthy mode with the undamaged members. A member is removed and the healthy replica is located in its place, and data is again acquired. After that, the healthy replica member is removed in order to change it with a damaged member, and the data acquisition is performed. Later, the damaged member is removed so as to replace it with the undamaged member, and again, data is acquired. This way the repeatability of the structure is measured. The same procedure is repeated using another member.

Results shown in Figure 14 demonstrated that damages are correctly detected. All green square shape represent Damage Indicators of data from healthy structure, and the other colors belongs to damaged structure in different parts of the substructure.

Feature Selection and Extraction

The preprocessing solution using first the Feature selection for later extracting the features, has shown that only the half of the sensors are needed in order to detect damages correctly. It is clear that not all of the sensors are needed for good damage detection. If the most informative ones are picked out, the damage is well detected. Apart from that, if the features are extracted, the detection becomes easier.

In results with no preprocessing (see Figure 14), the damaged datasets from the upper height (D3.1 and D3.2) where classified as damaged but their values were almost the same as the threshold. On the contrary (see Figure 15), if the preprocessing is used, the gap between the damaged and undamaged is bigger and it is easier to detect the damage.

Figure 14. Damage detection results

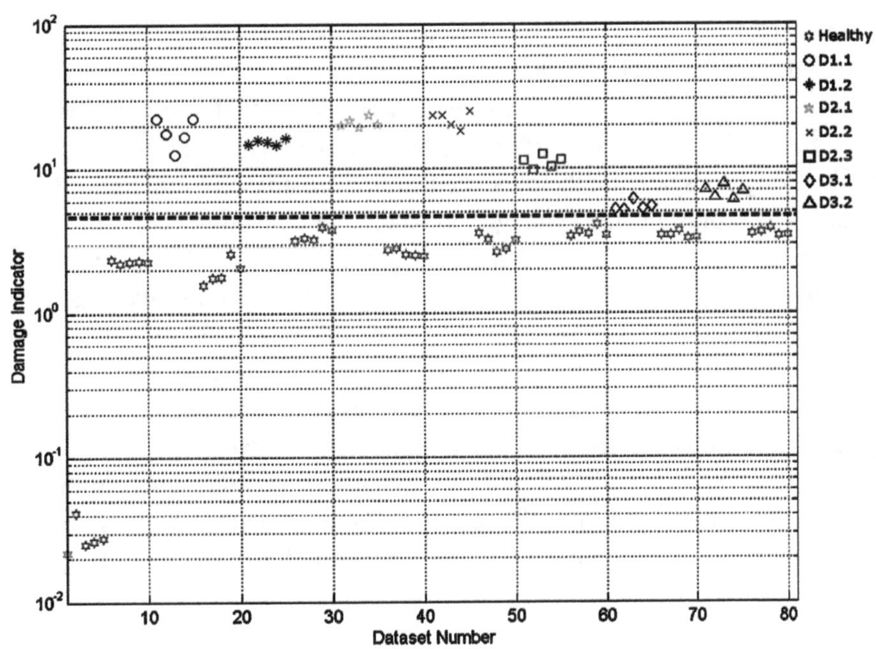

Figure 15. Damage detection results with 10 most important sensors

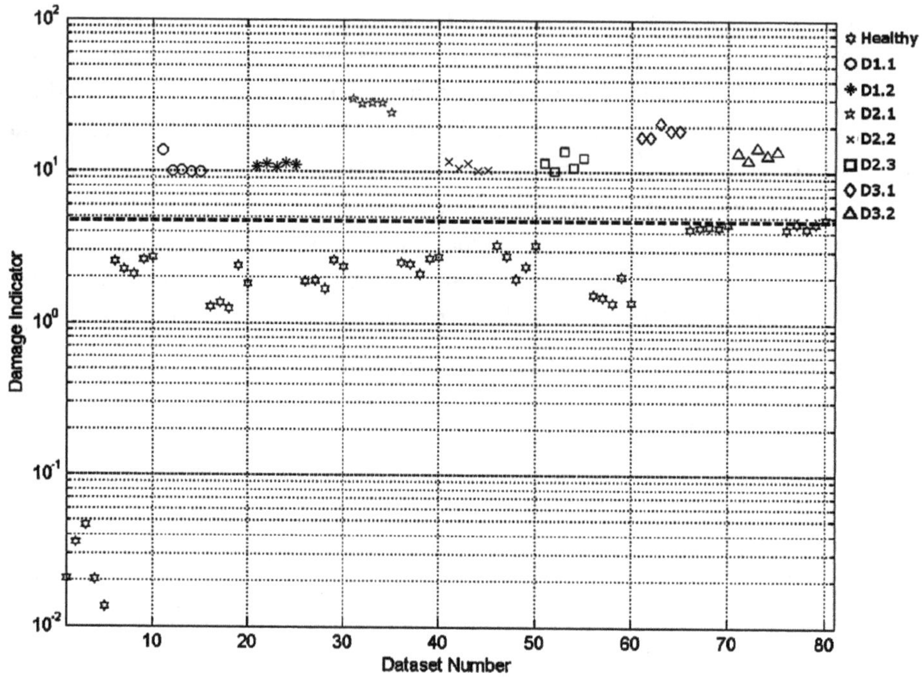

Environmental and Operational Changes Compensation

In this case, only one of the damage locations has been used for the test. This has been the D2.1. Again, the amount of experiments needed for the test, grow with the different EOCs. To simulate the different EOCs, the wind and nacelle orientations were changed. The wind was simulated by changing the amplitude of the shaker signal. For this test, 3 different amplitudes were used. The nacelle orientations (the top part of the structure) used were: 0º, 30º, and 90º. 9 clusters were used, one for each different condition, three winds and three orientations.

Figure 16 shows the Damage Indicator by each dataset using the NullSpace methodology without any type of compensation. The green and dark blue points belong to healthy structure. The green squares were used in the learning phase, while the dark blue stars were not. The red circles belong to damaged structure. Gray dash line represents the Damage Threshold, and the dotted color lines correspond to the mean value of the indicators in each case. The primitive NullSpace is not able to differentiate between the variations due to the EOC and the ones due to the damages. The method is also again more sensitive to the EOC than to the damage itself, therefore damaged and undamaged data are almost in the same level.

On the other hand, Figure 17 shows the damage indicators using the compensation method. The color and shape scheme is the same as in the previous figure, and results are much better than the unclustered method. Every damage is well detected and there are no false positives.

In the unclustered NullSpace, the baseline for every dataset was the same. In this case, in order to compensate the environmental conditions, the baselines for each dataset are different. That is why the method works, but also it is the reason of having big dispersion for the same damage.

Figure 16. Unclustered NullSpace result with EOCs

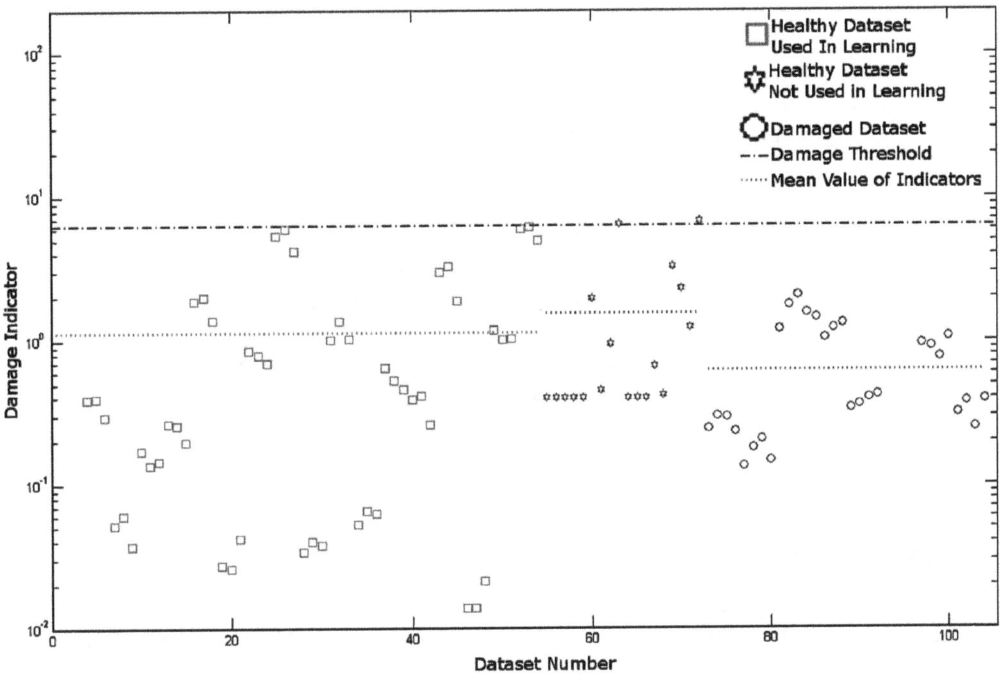

Figure 17. Clustered NullSpace results

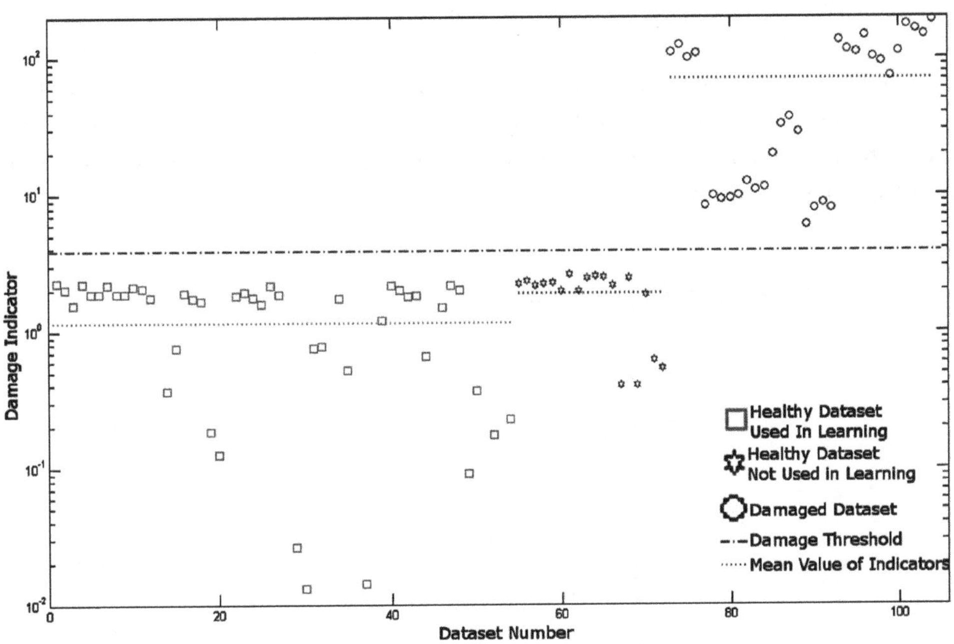

Figure 18. Sensor fault indicator with no faulty sensor

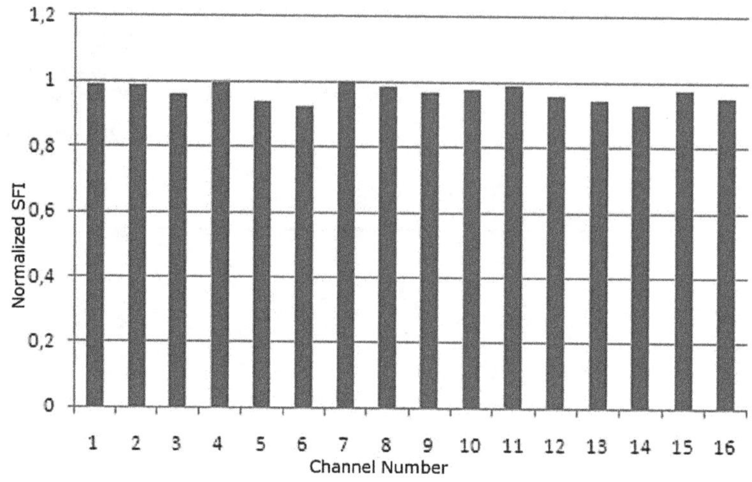

Sensor Fault Detection

To simulate the sensor fault, a reusable putty-like pressure-sensitive adhesive was attached between the sensor and the structure, that way the damping of the sensor increases, and simulates a faulty sensor. The structure was equipped with 16 sensors channels. 5 triaxial sensors and one uniaxial sensor that was placed in the 4th channel. All the damages where seeded in this sensor, as were the only one using only one single channel of the data acquisition system.

Figure 18 shows the result of the Sensor Fault Detection algorithm when all the sensors work correctly. As the SFI value is normalized, when no fault is present all the values are close to one. In other words, the difference between the SFIs of the sensors is really small. In this case, also the uniaxial sensor (4th channel) was attached with a special adhesive. As the Figure 19 shows, the results were satisfactory. When 5mm thick adhesive is placed under the sensor the fault indicator on the faulty sensor is 4 times bigger than the others. More tests were also made attaching thicker and thinner adhesives, and the error gets bigger and smaller, respectively.

Figure 19. Sensor fault indicator, damaged sensor

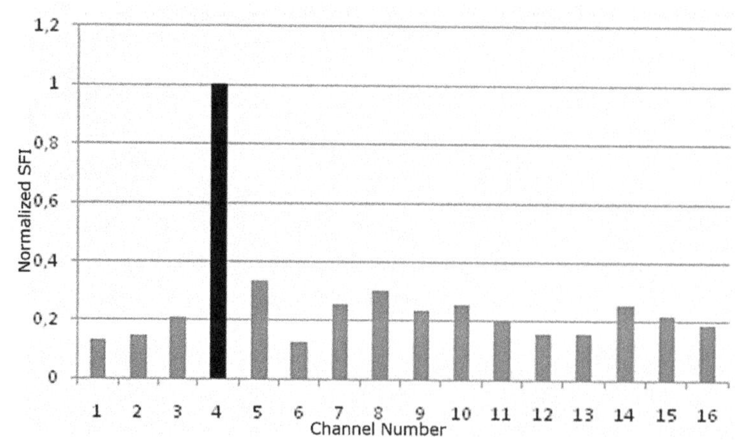

CONCLUSION

The main objective of this work was to develop a complete Smart System for wind turbines damage detection. This objective was achieved and the damage detection methodology has been put into test, with good results. This statement is supported by the fact that they can work properly with any type of structure regardless of whether it is a simple structure or a more complex one, as a wind turbine.

The different subtasks include damage detection, data preprocessing, Environmental and Operational Conditions compensation and sensor fault detection. All subtasks were correctly addressed, and good results were obtained for all of them. Although the inspection was performed under controlled laboratory conditions, the results showed a potential use in real applications. The applicability of the system is enhanced by the following elements:

- **The System Itself:** The methodology can be applied on an operating wind turbine following the steps needed for the damage assessment.
- **The Data Collecting Autonomous System:** It is able to transmit data from anywhere with 3G connections.
- **The Preprocessing Step:** This step allows normalizing, organizing, extracting and selecting the important information of the signals, in order to perform an adequate analysis of the data.
- **The Environmental and Operational Conditions Compensation Step:** The use of Fuzzy clustering methods to classify different statistical models allows obtaining and estimating a baseline model for different operating scenarios.
- **The Use of Damage Indicators:** These indicators allow identifying drifts in the structural behavior to define the presence of damages by comparing the data from different structural states with the undamaged state.
- **The Sensor Checking:** This step checks if the sensors that are used for damage detection are working properly or not. This implementation provides a high level of reliability to the methodology.

This work has contributed to SHM and especially to the data preprocessing area, but many open issues still remain. The following subjects are outlined as future works specifically related to the system developed in this work. Tests with real wind turbines in real conditions: Next step should consist of gathering data from a real operating wind turbine and use it to validate the methodology and confirm whether the results are as good as the ones presented in this work. The most difficult part in this area, besides of the natural challenge of going from laboratory conditions to the real world, is obtaining real data in damaged conditions; EOC: finding the optimum number of clusters: This is a very important phase for a real implementation of solution. It could be possible to make as much as clusters possible, and the results are going to be good, but it wouldn't be the most efficient computationally speaking. The computation in the outer field is not going to be the best, so this could be taken into account, and a way to find the optimum number of clusters should be worked out. Evaluation of different statistical methods: The damage detection method used in this work was the NullSpace damage detection method. In order to have a more secure damage detection system, it could be possible to implement different damage detection methods and integrate them in the methodology, in that way, the solution could have different Damage Indicators coming from different ways of calculating them; Evaluation of preprocessing step with different damage detection methods: The preprocessing step presented is a good way to find the optimum data that enters the damage detection block. This helps to have a solution that is more efficient computationally speaking. In this work, the results are obtained only by using the NullSpace damage detection method. It would be interesting to test the preprocessing solution with other damage detection algorithms and see if the results are still better than the RAW implementation of the algorithm.

REFERENCES

Adams, D., White, J., Rumsey, M., & Farrar, C. (2011). Structural health monitoring of wind turbines: Method and application to a HAWT. *Wind Energy (Chichester, England)*, *14*(January), 603–623. doi:10.1002/we.437

Basseville, M., Abdelghani, M., & Benveniste, A. (2000). Subspace-based fault detection algorithms for vibration monitoring. *Automatica*, *36*(1), 101–109. doi:10.1016/S0005-1098(99)00093-X

Bendat, J., & Piersol, A. (1986). *Random Data - Analysis and Measurement Procedures*. John Wiley & Sons.

Brincker, R. (2006). Understanding stochastic subspace identification. In *Proc. 24th IMAC*.

Chen, R. B., & Wu, Y. N. (2002). *A null-space algorithm for overcomplete independent component analysis*. Learning.

Dunn, J. C. (1973). A Fuzzy Relative of the ISODATA Process and Its Use in Detecting Compact Well-Separated Clusters. *Cybernetics and Systems*, *3*(3), 32–57.

Friedmann, A., Mayer, D., Koch, M., & Siebel, T. (2011). *Monitoring and Damage Detection in Structural Parts of Wind Turbines*. Intechopen. doi:10.5772/21456

Fritzen, C. P. (2005). Vibration - Based Structural Health Monitoring - Concepts and Applications. *Key Engineering Materials*, *293-294*, 3–20. doi:10.4028/www.scientific.net/KEM.293-294.3

Fritzen, C. P. (2005). Vibration - Based Structural Health Monitoring - Concepts and Applications. *Key Engineering Materials*, *293–294*, 3–20. doi:10.4028/www.scientific.net/KEM.293-294.3

Fritzen, C. P., Kraemer, P., & Klinkov, M. (2011). An Integrated SHM Approach for Offshore Wind Energy Plants. *Structural Dynamics*, *3*, 727–740.

Greiner. (2009). *Operational modal analysis and its application for SOFIA telescope assembly vibration measurements*. Academic Press.

Klinkov, M., & Fritzen, C. P. (2011). Wind Load Observer for a 5MW Wind Energy Plant. *Structural Dynamics*, *3*, 719–726.

Kraemer, P. (2011) *Schadensdiagnoseverfahren fr die Zustandsberwachung von Offshore- Windenergieanlagen*. (PhD thesis). Siegen.

Kraemer, P., & Fritzen, C. P. (2011). Aspects of Operational Modal Analysis for Structures of Offshore Wind Energy Plants. *Structural Dynamics and Renewable Energy*, *1*, 145–152. doi:10.1007/978-1-4419-9716-6_14

Kraemer, P., & Fritzen, C. P. (2008). *Damage identification of structural components of offshore wind energy plants.* Paper presented at the 9th German wind energy conference (DEWEK2008).

Mevel, L., Goursat, M., & Basseville, M. (2003, January). Stochastic Subspace-Based Structural Identification and Damage Detection and Localisation-Application To the Z24 Bridge Benchmark. *Mechanical Systems and Signal Processing*, *17*(1), 143–151. doi:10.1006/mssp.2002.1552

Oberholster, A. J., & Heyns, P. S. (2006). On-line fan blade damage detection using neural networks. *Mechanical Systems and Signal Processing*, *20*(1), 78–93. doi:10.1016/j.ymssp.2004.09.007

Pastor, M., Binda, M., & Harcarik, T. (2012, January). Modal Assurance Criterion. *Procedia Engineering*, *48*, 543–548. doi:10.1016/j.proeng.2012.09.551

Rolfes, R., Zerbst, S., & Haake, G. (2007). Integral SHM-System for Offshore Wind Turbines Using Smart Wireless Sensors. *Proceedings of the 6th.*

Shannon, C., & Weaver, W. (1948). The mathematical theory of communication. *Univ. Illinois Press*, *14*(4), 306–317. PMID:9230594

Smarsly, K., Law, K. H., & Hartmann, D. (2011) Implementing a Multiagent-Based Self-Managing Structural Health Monitoring System on a Wind Turbine. In N*SF (National Science Foundation),* 2011 NSF Engineering Research and Innovation Conference. Atlanta, GA.

Toivola, J., & Hollmén, J. (2009). Feature extraction and selection from vibration measurements for structural health monitoring. *Adv. Intelligent Data Analysis*, *8*, 213–224.

Van Overschee & De Moor. (1996). *Subspace Identification for Linear Systems: Theory-Implementation-Applications.* Kluwer Academic Publishers.

Yan, A.-M., & Golinval, J.-C. (2006). Null subspace-based damage detection of structures using vibration measurements. *Mechanical Systems and Signal Processing*, *20*(3), 611–626. doi:10.1016/j.ymssp.2005.04.010

Zugasti, E., Gonzalez, A. G., Anduaga, J., Arregui, M. A., & Martínez, F. (2012, August). NullSpace and AutoRegressive damage detection: A comparative study. *Smart Materials and Structures*, *21*(8), 085010. doi:10.1088/0964-1726/21/8/085010

Zugasti, E., Mujica, L. E., Anduaga, J., & Martínez, F. (2013). *Feature Selection - Extraction Methods based on PCA and Mutual Information to improve Damage Detection problem in Offshore Wind Turbines* (Vol. 579–570, pp. 620–627). Key Eng. Mater.

Compilation of References

Achenbach, J. (2000). Quantitative nondestructive evaluation. *International Journal of Solids and Structures*, *37*(1-2), 13–27. doi:10.1016/S0020-7683(99)00074-8

Adams, D., White, J., Rumsey, M., & Farrar, C. (2011). Structural health monitoring of wind turbines: Method and application to a HAWT. *Wind Energy (Chichester, England)*, *14*(January), 603–623. doi:10.1002/we.437

Aickelin, U., & Dasgupta, D. (2005). Artificial Immune System. In E. Burke & G. Kendall (Eds.), *Search Metodologies: Introductory Tutorials in Optimization and Decision Support Techniques* (pp. 375–399). New York, NY: Springer. doi:10.1007/0-387-28356-0_13

Ainsworth, M., & Oden, J. T. (1997). A posteriori error estimation in finite element analysis. *Computer Methods in Applied Mechanics and Engineering*, *142*(1), 1–88. doi:10.1016/S0045-7825(96)01107-3

American Association of State and Highway Transportation Officials (AASHTO). (2005). *T 277: Standard Method of Test for Electrical Indication of Concrete's Ability to Resist Chloride Ion Penetration*. Washington, DC: American Association of State and Highway Transportation Officials.

American Society for Testing and Materials (ASTM). (1997). *C 1202: Test method for electrical indication concrete's ability to resist chloride ion penetration*. American Society for Testing and Materials.

Anaya, M., Tibaduiza, D., & Pozo, F. (2014). Data driven methodology based on artificial immune systems for damage detection. In *Proceedings of the 7th European Workshop on Structural Health Monitoring*. Nantes, France: Academic Press.

An, Y., & Sohn, H. (2010). Instantaneous crack detection under varying temperature and static loading conditions. *Structural Control and Health Monitoring*, *17*(7), 730–741. doi:10.1002/stc.394

Armitage, P., & Wright, C. (2013). Design, development and testing of multi-functional non-linear ultrasonic instrumentation for the detection of defects and damage in CFRP materials and structures. *Composites Science and Technology*, *87*, 149–156. doi:10.1016/j.compscitech.2013.07.014

Arulampalam, M. S., Maskell, S., Gordon, N., & Clapp, T. (2002). A tutorial on particle filters for online nonlinear/non-Gaussian Bayesian tracking. *Signal Processing. IEEE Transactions on*, *50*(2), 174–188. doi:10.1109/78.978374

Au, S. K., & Beck, J. L. (2001). Estimation of small failure probabilities in high dimensions by Subset Simulation. *Probabilistic Engineering Mechanics*, *16*(4), 263–277. doi:10.1016/S0266-8920(01)00019-4

Aymerich, F., & Staszewski, W. J. (2010). Impact damage detection in composite laminates using nonlinear acoustics. *Composites. Part A, Applied Science and Manufacturing*, *41*(9), 1084–1092. doi:10.1016/j.compositesa.2009.09.004

Babuska, I., & Rheinboldt, W. C. (1978). A posterori error estimates for the finite element method. *International Journal for Numerical Methods in Engineering*, *12*(10), 1597–1615. doi:10.1002/nme.1620121010

Babuska, R. (2012). *Fuzzy Modeling for control*. Netherlands: Springer.

Baby, S., Balassubramanian, T., Pardikar, R. J., Palaniappan, M., & Subbaratnam, R. (2007). *Time of Flight Diffraction Technique for Accurate Sizing of Surface Breaking Cracks*. Institute of Materials and Minig.

Balasko, B., Abonyi, J., & Feil, B. (Eds.). (2005). *Fuzzy Clustering and Data Analysis Toolbox (for Use with Matlab)*. Academic Press.

Baravelli, E., Senesi, M., Ruzzene, M., & De Marchi, L. (2013). Fabrication and Characterization of a Wavenumber-Spiral Frequency-Steerable Acoustic Transducer for Source Localization in Plate Structures. *IEEE Transactions on Instrumentation and Measurement*, *62*(8), 2197–2204. doi:10.1109/TIM.2013.2255992

Barbieri, E., Meo, M., & Polimeno, U. (2009). Nonlinear wave propagation in damaged hysteretic materials using a frequency domain-based PM space formulation. *International Journal of Solids and Structures*, *46*(1), 165–180. doi:10.1016/j.ijsolstr.2008.08.025

Barnard, D. J., Dace, G. E., & Buck, O. (1997). Acoustic Harmonic Generation Due to Thermal Embrittlement of Inconel 718. *Journal of Nondestructive Evaluation*, *16*(2), 67–75. doi:10.1007/BF02683879

Barnard, D. J., Dace, G. E., Rehbein, D., & Buck, O. (1997). Acoustic harmonic generation at diffusion bonds. *Journal of Nondestructive Evaluation*, *16*(2), 77–89. doi:10.1007/BF02683880

Basseville, M., Abdelghani, M., & Benveniste, A. (2000). Subspace-based fault detection algorithms for vibration monitoring. *Automatica*, *36*(1), 101–109. doi:10.1016/S0005-1098(99)00093-X

Beck, J. (1989). Statistical system identification of structures. In *Fifth International Conference on Structural safety and reliability*, (pp. 1395–1402). New York: ASCE.

Beck, J. (2010). Bayesian system identification based on probability logic. *Structural Control and Health Monitoring*, *17*(7), 825–847. doi:10.1002/stc.424

Begambre, O. J. (2004). *Detecção de dano a partir da resposta dinâmica da estrutura: estudo analítico com aplicação a estruturas do tipo viga*. (Master dissertation). Retrieved from http://web.set.eesc.usp.br/producao/ \

Bendat, J., & Piersol, A. (1986). *Random Data - Analysis and Measurement Procedures*. John Wiley & Sons.

Bender, F., Kim, J.-Y., Jacobs, L., & Qu, J. (2013). The generation of second harmonic waves in an isotropic solid with quadratic nonlinearity under the presence of a stress-free boundary. *Wave Motion*, *50*(2), 146–161. doi:10.1016/j.wavemoti.2012.08.009

Bentley, J. P. (1993). *Sistemas de Medición. Compañía Editorial Continental*. S.A. Mexico.

BERRAIS, A. (1999). Artificial Neural Networks in Structural Engineering: Concept and Applications. *Engineering and Science*, *12*(1), 53–67.

Best, S., Croxford, A., & Neild, S. (2013). Pulse-Echo Harmonic Generation Measurements for Non-destructive Evaluation. *Journal of Nondestructive Evaluation*, *32*(4), 1–11.

Binitha, S., & Siva Sathya, S. (2012). A survey of bio-inspired optimization algorithms. *International Journal of Soft Computing and Engineering*, *2*, 137-151.

Blaise, E. J., & Chang, F. K. (2001). Built-in Damage Detection System for Sandwich Structures under Cryogenic Temperatures. In *Proceedings ofThird International Workshop on Structural Health Monitoring*, (pp. 154–63). Academic Press.

Bochud, N. (2014). *Signal processing-based identification of pathology using ultrasonics.* (Unpublished doctoral dissertation). University of Granada, Granada, Spain.

Boer F., Bonsague, M., & Van Buuren R. (2007). *Concepts for Architectural Description.* Telematica Instituut / Archimate Consortium.

Bossuat, B., Walaszeck, & Falvenot, J. (2006). *Non Destructive Evaluation by Time of Flight Diffraction.* Method Mechanical Applications, ECNDT.

Breazeale, M., & Thompson, D. (1963). Finite-Amplitude ultrasonic waves in aluminum. *Applied Physics Letters, 3*(5), 77–78. doi:10.1063/1.1753876

Brincker, R. (2006). Understanding stochastic subspace identification. In *Proc. 24th IMAC.*

Broda, D., Staszewski, W. J., Martowicz, A. U., & Silberschmidt, V. V. (2014). Modelling of nonlinear crack-wave interactions for damage detection based on ultrasound-A review. *Journal of Sound and Vibration, 333*(4), 1097–1118. doi:10.1016/j.jsv.2013.09.033

Bruno, C. L., Gliozzi, A. S., Scalerandi, M., & Antonaci, P. (2009). Analysis of elastic nonlinearity using the scaling subtraction method. *Physical Review B: Condensed Matter and Materials Physics, 79*(6), 1–13. doi:10.1103/PhysRevB.79.064108

Bucher, C. G. (1988). Adaptive sampling—an iterative fast Monte Carlo procedure. *Structural Safety, 5*(2), 119–126. doi:10.1016/0167-4730(88)90020-3

Caflisch, R. E. (1998). Monte carlo and quasi-monte carlo methods. *Acta Numerica, 1998*, 1–49. doi:10.1017/S0962492900002804

Caicedo, J. M. (2011). Practical guidelines for the natural excitation technique (NexT) and the eigensystem realization algorithm (ERA) for modal identification using ambient vibration. *Experimental Techniques, 35*(4), 52–58. doi:10.1111/j.1747-1567.2010.00643.x

Calvetti, D., & Somersalo, E. (2007). *Introduction to Bayesian Scientific Computing: Ten Lectures on Subjective Computing*. Springer.

Camacho, J. (2010). *Sistema experto para la monitorización de salud estructural mediante el reconocimiento de patrones: adaptación y validación numérica.* (Master dissertation). Retrieved from http://tangara.uis.edu.co/biblioweb/tesis/2010/136485.pdf

Camacho, J., & Villamizar, R. (2010). Adaptación del razonamiento basado en casos al problema de monitorización de salud estructural. Memorias Semana Técnica de Ingenierías eléctrica, electrónica y afines STIEEA.

Cantrell, J. H., & Yost, W. T. (1994). Acoustic harmonic generation from fatigue-induced dislocation dipoles. *Philosophical Magazine A, 69*(2), 315–326. doi:10.1080/01418619408244346

Cantrell, J., Thompson, D. O., & Chimenti, D. E. (2009). Nondestructive evaluation of metal fatigue using nonlinear acoustics. *AIP Conference Proceedings, 1096*(1), 19–32. doi:10.1063/1.3114205

Cappé, O., Godsill, S. J., & Moulines, E. (2007). An overview of existing methods and recent advances in sequential Monte Carlo. *Proceedings of the IEEE, 95*(5), 899–924. doi:10.1109/JPROC.2007.893250

Carden, E. P., & Fanning, P. (2004). Vibration based condition monitoring: A review. *Structural Health Monitoring, 3*(4), 355–377. doi:10.1177/1475921704047500

Celaya, J., Saxena, A., & Goebel, K. (2012). *Uncertainty Representation and Interpretation in Model-based Prognostics Algorithms based on Kalman Filter Estimation.* Paper presented at the Proceedings of the Annual Conference of the PHM Society.

Chakrapani, S. K., Barnard, D. J., & Vinay, D. (2015). Influence of fiber orientation on the inherent acoustic nonliearity in carbon fiber reinforced composites. *The Journal of the Acoustical Society of America*, *137*(2), 617–624. doi:10.1121/1.4906165 PMID:25697996

Chang, F.-K., Markmiller, J. F. C., Ihn, J.-B., & Cheng, K. Y. (2007). A Potential Link from Damage Diagnostics to Health Prognostics of Composites Through Built-in Sensors. *Transaction of the ASME*, *129*(December). doi:10.1115/1.2730530

Chang, O., & Sohn, H. (2009). Damage diagonosis under environmental and operational variations using unsupervised support vector machine. *Journal of Sound and Vibration*, *325*(1-2), 224–239. doi:10.1016/j.jsv.2009.03.014

Charlesworth, J. P. (2001). *Engineering Applications of Ultrasonic Time of Flight Diffraction.* Research Studies Press LTD.

Chen, B. (2010). Agent-based artificial immune system approach for adaptative damage detection in monitoring network. *Journal of Network and Computer Applications*, *33*(6), 633–645. doi:10.1016/j.jnca.2010.03.011

Chen, B., & Zang, C. (2009). Artificial immune pattern recognition for structure damage classification. *Computers & Structures*, *87*(21-22), 1394–1407. doi:10.1016/j.compstruc.2009.08.012

Chen, B., & Zang, C. (2011). Emergent damage pattern recognition using immune network theory. *Smart Structures and Systems*, *8*(1), 69–92. doi:10.12989/sss.2011.8.1.069

Chen, B., Zhao, S. L., & Li, P. Y. (2014). Application of Hilbert-Huang Transform in Structural Health Monitoring: A State-of-the-Art Review. *Mathematical Problems in Engineering.*

Chen, R. B., & Wu, Y. N. (2002). *A null-space algorithm for overcomplete independent component analysis.* Learning.

Cheon, Y.-M., Alam, M., & Kim, C. (2010). Nonlinear parameters for diagnosis of microscale cracks using a nonlinear resonant ultrasound spectroscopy (NRUS). *AIP Conference Proceedings*, *1211*(1), 1439–1444. doi:10.1063/1.3362237

Chiachio, M., Chiachio, J., Saxena, A., & Goebel, K. (2014). An efficient simulation framework for prognostics of asymptotic processes- a case study in composite materials. *2nd European Conference of the Prognostics and Health Management Society, Nantes, France2014. EPHM 20141.*

Chiachío, J., Chiachío, M., Saxena, A., Shankararaman, S., & Goebel, K. (2015). Bayesian model selection and parameter estimation for fatigue damage progression models in composites. *International Journal of Fatigue*, *70*, 361–373. doi:10.1016/j.ijfatigue.2014.08.003

Chilengue, Z., Dente, J., & Costa Branco, P. (2011). An artificial immune system approach for fault detection in the stator and rotor circuits of induction machines. *Electric Power Systems Research*, *81*(1), 158–169. doi:10.1016/j.epsr.2010.08.003

Ching, J., Beck, J., & Porter, K. (2006). Bayesian state and parameter estimation of uncertain dynamical systems. *Probabilistic Engineering Mechanics*, *21*(1), 81–96. doi:10.1016/j.probengmech.2005.08.003

Chou, J. H., & Ghaboussi, J. (2001). Genetic algorithm in structural damage detection. *Computers & Structures*, *79*(14), 1335–1353. doi:10.1016/S0045-7949(01)00027-X

Chu, T. Y. (1978). *Radiant heat evaluation of concrete—A study of the erosion of concrete due to surface heating.* Research paper SAND 77-0922. Sandia Laboratories.

Ciampa, F., & Meo, M. (2010). A new algorithm for acoustic emission localization and flexural group velocity determination in anisotropic structures. *Composites. Part A, Applied Science and Manufacturing*, *41*(12), 1777–1786. doi:10.1016/j.compositesa.2010.08.013

Ciampa, F., Meo, M., & Barbieri, E. (2012). Impact localization in composite structures of arbitrary cross section. *Structural Health Monitoring*, *11*(6), 643–655. doi:10.1177/1475921712451951

Cleveland, R., Hamilton, M., & Blackstock, D. (1996). Time-domain modeling of finite-amplitude waves in relaxing fluids. *The Journal of the Acoustical Society of America*, *99*(6), 3312–3318. doi:10.1121/1.414983

Coelho, C. K., Hiche, C., & Chattopadhyay, A. (2010). Impact Localization and Force Estimation on a Composite Wing using Fiber Bragg Gratings Sensors. In *Proceedings of 51st AIAA/ASME/ASCE/AHS/ASC Structures, Structural Dynamics, and Materials Conference*. AIAA.

Cohen, L. (1995). *Time-Frequency Analisys*. Prantice Hall.

Contreras, M., Vergara, C., Pereira, M., Colonia, J. D., & García, H. (2013). Pipeline integrity assessment and mitigation techniques in unstable soil conditions based on comparison of in line inspection results. *ASME 2013 International Pipeline Geotechnical Conference*. Pipeline Systems Division.

Costa Branco, P. J., Dente, J. A., & Vilela Mendes, R. (2003). Using immunology principles for fault detection. *Industrial electronics. IEEE Transactions on*, *50*, 362–373.

Courtney, C. R., Drinkwater, B. W., Neild, S. A., & Wilcox, P. D. (2008). Factors affecting the ultrasonic intermodulation crack detection technique using bispectral analysis. *NDT & E International*, *41*(3), 223–234. doi:10.1016/j.ndteint.2007.09.004

Cross, E., & Worden, K. (2012). Cointegration and why it works for SHM. *Journal of Physics: Conference Series*, 382.

Cross, E., Worden, K., & Chen, Q. (2011). Cointegration: A novel approach for the removal of environmental trends in structural health monitoring data. *Proceedings of the Royal Society of London. Series A*, *467*(2133), 2712–2732. doi:10.1098/rspa.2011.0023

Crouch, L. K., Hewitt, R., & Byard, B. (2007). *High Volume Fly Ash Concrete, Proceedings of World of Coal Ash* (pp. 1–14). WOCA.

Croxford, A. J., Wilcox, P. D., Drinkwater, B. W., & Nagy, P. B. (2009). The use of non-collinear mixing for nonlinear ultrasonic detection of plasticity and fatigue. *The Journal of the Acoustical Society of America*, *126*(5), 117–122. doi:10.1121/1.3231451 PMID:19894785

Cruz Cortes, N. (2004). *Sistema Inmune Artificial para Solucionar Problemas de Optimización.* (PhD thesis). Cinvestav.

Daigle, M., Saxena, A., & Goebel, K. (2012). *An Efficient Deterministic Approach to Model-based Prediction Uncertainty Estimation*. Paper presented at the Annual Conference of the Prognostics and Health Management Society.

Daigle, M. J., & Goebel, K. (2013). Model-based prognostics with concurrent damage progression processes. *Systems, Man, and Cybernetics: Systems. IEEE Transactions on*, *43*(3), 535–546.

Daigle, M., Sankararaman, S., & Kulkarni, C. (2015). Stochastic Prediction of Remaining Driving Time and Distance for a Planetary Rover. In *2015 IEEE Aerospace Conference*, Big Sky, MT. doi:10.1109/AERO.2015.7119144

Dalai, M., Weyer, E., & Campi, M. (2005). Parametric Identification of Nonlinear Systems : Guaranteed Confidence Regions. In *44th IEEE Conference on Decision and Control, and the European Control Conference*. Seville, Spain: IEEE. doi:10.1109/CDC.2005.1583191

De Carvalho, J., Tita V.,& Lirani, J. (2003).Theoretical and experimental dynamic analysis of fiber reinforced composite beams. *Journal of the Brazilian Society of Mechanical Sciences and Engineering, 25*(3), 306-310.

de Castro, L. N., & Von Zuben, F. J. (1999). Artificial immune systems: Part I basic theory and applications. Campina, Brazil: Technical report. School of Computing and Electrical Engineering. State University of Campinas. Brazil, No. DCA-RT 01/99.

de Castro, L. N., & Timmis, J. (2002). Artificial Immune Systems: A Novel Approach to Pattern Recognition. In J. Corchado, L. Alonso, & C. Fyfe (Eds.), *Artificial Neural Networks in Pattern Recognition* (pp. 67–84). Paisley: University of Paisley.

Dehghan Niri, E., Farhidzadeh, A., & Salamone, S. (2012). Adaptive multisensor data fusion for acoustic emission source localization in noisy environment. *Structural Health Monitoring, 12*(1), 59–77. doi:10.1177/1475921712462937

Delsanto, P., Hirsekorn, S., Agostini, V., Loparco, R., & Koka, A. (2002). Modeling the propagation of ultrasonic waves in the interface region between two bonded elements. *Ultrasonics, 40*(1-8), 605–610. doi:10.1016/S0041-624X(02)00183-X PMID:12160009

Delves, P. J., Martin, S. J., Burton, D. R., & Roitt, I. M. (2011). *Roitt's essential immunology.* Wiley-Blackwell.

Der Kiureghian, A., Lin, H.-Z., & Hwang, S.-J. (1987). Second-order reliability approximations. *Journal of Engineering Mechanics, 113*(8), 1208–1225. doi:10.1061/(ASCE)0733-9399(1987)113:8(1208)

Deraemaeker, A., Reynders, E., De Roeck, G., & Kullaa, J. (2008). Vibration based structural health monitoring using output only measurements under changing environment. *Mechanical Systems and Signal Processing, 22*(1), 34–56. doi:10.1016/j.ymssp.2007.07.004

Doebling, S., Farrar, C., & Prime, M. (1998). A summary review of vibration-based damage identification methods. *The Shock and Vibration Digest, 30*(2), 91–105. doi:10.1177/058310249803000201

Dolinski, K. (1983). First-order second-moment approximation in reliability of structural systems: Critical review and alternative approach. *Structural Safety, 1*(3), 211–231. doi:10.1016/0167-4730(82)90027-3

Donskoy, D. M., & Sutin, A. M. (1998). Vibro-acoustic modulation nondestructive evaluation technique. *Journal of Intelligent Material Systems and Structures, 9*(9), 765–771. doi:10.1177/1045389X9800900909

Donskoy, D., Sutin, A., & Ekimov, A. (2001). Nonlinear acoustic interaction on contact interfaces and its use for non-destructive testing. *NDT & E International, 34*(4), 231–238. doi:10.1016/S0963-8695(00)00063-3

Döring, C., Lesot, M. J., & Kruse, R. (2006). Data Analysis with Fuzzy Clustering Methods. *Computational Statistics & Data Analysis, 51*(1), 192–214. doi:10.1016/j.csda.2006.04.030

Doucet, A., De Freitas, N., & Gordon, N. (2001). *Sequential Monte Carlo methods in practice.* Springer Verlag. doi:10.1007/978-1-4757-3437-9

Drewry, M., Smith, R., Phang, A., Yan, D., Wilcox, P., & Roach, D. (2009). Ultrasonic techniques for detection of weak adhesion. *Materials Evaluation, 67*(9), 1048–1058.

Dubuget, M., El Guerjouma, R., Dubois, S., Baboux, J., & Vincent, A. (1996). Characterization of the non-linear elastic properties of aluminium alloys using ultrasonic evaluation under load. *Materials Science Forum, 217-222*, pp. 951-956.

Dunn, J. C. (1973). A Fuzzy Relative of the ISODATA Process and Its Use in Detecting Compact Well-Separated Clusters. *Cybernetics and Systems, 3*(3), 32–57.

Dwight, R. P., & Han, Z.-H. (2009). *Efficient uncertainty quantification using gradient-enhanced kriging*. Paper presented at the 11th AIAA Non-Deterministic Approaches Conference, Reston, VA.

Dyke, S. J., Bernal, D., Beck, J., & Ventura, C. (2003). Experimental phase II of the structural health monitoring benchmark problem. In *Proceedings of the 16th ASCE engineering mechanics conference*.

Dyke, S., & Caicedo, J. (2000). *Monitoring of a Benchmark Structure for Damage Identification. Department of Civil Engineering*. St Louis, MO: Washingtong University.

Edvardsen, C. (1999). Water permeability and Autogenous Healing of Cracks in Concrete. *ACI Materials Journal, 96*, 448–455.

Edwin, F., & Rodríguez, J. A. (2008). Procesamiento Digital de Señales TOFD. *Journal, STSIVA, 1*, 93–97.

Ekimov, A., Didenkulov, I., & Kazakov, V. (1999). Modulation of torsional waves in a rod with a crack. *Journal of the Acoustical Society of America, 106*(3–1), 1289.

Ellwood, R., Stratoudaki, T., Sharples, S., Clark, M., & Somekh, M. (2014). Determination of the acoustoelastic coefficient for surface acoustic waves using dynamic acoustoelastography: An alternative to static strain. *The Journal of the Acoustical Society of America, 135*(3), 1064–1070. doi:10.1121/1.4864308 PMID:24606250

Elsadig, M., Adbullah, A., & Samir, B. (2010). Intrusion prevention and self-healing algorithms inspired by danger theory. In *Proceedings of 2nd international conference on computer and automation engineering (ICCAE)*. Singapore, Singapore: IEEE Press.

Engel, S. J., Gilmartin, B. J., Bongort, K., & Hess, A. (2000). *Prognostics, the real issues involved with predicting life remaining*. Paper presented at the Aerospace Conference Proceedings. IEEE. doi:10.1109/AERO.2000.877920

Eroglu, A., Erden, Z., & Erden, A. (2011). Bioinspired conceptual design (bicd) approach for hybrid bioinspired robot design process. In *Proceedings of Mechatronics (ICM), 2011 IEEE International Conference on* (pp. 905-910). IEEE.

Escobar-Ruiz, E., Ruiz, A., Hassan, W., Wright, D. C., Collison, I. J., Cawley, P., & Nagy, P. B. (2014). Non-linear Ultrasonic NDE of Titanium Diffusion Bonds. *Journal of Nondestructive Evaluation, 33*(2), 187–195. doi:10.1007/s10921-013-0217-5

Fan, W., & Qiao, P. (2011). Vibration-based damage identification methods: A review and comparative study. *Structural Health Monitoring, 10*(1), 83–111. doi:10.1177/1475921710365419

Farim, A., Gallego, R., Bochud, N., & Rus, G. (2013). Model-based damage reconstructionin compositesfrom ultrasound transmission. *Composites. Part B, Engineering, 45*(1), 50–62. doi:10.1016/j.compositesb.2012.09.003

Farrar, C. R., & Lieven, N. A. (2007). Damage prognosis: the future of structural health monitoring. *Philosophical Transactions of the Royal Society A: Mathematical, Physical and Engineering Sciences, 365*(1851), 623-632.

Farrar, C. R., & Worden, K. (2007). An introduction to structural health monitoring. *Philosophical Transactions of the Royal Society A: Mathematical, Physical and Engineering Sciences, 365*(1851), 303–315.

Farrar, C. R., Doebling, S. W., & Nix, D. A. (2001). Vibration-based structural damage identification. *Philosophical Transactions: Mathematical, Physical & Engineering Sciences, 359*(1778), 131–149.

Fawaz, S. A. (2003). Equivalent initial flaw size testing and analysis of transport aircraft skin splices. *Fatigue & Fracture of Engineering Materials & Structures, 26*(3), 279–290. doi:10.1046/j.1460-2695.2003.00637.x

Fayyadh, M. M., & Razak, H. A. (2011). Modified damage location indices in beam-like structure: Analytical study. *Scientific Research and Essays, 6*(32), 6605–6621.

Figuiredo, E., Todd, M., Farrar, R., & Flynn, E. (2010). Autoregressive modeling with state-space embedding vectors for damage detection Ander operacional variability. *International Journal of Engineering Science*, *48*(10), 822–834. doi:10.1016/j.ijengsci.2010.05.005

Fink, M. (1997). Time reversed acoustics. *Physics Today*, *50*(3), 34–40. doi:10.1063/1.881692

Friedmann, A., Mayer, D., Koch, M., & Siebel, T. (2011). *Monitoring and Damage Detection in Structural Parts of Wind Turbines*. Intechopen. doi:10.5772/21456

Fritzen, C. P. (2005). Vibration - Based Structural Health Monitoring - Concepts and Applications. *Key Engineering Materials*, *293-294*, 3–20. doi:10.4028/www.scientific.net/KEM.293-294.3

Fritzen, C. P., Kraemer, P., & Klinkov, M. (2011). An Integrated SHM Approach for Offshore Wind Energy Plants. *Structural Dynamics*, *3*, 727–740.

Fu, H., & Xu, Q. (2013). Locating impact on structural plate using principal component analysis and support vector machines. *Mathematical Problems in Engineering*, article ID 352149. doi: .10.1155/2013/352149

Fukuda, M., Nishihira, M., & Imano, K. (2007). Real Time Detection of Second-Harmonic Components Generated from Plastic-Deformed Metal Rod Using Double-Layered Piezoelectric Transducer. *Japanese Journal of Applied Physics*, *46*(7B), 4529–4531. doi:10.1143/JJAP.46.4529

Gang, T., & Chi, D. (2007). Nobel Approach to Enhancement of Ultrasonic, TOFD B-Scan, image for measurement of weld crack. *Science and Technology of Welding and Joining*, *12*(1), 87–92. doi:10.1179/174329307X159847

García, A., Desimone, C., & Katchadjian, P. (2007). *Caracterización de equipos de END. In IV Conferencia Panamericana de END*. Buenos Aíres: AAENDE.

Garrett, K., & Bailey, J. (1977). Multiple transverse fracture in Latex translation cross-ply laminates of a glass fibre-reinforced polyester. *Journal of Materials Science*, *12*(1), 157–168. doi:10.1007/BF00738481

Gentile, A., & Messina, A. (2003). On the continuous wavelet transform applied to discrete vibrational data for detecting open cracks in damaged beams. *International Journal of Solids and Structures*, *40*(2), 295–315. doi:10.1016/S0020-7683(02)00548-6

Georgali, B., & Tsakiridis, P. E. (2005). Microstructure of fire-damaged concrete. A case study. *Cement and Concrete Composites*, *27*(2), 255–259. doi:10.1016/j.cemconcomp.2004.02.022

Gharibnezhad, F., Mujica, L. E., & Rodellar, J. (2011). Damage Detection Using Andrew Plots. In *Proceedings of the 8th International Workshop on Structural Health Monitoring*. Stanford University.

Gharibnezhad, F., Mujica, L. E., & Rodellar, J. (2014). Applying Robust Variant of Principal Component Analysis as a Damage Detector in Presence of Outliers. *Mechanical Systems and Signal Processing*, *50-51*, 467–479. doi:10.1016/j.ymssp.2014.05.032

Ginsberg, J. H. (2008). Perturbation Methods. In M. H. Hamilton & D. T. Blackstock (Eds.), *Nonlinear Acoustics* (pp. 279–308). San Diego, CA: Academic Press.

Gliozzi, A., Griffa, M., & Scalerandi, M. (2006). Efficiency of time-reversed acoustics for nonlinear damage detection in solids. *The Journal of the Acoustical Society of America*, *120*(5), 2506–2517. doi:10.1121/1.2345955

Glynn, P. W., & Iglehart, D. L. (1989). Importance sampling for stochastic simulations. *Management Science*, *35*(11), 1367–1392. doi:10.1287/mnsc.35.11.1367

Goebel, K. (2014). PHM Design Process. NASA Ames Research Center.

Gordon, N., Salmond, D., & Smith, A. (1993). Novel approach to nonlinear/non-Gaussian Bayesian state estimation. *IEEE-Proceedings-F*, *140*, 107–113.

Goursolle, T., Callé, S., Dos Santos, S., & Bou Matar, O. (2007). A two-dimensional pseudospectral model for time reversal and nonlinear elastic wave spectroscopy. *The Journal of the Acoustical Society of America*, *122*(6), 3220–3229. doi:10.1121/1.2799900 PMID:18247734

Greiner. (2009). *Operational modal analysis and its application for SOFIA telescope assembly vibration measurements*. Academic Press.

Grell, W. A., & Laz, P. J. (2010). Probabilistic fatigue life prediction using AFGROW and accounting for material variability. *International Journal of Fatigue*, *32*(7), 1042–1049. doi:10.1016/j.ijfatigue.2009.12.001

Gudmundson, P., & Weilin, Z. (1993). An analytic model for thermoelastic properties of composite laminates containing transverse matrix cracks. *International Journal of Solids and Structures*, *30*(23), 3211–3231. doi:10.1016/0020-7683(93)90110-S

Guise, S. E. (1999). Petrographic and color analysis for assessment of fire damaged concrete. In *Proceedings of the 19th International Conference on Cement Microscopy*.

Gusev, V. (2000). Propagation of acoustic pulses in material with hysteretic nonlinearity. *The Journal of the Acoustical Society of America*, *107*(6), 3047–3058. doi:10.1121/1.429333 PMID:10875350

Gustafero, A. H. (1983). *Experiences from evaluating fire-damaged concrete structures – Fire safety of concrete structures*. American Concrete Institute.

Guyan, R. J. (1965). Reduction of stiffness and mass matrices. AIAA Journal, 3(2), 380.

Guyer, R., & Johnson, P. (1999). Nonlinear mesoscopic elasticity: Evidence for a new class of materials. *Physics Today*, *52*(4), 30–36. doi:10.1063/1.882648

Guyer, R., McCall, K., & Van Den Abeele, K. (1998). Slow elastic dynamics in a resonant bar of rock. *Geophysical Research Letters*, *25*(10), 1585–1588. doi:10.1029/98GL51231

Haldar, A., & Mahadevan, S. (2000). *Probability, reliability, and statistical methods in engineering design*. John Wiley & Sons, Incorporated.

Hallaj, I. M., & Cleveland, R. O. (1999). FDTD simulation of finite-amplitude pressure and temperature fields for biomedical ultrasound. The Journal of the Acoustical Society of America. *The Journal of the Acoustical Society of America*, *105*(5), L7–L12. doi:10.1121/1.426776 PMID:10335650

Hamilton, M. F., & Blackstock, D. T. (1998). *Nonlinear acoustics* (Vol. 237). San Diego, CA: Academic press.

Handoo, S. K., Agarwal, S., & Agarwal, S. K. (2002). Physicochemical, mineralogical, and morphological characteristics of concrete exposed to elevated temperatures. *Cement and Concrete Research*, *32*(7), 1009–1018. doi:10.1016/S0008-8846(01)00736-0

Harley, J. B., & Moura, J. M. F. (2012). Scale Transform Signal Processing for Optimal Ultrasonic Temperature Compensation. *Ultrasonics, Ferroelectrics and Frequency Control. IEEE Transactions on*, *59*(10), 2226–2236.

Hashin, Z. (1985). Analysis of cracked laminates: A variational approach. *Mechanics of Materials*, *4*(2), 121–136. doi:10.1016/0167-6636(85)90011-0

Hatch, M. R. (2001). *Vibration simulation using Matlab and Ansys*. Washington, DC: Chapman & Hall/CRC.

Haupert, S., Renaud, G., Riviere, J., Talmant, M., Johnson, P. A., & Laugier, P. (2011). High-accuracy acoustic detection of nonclassical component of material nonlinearity. *The Journal of the Acoustical Society of America*, *130*(5), 2654–2661. doi:10.1121/1.3641405 PMID:22087892

Haupert, S., Rivière, J., Anderson, B., Ohara, Y., Ulrich, T., & Johnson, P. (2014). Optimized dynamic acousto-elasticity applied to fatigue damage and stress corrosion cracking. *Journal of Nondestructive Evaluation*, *33*(2), 226–238. doi:10.1007/s10921-014-0231-2

Haywood, J., Coverley, P. T., Staszewski, W. J., & Worden, K. (2005). An automatic impactmonitor for a composite panel employing smart sensor technology. *Smart Materials and Structures*, *14*(1), 265–271. doi:10.1088/0964-1726/14/1/027

Highsmith, A., & Reifsnider, K. (1982). Stiffness-reduction mechanisms in composite laminates. *Damage in composite materials. ASTM Special Technical Publication*, *775*, 103–117.

Hirsekorn, S., Rabe, U., & Arnold, W. (2006). Characterization and evaluation of composite laminates by nonlinear ultrasonic transmission measurements.*Proceedings of the 9th European Conference on NDT (ECNDT 2006), CD-ROM BB-103.*Berlin, Germany.

Hohenbichler, M., & Rackwitz, R. (1983). First-order concepts in system reliability. *Structural Safety*, *1*(3), 177–188. doi:10.1016/0167-4730(82)90024-8

Hohenbichler, M., & Rackwitz, R. (1988). Improvement of second-order reliability estimates by importance sampling. *Journal of Engineering Mechanics*, *114*(12), 2195–2199. doi:10.1061/(ASCE)0733-9399(1988)114:12(2195)

Holcomb, D. J. (1981). Memory, relaxation, and microfracturing in dilatant rock. *Journal of Geophysical Research: Solid Earth (1978-2012)*, *86*(B7), 6235-6248.

Hombal, V. K., Ling, Y., Wolfe, K. A., & Mahadevan, S. (2012). Two-stage planar approximation of non-planar crack growth. *Engineering Fracture Mechanics*, *96*, 147–164. doi:10.1016/j.engfracmech.2012.07.017

Hombal, V. K., & Mahadevan, S. (2013). Surrogate modeling of 3D crack growth. *International Journal of Fatigue*, *47*, 90–99. doi:10.1016/j.ijfatigue.2012.07.012

Horn, B. (1986). *Robot Vision*. MIT Press.

Huang, G.-B., Wang, D. H., & Lan, Y. (2011). Extreme learning machine: A survey. *International Journal of Machine Learning and Cybernetics*, *2*(2), 107–122. doi:10.1007/s13042-011-0019-y

Huang, G.-B., Zhu, Q.-Y., & Siew, C.-K. (2006). Extreme learning machine: Theory and applications. *Neurocomputing*, *70*(1-3), 489–501. doi:10.1016/j.neucom.2005.12.126

Hwang, W., & Han, K. S. (1986). Fatigue of composites—fatigue modulus concept and life prediction. *Journal of Composite Materials*, *20*(2), 154–165. doi:10.1177/002199838602000203

Hyvarinen, A., Kahunen, J., & Oja, E. (2001). *Independent Component Analysis*. John Wiley & Sons. doi:10.1002/0471221317

Ii'kaev, R. I., Zelesnev, V. E., Aleshin, V. V., & Klishin, G. S. (2005). *Numerical simulation of gas pipeline networks. Theory, computational implementation, an industrial application.* Moscow: Editorial URSS.

Indian Standard (IS). (1991). *1489 (Part 1): Portland pozzolona cement- specification (Part 1- Fly ash based).* New Delhi, India: Bureau of Indian Standards.

Indian Standard (IS). (1992). *1331 (Part 1). Nondestructive testing of concrete: Part 1- Ultrasonic pulse velocity.* New Delhi: Bureau of Indian Standards.

Inman, D. J. (2001). *Engineering vibration*. Prentice Hall Inc.

Inman, D. J., Farrar, C., Lopes, C., & Steffen, V. (2005). *Damage Prognosis: For Aerospace, Civil and Mechanical Systems*. John Wiley & Sons Inc. doi:10.1002/0470869097

Jaynes, E. T. (1989). Papers on Probability, Statistics and Statistical Physics. (R. Rosenkrantz, Ed.). Dordrecht, The Nederlands: Kluwer Academic Publishers - Springer.

Jaynes, E. T. (1983). *Papers on Probability, Statistics and Statistical physics* (R. D. Rosenkrantz, Ed.). Kluwer Academic Publishers.

Jaynes, E. T. (2003). *Probability Theory: The Logic of Science* (G. Bretthorst, Ed.). New York, NY: Cambridge University Press. doi:10.1017/CBO9780511790423

Jeong, H., & Cho, S. (2013). Impact source location of composites using a single sensor and time reversal technique. In *AIP Conf. Proc.* (vol. 1511, pp. 246-253). doi:10.1063/1.4789055

Jhang, K. Y. (2009). Nonlinear ultrasonic techniques for nondestructive assessment of microdamage in material: A review. *International Journal of Precision Engineering and Manufacturing*, *10*(1), 123–135. doi:10.1007/s12541-009-0019-y

Jiang, X., & Mahadevan, S. (2008a). Bayesian Probabilistic Inference for Nonparametric Damage Detection of Structures. *Journal of Engineering Mechanics*, *134*(10), 820–831. doi:10.1061/(ASCE)0733-9399(2008)134:10(820)

Jiang, X., & Mahadevan, S. (2008b). Bayesian Wavelet Methodology for Structural Damage Detection. *Structural Control and Health Monitoring*, *15*(7), 974–991. doi:10.1002/stc.230

Joffe, R., & Varna, J. (1999). Analytical modeling of stiffness reduction in symmetric and balanced laminates due to cracks in 90 layers. *Composites Science and Technology*, *59*(11), 1641–1652. doi:10.1016/S0266-3538(99)00025-1

Johnson, P. A., Zinszner, B., & Rasolofosaon, P. N. (1996). Resonance and elastic nonlinear phenomena in rock. *Journal of Geophysical Research: Solid Earth (1978-2012)*, *101*(B5), 11553-11564.

Johnson, E. A., Lam, H. F., Katafygiotis, L. S., & Beck, J. L. (2001). In F. Casciati & G. Magonette (Eds.), *A benchmark problem for structural health monitoring and damage detection. Structural Control for Civil and Infrastructure Engineering* (pp. 317–324). Singapore: World Scientific. doi:10.1142/9789812811707_0028

Johnson, G. R., & Cook, W. H. (1985). Fracture characteristics of three metals subjected to various strains, strain rates, temperatures and pressures. *Engineering Fracture Mechanics*, *21*(1), 31–48. doi:10.1016/0013-7944(85)90052-9

Johnson, P. A. (2006). Nonequilibrium Nonlinear Dynamics in Solids: State of the Art. In P. P. Delsanto (Ed.), *Universality of Nonclassical Nonlinearity* (pp. 49–69). New York, NY: Springer. doi:10.1007/978-0-387-35851-2_4

Jolliffe, I. T. (1972). Discarding Variables in a Principal Component Analysis I: Artificial Data. *Applied Statistics*, *21*(2), 160–173. doi:10.2307/2346488

Jolliffe, I. T. (2002). *Principal Component Analysis*. John Wiley & Sons, Ltd.

Jones D.E. & Holtz W.G. (1973). Expansive soils - The hidden disaster. *American Society of Civil Engineers, 43*(8), 49-51.

Jones, R. T., Sirkis, J. S., & Friebele, E. J. (1997). Detection of impact location and magnitude for isotropic plates using neural networks. *Journal of Intelligent Material Systems and Structures*, *7*(1), 90–99. doi:10.1177/1045389X9700800109

Kaiser, G. (1994). *Friendly Guide to Wavelets*. Boston, MA: Birkhäuser.

Kalman, R. E. (1960). A new approach to linear filtering and prediction problems. *Journal of Fluids Engineering*, *82*(1), 35–45.

Kalman, R. E., & Bucy, R. S. (1961). New results in linear filtering and prediction theory. *Journal of Fluids Engineering*, *83*(1), 95–108.

Karamchandani, A., Bjerager, P., & Cornell, C. (1989). *Adaptive importance sampling*. Paper presented at the International Conference on Structural Safety and Reliability.

Karim, M. E., Yun, F., & Siva Krishna Madani, S. P. (2010). *Fuzzy Clustering Analysis*. (Master Thesis). Blekinge Institute of Technology.

Kasap, S. O. (2000). *Principles of Electrical Engineering Materials and Devices*. McGraw-Hill.

Kaufman, L., & Rousseeuw, P. J. (2005). *Finding Groups in Data: An Introduction to Cluster Analysis*. Wiley.

Kawashima, K., Murase, M., Yamada, R., Matsushima, M., Uematsu, M., & Fujita, F. (2006). Nonlinear ultrasonic imaging of imperfectly bonded interfaces. *Ultrasonics*, *44*(Supplement), e1329–e1333. doi:10.1016/j.ultras.2006.05.011 PMID:16782154

Khoo, L. M., Mantena, P. R., & Jadhav P (2004). Structural damage assessment using vibration modal analysis. *Structural Health Monitoring*, *3*(2), 177–194.

Kidder, R. L. (1973). Reduction of structural frecuency equations. AIAA Journal, 11(6), 892.

Kim, S., & Sohn, H. (2007). Instantaneous reference free crack detection based on polarization characteristics of piezoelectric materials. *Smart Materials and Structures*, *16*(6), 2375–2387. doi:10.1088/0964-1726/16/6/042

Kirikera, G. R., Balogun, O., & Krishnaswamy, S. (2011). Adaptive Fiber Bragg Grating Sensor Network for Structural Health Monitoring: Applications to Impact Monitoring. *Structural Health Monitoring*, *10*(1), 5–12. doi:10.1177/1475921710365437

Klinkov, M., & Fritzen, C. P. (2011). Wind Load Observer for a 5MW Wind Energy Plant. *Structural Dynamics*, *3*, 719–726.

Kohonen, T. (2001). Self-Organizing maps (3rd ed.). Berlin: Springer-Verlag. doi:10.1007/978-3-642-56927-2

Köjl, M., Hurlebaus, S., & Gaul, L. (2004). Finite element simulation of non-destructive damage detectionwith higher harmonics. *NDT & E International*, *37*(3), 195–205. doi:10.1016/j.ndteint.2003.09.003

Kong, A., Liu, J. S., & Wong, W. H. (1994). Sequential imputations and Bayesian missing data problems. *Journal of the American Statistical Association*, *89*(425), 278–288. doi:10.1080/01621459.1994.10476469

Konstantinidis, G., Wilcox, P. D., & Drinkwater, B. W. (2007). An Investigation into the Temperature Stability of a Guided Wave Structural Health Monitoring System Using Permanently Attached Sensors. *Sensors Journal, IEEE*, *7*(5), 905–912. doi:10.1109/JSEN.2007.894908

Kraemer, P. (2011) *Schadensdiagnoseverfahren fr die Zustandsberwachung von Offshore- Windenergieanlagen*. (PhD thesis). Siegen.

Kraemer, P., & Fritzen, C. P. (2008). *Damage identification of structural components of offshore wind energy plants*. Paper presented at the 9th German wind energy conference (DEWEK2008).

Kraemer, P., & Fritzen, C. P. (2011). Aspects of Operational Modal Analysis for Structures of Offshore Wind Energy Plants. *Structural Dynamics and Renewable Energy*, *1*, 145–152. doi:10.1007/978-1-4419-9716-6_14

Krohn, N., Stoessel, R., & Busse, G. (2002). Acoustic non-linearity for defect selective imaging. *Ultrasonics*, *40*(1–8), 633-637.

Kumar Mehta, P. (2004) *High-performance, high-volume fly ash concrete for sustainable development*. Paper presented at the International Workshop on Sustainable Development and Concrete Technology, Beijing, China.

Kyprianou, A., & Staszewski, W. (1999). On the cross wavelet: Interpretation and application to Duffing oscillator. *Journal of Sound and Vibration*, 228(1), Lim, J., Kim, M., & Sohn, H. (2011) Impedance based damage detectionunder varying temperature and loading conditions. *NDT & E International*, *44*, 740–750.

Landau, L. D., & Lifshitz, E. M. (1986). *Theory of Elasticity*. Oxford, UK: Elsevier Butterworth-Heinemann.

Larrosa, C., & Chang, F. (2012). Real time in-situ damage classification, quantification and diagnosis for composite structures*Proceedings of the 19th International Congress on Sound and Vibration* (Vol. 15).

LeClerc, J. R., Worden, K., Staszewski, W. J., & Haywood, J. (2007). Impact detection in an aircraft composite panel—a neural network approach. *Journal of Sound and Vibration*, *299*(3), 672–682. doi:10.1016/j.jsv.2006.07.019

Lee, Y., & Hamilton, M. (1995). Time-domain modeling of pulsed finite-amplitude sound beams. *The Journal of the Acoustical Society of America*, *97*(2), 906–917. doi:10.1121/1.412135

Li, R., Mita, A., & Zhou, J. (2013). Hybrid methodology for structural health monitoring based on immune algorithms and symbolic time series analysis. *Journal of Intelligent Learning System and Applications*, 48-56.

Liang, D., Yuan, S. F., & Liu, M. L. (2013). Distributed coordination algorithm for impact location of preciseness and real-time on composite structures. *Measurement*, *46*(1), 527–536. doi:10.1016/j.measurement.2012.08.011

Li, H. N., Yi, T. H., Ren, L., Li, D. S., & Huo, L. S. (2014). Reviews on innovations and applications in structural health monitoring for infrastructures. *Structural Monitoring and Maintenance*, *1*(1), 1–45. doi:10.12989/smm.2014.1.1.001

Limongelli, M. (2010). Frequency response function interpolation for damage detection under changing environment. *Mechanical Systems and Signal Processing*, *24*(8), 2898–2913. doi:10.1016/j.ymssp.2010.03.004

Lin, N., & Qun, C. (2013, January). Structural Health Monitoring and Damage Detection Using Neural Networks. In *Proceedings of the 2013 Third International Conference on Intelligent System Design and Engineering Applications* (pp. 1302-1304). IEEE Computer Society. doi:10.1109/ISDEA.2012.307

Liu, J., Qiu, J., Chang, W., Ji, H., & Zhu, K. (2010). Metal core piezoelectric ceramic fiber rosettes for acoustoultrasonic source localization in plate structures. *International Journal of Applied Electromagnetics and Mechanics*, *33*, 865–873.

Liu, J., & West, M. (2001). *Combined Parameter and State Estimation in Simulation-Based Filtering* (pp. 197–223). Sequential Monte Carlo Methods in Practice. doi:10.1007/978-1-4757-3437-9_10

Liu, Y., Liu, L., & Mahadevan, S. (2007). Analysis of subsurface crack propagation under rolling contact loading in railroad wheels using FEM. *Engineering Fracture Mechanics*, *74*(17), 2659–2674. doi:10.1016/j.engfracmech.2007.02.012

Liu, Y., & Mahadevan, S. (2005). Multiaxial high-cycle fatigue criterion and life prediction for metals. *International Journal of Fatigue*, *27*(7), 790–800. doi:10.1016/j.ijfatigue.2005.01.003

Liu, Y., & Mahadevan, S. (2007). Threshold stress intensity factor and crack growth rate prediction under mixed-mode loading. *Engineering Fracture Mechanics*, *74*(3), 332–345. doi:10.1016/j.engfracmech.2006.06.003

Liu, Y., & Mahadevan, S. (2009). Probabilistic fatigue life prediction using an equivalent initial flaw size distribution. *International Journal of Fatigue*, *31*(3), 476–487. doi:10.1016/j.ijfatigue.2008.06.005

Liu, Y., Mohanty, S., & Chattopadhyay, A. (2010). Condition Based Structural Health Monitoring and Prognosis of Composite Structures under Uniaxial and Biaxial Loading. *Journal of Nondestructive Evaluation*, *29*(3), 181–188. doi:10.1007/s10921-010-0076-2

Loh, W.-L. (1996). On Latin hypercube sampling. *Annals of Statistics*, *24*(5), 2058–2080. doi:10.1214/aos/1069362310

Lukasiewicz, J. (1970). *Selected Works*. Cambridge, UK: Cambridge University Press.

Lundmark, P., & Varna, J. (2005). Constitutive relationships for laminates with ply cracks in in-plane loading. *International Journal of Damage Mechanics*, *14*(3), 235–259. doi:10.1177/1056789505050355

Luukka, P., Saastamoinen, K., & Kononen, V. (2001). A Classifier Based on the Maximal Fuzzy Similarity in the Generalized Lukasiewicz-Structure. In *Proceedings of Fuzzy Systems, 2001. The 10th IEEE International Conference On* (pp. 195–98). IEEE. doi:10.1109/FUZZ.2001.1007281

Luukka, P. (2009). Classification Based on Fuzzy Robust PCA Algorithms and Similarity Classifier. *Expert Systems with Applications*, *36*(4), 7463–7468. doi:10.1016/j.eswa.2008.09.015

Luukka, P., & Leppälampi, T. (2006). Similarity Classifier with Generalized Mean Applied to Medical Data. *Computers in Biology and Medicine*, *36*(9), 1026–1040. doi:10.1016/j.compbiomed.2005.05.008 PMID:16159657

Mallat, S. (2009). *A Wavelet Tour of Signal Processing*. Burlington, MA: Elsevier.

Mallat, S., & Hwang, L. (1992). Singularity detection and processing with wavelets. *IEEE Transactions on Information Theory*, *38*(2), 617–642. doi:10.1109/18.119727

Martin, E., Kriegel, H. P., Sander, J., & Xiaowei, X. (1996). A Density-Based Algorithm for Discovering Clusters in Large Spatial Databases with Noise. In *Proceedings of the 2nd International on KDD* (vol. 96, pp. 226–31). Academic Press.

Matikas, T. E., & Aggelis, D. G. (2011). New Trends in Materials Nondestructive Characterization Using Surface Acoustic Wave Methodologies. In *Ultrasonic Waves* (pp. 251-282). In Tech. Retrieved from http://www.intechopen.com/books/ultrasonic-waves/new-trends-in-materials-nondestructive-characterization-using-surface-acoustic-wave-methodologies

Matikas, T. (2010). Damage Characterization and Real-Time Health Monitoring os Aerospace Materials Using Innovative NDE Tools. *Journal of Materials Engineering and Performance*, *19*(5), 751–760. doi:10.1007/s11665-009-9532-5

Matlab. (2014). *Fuzzy Logic Toolbox User`s Guide*. MathWorks.

Matsuda, N., & Shiro, B. (2014). Frequency Dependence of Second-Harmonic Generation in Lamb Waves. *Journal of Nondestructive Evauation*, *33*(2), 169–177. doi:10.1007/s10921-014-0227-y

Matz, V., Kreidl, M., & Smíd, R. (2004). *Signal to Noise Ratio Improvement Based of the Discrete Wavelet*. Academic Press.

Maymon, G. (2005). Probabilistic crack growth behavior of aluminum 2024-T351 alloy using the "unified"approach. *International Journal of Fatigue*, *27*(7), 828–834. doi:10.1016/j.ijfatigue.2004.12.003

McCall, K. (1994). Theoretical study of nonlinear elastic wave propagation. *Journal of Geophysical Research*, *99*(B2), 2591–2600. doi:10.1029/93JB02974

McCall, K., & Guyer, R. (1994). Equation of state and wave propagation In hysteretic nonlinear elastic materials. *Journal of Geophysical Research*, *99*(B12), 23887–23897. doi:10.1029/94JB01941

McCall, K., & Guyer, R. (1999). A new theoretical paradigm to describe hysteresis, discrete memory and nonlinear elastic wave propagation in rock. *Nonlinear Processes in Geophysics*, *3*(2), 89–101. doi:10.5194/npg-3-89-1996

McClung, R. C., Enright, M. P., Millwater, H. R., Leverant, G. R., & Hudak, S. J. Jr. (2004). A software framework for probabilistic fatigue life assessment of gas turbine engine rotors. *Journal of ASTM International*, *1*(8), 11563. http://www.astm.org/cgi-bin/scholar.cgi?JAI11563 doi:10.1520/JAI11563

McKeeman, M. (1962). Algorithm 145: Adaptive numerical integration by Simpson's rule. *Communications of the ACM*, *5*(12), 604. doi:10.1145/355580.369102

Meo, M., & Zumpano, G. (2005). Nonlinear elastic wave spectroscopy identification of impact damage on a sandwich plate. *Composite Structures*, *71*(3), 469–474. doi:10.1016/j.compstruct.2005.09.027

Merati, A., & Eastaugh, G. (2007). Determination of fatigue related discontinuity state of 7000 series of aerospace aluminum alloys. *Engineering Failure Analysis*, *14*(4), 673–685. doi:10.1016/j.engfailanal.2006.02.016

Meurer, T., Qu, J., & Jacobs, L. J. (2002). Wave propagation in nonlinear and hysteretic media - a numerical study. *International Journal of Solids and Structures*, *39*(21-22), 5585–5614. doi:10.1016/S0020-7683(02)00366-9

Mevel, L., Goursat, M., & Basseville, M. (2003, January). Stochastic Subspace-Based Structural Identification and Damage Detection and Localisation-Application To the Z24 Bridge Benchmark. *Mechanical Systems and Signal Processing*, *17*(1), 143–151. doi:10.1006/mssp.2002.1552

Michaels, J. E. (2008). Detection, Localization and Characterization of Damage in Plates with an in Situ Array of Spatially Distributed Ultrasonic Sensors. *Smart Materials and Structures*, *17*(3), 35035. doi:10.1088/0964-1726/17/3/035035

Migon, H., & Gamerman, D. (1999). *Statistical Inference: An Integrated Approach*. London: Arnold.

Moreschi, H., Callé, S., Guerard, S., Mitton, D., Renaud, G., & Defontaine, M. (2011). Monitoring trabecular bone microdamage using a dynamic acousto-elastic testing method. *Proceedings of the Institution of Mechanical Engineers. Part H, Journal of Engineering in Medicine*, *225*(3), 282–295. PMID:21485329

Mosterman, P., & Biswas, G. (1999). Diagnosis of continuous valued systems in transient operating regions. *IEEE Transactions on Systems, Man, and Cybernetics. Part A, Systems and Humans*, *29*(6), 554–565. doi:10.1109/3468.798059

Moustafa, A., Mahadevan, S., Daigle, M., & Biswas, G. (2010). Structural and Sensor Damage Identification using the Bond Graph Approach. *Structural Control and Health Monitoring*, *17*(2), 178–197. doi:10.1002/stc.285

Muir, D. D. (2009, May). *One-sided ultrasonic determination of third order elastic constants using angle-beam acoustoelasticity measurements.* (Doctoral Dissertation). Georgia Institute of Technology. Retrieved from http://hdl.handle.net/1853/29766

Mujica, L. E. (2006). *A hybrid approach of knowledge-based reasoning for structural assessment.* (Doctoral dissertation). Retrieved from http://www.tdx.cat/bitstream/handle/10803/7737/tlemd.pdf?sequence=1

Mujica, L. E., Rodellar, J., Fernandez, A., & Guemes, A. (2010). Q-Statistic and T2-Statistic PCA-Based Measures for Damage Assessment in Structures. *Structural Health Monitoring*, *10*(5), 539–553. doi:10.1177/1475921710388972

Mujica, L. E., Rodellar, J., Fernandez, A., & Guemes, A. (2010). Q-statistic and T2-statistic PCA-based measures for damage assessment in structures. *Structural Health Monitoring*, 1475921710388972.

Mujica, L. E., Vehí, J., Ruiz, M., Verleysen, M., Staszewskic, W., & Worden, K. (2008). Multivariate Statistics Process Control for Dimensionality Reduction in Structural Assessment. *Mechanical Systems and Signal Processing*, *22*(1), 155–171. doi:10.1016/j.ymssp.2007.05.001

Mujica, L. E., Vehi, J., Staszewskic, W., & Worden, K. (2008). Impact damage detection in aircraft composites using knowledge-based reasoning. *Structural Health Monitoring*, *7*(3), 215–230. doi:10.1177/1475921708090560

Mujica, L., Vehí, J., & Rodellar, J. (2005). *Detección de impactos mediante razonamiento basado en conocimiento: Aplicación a una sección de ala de avión. 1Departament d'Electrònica, Informàtica i Automàtica*. Escola Politècnica Superior - Universitat de Girona.

Mukherjee, Mandal, & Adhikari. (2012). Study on the Autogenic Healing Property of High Volume Fly Ash Concrete. *International Journal of 3R's, 3*, 491-494.

Mukherjee, S., Mandal, S., & Adhikari, U. B. (2012). Study on the physical and mechanical property of ordinary portland cement and fly ash paste. *International Journal of Civil & Structural Engineering*, *2*(3), 731–736.

Mukherjee, S., Mandal, S., & Adhikari, U. B. (2013). Comparative study on physical and mechanical properties of high Slump and zero slump high volume fly ash concrete (HVFAC). *Global NEST Journal*, *20*(10), 1–7.

Muller, M., Sutin, A., Guyer, R., Talmant, M., Laugier, P., & Johnson, P. (2005). Nonlinear resonant ultrasound spectroscopy (NRUS) applied to damage assessment in bone. *The Journal of the Acoustical Society of America*, *118*(6), 3946–3952. doi:10.1121/1.2126917 PMID:16419838

Muller, M., Tencate, J., Darling, T., Sutin, A., Guyer, R., Talmant, M., & Johnson, P. (2006). Bone micro-damage assessment using non-linear resonant ultrasound spectroscopy (NRUS) techniques: A feasibility study. *Ultrasonics*, *44*, e245–e249. doi:10.1016/j.ultras.2006.06.043 PMID:16876843

Murnaghan, F. (1951). *Finite deformation of an elastic solid*. London, UK: Chapman & Hall.

Nairn, J., & Hu, S. (1992). The initiation and growth of delaminations induced by matrix microcracks in laminated composites. *International Journal of Fracture*, *57*(1), 1–24. doi:10.1007/BF00013005

Na, J. K., Cantrell, J. H., & Yost, W. T. (1996). Linear and Nonlinear Ultrasonic Properties of Fatigued 410Cb Stainless Steel. In D. O. Thompson & D. E. Chimenti (Eds.), *Review of Progress in Quantitative Nondestructive Evaluation* (pp. 1347–1352). New York, NY: Plenum Press. doi:10.1007/978-1-4613-0383-1_176

Najm, H. N. (2009). Uncertainty quantification and polynomial chaos techniques in computational fluid dynamics. *Annual Review of Fluid Mechanics*, *41*(1), 35–52. doi:10.1146/annurev.fluid.010908.165248

Nazarov, V. E., Radostin, A., Ostrovsky, L., & Soustova, I. (2003). Wave Processes in Media with Hysteretic Nonlinearity. Part I. *Acoustical Physics*, *49*(3), 344–353. doi:10.1134/1.1574363

Negnevitsky, M. (2005). *Artificial intelligence: A guide to intelligent systems* (2nd ed.). London: Addison-Wesley.

Nestleroth, J. B. (2006) Pipeline in lines inspection. Challenges to NDT. *ECNDT 2006. 9th European Conference on NTD*. Berlin, Germany: ECNDT.

Newman, J. C. Jr, Brot, A., & Matias, C. (2004). Crack-growth calculations in 7075-T7351 aluminum alloy under various load spectra using an improved crack-closure model. *Engineering Fracture Mechanics*, *71*(16), 2347–2363. doi:10.1016/j.engfracmech.2004.01.004

Nicholson, P., & Bouxsein, M. (2000). Quantitative ultrasound does not reflect mechanically induced damage in human cancellous bone. *Journal of Bone and Mineral Research*, *15*(12), 2467–2472. doi:10.1359/jbmr.2000.15.12.2467 PMID:11127211

Nikias, C., & Mendel, J. (1993). Signal processing with higher order spectra. *IEEE Signal Processing Magazine*, *10*(3), 10–37. doi:10.1109/79.221324

Norris, A. (1998). Finite Amplitude Waves in Solids. In M. F. Hamilton & D. T. Blackstock (Eds.), *Nonlinear Acoustics* (pp. 263–277). San Diego, CA: Academic Press.

Oberholster, A. J., & Heyns, P. S. (2006). On-line fan blade damage detection using neural networks. *Mechanical Systems and Signal Processing*, *20*(1), 78–93. doi:10.1016/j.ymssp.2004.09.007

Ohtani, T., & Ishii, Y. (2012). Nonlinear Resonant Ultrasound Spectroscopy (NRUS) applied to fatigue damage evaluation in a pure copper. *AIP Conference Proceedings*, *1474*, 203–206. doi:10.1063/1.4749331

Orchard, M., Kacprzynski, G., Goebel, K., Saha, B., & Vachtsevanos, G. (2008). *Advances in uncertainty representation and management for particle filtering applied to prognostics*. Paper presented at the Prognostics and Health Management, 2008. PHM 2008. International Conference on. doi:10.1109/PHM.2008.4711433

Ostrovsky, L. A., & Johnson, P. A. (2001). Dynamic nonlinear elasticity in geomaterials. *Rivista del Nuovo Cimento*, *24*(7), 1–46.

Paris, P. C., Gomez, M. P., & Anderson, W. E. (1961). A rational analytic theory of fatigue. *The Trend in Engineering*, *13*(1), 9–14.

Park, B., Sohn, H., Olson, S. E., DeSimio, M. P., Brown, K. S., & Derriso, M. M. (2012). Impact localization in complex structures using laser-based time reversal. *Structural Health Monitoring*, *11*(5), 577–588. doi:10.1177/1475921712449508

Parra dos Anjos Lima, F., Chavarette, F., dos Santos e Souza, A., Silva Frutuoso de Souza, S., & Martins, M. L. (2013). Artificial immune systems with negative selection applied to health monitoring of aeronautical structures. *Advanced Materials Research*, *871*, 283-289.

Pasqualini, D., Heitmann, K., TenCate, J., Habib, S., Higdon, D., & Johnson, P. (2007). Nonequilibrium and nonlinear dynamics in Berea and Fontainebleau sandstones: Low-strain regime. *Journal of Geophysical Research. Solid Earth*, *112*(B01204), 1–16.

Pastor, M., Binda, M., & Harcarik, T. (2012, January). Modal Assurance Criterion. *Procedia Engineering*, *48*, 543–548. doi:10.1016/j.proeng.2012.09.551

Patrick, R., Orchard, M. E., Zhang, B., Koelemay, M. D., Kacprzynski, G. J., Ferri, A. A., & Vachtsevanos, G. J. (2007). An integrated approach to helicopter planetary gear fault diagnosis and failure prognosis. In *Autotestcon, 2007 IEEE* (pp. 547–552). IEEE. Retrieved from http://ieeexplore.ieee.org/xpls/abs_all.jsp?arnumber=4374266

Pawitan, Y. (2001). *In All Likelihood: Statistical Modelling and Inference Using Likelihood* (1st ed.). New York: Oxford University Press.

Payan, C., Garnier, V., Moysan, J., & Johnson, P. (2007). Applying nonlinear resonant ultrasound spectroscopy to improving thermal damage assessment in concrete. *The Journal of the Acoustical Society of America*, *121*(4), EL125–EL130. doi:10.1121/1.2710745 PMID:17471756

Pecorari, C. (2003). An extension of the spring model to nonlinear interfaces. *AIP Conference Proceedings*, *657*, 49–56. doi:10.1063/1.1570118

Peng, K., Chu, L., & Tse, P. (2005). Singularity analysis of the vibration signals by means of wavelet modulus maxima method. *Mechanical Systems and Signal Processing*, *21*(2), 780–794. doi:10.1016/j.ymssp.2005.12.005

Peng, T., Saxena, A., Goebel, K., Xiang, Y., Sankararaman, S., & Liu, Y. (2013). A novel Bayesian imaging method for probabilistic delamination detection of composite materials. *Smart Materials and Structures*, *22*(12), 125019. doi:10.1088/0964-1726/22/12/125019

Perelli, A., De Marchi, L., Marzani, A., & Speciale, N. (2014). Frequency warped cross-wavelet multiresolution analysis of guided waves for impact localization. *Signal Processing*, *96*, 51–62. doi:10.1016/j.sigpro.2013.05.008

Peterka, V. (1981). Bayesian Approach to System Identification. In *Trends and Progress in System Identification*. Oxford, UK: Pergamon Press. doi:10.1016/B978-0-08-025683-2.50013-2

Polikar, R., Upda, L., Upda, S., & Taylor, T. (1998). Frequency Invariant Classification of Ultrasonic Weld Inspection Signals. *IEEE Transactions on Ultrasonics, Ferroelectrics, and Frequency Control*, *45*(3), 614–624. doi:10.1109/58.677606 PMID:18244213

Polimeno, U., & Meo, M. (2009). Detecting barely visible impact damage detection on aircraft composites structures. *Composite Structures*, *91*(4), 398–402. doi:10.1016/j.compstruct.2009.04.014

Powers-Couche, L. (1992). Fire damaged concrete-up close. *Concrete Repair Digest*, *1*, 241–248.

Powers-Couche, L. (1994). Observations of concrete exposed to very high temperature. In *Proceedings of the 16th International Conference on Cement Microscopy.*

Proakis, M. (2007). *Tratamiento Digital de Señales*. Prentice Hall.

Qiu, L., Yuan, S., Liu, P., & Qian, W. (2013, July). Design of an All-Digital Impact Monitoring System for Large-Scale Composite Structures. *IEEE Transactions on Instrumentation and Measurement*, *62*(7), 1990–2002. doi:10.1109/TIM.2013.2247491

Quek, S. T., Tua, P. S., & Wang, Q. (2003). Detecting anomalies in beams and plate based on the Hilbert-Huang transform of real signals. *Smart Materials and Structures*, *12*(3), 447–460. doi:10.1088/0964-1726/12/3/316

Raghavan, A., & Cesnik, E. (2008). Effects of Elevated Temperature on Guided-Wave Structural Health Monitoring. *Journal of Intelligent Material Systems and Structures*, *19*(12), 1383–1398. doi:10.1177/1045389X07086691

Ramm, W., & Biscoping, M. (1998). Autogenous healing and reinforcement corrosion of water-penetrated separation cracks in reinforced concrete. *Nuclear Engineering and Design*, *179*(2), 191–200. doi:10.1016/S0029-5493(97)00266-5

Ramos, P., Sanz, F., & Montero, E. (1986). *Riera*. Mundo Electrónico.

Rasmussen, C. E. (2004). Gaussian processes in machine learning. In Advanced Lectures on Machine Learning (pp. 63–71). Springer. Retrieved from http://link.springer.com/chapter/10.1007/978-3-540-28650-9_4 doi:10.1007/978-3-540-28650-9_4

Renaud, G., Callé, S., Remenieras, J. P., & Defontaine, M. (2008). Exploration of trabecular bone nonlinear elasticity using time-of-flight modulation. *IEEE Transactions on Ultrasonics, Ferroelectrics, and Frequency Control*, *55*(7), 1497–1507. doi:10.1109/TUFFC.2008.825 PMID:18986939

Renaud, G., Riviere, J., Haupert, S., & Laugier, P. (2013). Anisotropy of dynamic acoustoelasticity in limestone, influence of conditioning, and comparison with nonlinear resonance spectroscopy. *The Journal of the Acoustical Society of America*, *133*(6), 3706–3718. doi:10.1121/1.4802909 PMID:23742326

Renaud, G., Talmant, M., Callé, S., Defontaine, M., & Laugier, P. (2011). Nonlinear elastodynamics in micro-inhomogeneous solids observed by head wave based dynamic acoustoelastic testing. *The Journal of the Acoustical Society of America*, *130*(6), 3583–3589. doi:10.1121/1.3652871 PMID:22225015

Riahi, M., & Abolhasany, M. (2006). Substitution of the Time-of-Flight Diffraction Technique for Nondestructive Testing of Welds and Thick Layers of Steel. *Russian Journal of Nondestructive Testing*, *42*(12), 794–801. doi:10.1134/S1061830906120047

Richardson, J. M. (1979). Harmonic generation at an unbonded interface-I. Planar interface betwwen semi-infinite elastic media. *International Journal of Engineering Science*, *17*(1), 73–85. doi:10.1016/0020-7225(79)90008-9

Richards, S. A. (1997). Completed Richardson extrapolation in space and time. *Communications in Numerical Methods in Engineering*, *13*(7), 573–582. doi:10.1002/(SICI)1099-0887(199707)13:7<573::AID-CNM84>3.0.CO;2-6

Rivière, J., Remillieux, M., Ohara, Y., Anderson, B., Haupert, S., Ulrich, T., & Johnson, P. (2014). Dynamic Acousto-Elasticity in a Fatigue-Cracked Sample. *Journal of Nondestructive Evaluation*, *33*(2), 216–225. doi:10.1007/s10921-014-0225-0

Rivière, J., Renaud, G., Guyer, R., & Johnson, P. (2013). Pump and probe waves in dynamic acousto-elasticity: Comprehensive description and comparison with nonlinear elastic theories. *Journal of Applied Physics*, *114*(5), 054905. doi:10.1063/1.4816395

Rodríguez, J. A., Sandoval, S., & Vitola, J. (2007). Fundamentos Teórico-Prácticos de Ultrasonido. *Revista Tecnura*, *10*(20), 4–18.

Rodríguez, J. A., Vitola, J., & Sandoval, S. (2009). Diseño y Construcción de un Sistema de Ultrasonido para la detección de Discontinuidades en Soldadura. *Revista Colombiana de Física*, *41*(1), 159–161.

Rodríguez, J. A., Vitola, J., Sandoval, S., & Forero, E. (2009). Diseño y Construcción de un Sistema para Examen no Destructivo de Fallas y Defectos en Metales Utilizando Señales Ultrasónicas. *Revista EIA*, *12*, 9–21.

Rokach, L., & Maimon, O. (2005). Clustering Methods. In O. Maimon, & L. Rokach (Eds.), Data Mining and Knowledge Discovery Handbook (pp. 321-352). Springer.

Rokach, L., & Maimon, O. (2010). Distance Measures for Numeric Attributes. In Data Mining and Knowledge Discovery Handbook, (pp. 322–52). Academic Press.

Rolfes, R., Zerbst, S., & Haake, G. (2007). Integral SHM-System for Offshore Wind Turbines Using Smart Wireless Sensors.*Proceedings of the 6th.*

Rosenblatt, M. et al. (1956). Remarks on some nonparametric estimates of a density function. *Annals of Mathematical Statistics*, *27*(3), 832–837. doi:10.1214/aoms/1177728190

Rus, G., Wooh, S., & Gallego, R. (2007). Processing of ultrasonic array signals for characterizing defects. Part I: signal synthesis. Part II: Experimental work. *IEEE Transactions on Ultrasonics, Ferroelectrics, and Frequency Control*, *54*(10), 2129–2145. doi:10.1109/TUFFC.2007.508 PMID:18019251

Rytter, A. (1993). *Vibration Based Inspection of Civil Engineering Structures.* (PhD Thesis). Aalborg University, Aalborg, Denmark.

Sagar, S. P., Das, S., Paridaa, N., & Bhattacharya, D. (2006). Non-linear ultrasonic technique to assess fatigue damage in structural steel. *Scripta Materialia*, *55*(2), 199–202. doi:10.1016/j.scriptamat.2006.03.037

Saha, B., & Goebel, K. (2008). *Uncertainty management for diagnostics and prognostics of batteries using Bayesian techniques.* Paper presented at the Aerospace Conference. doi:10.1109/AERO.2008.4526631

Şahmaran, M., Keskin, S. B., Ozerkan, G., & Yaman, I. O. (2008). Self-healing of mechanically-loaded self consolidating concretes with high volumes of fly ash. *Cement and Concrete Composites*, *30*(10), 872–879. doi:10.1016/j.cemconcomp.2008.07.001

Saltelli, A., Ratto, M., Andres, T., Campolongo, F., Cariboni, J., Gatelli, D. … Tarantola, S. (2008). Global Sensitivity Analysis : The Primer. John Wiley and Sons.

Saltelli, A., Ratto, M., Andres, T., Campolongo, F., Cariboni, J., Gatelli, D., & Tarantola, S. (2008). *Global Sensitivity Analysis: The Primer.* Wiley-Interscience.

Sandoval, W. A. (2010). *Sistema experto para la evaluación del estado de deformación en tuberías de gas: diseño y validación numérica.* (Master dissertation). Retrieved from http://tangara.uis.edu.co/biblioweb/tesis/2010/134415.pdf

Sankararaman, S., Daigle, M., Saxena, A., & Goebel, K. (2013). *Analytical algorithms to quantify the uncertainty in remaining useful life prediction.* Paper presented at the Aerospace Conference. doi:10.1109/AERO.2013.6496971

Sankararaman, S., Ling, Y., Shantz, C., & Mahadevan, S. (2009). Uncertainty quantification in fatigue damage prognosis. In *Annual conference of the prognostics and health management society* (pp. 1–13). Retrieved https://www.phmsociety.org/sites/phmsociety.org/files/phm_submission/2009/phmc_09_51.pdf

Sankararaman, S., Ling, Y., Shantz, C., & Mahadevan, S. (2009). *Uncertainty Quantification in Fatigue Damage Prognosis.* Paper presented at the Annual Conference of the PHM Society, PHM09, San Diego, CA.

Sankararaman, S. (2014). (Already published). Significance, interpretation, and quantification of uncertainty in prognostics and remaining useful life prediction. *Mechanical Systems and Signal Processing*. (52)228-247

Sankararaman, S., Daigle, M. J., & Goebel, K. (2014). Uncertainty Quantification in Remaining Useful Life Prediction Using First-Order Reliability Methods. *Reliability. IEEE Transactions on*, *63*(2), 603–619.

Sankararaman, S., Ling, Y., & Mahadevan, S. (2010). Statistical inference of equivalent initial flaw size with complicated structural geometry and multi-axial variable amplitude loading. *International Journal of Fatigue*, *32*(10), 1689–1700. doi:10.1016/j.ijfatigue.2010.03.012

Sankararaman, S., Ling, Y., & Mahadevan, S. (2011). Uncertainty quantification and model validation of crack growth prediction. *Engineering Fracture Mechanics*, *78*(7), 1487–1504. doi:10.1016/j.engfracmech.2011.02.017

Sankararaman, S., Ling, Y., Shantz, C., & Mahadevan, S. (2011). Inference of equivalent initial flaw size under multiple sources of uncertainty. *International Journal of Fatigue*, *33*(2), 75–89. doi:10.1016/j.ijfatigue.2010.06.008

Sankararaman, S., & Mahadevan, S. (2011). Likelihood-based representation of epistemic uncertainty due to sparse point data and/or interval data. *Reliability Engineering & System Safety*, *96*(7), 814–824. doi:10.1016/j.ress.2011.02.003

Sankararaman, S., & Mahadevan, S. (2011). Uncertainty quantification in structural damage diagnosis. *Structural Control and Health Monitoring*, *18*(8), 807–824. doi:10.1002/stc.400

Sankararaman, S., & Mahadevan, S. (2013). Bayesian methodology for diagnosis uncertainty quantification and health monitoring. *Structural Control and Health Monitoring*, *20*(1), 88–106. doi:10.1002/stc.476

Sankararaman, S., McLemore, K., Mahadevan, S., Bradford, S. C., & Peterson, L. D. (2013). Test resource allocation in hierarchical systems using Bayesian networks. *AIAA Journal*, *51*(3), 537–550. doi:10.2514/1.J051542

Saxena, A., Celaya, J., Saha, S., Saha, B., Roychoudhury, I., & Goebel, K. (2010). *Requirements Specification for Prognostics Performance – An Overview.* Paper presented at the AIAA Infotech @ Aerospace, Atlanta, GA.

Saxena, A., Goebel, K., Larrosa, C. C., & Chang, F.-K. (2013). *CFRP Composites Dataset.* Retrieved from http://ti.arc.nasa.gov/project/prognostic-data-repository

Saxena, A., Roychoudhury, I., Celaya, J., Saha, B., Saha, S., & Goebel, K. (2012). *Requirement Flowdown for Prognostics Health Management.* Paper presented at the AIAA Infotech@Aerospace, Garden Grove, CA. doi:10.2514/6.2012-2554

Saxena, A., Shankararaman, S., & Goebel, K. (2014, July, 2014). *Performance Evaluation for Fleet-based and Unit-based Prognostic Methods.* Paper presented at the 2nd European Conference of the PHM Society, Nantes, France.

Saxena, A., Celaya, J., Saha, B., Saha, S., & Goebel, K. (2010a). Metrics for Offline Evaluation of Prognostic Performance. *International Journal of the PHM Society*, *1*(1), 20.

Saxena, A., Celaya, J., Saha, B., Saha, S., & Goebel, K. (2010b). Metrics for Offline Evaluation of Prognostic Performance. *International Journal of Prognostics and Health Management*, *1*(1), 21.

Seber, G., & Wild, C. (1989). *Nonlinear Regression.* New York: John Wiley and Sons. doi:10.1002/0471725315

Segura D., A., & Rodríguez J., A. (2011). Interfase para el Control de Tarjeta de Ultrasonido. *CIMM*, 74.

Segura, D., & Rodríguez, J. A. (2013). Algoritmo que Reduce Distorsión de Imágenes Ultrasónicas en Ensayos no Destructivos. *CIMM*, 65.

Segura, D. A., & Rodríguez, J. A. (2012). Sistema de Formación de Imágenes TOFD para Ensayos no Destructivos. *Revista EIA*, *17*, 139–149.

Shannon, C., & Weaver, W. (1948). The mathematical theory of communication. *Univ. Illinois Press*, *14*(4), 306–317. PMID:9230594

Sheikholeslami, G., Chatterjee, S., & Zhang, A. (1998). Wavecluster: A Multi-Resolution Clustering Approach for Very Large Spatial Databases. In *Proceedings of24th VLDB Conference*. New York, NY: VLDB.

Sheikholeslami, G., Chatterjee, S., & Zhang, A. (2000). WaveCluster: A Wavelet-Based Clustering Approach for Spatial Data in Very Large Databases. *The VLDB Journal The International Journal on Very Large Data Bases*, *8*(3-4), 289–304. doi:10.1007/s007780050009

Shin, K., Yang, H., Lee, S.-K., & Lee, Y.-S. (2013). Group delay based location template matching method for the identification of the impact location on a plate. *Journal of Sound and Vibration*, *332*(8), 2111–2117. doi:10.1016/j.jsv.2012.12.007

Shkerdin, G., & Glorieux, C. (2010). Nonlinear clapping Modulantion of lamb modes by normally closed delamination. *IEEE Transactions on Ultrasonics, Ferroelectrics, and Frequency Control*, *57*(6), 1426–1433. doi:10.1109/TUFFC.2010.1561 PMID:20529717

Sinou, J. J. (2009). A review of damage detection and health monitoring of mechanical systems from changes in the measurement of linear and non-linear vibrations. *Mechanical Vibrations: Measurement, Effects and Control*, 643-702.

Sivasundaram, V., Carette, G. G., & Malhotra, V. M. (1990). Long-term strength development of high-volume fly ash concrete. *Cement and Concrete Composites*, *12*(4), 263–270. doi:10.1016/0958-9465(90)90005-I

Smarsly, K., Law, K. H., & Hartmann, D. (2011) Implementing a Multiagent-Based Self-Managing Structural Health Monitoring System on a Wind Turbine. In N*SF (National Science Foundation),* 2011 NSF Engineering Research and Innovation Conference. Atlanta, GA.

Smith, S. (1966, May). Random-loading fatigue crack growth behavior of some aluminum and titanium alloys. *Fifth Pacific Area National Meeting of the American Society for Testing and Materials: Structural Fatigue in Aircraft. STP 404*, (pp. 74-100). American Society for Testing and Materials (ASTM). doi:10.1520/STP48341S

Smith, L. (2002). *A Tutorial on Principal Components Analysis*. Academic Press.

Sohh, H., Dutta, D., & An, Y. (2011). Temperature independent damage detction in plates using redundant signal measurements. *Journal of Nondestructive Evaluation*, *30*(2), 106–116. doi:10.1007/s10921-011-0096-6

Sohn, H. (2007). Effects of Environmental and Operational Variability on Structural Health Monitoring. *Philosophical Transactions Series A: Mathematical, Physical, and Engineering Sciences, 365*(1851).

Sohn, H. (2007). Effects of environmental and operational variability on structural health monitoring. *Philosophical Transactions of the Royal Society*, *A365*, 539–556. PMID:17255051

Sohn, H. (2007). Effects of environmental and operational variability on structural health monitoring. *Phil. Trans. R. Soc. A*, *365*(1851), 539–560. doi:10.1098/rsta.2006.1935 PMID:17255051

Sohn, H., & Farrar, R. (2001). Damage diagnosis using time series analysis of vibration signals. *Smart Materials and Structures*, *10*(3), 446–451. doi:10.1088/0964-1726/10/3/304

Sohn, H., Worden, K., & Farrar, R. (2002). Novelty detection in a changing environment: Regression and interpolation approaches. *Journal of Sound and Vibration*, *285*(4), 741–761.

Solodov, I., Krohn, N., & Busse, G. (2002). CAN: An example of nonclassical acoustic nonlinearity in solids. *Ultrasonics*, *40*(1), 621–625. doi:10.1016/S0041-624X(02)00186-5 PMID:12160013

Song, P.-S., Sheu, B.-C., & Chang, L. (2001). A Modified Wheeler Model to Improve Predictions of Crack Growth Following a Single Overload. *JSME International Journal Series A*, *44*(1), 117–122. doi:10.1299/jsmea.44.117

Stobbe, D. M. (2005, August). *Acoustoelasticity in 7075-T651 Aluminum and Dependence of Third Order Elastic Constants on Fatigue Damage.* (Master's thesis). Georgia Institute of Technology. Retrieved from http://hdl.handle.net/1853/7184

Stratoudaki, T., Ellwood, R., Sharples, S., Clark, M., Somekh, M. G., & Collison, I. J. (2011). Measurement of material nonlinearity using surface acoustic wave parametric interaction and laser ultrasonics. *The Journal of the Acoustical Society of America*, *129*(4), 1721–1728. doi:10.1121/1.3560945 PMID:21476629

Su, Z., & Ye, L. (2009). Heat and Mass Transfer. In Identification of Damage Using Lamb Waves: From Fundamentals to Applications. Springer-Verlag Berlin Heidelberg.

Subbaratnam, R., Venkatraman, B., & Raj, B. B. (2006). Time of Flight Difracttion- an Alternative to Radiography Examination of Thick Walled Stainless Weldmentes. In *Proceedings of A-PCNDT -Asia-Pacific Conference on NDT.* A-PCNDT.

Surace, C., & Worden, K. (2010). Novelty detection in changing environment: A negative selection technique. *Mechanical Systems and Signal Processing*, *24*(4), 1114–1128. doi:10.1016/j.ymssp.2009.09.009

Sutin, A., & Nazarov, V. (1995). Nonlinear acoustic methods of crack diagnostics. *Radiophysics and Quantum Electronics, 38*(3-4), 109-120.

Suykens, J. A. K. (2001). Support vector machines: A nonlinear modeling and control perspective. *European Journal of Control*, *7*(2-3), 311–327. doi:10.3166/ejc.7.311-327

Taha, M. R., Noureldin, A., Lucero, J. L., & Baca, T. J. (2006). Wavelet transform for structural health monitoring: A compendium of uses and features. *Structural Health Monitoring*, *5*(3), 267–295. doi:10.1177/1475921706067741

Talreja, R., & Singh, C. V. (2012). *Damage and failure of composite materials*. Cambridge University Press. doi:10.1017/CBO9781139016063

Tang, L., Kacprzynski, G. J., Goebel, K., & Vachtsevanos, G. (2009). *Methodologies for uncertainty management in prognostics*. Paper presented at the Aerospace conference.

Tanizaki, H., & Mariano, R. S. (1998). Nonlinear and non-Gaussian state-space modeling with Monte Carlo simulations. *Journal of Econometrics*, *83*(1), 263–290. doi:10.1016/S0304-4076(97)80226-6

Tarantola, A. (2005). *Inverse Problem Theory and Methods for Model Parameters Estimation*. Philadelphia, PA: SIAM. doi:10.1137/1.9780898717921

TenCate, J. A. (2011). Slow Dynamics of Earth Materials: An Experimental Overview. *Pure and Applied Geophysics*, *168*(12), 2211–2219. doi:10.1007/s00024-011-0268-4

Ter Heide, N. (2005). *Crack healing in hydrating concrete*. (Master thesis). TU Delft, Delft, The Netherlands.

Termkhajornkit, P., Nawa, T., Yamashiro, Y., & Saito, T. (2009). Self-healing ability of fly ash–cement systems. *Cement and Concrete Composites*, *31*(3), 195–203. doi:10.1016/j.cemconcomp.2008.12.009

Tibaduiza Burgos, D., Mujica, L., Guemes, A., & Rodellar, J. (2010). Active piezoelectric system using PCA.*Proceedings of the Fifth European Workshop on Structural Health Monitoring*, (pp. 164-169). Sorrento.

Tibaduiza, D. (2013). *Design and validation of a structural health monitoring system for aeronautical structures.* (PhD Thesis). Universitat Politècnica de Catalunya.

Tibaduiza, D. A. (2013). *Design and Validation of a Structural Health Monitoring System for Aeronautical Structures.* (PhD. Thesis). Technical University of Catalunya.

Tibaduiza, D. A., Mujica, L. E., & Rodellar, J. (2013). Damage classification in structural health monitoring using principal component analysis and self-organizing maps. *Structural Control and Health Monitoring*, *20*(10), 1303–1316. doi:10.1002/stc.1540

Tibaduiza, D. A., Torres-Arredondo, M.-A., Mujica, L. E., Rodellar, J., & Fritzen, C.-P. (2013). A study of two unsupervised data driven statistical methodologies for detecting and classifying damages in structural health monitoring. *Mechanical Systems and Signal Processing*, *41*(1-2), 467–484. doi:10.1016/j.ymssp.2013.05.020

Tiratsoo, J. N. H. (1999). *Pipeline pigging technology* (2nd ed.). Houston, TX: Butterworth Heinemann.

Tjirkalis, A. (2013). *Continuous Structural Health Monitoring Under Varying Environmental and Operational Conditions using Wavelet Transform Modulus Maxima Decay Lines Similarity*. PhD Thesis.

Toivola, J., & Hollmén, J. (2009). Feature extraction and selection from vibration measurements for structural health monitoring. *Adv. Intelligent Data Analysis*, *8*, 213–224.

Toyama, N., & Takatsubo, J. (2004). An investigation of non-linear elastic behavior of CFRP laminates and strain measurement using Lamb waves. *Composites Science and Technology*, *64*(16), 2509–2516. doi:10.1016/j.compscitech.2004.05.007

Trejo Perez, D. (2005). *Optimización global en espacios restringidos mediante un sistema inmune artificial.* (Master Thesis). Cinvestav.

Tryon, R. G., Cruse, T. A., & Mahadevan, S. (1996). Development of a reliability-based fatigue life model for gas turbine engine structures. *Engineering Fracture Mechanics*, *53*(5), 807–828. doi:10.1016/0013-7944(95)00138-7

Tsa, M., & Chiang, H. (2002). Rotation-invariant pattern matching using wavelet decomposition. *Pattern Recognition Letters*, *23*(1-3), 191–201. doi:10.1016/S0167-8655(01)00099-X

Tvedt, L. (1989). *Second order reliability by an exact integral.* Springer. doi:10.1007/978-3-642-83828-6_26

Ulrich, T. J., Sutin, A. M., Guyer, R. A., & Johnson, P. A. (2008). Time reversal and non-linear elastic wave spectroscopy (TR NEWS) techniques. *International Journal of Non-linear Mechanics*, *43*(3), 209–216. doi:10.1016/j.ijnonlinmec.2007.12.017

Urquhart, R. (1982). Graph Theoretical Clustering Based on Limited Neighbourhood Sets. *Pattern Recognition*, *15*(3), 173–187. doi:10.1016/0031-3203(82)90069-3

Van Den Abeele, K. E. (1996). Elastic pulsed wave propagation in media with second or higher order nonlinearity. Part I. Theoretical framework. *The Journal of the Acoustical Society of America*, *99*(6), 3334–3345. doi:10.1121/1.414890

Van Den Abeele, K. E., & Johnson, P. A. (1996). Elastic pulsed wave propagation in media with second or higher order nonlinearity. Part II. Simulation of experimental measurements on Berea sandstone. *The Journal of the Acoustical Society of America*, *99*(6), 3346–3352. doi:10.1121/1.414891

Van Den Abeele, K. E., Johnson, P. A., Guyer, R. A., & McCall, K. R. (1997). On the quasi-analytic treatment of hysteretic nonlinear response in elastic wave propagation. *The Journal of the Acoustical Society of America*, *101*(4), 1885–1898. doi:10.1121/1.418198

Van Den Abeele, K., Johnson, P., & Sutin, A. (2000). NEWS techniques to discern material damage, Part I: NWMS (Nonlinear Wave Modulation Spectroscopy). *Research in Nondestructive Evaluation*, *12*, 17–30. doi:10.1080/09349840009409646

Van Den Abeele, K., Katkowski, T., & Mattei, C. (2006). Nonlinear Acoustic and Ultrasonic NDT of Aeronautical Components. *AIP Conference Proceedings*, *838*(1), 75–78. doi:10.1063/1.2210321

Van Den Abeele, K., Le Bas, P. Y., Van Damme, B., & Katkowski, T. (2009). Quantification of material nonlinearity in relation to microdamage density using nonlinear reverberation spectroscopy: Experimental and theoretical study. *The Journal of the Acoustical Society of America*, *126*(3), 963–972. doi:10.1121/1.3184583 PMID:19739709

Van Den Abeele, K., Sutin, A., Carmeliet, J., & Johnson, P. (2001). Microdamage diagnostics using nonlinear elastic wave spectroscopy (NEWS). *NDT & E International*, *34*(4), 239–248. doi:10.1016/S0963-8695(00)00064-5

Van Overschee & De Moor. (1996). *Subspace Identification for Linear Systems: Theory-Implementation-Applications*. Kluwer Academic Publishers.

Van Zandt, J. R. (2001). *A more robust unscented transform*. Paper presented at the International Symposium on Optical Science and Technology.

Vander Meulen, F., & Haumesser, L. (2008). Evaluation of B/A nonlinear parameter using an acoustic self-calibrated pulse-echo. *Applied Physics Letters*, *92*(21), 214106. doi:10.1063/1.2937838

Vanik, M., Beck, J., & Au, S. (2000). Bayesian probabilistic approach to structural health monitoring. *Journal of Engineering Mechanics*, *126*(7), 738–745. doi:10.1061/(ASCE)0733-9399(2000)126:7(738)

Vila, M., Vander Meulen, F., Dos Santos, S., Haumesser, L., & Bou Matar, O. (2004). Contact phase modulation method for acoustic nonlinear parameter measurement in solid. *Ultrasonics*, *42*(1-9), 1061–1065. doi:10.1016/j.ultras.2003.12.024 PMID:15047429

Villamizar, R. (2005). *Robust control of systems subjected to uncertain disturbances and actuator dynamics*. (Doctoral dissertation). Retrieved from http://www.tdx.cat/bitstream/handle/10803/7729/Trvm.pdf?sequence=1

Viswanath, A., Purna Chandra Rao, B., Mahadevana, S., Parameswaran, P., Jayakumar, T., & Raj, B. (2011). Nondestructive assessment of tensile properties of cold worked AISI type 304 stainless steel using nonlinear ultrasonic technique. *Journal of Materials Processing Technology*, *211*(3), 538–544. doi:10.1016/j.jmatprotec.2010.11.011

Vuerinckx, R., Pintelon, R., Schoukens, J., & Rolain, Y. (2001). Obtaining Accurate Confidence Regions for the Estimated Zeros and Poles in System Identification Problems. *IEEE Transactions on Automatic Control*, *46*(4), 656–659. doi:10.1109/9.917672

Wahab, A. M. M., & De Roeck, G. (1999). Damage Detection in Bridges using Modal Curvatures: Application to a Real Damage Scenario. *Journal of Sound and Vibration*, *226*(2), 217–235. doi:10.1006/jsvi.1999.2295

Wang, Q., & Deng, X. (1999). Damage detection with spatial wavelets. *International Journal of Solids and Structures*, *36*(23), 3443–3468. doi:10.1016/S0020-7683(98)00152-8

Ward, J., Croxford, A., & Paget, C. (2012). Passive impact localisation for the structural health monitoring of new airframe materials. 11th Anglo-French Physical Acoustics Conference. *Journal of Physics: Conference Series*, *457*, 012010. doi:10.1088/1742-6596/457/1/012010

Watanabe, Fujiwara, Hashimoto, & Ishimaru. (2011). Evaluation of self healing effect in fly-ash concrete by ultrasonic test method. *International Journal of Modern Physics B, 25.*

Wheeler, O. E. (1972). Spectrum loading and crack growth. *Journal of Fluids Engineering*, *94*(1), 181–186.

Wijewickreme, D. (2013). *Role of geotechnical engineering in assuring the integrity of buried pipeline systems*. Retrieved from http://dl.lib.mrt.ac.lk/bitstream/handle/123/8879/SBE-KN_1.pdf?sequence=1

Williams, R. B., Inman, D. J., & Wilkie, W. K. (2004). Temperature-Dependent Thermoelastic Properties for Macro Fiber Composite Actuators. *Journal of Thermal Stresses*, *27*(10), 903–915. doi:10.1080/01495730490498386

Winkler, K., & Liu, X. (1996). Measurements of third-order elastic constants in rocks. *The Journal of the Acoustical Society of America*, *100*(3), 1392. doi:10.1121/1.415986

Wong, P. K., Xu, Q., Vong, C. M., & Wong, H. C. (2012). Rate-dependent hysteresis modeling and control of a piezostage using online support vector machine and relevance vector machine. *IEEE Transactions on Industrial Electronics*, *59*(4), 1988–2001. doi:10.1109/TIE.2011.2166235

Worden, K., & Staszewski, W. J. (2000). Impact location and quantification on a composite panel using neural networks and a genetic algorithm. *Strain*, *36*(2), 61–70. doi:10.1111/j.1475-1305.2000.tb01175.x

Wu, Y.-T. (1994). Computational methods for efficient structural reliability and reliability sensitivity analysis. *AIAA Journal*, *32*(8), 1717–1723. doi:10.2514/3.12164

Wu, Y.-T., Enright, M. P., & Millwater, H. R. (2002). Probabilistic methods for design assessment of reliability with inspection. *AIAA Journal*, *40*(5), 937–946. doi:10.2514/3.15143

Xiao, W. (2012). *Structural health monitoring and fault diagnosis based on artificial immune system.* (PhD Thesis). Worcester Polytechnic Institute.

Xu, X., & Martin, E. (1998). A Distribution-Based Clustering Algorithm for Mining in Large Spatial Databases. In *Proceedings of Data Engineering*. IEEE.

Xu, Q. (2014). Impact detection and location for a plate structure using least squares support vector machines. *Structural Health Monitoring*, *13*(1), 5–18. doi:10.1177/1475921713495083

Xu, Q., & Wong, P. K. (2011). Hysteresis modeling and compensation of a piezostage using least squares support vector machines. *Mechatronics*, *21*(7), 1239–1251. doi:10.1016/j.mechatronics.2011.08.006

Xu, Y., Wong, Y. L., Poon, C. S., & Anson, M. (2003). Influence of PFA on cracking of concrete and cement paste after exposure to high temperatures. *Cement and Concrete Research*, *33*(12), 2009–2016. doi:10.1016/S0008-8846(03)00216-3

Yagawa, G., Kanto, Y., Yoshimura, S., Machida, H., & Shibata, K. (2001). Probabilistic fracture mechanics analysis of nuclear structural components: A review of recent Japanese activities. *Nuclear Engineering and Design*, *207*(3), 269–286. doi:10.1016/S0029-5493(01)00337-5

Yan, A.-M., & Golinval, J.-C. (2006). Null subspace-based damage detection of structures using vibration measurements. *Mechanical Systems and Signal Processing*, *20*(3), 611–626. doi:10.1016/j.ymssp.2005.04.010

Yang, M. S., Hwang, P. Y., & Chen, D. (2004). Clustering Algorithms for Mixed Feature Variables. *Fuzzy Sets and Systems*, *141*(2), 301–317. doi:10.1016/S0165-0114(03)00072-1

Yan, Y. J., Cheng, L., Wu, Z. Y., & Yam, L. H. (2007). Development in vibration-based structural damage detection technique. *Mechanical Systems and Signal Processing*, *21*(5), 2198–2211. doi:10.1016/j.ymssp.2006.10.002

Yinghui, L., & Michaels, J. E. (2005). A Methodology for Structural Health Monitoring with Diffuse Ultrasonic Waves in the Presence of Temperature Variations. *Ultrasonics*, *43*(9), 717–731. doi:10.1016/j.ultras.2005.05.001 PMID:15992847

Yolken, H. T., & Matzkanin, G. A. (2008). Recent trends in structural health monitoring technologies. *The AMMTIAC Quarterly*, *3*(4), 3-6.

Yuan, S., Liu, P., & Qiu, L. (2012). A miniaturized composite impact monitor and its evaluation research. *Sensors and Actuators. A, Physical*, *184*, 182–192. doi:10.1016/j.sna.2012.07.011

Yue, Z., Shi, T., Xuesong, J., Dongwei, Z., & Chuanzhi, Z. (2013). Structural damage detection and classification algorithm based on artificial immune pattern recognition. *Journal of Shenyang Jianzhu University*, 2.

Yu, E., Kim, S. N., Park, T., & Lee, S. H. (2014). Detection of earthquake-induced damage in a framed structure using a finite element model updating procedure. *TheScientificWorldJournal*. PMID:24574888

Yuen, K. (2010). Recent developments of Bayesian model class selection and applications in civil engineering. *Structural Safety*, *32*(5), 338–346. doi:10.1016/j.strusafe.2010.03.011

Yun, Y., Miao, G. Q., Zhang, P., Huang, K., & Wei, R. J. (2005). Nonlinear acoustic wave propagation in one-dimensional layered system. *Physics Letters. [Part A]*, *343*(5), 351–358. doi:10.1016/j.physleta.2005.05.092

Zagrai, A., Donskoy, D., Chudnovsky, A., & Golovin, E. (2008). Micro- and macroscale damage detection using the nonlinear acoustic vibro-modulation technique. *Research in Nondestructive Evaluation*, *19*(2), 104–128. doi:10.1080/09349840801931817

Zaitsev, V. Y., & Matveev, L. (2006). Microinhomogeneous medium, elastic nonlinearity, linear dissipation, amplitude dependent dissipation. *Russian Geology and Geophysics*, *47*(5), 694–709.

Zaitsev, V. Y., Matveev, L., & Matveyev, A. (2011). Elastic-wave modulation approach to crack detection: Comparison of conventional modulation and higher-order interactions. *NDT & E International*, *44*(1), 21–31. doi:10.1016/j.ndteint.2010.09.002

Zaitsev, V., Gusev, V., & Castagnede, B. (2002). Luxemburg-Gorky effect retooled for elastic waves: A mechanism and experimental evidence. *Physical Review Letters*, *89*(10), 105502. doi:10.1103/PhysRevLett.89.105502 PMID:12225205

Zarembo, L., & Krasil'Nikov, V. (1971). Nonlinear phenomena in the propagation of elastic waves in solids. *Physics-Uspekhi*, *13*(6), 778–797. doi:10.1070/PU1971v013n06ABEH004281

Zhang, J., Worden, K., & Staszewski, W. (2008). Sensor optimization using an immune system metaphor. In *Proceedings of 26th International Modal Analysis Conference (IMAC)* (pp. 1-10). IMAC.

Zhang, T., Ramakrishnan, R., & Livny, M. (1996). An Efficient Data Clustering Method for Very Large Databases. In *Proceeding SIGMOD '96*. ACM.

Zhang, R., & Mahadevan, S. (2001). Fatigue reliability analysis using nondestructive inspection. *Journal of Structural Engineering*, *127*(8), 957–965. doi:10.1061/(ASCE)0733-9445(2001)127:8(957)

Zheng, Y., Maev, R., & Solodov, I. (1999). Nonlinear acoustic applications for material characterization: A review. *Canadian Journal of Physics*, *77*(12), 927–967. doi:10.1139/cjp-77-12-927

Zhou, Y., Tang, S., Zang, C., & Zhou, R. (2012). An artificial immune pattern recognition approach for damage classification in structures. In D. Zeng (Ed.), *Advances in information technology and industry applications, LNEE 136* (pp. 11–17). Berlin: Springer. doi:10.1007/978-3-642-26001-8_2

Zhou, Z. H., Li, Z. Q., Xu, D. Y., & Yu, J. H. (2011). Influence of Slag and Fly Ash on the Self-Healing Ability of Concrete, Advanced. *Materials Research*, *306-307*, 1020–1023. doi:10.4028/www.scientific.net/AMR.306-307.1020

Zugasti, E., Gonzalez, A. G., Anduaga, J., Arregui, M. A., & Martínez, F. (2012, August). NullSpace and AutoRegressive damage detection: A comparative study. *Smart Materials and Structures*, *21*(8), 085010. doi:10.1088/0964-1726/21/8/085010

Zugasti, E., Mujica, L. E., Anduaga, J., & Martínez, F. (2013). *Feature Selection - Extraction Methods based on PCA and Mutual Information to improve Damage Detection problem in Offshore Wind Turbines* (Vol. 579–570, pp. 620–627). Key Eng. Mater.

About the Contributors

Diego Alexander Tibaduiza Burgos received his Electronic Engineer Degree from the Industrial University of Santander, Colombia in 2003, the M.Sc. degree in 2006 from the same university and his Doctoral Degree in 2013 from the Universitat Politècnica de Catalunya in Barcelona, Spain. He is the author and coauthor of more than 80 publications in journals, conferences, books and chapter books. Currently, Dr. Tibaduiza is a titular professor in the Faculty of Electronics Engineering at the Universidad Santo Tomás in Bogotá, Colombia. His research interest includes Structural Health Monitoring, Artificial Intelligence, Control Systems, Robotics and Signal Processing.

Luis Eduardo Mujica was born in Colombia in 1976. He graduated with bachelor degrees in electrical engineering and electronic engineering from the Universidad Industrial de Santander, Colombia in 2000. His doctoral degree in information technologies was obtained in 2006 from the University of Girona, Spain. Since 2007, he has been a post-doctoral researcher and assistant professor at the Technical University of Catalonia –UPC- (Barcelona, Spain). In recent years, he has been a visiting researcher at the University of Sheffield (UK), Ghent University (Belgium), the Catholic University of Louvain (Belgium) and the Universidad Politécnica de Madrid (Spain). His present research interests include intelligent and advanced signal processing for damage assessment in structural health monitoring.

José Rodellar, born in 1954 in Spain, is Professor in the Department of Applied Mathematics III, where he has Director in the period 2007-2013, in the School of Civil Engineering and the School of Industrial Engineering of the Technical University of Catalonia (Barcelona, Spain). His research interests include control systems, from theory to applications, including passive/semi-active/active control of structures and structural health monitoring using statistical methods. The outcomes of his research are summarized in three books, over 100 papers in peer-reviewed journals and over 200 presentations in the major international conferences on the above research topics. He has been principal investigator of 25 funded research projects and has supervised 20 PhD Thesis. From 2012, he is the President of the International Association for Structural Control and Monitoring.

* * *

Maribel Anaya Vejar received in 2005 her Electronic engineer degree from the Industrial University of Santander in Colombia and the M.Sc. degree in 2011 in electronic engineering from the same University. She is currently a Ph.D. candidate at Earthquake engineering and structural dynamics program in the "Universitat Politècnica de Catalunya" in Barcelona-Spain. In addition, she has experience in teaching and research in different fields of Electronics engineering and currently is an associated professor in the Faculty of Electronics Engineering at the "Universidad Santo Tomas" in Bogotá-Colombia. Her research interest includes Structural Health Monitoring, Control, Robotics and Biomedical Systems.

Daniel J. Barnard is Staff Associate Scientist at CNDE, Iowa State University since 1996, with primary focus on linear and nonlinear ultrasonic methods for engineering materials. Previous work included ultrasonics and mechanical testing of metals and alloys at the Ames Laboratory (USDOE) at Iowa State University, and fossil and nuclear fueled power plant construction projects engineering.

Nicolas Bochud holds a PhD in Structural Mechanics from the University of Granada, Spain, in 2014. His research focuses on the signal modeling-based identification of pathologies in advanced materials and biomaterials using ultrasonics. He received a MsC Degree in Mechanical Engineering from the Swiss Federal Institute of Technology (ETH), Zurich, in 2008, and a MsC Degree in Information Technology from the University of Granada, in 2010. He is actually continuing his research at the Laboratoire d'Imagerie Biomédicale, UPMC-CNRS, Paris, where he works on the assessment of the geometrical and elastic properties of cortical bone using ultrasonic guided waves.

Leonard J. Bond received his B.Sc. (Applied Physics) and PhD. (Physics) Degrees from the City University London. His graduate work contributed to the science base for quantitative nondestructive evaluation and the first British Gas "Ultrasonic Pigs," used for pipeline inspection. With the support of the Ministry of Defense (UK) he then analyzed ultrasonic inspection applied to aging solid rocket motors and an increasingly diverse range of defense systems. He joined University College London (UCL), University of London and as a member of faculty developed a research group focused on QNDE. In recognition of his contribution to the field he was promoted to "Reader in Ultrasonics." Dr. Bond moved to the United States in 1990, and initially worked at NIST in Boulder, CO and held Research Professorships with both the University of Colorado at Boulder and University of Denver. He then joined PNNL in 1998, and was appointed as a Laboratory Fellow. In 2005-6 he was for two years the founding Director of the Center for Advanced Energy Studies, Idaho National Laboratory. He returned to PNNL in January 2007, as a Laboratory Fellow, leading activities in advanced diagnostics/prognostics. This work had a history that goes back to UK projects, in particularly those looking at aging for turbine disks and aging for solid rocket motors. His supported various US Nuclear Regulatory Commission (NRC) and several DOE – Nuclear Energy programs. He was appointed as Director of CNDE and as a Professor in July 2012. CNDE is an NSF IU CRC, with a 29 history that has 18/19 sponsors. Dr. Bond leads the 12 research professional staff, 5 administrative staff and 20 center affiliated faculty, and his own group has currently has 7 graduate students. He now has more than 40 years R&D experience in academia, government laboratories, and the private sector. His current research is focused on applications of ultrasonics including to pipelines and high temperature inspection of modular nuclear power reactors. He is author or co-author of more than 300 publications, including one book: "Ultrasonics" – Ensminger and Bond (2011). He is a Fellow of both the Institute of Physics (UK) and the American Association for the Advancement of Science. He holds 10 patents.

Antonio M. Callejas holds a degree in Civil Engineering and a MSc degree in Structural Engineering from University of Granada (Spain). His research started with Dr. Rus on 10/2013 while he was studying the MSc degree in Structural Engineering. Nowadays, he is focused on nonlinear ultrasonic techniques for early nondestructive damage monitoring, both experimentally and computationally.

Jhonatan Camacho is an Electronics Engineer (2008) and he obtained his master Degree in Electronics Engineering at 2010 from Universidad industrial de Santander (UIS) – Colombia. Currently Msc Camacho is a doctoral student from universitat politecnica de catalunya (UPC), Spain in topics related to structural health monitoring by means of piezodiagnostics approach. He is an active member of the CEMOS research group at UIS and CoDAlab center at UPC. Research interests include the implementation of machine learning algorithms for application in structural condition monitoring and medical areas. Some of his works are related to the application of mixed integer nonlinear optimization methods, nonlinear predictive control algorithms and digital signal processing techniques.

Sergio Cantero holds a degree in Civil Engineering and a MSc degree in Structural Engineering from University of Granada (Spain). His research started with Dr. Rus on 10/2012 under an initial researcher grant on nanotechnology for improving matrix-fiber adhesion of carbon fiber composites. During MSc, he was focused on nonlinear ultrasonic techniques for early nondestructive damage monitoring, both experimentally and computationally. Nowadays, He is currently working in an engineering company in Madrid, which is called SENER.

Juan Chiachío holds a PhD on Structural Engineering by the University of Granada, Spain. His research focuses on uncertainty quantification for fatigue damage assessment in composites materials using Bayesian methods. He received his Master's Degree in Civil Engineering in 2007 and in Structural Engineering in 2011, both by University of Granada. His research has been awarded by the National Council of Education through one of the FPU annual fellowships, by the Andalusian Society of Promotion of the Talent, by the Prognostics and Health Management Society with a Best Paper Award, and by the European Council of Civil Engineers (ECCM) with the Silver Medal prize in the 1st European Contest of Structural Design. During the course of his research work, Juan has worked as guest scientist at NASA Ames Research Center (USA), California Institute of Technology (Caltech) and Technical University of Hamburg (Germany).

Manuel Chiachío holds a PhD (international mention) on Structural Engineering by the University of Granada, Spain. His research focuses on prognostics methods and uncertainty quantification methodologies in application to (but not limited to) a wide range of mechanical engineering problems. He received his Masters Degree in Civil Engineering in 2007 and in Structural Engineering in 2011, both by University of Granada. After several years in the industry, Manuel came back to the research by means of one of the national FPU fellowships (Spanish Ministry of Science and Education). His research has been also awarded by the Andalusian Society of Promotion of the Talent, by the Prognostics and Health Management Society with a Best Paper Award, and by the European Council of Civil Engineers (ECCM) with the Silver Medal prize in the 1st European Contest of Structural Design. During the course of his research work, Manuel has worked as guest scientist at NASA Ames Research Center (USA), California Institute of Technology (Caltech) and Technical University of Hamburg (Germany).

Edwin Forero received his Electrical Engineer and Master Graduated (MIE) with experience in: Electrical and electronic research field,University Education, Automation and Instrumentation, Rational and Efficiency Use of Energy, Building of electrical facilities, Coordination of contractors related to control of electricity areas in an Electric Company. Supervision of contractors and personnel.

Fahit Gharibnezhad graduated with bachelor degrees in electrical and control engineering from the Tabriz University, Iran in 2005. He did his master in Iran University of Science and Technology in 2008. His doctoral degree in Dynamic Structures was obtained in 2014 from the Technical University of Catalunya, Spain. He has been a visiting researcher at the University of Siegen (Germany), University of California San Diego (USA) and Technical University of Madrid (Spain). His present research interests include techniques and advanced signal processing for robust damage detection in structural health monitoring.

Kai Goebel works at NASA Ames Research Center where he is the Area Lead for Discovery and Systems Health in the Intelligent Systems Division. He received the degree of Diplom-Ingenieur from Technische Universitaet Muenchen in 1990 and the Ph.D. from the University of California at Berkeley in 1996. Dr. Goebel worked at General Electric's Corporate Research Center in upstate New York where he was also an adjunct professor at Rensselear Polytechnic Institute. He has been on the dissertation committee of eight Ph.D. students at RPI, Syracuse University, University of Cincinnati, Vanderbilt University, Georgia Institute of Technology, London Metropolitan University and Stanford University. He is co-founder of the PHM Society and Associate Editor of the International Journal of PHM. He is currently guest professor at University of Cincinnati. Dr. Goebel holds eighteen patents and he has co-authored more than 300 technical papers.

Andreas Kyprianou is an assistant professor at the Mechanical Engineering and Manufacturing department of the University of Cyprus. He studied Mechanical and Process engineering at the University Sheffield from where he graduated in 1996. He obtained his PhD in 1999 from the same University and MSc in Mathematics and the foundations of computer science from Oxford University. During the years of 2000-2001 he worked as research associate at the Department of Public and Business Administration of University of Cyprus on bankruptcy prediction using neural networks. During the course of this work, he has been exposed to the development and application of probabilistic neural networks that utilized information inferred from the theory of derivatives pricing (Black and Scholes) in order to assign a probability of failure to a specific company. In October 2001 he joined the dynamics group of university of Liverpool to undertake research in structural modification and inverse problems. The main field of his research interest is the theoretical and experimental characterization of mechanical uncertain vibrating structures and their health condition assessment. His secondary research interests lie in the field of nanotechnology and its applications to sensing, particle manipulation and atomic force microscopy based methodologies for imaging various processes at nanoscale and energy.

You Ling is a mechanical engineer in the probabilistics lab at General Electric Global Research Center. He obtains his Ph.D. degree from Vanderbilt University, and bachelor's degree from Shanghai Jiao Tong University. His research interest includes uncertainty quantification in structural and multi-disciplinary reliability analysis and life modeling, with extensive use of probability theory, Bayesian statistics, surrogate modeling, and advanced sampling techniques. He has worked on projects funded by

NASA, FAA, DOE, and USAF, with applications to fatigue/creep modeling of aerospace structures and MEMS devices. He has 20+ publications on peer-reviewed journals and conference proceedings, and has served as reviewer for 7 journals, including SMO, AIAA journal, IJPHM, SHM, and ASCE JEM.

Sankaran Mahadevan is Professor of Civil and Environmental Engineering at Vanderbilt University, Nashville, Tennessee, where he has served since 1988. He also has a joint appointment as Professor of Mechanical Engineering. His research interests are in reliability and uncertainty analysis methods, material degradation, structural health monitoring, design optimization, and model uncertainty. The methods have been applied to civil, mechanical and aerospace systems. This research has been funded by NSF, NASA (Glen, Marshall, Langley, Ames), FAA, U. S. DOE, U. S. DOT, Nuclear Regulatory Commission, U. S. Army Research Office, U.S. Air Force, U. S. Army Corps of Engineers, General Motors, Chrysler, Union Pacific, Transportation Technology Center, and the Sandia, Los Alamos, Idaho and Oak Ridge National Laboratories. Professor Mahadevan has developed and directs an NSF-IGERT multidisciplinary graduate program in Reliability and Risk Engineering and Management at Vanderbilt University, since 2001. He has directed 30 Ph.D. dissertations and 20 M.S. theses, taught several industry short courses on reliability methods, and authored more than 300 technical publications, including two textbooks and 120 peer-reviewed journal articles.

Saroj Mandal is currently Professor in the Department of Civil Engineering at the Jadavpur University, Kolkata, India. He has been involved in teaching, research, design and consultancy for over twenty three years. Prof. Mandal joined the Civil Engineering Department, Jadavpur University in June 1991. He has guided 12 Master thesis and 3 Ph.d Thesis and hold 5 Indian/ US patents. Prof. Mandal has successfully completed several consultancy/project works. He has several national and international journals and conferences articles published as well as he is reviewer of several national and international journal. He has chaired different international conferences and delivered different invited lectures. Prof. Mandal was the finalist of Lafarge Innovation Award, 2012. He is member of The Institution of Engineers, Indian Concrete Institute, International Centre for Fibre Reinforced Concrete and other scientific societies.

Fernando Martínez has received his B. Eng in Industrial Engineering at Mondragon University (Spain) in 1987, his M.Sc.in Automation and Electronics Engineering at University Paul Sabatier (France) in 1988 and his PhD in Industrial Engineering in 1993 at Institut National Polytechnique de Toulouse (France) in 1993. Currently, he is the head of the Sensors Department at IK4-IKERLAN. His research activities include design and development of sensors and their application to the monitorization of products and industrial processes. He has participated and coordinated several national and international research projects in domains related to: structural health monitoring and smart structures, sensor-actuator design based on smart materials, autonomous sensor integration, optoelectronic systems and mechatronic systems. He has published many research papers in journals and international conferences.

Juan Melchor holds a degree in Mathematics and master degree in Biomathematics from University of Granada (Spain) and worked as biostatician and mathematician for EASP (Andalusian of Public Health School) during three years. On 01/05/2008, he started to work with RARECARE team (supported by the Executive Agency for Health and Consumers (EAHC) of the European Commission) contributing to definition of rare cancers and developing a data mining study of quality control process. In parallel to this, he collaborated with the University of Granada in a project of control and tumour growth, using

PDE's and evolution models supervised by Dr. Soler. In 01/10/2010, he started his PhD in Structural Engineering at the Department of Structural Mechanics of University of Granada, under the supervision of Dr. Rus, and focusing on tissular engineering and their applications to biomedical fields. Finite elements methods, optimization, inverse problem, biomechanical structures and damage modelling using ultrasound are topics they are working on in this team.

Shaswata Mukherjee is currently research scholar in the Department of Civil Engineering at the Jadavpur University. He is basically a Ceramic Engineer and initially he worked as JRF at Central Glass and Ceramic Research Institute (CSIR), where he worked on Cf/SiC and Reaction bonded Silicon Nitride. Later in the year 2010 he moved to Jadavpur University as JRF and enrolled for Ph.D in Advanced Concrete Technology in the year 2012 from Department of Civil Engineering of the same University. He has been selected and invited as Visiting Research Scholar from Curtin University, Perth, WA for conducting research work on Application of Nanotechnology in Construction Industry. He has published several national and international journal and conferences articles.

Rafael Muñoz was born in San Sebastián (Spain) and graduated at the University of Navarra as Industrial Engineer (San Sebastián, Spain), with the speciality on Electricity, Electronics and Automation. He did his doctorate in the R&D dept. of the company Tecnatom S.A. (Madrid, Spain), with the academic guidance of the University of Navarra, about propagation models based on the Rayleigh Integral. Once he finished his granted time in Tecnatom, he continued there as a Project Manager in the UT Lab, where the main target was to prepare new UT automated inspections for different industries like nuclear, steel, chemical and aeronautical industry, from the design of the inspection, testing, generating a procedure, and putting it into exploitation. He also holds an Executive MBA degree by IE Business School (Madrid, Spain). Currently, he is an adjunct professor at the Sutructural Mechanics Department of the University of Granada (Granada, Spain), and Business Consultant for companies with a technological base. He collaborates with the NDE Lab of the department, working on nonlinear ultrasonics, Bayesian inverse problem and prognosis.

Laura Peralta is Civil Engineer, and FPI predoctoral fellow. She received her Msc in Structural Mechanics and is currently pursuing her PhD on a Model Micro-Meso-Macro-mechanical cervical ultrasound. She has been working in her PhD at UCL (London) and Langevin Institut (Paris) during a short-term as invited researcher, and they currently collaborate closely.

Francesc Pozo received the degree in Mathematics from the University of Barcelona (UB), Spain, in 2000 and the PhD in Applied Mathematics in 2005 from the Technical University of Catalonia (UPC), Spain. He is currently with the Department of Applied Mathematics III and the College of Industrial Engineering of Barcelona (EUETIB) (since 2000) at the Technical University of Catalunya, Barcelona, Spain, where he belongs to the research group Control, Dynamics and Applications (www-ma3.upc.edu/codalab). He is also a teaching collaborator at the Open University of Catalonia (UOC). His research interests include modelling, dynamics, control theory and applications, particularly in civil engineering, like vibration control of structures and structural health monitoring. Dr. Pozo, which is currently an Associate Professor, has published over twenty-five papers in these topics.

Jairo Alejandro Rodriguez graduated in Physics in the Universidad Nacional de Colombia. Post-graduates studies in Telecomunications Engineering in the Universidad Politécnica de Valencia (Spain). His research area lies in non-destructive testing, particularly: electromagnetism, solid state, DSP, semi-conductors and structural dynamics.

Guillermo Rus started his research on computational mechanics at the University of Granada (UGR, 1995), where he disputed the PhD thesis on Numerical Methods for Nondestructive Identification of Defects (2001), providing defect search algoritms and sensitivity computation with Boundary Elements. He applied these experimentally at the NDE Lab at MIT (USA) as a Fulbright Postdoctoral Fellow, rendering novel robust quantitative approaches to ultrasonics monitoring. He started up the NDE Lab at the UGR (www.ugr.es/~endlab) as assistant professor in 2003, focusing on bioengineering applications such as orthopaedic implants debonding in collaboration with University College London, bone quality diagnosis with ultrasound with Université Paris VI and biomaterials characterization with the Nanomaterials Technology Lab (Spain). He is also transferring this diagnosis technology to civil engineering for monitoring structural health of advanced materials, such as FRP damage state monitoring, in collaboration with NASA, Caltech (USA), T. U. Hamburg - Airbus (Germany) and Andong National University (Korea). Rus tenured as associate professor in 2009 at UGR, is the author of 40 SCI papers (H-factor 10, over 400 citations), 9 books, 3 patents and 14 invited seminars. His research career has been awarded by the Juan Carlos Simo prize for young researchers (Spain, 2007), the Honorary Fellowship of the Wessex Institute of Technology (UK, 2005), Fulbright Fellowship (USA, 2002) and the Excellence PhD award (Granada, 2001).

Wilmer Alexis Sandoval Caceres was born in Malaga, Santander/ Colombia in 1978. Wilmer attended Instituto Tecnico Industrial Malaga, which he graduated as technician in electricity in 1995. Later, Wilmer moved to Bucaramanga, Santander/ Colombia where he began his collage studies at Universidad Pontificia Bolivariana (UPB), which he graduated with a degree in Electronics Engineering. While he was in college, he participated in different Mobil robotics' contests. Additionally, Wilmer attended the Universidad Industrial de Santander (UIS), where he received a Masters in Electronics Engineering in 2010. During his graduate studies he belonged to a research group call CMOS. Since 2006, He was working for six years at Universidad Santo Tomas de Aquino (USTA) as a part-time and full-time professor, which he belonged to research group call GRAM, and participated in various committees such as thesis, research and curriculum committee. Additionally, he was working as part time professor with Universitaria de Investigacion y desarrollo (UDI) in 2006, and Unidades Tecnologicas de Santander (UTS) during 2007-2008. Wilmer has participated in different national and international scientific events such as: VIII International convention on environment and development held in La Habana, Cuba 2011, XIV Symposium on Signal Processing, Images, and Artificial Vision held in Pereira, Colombia 2009 in many others. His interest research areas are relates with robotics, control design, digital signal processing and structural health monitoring using expert systems.

Abhinav Saxena is a Research Scientist with SGT Inc. at the Prognostics Center of Excellence of NASA Ames Research Center, Moffett Field CA. His research focus lies in developing and evaluating prognostic algorithms for engineering systems using soft computing techniques. He has co-authored more than hundred technical papers including several book chapters on topics related to PHM. He is also a member of the SAE's HM-1 committee on Integrated Vehicle Health Management Systems and IEEE working group for standards on prognostics. Dr. Saxena is the editor-in-chief of International Journal of PHM and has led technical program committees in several PHM conferences. He is also a SGT technical fellow for prognostics. He has a PhD in Electrical and Computer Engineering from Georgia Institute of Technology, Atlanta. He earned his B.Tech in 2001 from Indian Institute of Technology (IIT) Delhi, and MS Degree in 2003 from Georgia Tech. He has been a GM manufacturing scholar and is also a member of several professional societies for PHM including PHM Society, SAE, IEEE, AIAA, and ASME.

Shankar Sankararaman received his B.S. degree in Civil Engineering from the Indian Institute of Technology, Madras in India in 2007 and later, obtained his Ph.D. in Civil Engineering from Vanderbilt University, Nashville, Tennessee, U.S.A. in 2012. His research focuses on the various aspects of uncertainty quantification, integration, and management in different types of aerospace, mechanical, and civil engineering systems. His research interests include probabilistic methods, risk and reliability analysis, Bayesian networks, system health monitoring, diagnosis and prognosis, decision-making under uncertainty, treatment of epistemic uncertainty, and multidisciplinary analysis. He is a member of the Non-Deterministic Approaches (NDA) technical committee at the American Institute of Aeronautics, the Probabilistic Methods Technical Committee (PMC) at the American Society of Civil Engineers (ASCE), and the Prognostics and Health Management (PHM) Society. Currently, Shankar is a researcher at NASA Ames Research Center, Moffett Field, CA, where he develops algorithms for uncertainty assessment and management in the context of system health monitoring, prognostics,and decision-making.

Andreas Tjirkallis is a researcher in the field of mechanical vibrations and structural health monitoring. He has obtained his Bachelor degree in Mechanical Engineering from Imperial College, London, and his Ph.D. degree in Mechanical Engineering from the University of Cyprus. His research interests are in the area of non-destructive damage identification using mechanical vibrations and signal processing including wavelet transform analysis. His research work has been published in conference and journal papers. Andreas is a scientific/research collaborator at the Department of Mechanical and Manufacturing Engineering of the University of Cyprus, and a visiting lecturer at the Department of Mechanical Engineering of Frederick University, Cyprus. He is the founder and director of a research and development organisation – ACE R&D.

Rodolfo Villamizar Mejía is an Electrical (1999) and electronics (2000) Engineer from Universidad Industrial de Santander - Colombia, and he obtained his Doctoral degree in information technologies from universitat de Girona - Spain, 2005. Currently, Phd Villamizar is assistant professor at universidad industrial de Santander and an active member of the Control, Electronics, Modelling and simulation (CEMOS) research group. His works and research interests cover topics related to Structural Health monitoring techniques, active/semiactive control with application to civil engineering buildings structures, modal intervals based techniques with application to stayed bridges, Electrical power quality and power compensation, supervised control systems, as well as robust control algorithms.

Qingsong Xu received his B.S. degree in mechatronics engineering (with honors) from Beijing Institute of Technology, Beijing, China, in 2002, and the M.S. and Ph.D. degrees in electromechanical engineering from the University of Macau, Macao, China, in 2004 and 2008, respectively. He is currently an Associate Professor of electromechanical engineering with the University of Macau, where he directs the Smart and Micro/Nano Systems Laboratory. He was a visiting scholar at the Swiss Federal Institute of Technology (ETH), Zurich, Switzerland, and the National University of Singapore, Singapore. Dr. Xu is the holder of two patents and has published over 160 papers in journals, conferences, and book chapters. He is a member of IEEE and ASME. His research interests involve MEMS/NEMS, micro/nano mechatronics, micro/nano manipulation/assembly, computational intelligence, advanced and intelligent control, prognosis and diagnosis, structural health monitoring, sensors and actuators, smart materials and structures, robotics and automation.

Ekhi Zugasti received his Telecommunication engineering degree in 2009 at Mondragon Unibersity, in 2010, Master in New Materials at University of the Basque Country. From 2010 to 2014 he focused on his PhD in developing algorithms for monitoring systems for Structural Health Monitoring. He did a stay in Siegen University in order to focus in Structural Health monitoring techniques for Wind Turbines. He received his PhD in Civil Engineering on January 2014 in the Politecnical University of Catalonia. Right now he is part of the Mechanical Department in IK4-Ikerlan, working on Structural Health and condition monitoring, focusing on Signal Analysis.

Index

L

M

N

O

P

R

S

T

U

W

www.ingramcontent.com/pod-product-compliance
Lightning Source LLC
LaVergne TN
LVHW060413300926
849162LV00006B/1415
9781466684904